INTERDISCIPLINARY SCIENCE REVIEWS

ESSAY ANNUAL 1981

· ISR ·

ISR Interdisciplinary Science Reviews

ESSAY ANNUAL 1981
First published in
ISR – Interdisciplinary Science
Reviews Volume 5

ISBN 0855014776

Printed in the United Kingdom by
J. W. Arrowsmith Ltd, Bristol.

HEYDEN

HEYDEN & SON LTD
Spectrum House, Hillview Gardens, London NW4 2JQ, UK (Tel: 01-203 5171; Telex: 28303)

HEYDEN & SON INC.
247 South 41st Street, Philadelphia, PA 19104, USA (Tel: (215)382 6673; Telex: 831679)

HEYDEN & SON GmbH
Münsterstrasse 22, 4440 Rheine, West Germany, (Tel: (05971)55111; Telex: 981630)

ISR

Interdisciplinary Science Reviews

ESSAY ANNUAL 1981

Comprising the reviews, comments, critiques and editorials from Volume 5 of Interdisciplinary Science Reviews

Editor: Dr Anthony R. Michaelis

HEYDEN

LONDON · PHILADELPHIA · RHEINE

ISR Contributions

Contributions

Foreword

Interdisciplinary work comes from the joint and continuously integrated effort of two or more specialists having a different disciplinary background and training. To record such work is the primary aim of the essays in this volume, but they also show how science and technology affect society and, occasionally, how the bridge can be built between the Two Cultures of C. P. Snow.

Interdisciplinary research in its formal sense goes back to the Operations Research between scientists and the Defence Services of World War II, although of course there have been many eminent men in the past whose scholarship and erudition extended over more than one discipline. The Honourable Robert Boyle, Benjamin Franklin, Alexander von Humboldt and Louis Pasteur readily spring to mind.

Interdisciplinary co-operation is commonplace in industry where the profit motive demands the most efficient use of available manpower. In universities on both sides of the Atlantic, it is still difficult to achieve. The often rigid divisions between academic departments do not offer enough incentives for the researcher to transgress them. Co-operation between scientists and politicians is now being widely urged as an 'understanding of science is an essential prerequisite for the politicians of tomorrow', as Walter Scheel, formerly the President of the Federal Republic of Germany, put it in the columns of *Interdisciplinary Science Reviews.*

In this 1981 volume of essays, examples of these interdisciplinary concepts can all be found. The effect of a relatively new interdisciplinary science, biotechnology, on society and its nutrition, is the single theme of the first 78 pages. Lord Todd, at the time of writing the President of the Royal Society, London, discusses how our traditional university education no longer fits into our present technological and interdisciplinary age and how a hierarchy of universities could lead to a solution (p. 85). A similar plea for reform comes from the United States, from Dr Simon Ramo (p. 173) and Professor Edwin Mansfield (p. 182), for increased interdisciplinary innovation for American industry.

The best bridge between the Two Cultures is the history of science and technology and in this volume two outstanding examples will be found. Claude Breguet (p. 149) tells his family history over two centuries, and how five previous generations of interdisciplinary scientists and engineers have contributed to the greatness of France. The second example comes from Dr Joseph Needham (p. 263), giving a brief progress report on his monumental work *Science and Civilization in China*, which has now reached nine impressive volumes. It is recognized as one of the greatest works of scholarship of this century.

It would have been impossible to assemble these pages without the enthusiastic support from all the contributors, the advice from the eminent international Editorial Board and the ever-generous attention from the publishers.

London, December 1980 — ANTHONY R. MICHAELIS

ISR Editorial

Interdisciplinary Biotechnology for Nutrition

As hunger is the most serious danger to peace, food must be placed immediately after human rights to a free life and security of the person, according to Pope John Paul II. No human being, past, present or future, of whatever religion or race, can disagree with this statement and neither could I write these sentences nor you read them, if gnawing hunger was our constant companion. Too many millions are still doomed to a life of malnutrition and hunger on this planet and this should not be so.

Malthus Again Disproved?

Will biotechnology be the third factor to disprove Malthus' theory that population growth always tends to outrun the food supply? After the geographical expansion of the American and Australian wheatfields and after the Haber–Bosch process of nitrogen fixation from the air for fertilizers was invented, the biotechnological production of single cell proteins and other substances might well prove the third and final balancing factor until population expansion finally comes to a halt.

Nutritional Prospects

On the following pages nutrition and biotechnology are examined in their widest context and perhaps not unexpectedly, some of the conclusions are more optimistic than others. Aquaculture and algaculture may have long-term prospects, but the biotechnology of single cell protein and of wine production – enology – have more immediate benefits to offer. The ever increasing consumption of wine has certainly become a vital need for many in the richer western countries. Our utter ignorance of the science of smell may be a revelation and a better understanding of it could greatly contribute to a successful introduction of novel synthetic foods, at present a very difficult task in some areas. And finally, it may come as a surprise that gastronomy and the politics of science can be combined for their mutual benefit.

Powerful Opinions

In his introductory review Sir Michael Swann, FRS, the Chairman of the world's greatest and best broadcasting system, the BBC, stresses correctly the powerful influence of public and of scientific opinion on political decision-makers who can balance or unbalance the biological scales. The time has now come for scientific opinion to add its weight to that of the Pope and to press for a mighty biotechnological effort to remove malnutrition and hunger from Earth. Only an interdisciplinary effort, ranging from biology through biotechnology to engineering, from psychology through sociology to politics and from efficient housekeeping through management to economics, can give us all our daily bread.

Anthony R. Michaelis.

The Biological Credit-Debit Balance

PROFESSOR SIR MICHAEL SWANN, FRS
Chairman of the British Broadcasting Corporation, London, UK

Through the media of print and broadcasting, the British Public has become aware during the last few decades of the problems studied by science. Here the extinction of species, the uses of herbicides and pesticides, genetic manipulation and the misuse of antibiotics are considered, and how these have become a matter of public concern.

The present multiplicity of BBC television programmes about Science, with audiences running into many millions, does not mean that the great British public is educated to any great depth in science. Those who aspire to greater understanding must graduate through talks on Radio 3, articles in journals and books, and by attending the British Association. For the tens of thousands who do so, there are, I suppose, millions who do not. Nevertheless, the effect has been beyond the wildest dreams of the early popularizers of science, and it can now be said truthfully that a large part of the British public has some appreciation of what science is about, and still more so, that society has as much awareness of the problems that science studies, or throws up, as of most other corners of human activity.

I want briefly to consider here some of these problems in the biological area before reflecting on how society can approach them. And I mean society, because most of the problems, perhaps all of them, arise not just out of thin air, but because society does something, believing it to be of benefit, only to discover that all sorts of undesirable consequences follow. Research can sometimes mitigate the undesirable consequences, and is almost always needed as a pointer for action. But in the last resort, it is society, local, national or world-wide that has to strike a balance between good and ill, between credit and debit, and decide, as best it may, what shall and shall not be done.

EXTINCTION OF SPECIES

Let us first take a rather simple case – the large problem of species reaching, or being threatened with, extinction. Whales in general, and various species in particular are threatened, though at least with the excuse of being used for food. Rhinoceroses are threatened because some people imagine that their horns have aphrodisiac properties, and elephants are threatened for their ivory.

CCC-0308-0188/80/0005-0002$02.00/1

But with modern methods of agriculture, whales are not necessary as a source of oil or meat. Ivory is elegant stuff, but there are chemical substitutes. And as for rhinoceros horn as an aphrodisiac, there is no evidence that it works at all. So there is very little indeed one can put in the credit side of the social balance for good and ill.

What does one put on the debit side? There is a trouble here in that species have become extinct in vast numbers throughout evolution – the dinosaurs for instance on a colossal scale, and in geological terms, rather suddenly, but the world went on. Fresh ecological balances were struck, and new species evolved.

And no doubt, if the present species threatened with extinction were to go, fresh balances would be struck again. Perhaps they would be less favourable to us. But in all probability it would not make much difference. The debit here is something quite different, best described in aesthetic, not to say moral, terms. It is simply that a number of people, from a variety of ethical standpoints, are not prepared to destroy something they cannot recreate. Some carry these feelings to lengths that not many of us share – pacifism, vegetarianism, anti-vivisection, resistance to the controlled culling of animals that have grown too numerous. Most people are prepared to strike some sort of a balance in these areas, mitigating unkindness when they can. But it seems, and I welcome it, that the prospect of extinguishing a species for ever has become increasingly repugnant to very large numbers of people and for them, the scales are weighted very firmly in one direction.

PESTICIDES AND HERBICIDES

I now want to turn to another facet of modern technology that may do much harm in an even more indiscriminate way – the use of pesticides and herbicides in modern agriculture. No-one buys these expensive chemicals for the fun of it. On the credit side of the balance one can immediately put a great increase in food production in a world that is chronically short of it, as well as great reductions in disease. It is not self-evident, for instance, that the virtual eradication of malaria by killing mosquitoes with DDT, was too high a price to pay for the harm DDT did by entering into food chains. And were there no other way of controlling the disease, we might have felt bound to compromise. In the event, pesticides and herbicides that do not persist have been developed. But would this have happened, or happened so quickly, if there had been no public pressure pointing a finger at the debit side of the equation? I doubt it – interested parties, naturally, point only to the credit side.

There is an even more indiscriminate source of harm to the living environment that comes as a product, or rather a by-product of modern technology – namely pollution. Rivers are spoiled by the effluent of paper mills, inland waters ruined by detergents or nitrogenous waste from intensive farming, or phosphates from domestic dishwashers, shores are polluted with oil, the list is endless. Once again no-one does it for fun, the credit side of the equation is obvious. But until public pressure started to point a finger to the debit side, there were no new inspectorates, no Royal Commissions, no serious attempts to mitigate the harm that was being done.

GENETIC MANIPULATION

My concerns so far stem from what one may call the more blunderbuss effects of modern science and technology. But there have been other developments, far more subtle and delicate, and still in the hands of scientists, that are also causing public concern: Genetic Manipulation. The cracking of the genetic code, by Crick and Watson 30 years ago must, I think, rank as the most significant piece of biological research of this century. And it soon led, of course, to a realization that in principle, one might so alter the DNA of higher organisms as to transform them into something alarmingly different. In practice that is something so far beyond understanding and expertise as to be in all probability, beyond our grasp for all time. Just as well perhaps.

But it has become possible recently to insert small amounts of DNA into micro-organisms. This offers the possibility of getting a bacterium to manufacture something we cannot make ourselves, like insulin, which can at present be obtained only from certain cells of higher animals, from which it can be extracted only with difficulty and expense.

But if one starts changing genetic material, putting bits in and taking bits out, one may, inadvertently alter the bacterium in some unexpected way. One may indeed create in effect a new bacterium, capable of causing a new disease, or resistant to present day treatment. It is this hazard, very remote perhaps, but because it cannot be assessed with any certainty, an alarming one, that has led to an elaborate control mechanism, to monitor experiments in this area – to satisfy the government and through it the public, that the safety precautions are sufficiently stringent.

If Genetic Manipulation had become possible 50, or even only 30 years ago, I doubt if society would have imposed any restraints on it. But because it has become possible only very recently, and because we now live in an age of public pressure, conditions are different. Perhaps the restraints are unnecessary. Some think they are, and that they are inhibiting important work. But whatever the truth, perhaps there is no harm in taking care in the first instance until one is in a position to assess the credits and debits with some certainty.

MISUSE OF ANTIBIOTICS

There is one further concern I want to touch on, the misuse of antibiotics. They are, of course, the greatest achievement of medicine of this century, comparable in their effect on death and disease with the advances in sanitation of the last century. But precisely because they are so effective, are so easily administered, and have so few undesirable effects, they are all too readily over-used or misused. And they do have their Achilles heel – small numbers of bacteria normally sensitive to an antibiotic may in fact be or become resistant. Repeated use of antibiotics can then build up much greater proportions of resistant strains simply because the sensitives are killed off, and the resistants are not.

This effect was realized very early on. Workers in penicillin factories, for instance, before stringent precautions were taken, used to develop intractable infections from resistant strains. It was even feared that all bacteria might some day become resistant, and antibiotics thus become useless. Fortunately, this

SIR MICHAEL SWANN, FRS, FRSE, has been chairman of the BBC since 1973. In his academic career in Edinburgh he served successively as Professor of Natural History, Dean of the Faculty of Science and as Principal and Vice Chancellor of the University. He has been Chairman of many public, educational and scientific Committees of Enquiry and Royal Commissions; best known perhaps became the Swann Report on Scientific Manpower. Address: BBC, Broadcasting House, London W1, UK.

has not happened, even though antibiotics have been used for more than 30 years, presumably because sensitive strains of bacteria have a slight advantage of some sort over resistant ones, and left to its own devices, natural selection tends to restore the original *status quo*. Nevertheless, resistant strains selected out by continuous use of antibiotics do cause trouble, and it looks as though periodic trouble with them will continue.

AGRICULTURAL CONCERN

There are other concerns. One stems from agricultural practice. A curious property of some antibiotics is that administered in feedstuffs in rather small doses over a long period, they promote better growth. This is good for the economics of agriculture, but it is also a recipe for building up resistant strains in the guts of farm animals. Since most gut bacteria are harmless this might be expected not to matter too much, but for a curious phenomenon discovered quite some time ago, known as 'infective transformation'. It amounts to naturally occurring genetic manipulation, and it can lead to antibiotic resistance being transferred from a harmless bacterium to a harmful one. And this is exactly what began to happen. Worse still, these newly resistant bacteria were occasionally passed on from animals to man, usually through food. Hence occasional outbreaks of enteritis, usually in babies or old people not readily treatable, and with a high mortality rate.

In due course a Government committee was set up, of which I was chairman, and to cut a long story short, we recommended that antibiotics in feedstuffs should be limited as far as possible to certain antibiotics that were of little use for man, and that these should only be available on a veterinary prescription. So far so good, but in fact there is in the United Kingdom something of a black market in antibiotics from abroad, so that control is not as good as it should be. Gradually though, other countries are falling into line, and I am hopeful for the future.

I am less hopeful, on the other hand, about another problem. One of the rather rare forms of gut bacterium is the typhoid bacillus, and there is only one antibiotic that treats it effectively, chloramphenicol. And chloramphenicol, we discovered, was used quite widely in veterinary medicine for a number of surface infections. What would happen, we wondered, if the typhoid bacillus was made resistant, either directly, or more probably by infective transformation, to the one antibiotic that will deal with it. Typhoid is a very serious disease, and we recommended stringent controls. But once again, many other countries are alarmingly easy-going, and chloramphenicol resistant bacteria have indeed now turned up; in some countries there have been very nasty outbreaks. We can only hope they do not find their way here, and that the rest of the world becomes as sensitive to the dangers as the United Kingdom.

PUBLIC CONCERN

I have now touched very briefly on several public concerns of a biological nature. How have they actually become public concerns? Not, it is clear, from the people who benefited in the first instance from the activity in question.

Whalers do not seem to have doubts about whaling. The eyes of sprayers of DDT were on the benefits of so doing. Industries have seldom mitigated pollution of their own free will. And it is doubtful if many farmers, vets, doctors or patients brood on the hazards of building up bacterial resistance to antibiotics. The credit side of these activities is taken for granted, and deemed justification enough.

Who then puts the debit side? Scientists? Sometimes yes, more now than a decade ago, but not so very often. More usually it is the observers of science and the joiners of pressure groups, and for that very reason the debit side is sometimes presented more stridently or more hysterically than the situation may warrant. Perhaps the scientific community has only itself to blame.

But because the pressures, humanitarian, commercial or otherwise, to continue administering antibiotics, or polluting the environment are powerful, it is seldom that persuasion has much effect. On the contrary, none of the concerns I have described are likely to be dealt with by anything short of Government intervention, usually legislation.

Here, however, we come up against a difficulty. Present-day Governments are assailed by so many pressure groups complaining of everything under the Sun, that they seldom take much note of a pressure group that does not visibly have wide and powerful support. To be fair to them, how could it be otherwise?

How then do people who discern a hazard succeed in getting Governments to do something about it? It is not easy. But in brief they must raise enough support to convince Governments that there will be more trouble doing nothing than doing something. And this means putting their views to society, by books, or periodicals or the Press, or Radio or Television – each of these media reaching, as a rule, more than the preceding one – in the hope that they convince people.

It can be done. Indeed I have always thought that the credit for such controls on the use of antibiotics in farming as we have, should go not to my Committee's Report, but to Dr E. S. Anderson, who waged an almost one-man campaign with the help of the *New Scientist*. But some campaigns need more and bigger loudspeakers, which means the Press, Radio and above all Television. Without them the credit–debit balance will not get through to Governments, harassed as they are by all the other concerns of society.

FUTURE OPTIMISTIC

Having now considered the hazards and concerns that the over-mighty organism Man manages to inflict on himself and his fellow creatures, and how society, slowly, is putting some restraints into the system, it remains only to take a glimpse into the future. I am not altogether pessimistic.

True, I fear that a combination of commercial pressures and Government inertias will render some species extinct. Most probably whales. With luck the zoos will preserve the endangered terrestrial species. More hopefully I think that the hazards of modern agricultural methods and pollution are on the way to being solved. They are more susceptible to scientific and technological solutions, which are not too hard for Governments to insist on. I believe that genetic manipulation is much more likely to be a gain than a loss. I am most pessimistic about antibiotic resistance, not for any scientific reason, but because the credit side of the equation here is the most heavily loaded by the preservation of human life. And being an essentially caring organism, the overwhelming likelihood is that we shall be prepared to snatch at present rather than future salvation of our kin. In short, I think the control and limitation of the use of antibiotics must in practice prove very difficult for any society, for wholly understandable reasons. But the result in the long run could well be more intractable outbreaks of disease. Perhaps the British Association in the year 2000 will have something to say about it.

This review is a shortened and edited version of the author's Presidential Address to the Zoology Section of the British Association for the Advancement of Science. It was delivered at Heriot Watt University, Edinburgh, on 4 September 1979.

Nutrition
An Interdisciplinary Survey

DR PAUL PRÄVE, DR GERT-WOLFHARD von RYMON LIPINSKI and DR WALTHER SAMBETH
Hoechst AG, Frankfurt am Main, Federal Republic of Germany

With a critical world food situation and an increasing protein gap between the rich and the poor, scientific and technological solutions exist, at least in principle, to extend food production and thus avoid world wide famine. Human food requirement, malnutrition, the world markets of supply and food habits are reviewed. Conventional developments of food production consist in expanding arable land, the Green Revolution, more fertilizers and irrigation, plant protection and the accumulation of reserves. But the efficient processing of animal and plant proteins is equally important and so is the production of new concentrated proteins. Amongst the more unconventional developments, algae, fungi, yeast and bacteria as single cell proteins, are only the first steps in the biotechnological production of natural substances for human consumption to remove the food crisis.

This review is dedicated to Professor Dr Rolf Sammet, Chairman of the Board of Hoechst AG, on the occasion of his 60th birthday.

The seed is rotten under their clods,
the garners are laid desolate,
the barns are broken down;
for the corn is withered.

How do the beast groan!
The herds of cattle are perplexed,
because they have no pasture;
yea, the flocks of sheep are made desolate.

JOEL I:17, 18
About 800 B.C.

At present there is much talk about crises and the future of the human race. Crisis means difficulty or moment of decision – in medicine it is the crux of an illness. The energy crisis, the population growth crisis and the food crisis are all interconnected and have common roots.

The Club of Rome's survey *Limits to Growth* turned man's conscience to future problems. Amongst the many reactions some foresaw a new social and policitical structure within the next 200 years.[1] However this required the development of new opportunities especially in technology, in fact, another industrial revolution. These interpretations imply that a crisis exists in our faith for the future.

Naturally, this type of discussion is not new; the Bible and, more recently, Malthus (1766–1843), saw the essential problem as one of hunger. Malthus proposed that food supply would always fall behind birth-rate. Malthus's *Essay on the Principle of Population* (1798) influenced Darwin who, when he read it, realized that his theory of evolution of animals and plants agreed with the evolution of man.

Animals and plants are not able to use material aids for qualitatively improving their food and certainly not for quantitatively increasing it – but man can! And, if it is proved that the main cause of hunger is poverty and ignorance,[2] then the converse should be stated, that the civilized powers must be, at least in theory, in a position to solve the problem.

ADVANCEMENT – HUNGER – SOLUTIONS TO PROBLEMS

The progress of the last hundred years, especially with respect to medicine and artificial fertilizers, has opened up a new dimension for mankind. Is man's food supply thereby secured? It is doubtful! Today, one can, however, specify many of the problems – land requirements, shortage of capital, distribution problems, lack of raw materials and lack of energy.

Extensive studies of nutrition have begun and a simple 'eat to repletion' concept is no longer accepted by scientists. Proof of progress lies in the fact that today we can feed almost 90% of the world's population of 4×10^9 people; this has become possible by intensive agriculture, fertilizers, plant protection, mechanization and preservation. The difficulty lies in the approximate number of 400–500 million starving people:[3] figures of this magnitude are being quoted more frequently and not only by UN organizations. It is obvious that further new opportunities must be created; for example new hybrid strains of grains, utilization of krill and the breeding of cows for milk, rich in protein content. New methods of processing can be envisaged whereby many useless products or side-products can be effectively treated.

CCC-0308-0188/80/0005-0006$09.00/1

The poor or wasteful utilization of nature's resources must be avoided; the high energy contents of many products and processing techniques in the food industry has been overlooked; and in this context biotechnology has a special meaning. It has the advantage of connecting both industrial preparations with the cycling and, to some extent, recycling, of biological products. Micro-organisms such as algae or bacteria can now be technically manipulated in a way impossible with other forms of life; genetic improvement is also feasible.

A wide field of new possibilities presents itself: Of primary importance, is man's food a basic requirement for all our existence – but reaction in this area is extremely conservative. One asks oneself if, in the first instance, sociological or physiological forces or political pressure must open the way towards these new possibilities. It is also within the bounds of imagination that success can be attained with market strategy and scientific-economic action. This part of the food problem is equally important to new development for better quality and sufficiency of food in human nutrition.

THE PRESENT SITUATION

To summarize the world's present food situation is to say that one works with yesterday's figures and, obviously, these are out-of-date. Therefore, computer investigations by the Club of Rome[4] or Bariloche[5] are important because they show trends without getting lost in detail. These very details are, however, often the beginning of very harsh criticism. Whether this criticism is well founded or not, the work of Meadows[4] as well as that of Mesarovic and Pestel[6] started a world-wide discussion with the result that, 'Considerably more than half the world's population can't sleep for hunger, and the rest from worry of what these starving people might do!', to quote Josue de Castro.[5]

It is certainly an historical fact that starvation has always been with us. But what is frightening is its escalation and the frequency of its occurrence in recent times. Today's communication media, the press, radio and television, make wide publicity possible. World statistics are available which give information according to year, continents and countries about birthrate, food production and food supply per person in the population. Population density – as a ratio of cultivated arable land – is increasing in South-East Asia and South America much faster than, for example, in Western Europe or North America.[6] It does not require complicated computer analyses to present some prospects for the near future. The poorer parts of the world, already with millions of undernourished people, will find themselves tomorrow with thousands of millions unable to obtain sufficient calories, proteins, vitamins or essential minerals. From FAO estimates the number of starving could grow from today's 400–500 million to 750 million by as early as 1985.

The concept of undernourishment, on the basis of today's medical and nutritional knowledge, is not adequate. Undernourishment would not be recognized by many of the people on Earth, because they have never experienced an adequacy of nourishment. To judge the world's food situation as it is today one needs first some knowledge of man's food requirements and the effects of malnutrition.

Food Requirements

We know of about 50 essential food substances which the body cannot synthesize and which must form part of our diet. If they are lacking, either in the long- or the short-term, deficiency effects follow. The results are particularly obvious when there is a shortage of essential amino acids due to lack of food protein intake or too one-sided composition. If, in addition, there is a calorie deficiency then the effects of any protein shortage are felt still more, as part of the protein is used for energy production and its inherent properties, as a protein, are lost (Fig. 1).

Recent discussions tend to the view that protein shortage should not be taken so seriously as today's requirements are estimated at about a third of the value of previous estimates. New standards based on

DR P. PRÄVE was born in 1930 and obtained his PhD in microbiology from the University of Göttingen for research on iron bacteria; he continued to work there on antibiotics. Since 1958 he has been employed in the Department of Biochemistry and Microbiology of Hoechst where his research interests included microbial tests and the screening for micro-organisms capable of producing active metabolites. During the last decade his main field has been the development of industrial processes for the production of single cell proteins.

DR G.-W. von RYMON LIPINSKI was born in 1947 and obtained his doctor's degree in food science from the University of Hamburg for research on fish extracts. Since 1975 he has been employed at Hoechst AG. His present interests are the application of food additives and bioproteins.

DR W. L. SAMBETH, born 1934, received his degrees from the Agricultural College of the University of Bonn, where in 1964 he finished his PhD thesis on biological protein evaluation. After two years as Research Associate in poultry nutrition at Cornell University, he has been working as nutritionist since 1966 with Hoechst in the Department of Veterinary Medical Research. His current main activities are research and development of feed additives and single cell proteins in animal experiments.

Address: HOECHST AG, Postfach 80 03 20, 6230 Frankfurt am Main 80, Federal Republic of Germany.

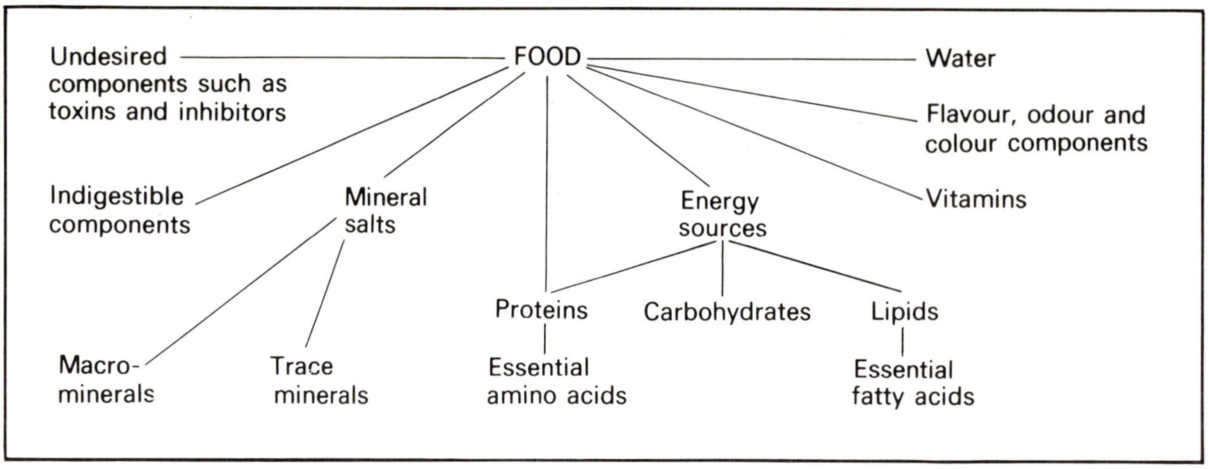

Figure 1. Components of food with respect to their use and function.

the results of many human metabolic experiments remove the apparent protein shortage and the calorie deficit assumes priority. Women, children and the sick are often overlooked with respect to protein requirement, as they do not belong to the statistical average but rather to the special 'vulnerable group'. Scrimshaw[7] and Latham[8] have forcibly focused attention to the fact that requirement figures following infections are certainly not valid. The requirement increases, for example, to about 150% if food utilisation is greatly reduced as a result of diarrhoea, invasion by parasites, and for similar reasons.

A further point of discussion arises about FAO's figures of protein requirement: here a global protein requirement is under discussion although what is really meant is the requirement for essential amino acids. Depending upon the type of basic food, the essential amino acid content varies considerably from country to country. One speaks, therefore, of differences in protein quality. In principle, a poor quality can be compensated by increasing the total protein intake which will avoid relative shortage of limiting essential amino acids. All other excess amino acids would then be used up and lead in effect to an increased protein requirement.

Ganzin[9] has calculated that if the protein quality drops around 10%, then the increase in the amount of protein required is in the region of 15%. The higher the living standard, the higher the demand for animal protein products.[10] In countries with a highly developed technology and strong purchasing power we are aware of an especially high intake of animal protein which lies well above the maximum physiological requirement figures. Here one can speak of a social or psychological requirement with FAO standard values again allowed to vary considerably, according to different regional considerations. However, in developing countries, too, the daily intake of animal protein and calories is directly dependent on family income (Fig. 2).[11]

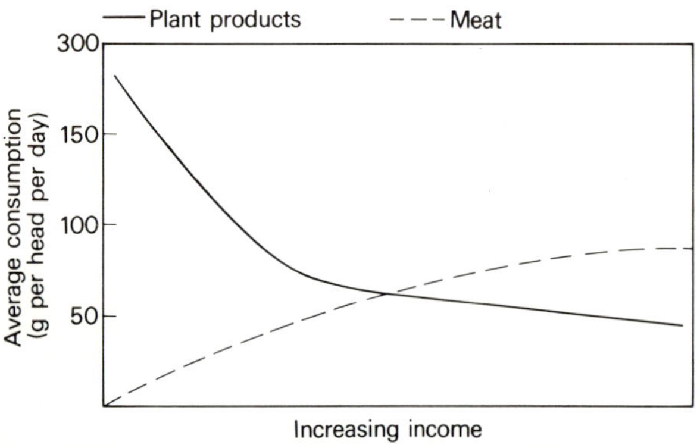

Figure 2. Cereal and meat consumption as a function of income.[17]

Malnutrition

Malnutrition is also found in industrialized countries. Though here the problems are due to overeating as a disease of civilization or a shortage of trace elements or of vitamins.[12] The worst problems in the developing countries are Kwashiorkor and Marasmus – diseases which are due to chronic protein and calorie shortages. These clinical cases of disease, running into hundreds of millions, have not dropped in number although the FAO and WHO revised their protein requirement figures; the result was that, theoretically, there should exist no shortage in these countries. What these average national figures are unable to show is the unequal numbers of people either starving or suffering from food shortages when the remainder live on a surplus thus considerably raising the average value of food consumption. If statistically evaluated averages for the total population are correct then, according to FAO figures, Sweden and Chile are nutritionally on the same level, both countries having about the same protein use *per capita*. In Chile protein–calorie malnutrition is a very serious problem, but in Sweden does not exist at all.[9]

Infections make malnutrition worse and poor nutrition increases the severity of infectious diseases. The death rate from measles is therefore about 200 times higher in the developing countries than in the

USA. Those who manage to survive a serious infection are often left with permanent psychiatric disturbances, reduction of physical capabilities and reduced life expectancy.[8] The situation is similar for children whose mothers suffered a chronic protein shortage during pregnancy or nursing, the central nervous system having been irreparably damaged.[12,13] These connections have been proved beyond any doubt in extensive animal experiments.

Within a genetically similar population, an infant's weight at birth is a reliable index for assessing the maternal nutrition. In 1970 WHO established that the birth weight in low caste Indians was 10–20% below that in comparable, wealthy families. In Africa and Indonesia the same observations could be made.[14]

Although hygienic conditions vary widely from country to country, the death-rate for children between 1–4 years is also used as an indicator of malnutrition. Information provided by FAO puts child death-rate in some African and South American countries 5–20 times higher than in Europe.[14]

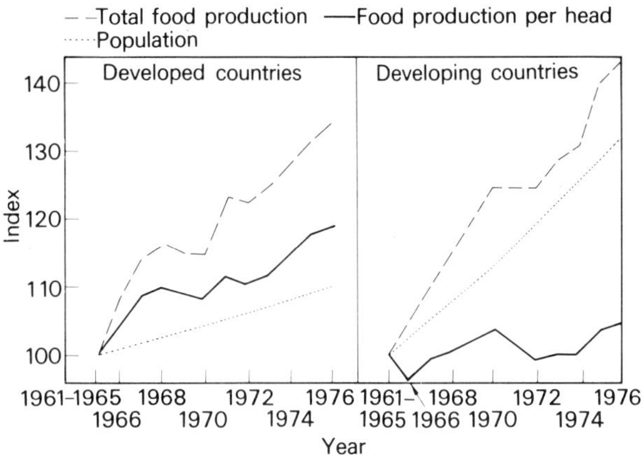

Figure 3. Food production and population 1961–1976.[24]

The Present Supply Position

Although FAO figures show that global food production has greatly increased in the last 15 years, the supply per person in the developing countries has remained practically the same as that in 1961 (Fig. 3), the reason for this being the population explosion. According to the latest reports it appears that the global birth-rate has passed its peak but such wide ranging statistics can be misleading. Different considerations, arranged according to groups of countries with similar socio-economic structure show that, up to now, undernourished people have no hope of quick relief and because of this, they could be a source of political instability. 'You can't build peace on empty stomachs!' stated Lord Boyd Orr, the first Director General of the FAO.

The social requirement for animal protein is reflected by the supply of meat, eggs and milk. The developing countries, about 50% of mankind, have today only 16–20% of the world's supply of animal protein. Thus, expressed in terms of supply per person per day for protein and energy, there is a drastic difference between the privileged world and the forgotten world (Table 1). On an average, all industrialized countries consumed more than five times as much animal protein per person than the developing countries. If one takes North America with 71 g protein per person per day and some African countries with 4.5 g protein per person per day as the two extremes, the consumption in North America is even 15 times greater.[15] The industrial and the oil-rich countries are in a position to buy available supplies of grain and meat on world markets whereas many developing countries, with infinitely poorer food supplies have a great deficit due to their rapidly increasing populations.

During the last 40 years North America, Australia and New Zealand have become the only grain exporting countries, all others import, as far as their foreign exchange allows it (Fig. 4). In fact, the United States are 'in a position to make life and death decisions in global nutrition.'[16] The large imports of the centrally planned countries are conspicuous; from 1934 to 1938 they were exporters. These movements on the world grain markets lead to economic and political dependency, so clearly evident within the EEC during the 1973 protein crisis. The EEC is not only the world's largest food importer of 8 million tonnes per year of protein for animal feed, which is 85% of its feed protein requirement;[17] the Community also put more animal products into storage: in 1976 the butter mountain in the EEC stood at

Table 1. Food and Nutrient Consumption; Present Status[24]

	Developed countries	Centrally planned countries	Developing countries
Nutrient consumption per head per day in 1974			
Protein from plants g	39.2	51.2	42.5
from animals g	56.2	24.6	11.6
Energy from plants MJ	9.27	9.32	8.42
from animals MJ	4.68	1.87	0.76
Total food availability in 1976			
Meat million tonnes	68.3	39.3	21.4
Eggs million tonnes	11.5	8.5	3.8
Milk million tonnes	213.9	135.8	85.0

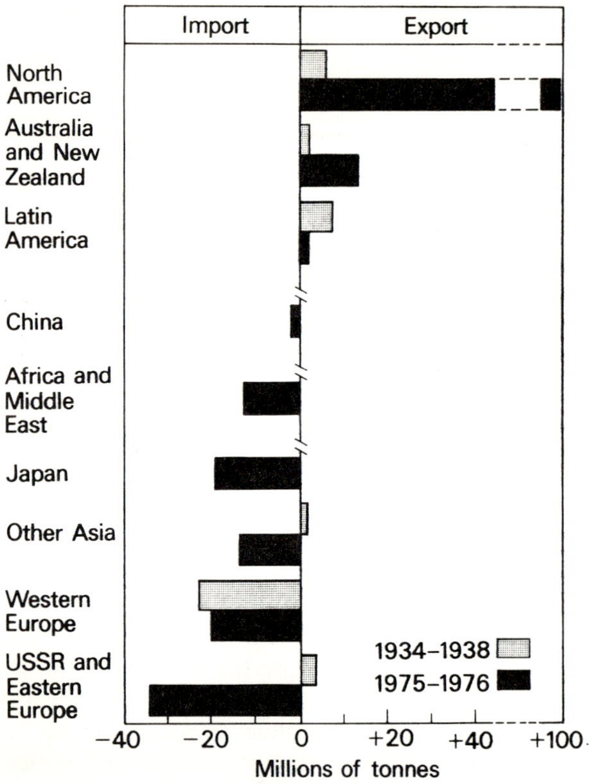

Figure 4. Development in world grain trade.[67]

400 000 tonnes, the storage of milk-powder at 1.3 million tonnes.[18]

The world market of animal feed protein, the average of which was about 30 million tonnes per year (1971–1977)[19] is composed mainly of oil seed-cake and fishmeal. Soy bean meal provides about 58% of the total protein offered, this establishing its special leading position. Soy beans are exported in increasing amounts almost exclusively from the USA and Brazil; the large areas under cultivation in Asia only just meet their own demands. Despite efforts for self-sufficiency in proteins, the EEC has not been successful in acclimatizing the soy bean in Europe because of its unsuitable climate.

Fishmeal production has declined about 30% in the last eight years and, depending on the catch, provides only 5–10% of the protein on the world market.[19]

FOOD HABITS

The choice of food is determined by two main factors, the most important being to satisfy the pangs of hunger. During a serious famine, habits and preferences are forgotten and, as a last resort, products which are normally regarded as inedible may be eaten. When ample food is available a choice is made. This choice depends on food habits.

Food habits are complicated and stem from a variety of interconnected factors of which one is often unaware. Basically, they are determined by the food available locally. Fish, for example, is a normal part of the diet on the coast, near lakes and rivers. When the climate is moderate, grain is the main source of carbohydrates and in tropical areas yams, tapioca or similar starch containing plants. Where food is plentiful only a few sources are used and, together with the food which is easy to collect, delicacies are sought after. When common foods become scarce, they may even be regarded as luxuries together with those which are not normally considered as appetizing. In Europe the use of the herring can be quoted as an example; a few years ago it formed part of the diet of the poorer parts of the population, but now with decreasing stocks and increasing prices it is highly valued. The converse is true when a luxury food becomes plentiful and its price drops.

The food habits formed at an early age in childhood are, in general, those of the parents and are taken for granted. Religious and cultural influences also play a large part in the formation of food habits. In the earliest cultures food was not only nourishment but also an integral part of important ceremonies. Even today restrictions based on religious customs can be found in almost all countries. In the religious texts of several religions rules are given as to the choice of food, in the Bible, the Koran and the Talmud. The prohibition of pork in several eastern countries has a religious basis just as the sanctity of the cow in Hindu countries and the prohibition of horse meat in some European countries. During the Christian time of Lent and during Ramadan there may be complete or partial periods for fasting and abstinence.[20]

As they are acquired so early in life, food habits tend to be very conservative. New foods and changes in agricultural methods may be hard to accept. The upper classes are more open to accept new foods and the introduction of novel products may be successful if they are offered as delicacies to them. But this would only work if the food offered was really new; and it is unlikely that a change in food habits in the upper classes could be achieved in a short period. The chance of changing the food habits of the majority of the population is probably impossible by this method.

When new foods compete with familiar ones, or are introduced to supplement them, they are often regarded as inferior. To achieve a basic change in food habits, therefore, strategies and technologies must be developed in such a way that, at least in the beginning, the proposed change is integrated as much as possible with local habits, with local religious customs and avoids any impression that the products are in any way inferior, or *Ersatz*.

CONVENTIONAL DEVELOPMENTS AND SOLUTIONS

'Present famine, coexisting with surplus, does not yet prove that Malthus was right; the Malthusian limits have not yet been reached'. This conclusion of Chaudrey[21] made in 1976 is based on existing opportunities for agriculture to expand in order to provide adequate food for mankind.

Expansion of Agricultural Land

The most important requirement for food production is arable land. On the continents of Australia, South America and in tropical Africa, for example, agricultural potential is still very great. On the other hand, in Asia with its very large population, over 80% of the possibly available arable land area is already under cultivation. On a world scale it has been estimated that the present available area of agricultural production could be more than doubled. This potential has, however, some economic limitations as the reserve lands consist in part of deserts, soils with high salt content, ice-covered regions or mountains.[6] How far these regions can be made useful is an economic decision with all its consequent social problems. One can take the thesis that the unit cost for opening up such new areas has risen exponentially to a maximum of $6000 per hectare.[5] By today's standards such costs would be uneconomical.

If one takes the most optimistic case and all potentially arable land will be fully used for food production then, according to Meadows,[4] the present rate of population explosion will ensure that in only 30 years time we will have the same shortage of land, and hence of food, as exists today.

Furthermore, the increase in population and use of large areas for housing development means that more land is being lost which would formerly have been used for agriculture. Pimentel and colleagues[22] estimated that the USA has developed 50 million hectares of arable land since 1900 but that 30 million hectares were lost for building towns and roads. Apart from this, 80 million hectares have been completely ruined by soil erosion. On a global scale it is estimated that the environmental destruction of land is even worse than in the USA.[22] The more intensive use of agricultural areas already under the plough has proved far more successful than the opening up of new agricultural land.

The Green Revolution

During the last ten years the successful use of new seed strains, intensive fertilizing, appropriate methods including artificial irrigation in many of the developing countries, has become known collectively as the 'Green Revolution.'

Especially in India and Pakistan the very successful introduction of these various measures has brought about a fivefold increase in the grain harvest.[23] The importance of this success is clear if one is aware that in these countries more than 80% of the population are dependent on agriculture.

The wheat geneticist N. E. Borlaug who received the 1970 Nobel Peace Prize for this success has however, warned, that the Green Revolution must not be seen too optimistically and raise too much hope, as difficult problems are hidden in the system.[23] The Green Revolution is not only a question of available capital but also of technological 'know-how' and this may, in some circumstances, lead to negative social side effects.[44] As the owners of large farms are usually better trained and have more capital to put into new technology this will obviously increase existing agricultural and social inequality, a phenomenon, however, that not only concerns the developing countries. Certainly, 'the Green Revolution has won a temporary success in man's fight against hunger and deprivation. It has given man a breathing spell.'[23]

With modern cultivation technology it is not only possible to increase the harvest yield but also to improve the quality of the food value through breeding disease-resistant strains as well as through crops whose composition has a better nutritional value than those of the original strains. The best known example is the cultivation of a strain of maize carrying the recessive gene 'opaque-2' which significantly raises its lysine and tryptophane content. As lysine is nutritionally the least of all the essential amino acids present in grain, the opaque-2 maize has biologically a higher value than normal maize. Another example is the cultivation of rape-seed, low in glucosinolate; its extraction residue can be used without difficulty as a protein feedstuff for animals.

Fertilizers

The new varieties have genetically more growth potential and usually react much better to intensive fertilization. Justus von Liebig's knowledge, large-scale industrial technology and the use of fossil energy have, especially in the last few decades, increased the yield per hectare several-fold. In the developing countries these increased yields can also be noted (Table 2).

The less the soil itself is able to provide mineral nutrients for plants and the fewer the organic fertilizers which can be used, the more important are

Table 2. Use of Fertilizer and Grain Yield (expressed as kg per ha cropland)[a]

Year	Fertilizer input grain yield	Developed countries	Centrally planned countries	Developing countries
1961–1965	N	23	9.2	3.1
	P_2O_5	23	6.4	1.4
	K_2O	18	6.3	0.7
	Total grain	2340	1340	1090
1974	N	35	34	10.6
	P_2O_5	35	17	5.2
	K_2O	29	18	2.8
	Total grain	2820	1860	1290

[a] Adapted from FAO.[24]

synthetic fertilizers. This applies particularly to poor soils in the tropical developing countries. The enormous reserves which could be mobilized by the application of more fertilizers is clearly demonstrated by FAO statistics.[24] From 1961 to 1965 the use of fertilizers per hectare in the developing countries was about 10 per cent of that in the industrialized world. Now the percentage for these countries has risen on average to 25% (for nitrogen). The gap between the 'forgotten world' and the 'privileged world' is shown by the extreme example of a single country: in the Netherlands 185 kg nitrogen per hectare is used, in the developing countries, however, only about 10 kg per hectare! Even so the optimum in Europe has not been reached and for example, leaf fertilizing or bringing ammonia directly to the root region would certainly have the effect of increasing production still further. The more efficient use of older methods could be another solution. Finally, it is especially important for countries dependent on imports of energy supplies and agricultural production material, to realize that the manufacture of nitrogen-containing fertilizers is very energy-demanding.

Plant Protection

Agriculture can only be profitable if it is based on the farming of large areas with monocultures; plant protection is therefore essential if great losses, both with respect to quantity as well as quality, are to be prevented. Chemical plant protection agents should be used in conjunction with the breeding of new types of disease-resistant strains of plants, with fertilizing and mechanical earth cultivating methods. One talks, therefore, of integrated plant protection, an important part of the Green Revolution.

It is estimated that there is a world-wide average loss of harvest of about 30% as a result of the action of malevolent pests.[25] This loss, according to Büchel,[26] would be at least double if chemical plant protection were completely stopped. Higher food prices, for the USA from four to fivefold[23] would follow and famines in economically weaker countries could not be avoided. On the other hand, if one were able to prevent completely the wastage caused by pests one could feed adequately about 900 million people with the crops so saved.

Losses of over 40% have been quoted for Africa and Asia;[25] this is explained by the small amounts of chemical plant protectives used there, in contrast to the USA, Canada and Western Europe. These huge losses give an idea what enormous reserves could be mobilized by an extended use of just the conventional herbicides, fungicides and insecticides which, at the present state of knowledge, cannot be substituted by biological plant protection methods. Research for such new plant protection materials, harmless both to men and animals, is being continued with increased intensity.

Irrigation

Although irrigation has been well known for about 4000 years, it is being utilized only on a small scale and FAO data[24] estimate that only 15% of all cropland in the world is being irrigated. Modern irrigation is mostly based on fresh water as the desalination of sea water would be economic only under very limited conditions. However, brackish waters can be utilized for special plants, as demonstrated in the Negev and Sinai deserts. Fossil water might also be more efficiently tapped there and this would bring about 1% of the total desert area under regular crop production.[27]

In many development projects the technology of irrigation plays an important role. In India alone during 1969–1970 about 7000 new wells were drilled, providing for irrigation of an additional 1.4 million hectares.[28] In Mexico, in 1975 nearly 30% of all cropland was irrigated and from this area 65% of the total plant production of the country was harvested.[29] To sum up, irrigation together with other measures can increase crop productivity enormously for a developing country.

Meat Production

On a global scale about 20–25% of man's protein and 10% of his energy intake is derived from food of animal origin.[30] In the conversion of animal feed into human food, about 70–90% of its nutritional value is lost.[31] However, about two thirds of the animal feedstuff is derived from substances which are either undesired or unfit for human consumption.[30] Several examples of by-products for livestock feed are: straw from wheat, oats, rye, barley, millet, rice, soy beans; hulls from grains, soy beans, peanuts; pulp and molasses from beet, citrus fruit, grapes and tomatoes.

Animals are irreplaceable as consumers of this type of waste which would otherwise cause environmental pollution. With grain it is different, as here man competes with animals. In the USA about 93% of all grain is given to animals,[32] in contrast to the developing countries where only 10% of grain is used in animal feed and 90% is used directly for human consumption.[21] If more grain could be replaced more efficiently by non-competitive feedstuffs the more would become available for man.

In this respect, ruminants such as cattle, buffalo, sheep, goats, camels, llamas and reindeer have a special role to play. As their intestinal microflora can break down cellulose and hemicellulose, useful products are formed and can be absorbed by the livestock as an energy source. These micro-organisms can also synthesize amino acids essential to the ruminants which makes these animals irreplaceable for the grazing of permanent pastures, rangelands and meadows; these make up about two-thirds of the agriculturally useful areas of the world.[30] In many countries improvement and enlargement of herds of beef cattle could, therefore, bring about a real increase of high quality meat.

Since the Second World War intensive research in animal husbandry, in genetics and about the physiology of nutrition has led to considerably improved conversion of animal feedstuffs to human food. As a result, there has been a large reduction of the losses occurring during such conversions. Feed additives e.g. growth promoters have also helped in this improvement and today their scientific significance, even in a well balanced feed mixture, is undisputed. Optimum maintenance and, as far as possible, optimum nutrition, together with genetically greatly improved animals are a potential which, for many countries, is far from being exhausted and research in animal production is also nowhere near its end. Above all, in the animal and veterinary sciences there are very many problems which are awaiting their interdisciplinary solutions.[30]

Fishing and Aquaculture

The ten year increase in world fish catch to 70 000 million tonnes per year was interrupted in 1973 when Peru's important anchovy catch failed to materialize.[33] Although many traditional fishing grounds have been overfished, it appears that the available fishing potential of the world's seas are a long way from being exhausted. The price per kilogram of fish from these potential fishing grounds, often distant from the consumer, will certainly be high. In Europe this increased cost will be unavoidable for some countries owing to the imposition of the 200 mile limit.

The cost for a given yield of fish appears in recent years to favour aquaculture; its share in the edible fish market is constantly increasing and production at present is about 5 million tonnes per year. (See O. Kinne, Aquaculture, p. 24.) The FAO estimates that the area under aquaculture will multiply tenfold by the end of the century. Depending upon its intensity aquaculture demands a high standard of scientific knowledge and technical equipment. The most simple form of fish farming is the use of net-cages in coastal areas. At the other extreme there are silos with an optimization of temperature, pH and oxygen content, a ready-mix feed having the ideal combination of nutrients where fish are farmed independent of season.[34,35] By this method up to 300 kg of fish per cubic metre water can be grown and harvested. Progress in the field of aquaculture is in a rapid stage of development and is now passing through the phases which the poultry industry underwent 25 years ago.

Grain Reserves and Emergency Aid

The bad harvests in 1972, especially in Russia, China and India, which followed consistently unfavourable weather conditions, greatly affected the world's grain reserves, their prices and the planning of further grain cultivating areas in the main exporting countries.[36] To reach stability for the future, acceptable to both producers and consumers, Borlaug[23] put forward the suggestion of a world buffer stock as presented to the World Food Conference of 1974. In his scheme the keeping of grain reserves is only partly an agricultural problem, it is much more political. Not only could amounts and prices be regulated by having grain reserves but provision could also be made for food emergencies becoming increasingly common.

The catastrophic famine in the African Sahel in 1975, probably to repeat itself, is only one dramatic example. Emergency aid can never be more than a temporary and not a lasting solution.[18] Even a permanent reduction in the EEC's surplus can never be reached by this method. In many receiving countries the appropriate infrastructure, administration and storage facilities are lacking for the undernourished population in rural areas. Above all, the transfer of processed products, such as milk-powder, is particularly problematical; in comparison with the traditional food in the receiving countries, it would inevitably be regarded as an unconventional food.[18]

PROCESSING

Improved agricultural methods, new strains of plants with increased crop yields and high protein content, intensive animal rearing and improved fishing methods – all offer possibilities for increasing protein production. But in addition to production these proteins have to be transported from the site of origin to the regions where they are needed. This is not difficult when grains and pulses which do not spoil easily have to be transported. But milk, meat and fish can only be distributed as fresh and chilled products where a well-developed distribution system exists.

In the industrialized countries milk, meat and fish are distributed as fresh products from refrigerator to refrigerator, but such refrigeration equipment demands considerable capital investment and furthermore the distribution of refrigerated goods demands a reliable supply of energy as most refrigerated foods can be kept fresh only for a few days. Refrigeration, therefore, offers only limited possibilities for storage. Freezing, which is much more energy-consuming than refrigeration, allows storage of foods for several months and, in a few cases, for several years. The long-term storage for several years of meat and fish is, however, impossible even after the normal freezing process. Lowering the storage temperature below the usual values increases the storage time but demands considerably more energy. Faults in refrigeration equipment or interruption of energy supply can cause spoilage when the temperature rises.

An increase in the production of agricultural products with high protein content is therefore especially desirable. To overcome the time and distance problems of protein deficiency, it will be necessary to develop protein products which can be transported easily, stored for a long time and produced cheaply. Ideally, it should be possible to transport and store these products without any special care in simple packings and in all climates. Concentrated proteins with their high efficiency ratio and favourable properties of direct addition to human food, may contribute quickly to man's protein requirements; they alone can avoid the inevitable losses of protein which occur when feeding animals (Fig. 5).

Animal Proteins

Several years ago the first attempts were made to produce storable protein products such as skim milk-powder from defatted milk; they can be stored for long periods. The milk fat can be used in several, storable animal fat preparations.

Skim milk-powder is produced in large amounts in several countries, but it has only a limited use as its lactose content is a drawback for human nutrition. In several parts of the world adults are not able to assimilate lactose and most skim milk-powder with its concentrated protein therefore is fed to animals. Enzymatic splitting of lactose to galactose and glucose could improve this situation. The production of casein, of lactalbumin or other protein preparations is possible but more expensive.[36]

Also during cheese production large quantities of milk proteins are wasted as the whey proteins remain in solution and therefore whey has been fed to animals as much as possible or, more often discharged as sewage. Whey proteins can be recovered by drying, precipitating with suitable chemicals and especially by ultra-filtration, under pressure through membranes, which retain the proteins. The ultra-filtration process is expensive but the recovered protein concentrate can be used in the industrialized countries, in the production of several foodstuffs; this production of whey protein has been increasing in recent years.[37,38]

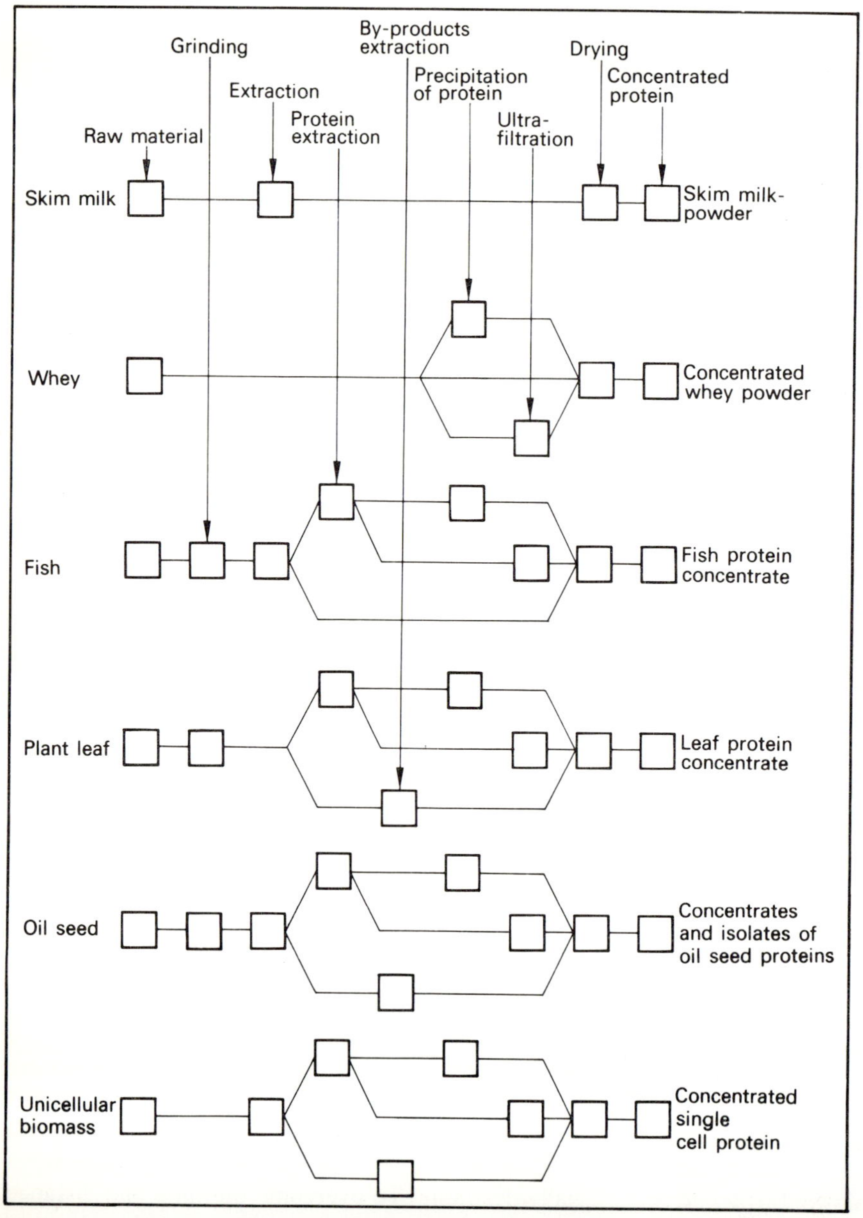

Figure 5. Preparation of concentrated protein.

Fish resources close to countries with high population densities are exploited extensively, even excessively, and fish stocks have been decreasing. [See also: Korringa, European fisheries mismanagement. *Interdiscip. Sci. Rev.* **3**, 335 (1978).] Fish caught in distant waters can be transported in a refrigerated condition. However, freezing is required from the time the fish are caught until they are distributed to the consumer and this is, of course, energy-consuming. To can fish is expensive. A storable protein meal could be prepared from fish which are not suitable for direct human consumption. Instead of preparing fish meal, which has been produced for animal feedstuffs for decades, fish which have been stored for several days or which are too small for direct use could be processed into protein preparations for human consumption. Fishmeal itself is not prepared under hygienic conditions and hence is unsuitable for human consumption.

Fish protein concentrates which can be used in human nutrition are prepared by extracting minced fish with water and certain solvents. By this method fats and low molecular compounds are removed from the fish, as these could cause undesirable changes in taste during storage. If only part of the fat is extracted, a typical fishy taste often remains and this restricts the use of these special concentrates to countries in which fish and dried fish are commonly eaten. In addition to muscle proteins, fish protein concentrates also contain connective tissue and minerals. The mineral content of fish protein concentrates, which originates mainly from their bones, is usually lowered by special pre-treatment. The minced fish is very susceptible to spoilage by bacteria and it has, therefore, to be processed quickly if the concentrates are to be used for human consumption.[39,40]

Fish protein concentrates are an example of the difficulties which occur when new proteins are introduced into human nutrition. In particular, it can be seen that during the processing of specialized products manufacturing plants need an efficient supply of raw material for economical running.[41]

The utilization of krill (*Euphasia superba*) brings problems similar to those of fish processing. Krill are small crustacea found in antarctic waters and can be a valuable source of protein if they are caught in sufficient amounts. They are only a few centimeters long and are, therefore, only of limited value for direct consumption, they spoil quickly and have to be processed at sea. Freezing is possible but, for unprocessed krill, this would be costly because of their high shell content and therefore is only possible for industrialized countries. Trials have been made to process krill into soups, pastes and several other products similar to those prepared from crabs. Krill can also be dried and may be processed into a storable protein concentrate similar to fish protein concentrate. The disadvantage of processing krill at sea is offset by the advantage of having a steady and very large supply of raw material hence the processing plants are well utilized.[24]

The blood of slaughtered cattle is very susceptible to spoilage if it is not immediately fully processed and then offers an excellent culture medium for many, even pathogenic, bacteria. Small quantities of blood are used directly in the manufacture of sausages. Blood protein is also produced industrially under strictest hygienic conditions for protein concentrates, but in many countries religious customs present obstacles to the use of this protein source.[43]

Other sources of animal protein, for example, slaughterhouse by-products and feathers, are of little importance. The quantities available are too small or processing is too complicated and hence too expensive when compared to other potential protein sources.

Plant Proteins

The preparation of protein powders or storable products is advisable when animal protein is considered. Processing is equally necessary for plant proteins when products for human consumption are required. Not only is the separation of proteins from other compounds desirable but it is also important to remove undesirable taste and flavour and, obviously, toxic compounds.

Soy proteins are the most common plant proteins. In the Far East many foods based on soy beans are esteemed. *Tofu*, for example, which is similar to white cheese, and the fermented *Miso*, are produced on a large scale. In these fermentation processes mainly carbohydrates are metabolized. In Western countries soy beans were initially grown for their fat content. The fat is extracted from the minced soy beans by solvents together with lecithin. The extracted soy beans are not normally used in foods for the special taste and flavour of the extracted material is rather disagreeable. However processes exist which can lower or remove this 'soy flavour.'

The protein of the extracted materials is, depending on the crude material, 40–50% dry matter. Extraction with water, acidified water or mixtures of water and ethanol, produces soy concentrates with about 70% crude protein. For removing the polysaccharides in the soy concentrates, the extracted soy material can be treated, at elevated temperatures, with dilute alkali, and the proteins precipitated by acidification. This process yields soy isolates with protein content. The various processes for preparing concentrated soy proteins also inactivate the diges tion inhibitors of the soy bean like trypsin inhibitors, the active form of which blocks the digestion of proteins. Defatted soy meal and soy grits and, especially, soy isolates and concentrates, are a valuable supplement in human nutrition and are used in the industrialized countries in a variety of foods.[44,45]

Other oil seeds which are grown for their fat content also contain substantial amounts of proteins and can be processed in a similar way to soy. The content of crude protein of cotton seed meal is 60–70%, about 40% for rape-seed meal, 60–70% for peanut meal, 40–50% for sesame meal and about 50% for sunflower seed meal. Defatted oil seed meals contain inhibitors or toxic substances, the gossypol of cotton seed, the glucosinolates of rape-seed or the chlorogenic acid of sunflower seeds; these can be removed by suitable processes.

The gossypol can be separated from cotton seeds by polar solvents together with the fat and can then be removed from the latter during the refining process: it can also be extracted. Oil seed meals which have not been heated during processing are reported to exhibit weak inhibitory activity against protein digesting enzymes. Depending on the processing, oil seeds can, as with soy, yield detoxified meals, give protein concentrates having a crude protein content of more than 70%, and protein isolates with a protein content of more than 90%. Basically, all those products may be used in human foods.[45,46]

Pulses are known for their rather high protein content. They may be used in a similar way to extracted soy meals, soy grits and other oil seeds, both as meal or as concentrated protein products. Growing pulses is advantageous even at low levels of N-fertilizing due to the symbiosis between them and N-fixing bacteria; seeds with high protein content can be harvested. The disadvantage of legumes, which include the soy bean, is that they usually contain enzyme inhibitors and other undesired compounds which have to be inactivated. This process can be combined with the concentration of the proteins in a method similar to that for soy.[47]

An almost unlimited source of proteins are leaves and other parts of green plants. The protein content of these parts is less than in defatted oil seeds. By means of special processes undesired compounds can be removed from these plants while concentrating the proteins. In practice plants with a rather high protein content should be used, for example, alfalfa. The first step in processing is to prepare a concentrated press-juice, or extract, which contains the dissolved proteins. By heating, ultra-filtration or other precipitating processes, the protein can then be obtained. The crude protein from the first stages of processing is not fully suitable for human consumption having a dark colour and a special flavour. By fractionation two different proteins can be obtained from the press-juice, one of which is almost colourless, has a weaker flavour and seems to be useful in food. The second, coloured fraction, can be used in animal feed as well as the residual leaf material.[48,49]

In addition to the plant materials mentioned above, there are such other sources as milling wastes and brewer's grains which could be processed into concentrated proteins. However, these seem to be of less importance than oil seeds, legumes and leaves.

Practical Application of Concentrated Proteins

The production of concentrated protein powders suitable for human consumption is only the first step in combatting protein malnutrition. In situations of direct privation and during famines, protein powders can be added to common foods which lack protein. Slight changes in appearance, flavour and taste of those foods would then not be so important. Very different is the situation when there is only an apparent lack of protein, when sufficient food is available but of poor nutritional quality.[50]

Attempts to produce substitutes for meat have been especially successful. Fibres resembling those of meat can be produced from proteins in solutions by spinning, by forcing them through nozzles or by texturizing and pressing dough through jets; the texturized soy proteins have especially favourable properties. Texturized proteins can be added to food containing minced meat, thus replacing part of the meat without negative effects. In many industrialized countries these preparations are accepted by consumers.[51]

A further possibility for substituting conventional proteins is egg white. Egg white is used in many fields of food technology because of its special properties. For example, it can be whipped to give a fine-structured, stable foam and it has special gelling and coagulating properties. Proteins of animal, plant or microbial origin can be given similar properties by special degradation or special modification and could play an important role as substitutes for egg white in foods. The inevitable protein losses due to production of eggs by feeding hens could thus be avoided.

The dough structure in baked goods is very much influenced by the protein of the meal. Protein preparations with swelling properties similar to those of gluten could improve the dough structure of the goods and, in addition to this, greatly improve the nutritional value when they are rich in essential amino acids. The preparation of protein-containing beverages, enrichment of farinaceous products and adding proteins to sweetmeat could all offer more solutions for increasing protein content of the diet.

Toxicological and Sensory Drawbacks to New Proteins

In addition to toxic substances and enzyme inhibitors in the crude material some other toxicological problems arise from the use of new proteins. Above all, the new proteins, as well as modified proteins, have to be completely safe for human consumption.[52]

When protein preparations containing carbohydrates are heated during processing, lysine, a very important essential amino acid, reacts with carbohydrates to form special compounds from which it cannot be recovered biologically. Treatment with alkali, a usual step during the processing of several plant proteins, can cause modifications of several amino acids even under mild conditions. Of special importance is the reaction between lysine and alanine to form lysinoalanine as some recent investigations revealed damage to several organs, especially the kidneys, when it was orally administered to animals. Modifications of other amino acids can, at least, cause losses in protein efficiency and heating under neutral or acid conditions can change protein structures by denaturation; this often improves digestibility and inactivates enzyme inhibitors.[53]

An unpleasant bitter taste can arise from degradation of high molecular proteins to low molecular compounds, as such peptides are responsible for this extremely bitter taste. Methods for removing compounds with a bitter taste from mixtures of low molecular peptides are known.

These considerations all show that availability, efficiency and toxicology of the concentrated proteins are more important than their origin. Therefore, the most important aspect for assessing new proteins should be the protein efficiency, absence of toxins due to processing, neutrality in taste and flavour and their usefulness as human foods. Economic factors should of course also be considered. As there is no basic difference between concentrated proteins from plants, animals or microbes, future developments must include microbial proteins.

UNCONVENTIONAL DEVELOPMENTS

Micro-organisms belong to the plant kingdom. Although they are a permanent part of our life, the fact that they are so small and one cannot normally see them without a microscope means that for most people they are 'non-existent.' Only when one comes in contact with mouldy bread or disease one realizes how deeply fungi and bacteria can affect our lives.

Micro-organisms play a very important role in nutrition extending from man's intestinal flora to the production of delicacies. The union between man and his intestinal flora is one of symbiosis and by taking certain drugs or by altering the pH, the micro-organisms in the alimentary canal can be killed or changed; the consequence is the abnormal break-down of nutrients leading for example, to diarrhoea.

In Table 3 a selection of microbially produced products is presented showing that techniques using micro-organisms are common and widely used in the food industry.[54] Many of these techniques belong to the most ancient modifications which man made – knowingly or unknowingly – to render his food edible, drinkable, appetizing, more digestible or to achieve a qualitatively higher nutritional value. These ancient techniques were similar to those used in the brewing industry today. A biochemical reaction, here fermentation, was brought about by using micro-organisms in urns, barrels, vats, tubs or kettles. This process, often under anaerobic conditions, lasted days or weeks and converted a basic raw material into

Table 3. Selection of Microbial Fermentation Products

Products	Micro-organism	Use of Products as:
Yoghurt	*Streptomyces thermophilus*	Food
Gouda	*Bacterium spec.*	Food
Beer	*Saccharomyces cerevisiae*	Drinks
Wine	*Saccharomyces ellipsoideus*	Drinks
Ethanol	*Saccharomyces cerevisiae*	Solvents, drinks
Citric acid	*Aspergillus niger*	Food, drinks
Lactic acid	*Rhizopus oryzae*	Food, pharmaceuticals
Lysine	*Micrococcus glutamicus*	Food supplement, feed
Vitamin B_{12}	*Propionibacterium freudenreichii*	Food additive, feed
Amylases	*Bacillus subtilis*	Starch processing
Dextran	*Leuconostoc mesenteroides*	Food stabilizer
Toxin	*Bacillus thuringiensis*	Insecticide
Penicillin	*Penicillium notatum*	Antibiotics

a new product, such as sugar to alcohol or milk to cheese. Ancient technology was simple and demanded no special know-how, careful controls or elaborate equipment. Duly specialized knowledge and modern equipment can today meet the increased demand for microbiologically produced foods and drinks.

With the discovery of antibiotics a new technology was required. Here an aerobic process is involved taking place in aerated and stirred tanks under strictly sterile conditions with one sole microbial strain in the reactor. Very simply expressed one can say that the older methods were fermentation processes and that the newer aerobic procedures involve an advanced technology where the aim is often to produce the micro-organisms as such, as biomass, the SCP–single cell protein. This procedure needs to be explained.

New Biotechnology

The substrates used in the fermentation of beer, enzymes or antibiotics are plant raw materials but animal products, chemicals and salts are also required. Most of the smallest of cells use sugar for their life synthesis and energy metabolism in the same way as higher forms. There are, however, some micro-organisms which are able to utilize most unusual substrates as carbon source, like methane, methanol, ethanol, paraffin and crude oil. When a culture substrate is prepared for this type of micro-organism, a clear water-based medium can be accurately and synthetically made up from constituents which can be precisely optimized for the subsequent use of the biomass, for example high protein, fat content or polysaccharide formation.

Table 4. Culture Medium for a Bacterium which uses Methanol as its Carbon Source

Carbon source	–Methanol
Nitrogen source	–Ammonium
Phosphorus source	–Phosphoric acid
	Salts
	Water

In the production of such a biomass two things are possible: first, it is relatively easy, due to the clear, defined medium, to analyse the contents of the culture medium at any stage during the growth of the micro-organism. Second, after analysis one can easily remedy any shortage by the addition of the required substrate. For this new biotechnology precise measurements, control technology and computers are now indispensible, in fact, it is now possible to control the life of the cell, to change within certain limits the composition of the cell mass, and to produce biomass continually over long periods of time.

Some continuous production techniques, already widely used in industry, are now being adapted to biological processes, allowing for the first time to apply industry's present 'know-how' to living cells. The advantages are: biomass is independent of climate and season, requires no large areas of land and can be produced when needed in a homogeneous and consistent quality. It should be mentioned here, that new energy-saving reactors can be designed and manufactured to conform to these conditions.

This new knowledge has been gained during the development of SCP production techniques and they are the key for the new science of biotechnology. With them, the advantages of industrial development are put to use for living matter.

Biomass

The term biomass can be applied to all organically produced substances containing the chemical element carbon, on the micro or macro scale. To the latter belongs the biomass which is quickly obtained when forest vegetation is fertilized, bioconversion, which is now tried experimentally in Germany and the USA. The object is not only to produce renewable biomass, such as trees and leaves for feed or food, but also for fuel.[55] Normally, however, the term biomass is used for SCP, single cell protein.

What then is the composition of a biomass of microbiological origin? In Table 5 the most important types of product are shown, beginning with the 'almost conventional' algae. The most recent possibility is the production of hydrogen.

Algae and Fungi

Before and during the Second World War, algae were grown and used as animal feed and a source of fat; there was also something called 'algal-butter'. More recently there has been intensive work on the production of algae in open ponds[56] and there is a project where the algae produced can be used as an addition to food; this is, however, at present only of local importance. This technology was first used in sewage treatment plants. (See A. Richmond and K. Preiss, Biotechnology of Algaculture, p. 60.)

Table 5. Biomass Products of SCP and Related Products

Products from different types of micro-organisms	Grown on	Composition in % Protein	Fat
Algae	CO_2	45–60	5
Fungi	Carbohydrate	34–50	5
Yeast	Paraffin	50–60	9
Yeast	Crude oil	60–65	2
Bacteria	Methanol	80–85	8
Comparable products			
Plant origin	Soy bean meal	45	1.5
Animal origin	Fishmeal	60–65	7

On the other hand, everyone is conversant with fungi which can be obtained by cultivation. The mushroom industry in Europe, but especially that in Asia, supplies tasty fungi of this type, although the nutritional value of this microbiological 'crop' is not very high. The culturing of other species is under way, such as truffles or the edible *Boletus*, though at the moment there are difficulties despite many experiments. Amongst the most interesting experiments are the growth of the oyster-mushroom on waste such as cellulose or straw.[57]

Yeasts and Bacteria

Yeasts are micro-organisms which have played an important role in human nutrition in many different ways. Yeasts are required for brewing purposes, beer, wine and mead, and their function in the baking of cake and bread is equally well known. Due to their nucleotide content, yeasts are further used for improving the flavour of many kinds of foods and as a source of vitamins. Many types of yeast can grow on 'waste substrates' such as molasses or sulphite waste liquor and this leads to the removal of ecologically unwanted products; they are purified into a yeast cell mass.

From waste substrates, it was only a short step to grow yeasts on crude oil, alkanes or methanol, a step, independent from the isolation of special strains using unusual substrates, instead of carbohydrate, as a carbon source. The advantages of the new science of biotechnology are here put to their greatest use. This type of micro-organism was first described in 1905[58] and studied in more detail in 1952;[59] however, it was the work of Champagnat[60] and British Petroleum[61] which put this knowledge to a useful application.

In Table 5 biomass products from SCP, single cell protein, are shown. Many animal experiments have

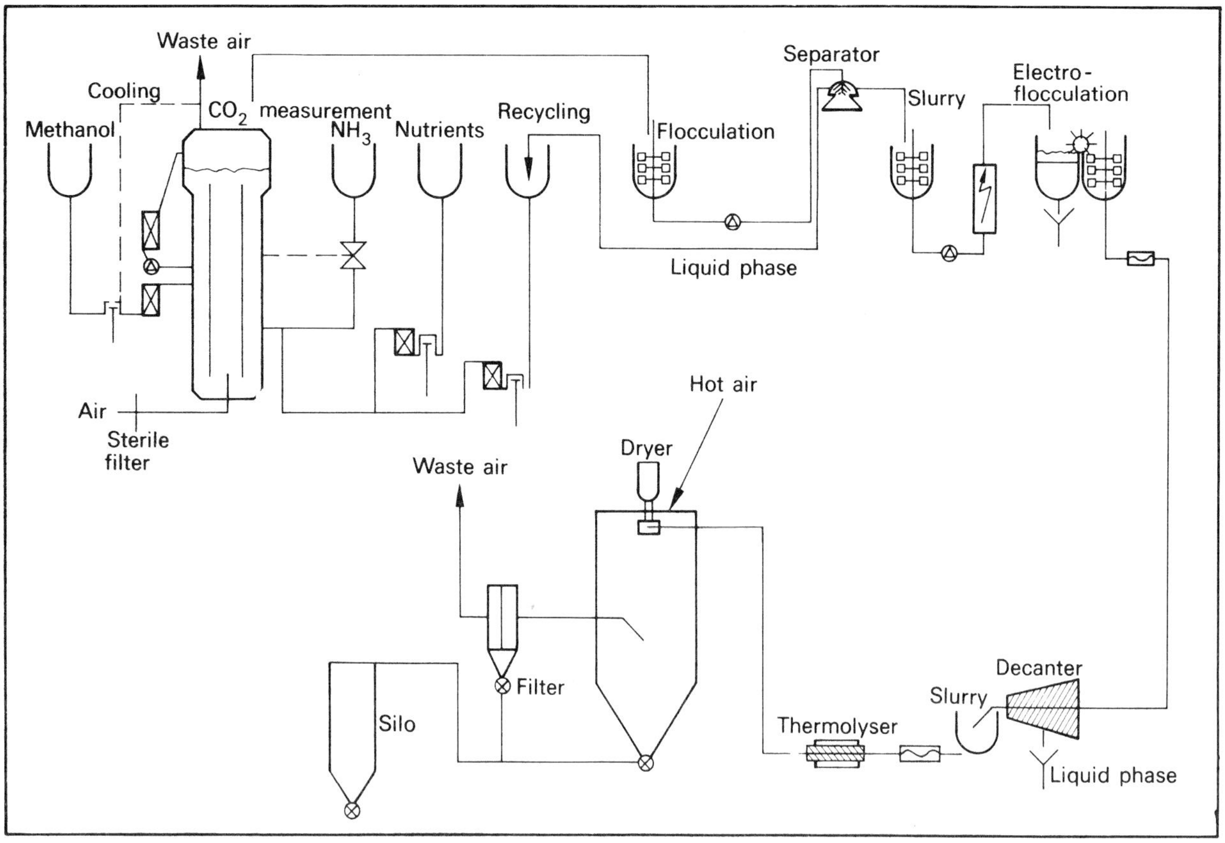

Figure 6. Production of SCP from methanol.[62]

demonstrated that these microbiological proteins in mixed feed can replace part of the plant protein normally used. Bacteria play a large role in the food chains of the nutrition industry, yoghurt, tripe and cheese for example. They are especially suitable for biotechnological techniques as they have a very high protein content of up to 85%. The unusually large numbers of different types and the special physiological qualities of these single cell organisms open up a world of possibilities. Their great usefulness to man will certainly be fully exploited in the future.

Figure 6 is a typical flow diagram of a process developed for the production of SCP using methanol as a carbon source. Here, with modern low energy technology, very simple substrates such as methanol, ammonia, phosphoric acid and salts are turned into high value single cell proteins. The process which here takes place synthetically, occurs much more rapidly than in nature where, due to external climate and specific substrate requirements, cell division requires a much longer time. The utilization of a continuous forward growth process in this new biotechnology leads to further economic advantages.

Energy and Economics

Although these bioindustrial developments have nothing in common with agriculture they create organic products similar to agricultural ones, thus extending an era begun in historic times with microorganisms refining and producing mead, sauerkraut, silage and many other desirable products. However, during the last century satisfying the world's food requirements became a direct challenge, as poor harvests or bad winters brought hunger in their wake. Furthermore, the industrial revolution with its intensive use of resources, fertilizers, machines and medicines leading to a rise in the world's population, all increased the interrelationship of industry and human food with energy and its economics.

However, the cost of progress has still to be answered: First with respect to the price to be paid for agricultural and bioindustrial food – and second the costs for the specific energy requirements to produce conventional and unconventional food supplies.

If a few important protein products now on the market are quoted, one notes that prices are very different indeed.

Soy bean meal	175– 450 \$ per tonne
Fishmeal	600 \$ per tonne
Casein	1250–2000 \$ per tonne
Meat	2000–4000 \$ per tonne

Consider also the different qualities of soy bean meal and meat, as well as their use in feed or food products; then SCP as a biomass product with its price range lying between 600 and 1000 \$ per tonne and its good quality of 80% crude protein, is even now an interesting addition to the available protein range. The advantages of its industrial production in countries relying heavily on imports or having poor soils and adverse climates, cannot be overlooked. The

Table 6. Selected SCP Plants and Pilot Plants

	Micro-organisms	Substrate	
USSR	Yeast	alkane	250 000 tonnes per year
UK (ICI)	Bacteria	methanol	50–70 000 tonnes per year
Romania (Roniprot)	Yeast	alkane	60 000 tonnes per year
USA (Amoco)	Yeast	ethanol	5000 tonnes per year
DDR	Yeast	alkane	60 000 tonnes per year (under construction)
FRG (Hoechst/Uhde)	Bacteria	methanol	1000 tonnes per year
Japan (Mitsubishi Gas Chem.)	Yeast Bacteria	methanol	500 tonnes per year

most important large plants and pilot plants for the production of SCP, which are either in the planning stage or already operating, are listed in Table 6.

These first steps in the bioindustrial production of natural substances, especially of SCP as protein source, are intended to produce a mix for animal feedstuffs. However, the high quality of this product immediately suggests its direct use as human food. Naturally many tests must be carried out first; whereas a drug is normally taken only for a short period of time a food product, in contrast, may be ingested during a complete lifetime. Three conditions must be fulfilled if an unconventional food product is to be acceptable: nutritive value, safety, palatability. Acceptability raises problems: strict controls of the substrate, process and products are essential and all additional production stages must be equally well tested and approved. Any introduction of new synthetic food is complicated and modern marketing is required to prove its quality.

The first project already underway is the yeast product Torutein, manufactured by Amoco in Indiana, USA. Five thousand tonnes per year of the yeast *Candida utilis*, grown on ethanol as a substrate, are produced there and sold on the food market. The same technique, yeast on ethanol, is working in Czechoslovakia and a Mitsubishi plant in Japan. In the United Kingdom, Rank Hovis McDougall work with a fungus growing on carbohydrates. Some processes with algae, light and CO_2, produce amounts running into hundreds of tonnes per year, at biological research centres in Germany, Peru and Thailand; at Sosa Texaco in Mexico and at Dainippon in Kuwait.

In addition to these first production steps there have been plans, innumerable experiments and tests prior to pilot plant and industrial manufacture.[63] The digestibility of bacteria and yeasts lies between 80 and 96% but to reach this value it is not as a rule sufficient to take the cell mass directly from the reactor as dried. Normally, a definite drying temperature must be used or a breakdown of the cells effected, a procedure not unusual for the food industry. Almost all food products intended for human consumption must be pre-treated either heated, smoked or preserved and the ultimate consumer may boil, mix or roast his raw food to improve its flavour and digestibility.

These processing qualities, the functionality, of nutritional products are extremely important – the possibility of spinning a food product, drawing it out into sheets or making foams. And it is by this very control of the process of growing cells for SCP that the composition as well as the chemical and physical structure is always known and can be controlled. Here, then, can the scientific know-how of industry cover new ground, not based on the experience of centuries, but on scientific principles.

Agriculture reached a level of productivity in the last decades that would have been thought impossible one or two generations ago, but a higher price in energy had to be paid. One can state, for example, that in coastal fishing to obtain food containing one unit of energy, one unit of fossil energy had to be invested. Our forefathers lived much more favourably, as hunters and gatherers required one-tenth or less of the requisite unit to obtain one energy unit. Our present system of acquiring food is, in contrast, very uneconomic. Intensive agriculture requires 2–5 and sometimes 8–10 energy units, deepsea fishing and similar systems such as feed-lot beef production 10, 20 or more energy units to provide one nutritional energy unit.[64–66]

For the production of 1 nutritional calorie of a bacterial cell biomass from methanol about 5–6 energy units are today expended. The yield of this new biotechnology lies, therefore, on the upper limit of the energy expenditure for the production of plant products and at the lower level of energy expenditure required for food products of animal origin. Naturally, these considerations cover only a few aspects of the problem. The quality of the food products is obviously of prime importance. Very simply stated one assumes that carbohydrates and fats are cheaper than protein. Not only are they lower in price but also in terms of energy expenditure their production in the cell is very much cheaper than the synthesis of protein.

It is, therefore, not unusual to find that priority of food production is focused on proteins, nutritionally and physiologically so important and so expensive to produce. One can imagine a future food policy using the biotechnological production and aiming at cheap agricultural bulk products, as well as vital nutritional components such as vitamins, proteins and hormones.

Synthetic production of Components for Foods and Feedstuffs

Amongst the foods and feedstuffs which can be produced synthetically, the synthesis of the amino acid methionine is of practical importance as it is an additive in those animal feeds which are short in protein and low in methionine. In contrast to other amino acids, methionine can be used as the racemic synthetic product. Other essential amino acids could also be of interest for upgrading protein efficiency, as for example, lysine. These amino acids are usually synthesized in racemic form, of which only the L-compound can be utilized. The stereospecific synthesis of the pure L-isomer is complicated and expensive and no use has been found for the D-compounds, half the racemic products. Even if a process could be found to remove the physiologically undesirable D-amino acids, either by microbiological or enzymatical methods this removal would be expensive.

As energy sources several easily synthesized polyols and fatty acids could be used, as these products can be physiologically metabolized even in high quantities without any harmful effects. At present unfortunately, they cannot be produced cheaper than agricultural products rich in carbohydrates.[52]

CONCLUSIONS

'We will have to find in the next 25 years food for as many people again as we will have been able to develop in the whole history of men till now.' This statement by J. Mayer expresses in simple terms the size of the task required to solve mankind's nutritional problems. The projection (Table 7) shows that the gap between demand for food and its production in the developing countries will remain or even increase.

Table 7. Food Production and Food Demand, 1970 and 1985[a] (average changes in percent per year)

	Developed countries	Developing countries	World total
Food production per head of population	+1.9	+0.2	+0.7
Demand for food	+1.5	+3.4	+2.4
Surplus or deficit	+1.3	−0.8	+0.3

[a] Adapted from FAO[24] and FAO Provisional Indicative World Plan 1970.

Total world-wide protein production as it stands at present could eventually be sufficient to cover all physiological requirements, provided that a better distribution could be achieved world-wide and that all the protein produced is acceptable to all consumers. Unfortunately, however, rapid population growth will bring about an ever increasing demand for food and the threat of protein shortage will be severely aggravated by an ever higher demand for animal protein as standards of living rise in the wealthier sections of all communities.

The potential for increasing conventional food production is still great. Obviously, further growth of agricultural production would be advantageous as additional food would be easily accepted and consumed without any change in cherished traditions. However, it is more than doubtful if agricultural expansion is the only solution to the urgent problem. Major obstacles would be the lack of experienced personnel; the huge capital demands for fertilizers and the pesticides needed for intensive management and, finally, the vast demands for water for irrigation. An additional disadvantage could be the necessity to cultivate virgin regions with serious ecological consequences, such as changes in climate or extermination of genetically important plant and animal species.

Agriculture will remain the most important source of food and feed and great efforts should be made to protect plants during growth and storage. Moreover, a careful evaluation is required to determine which alternative production resources for food could and should be developed. In particular the processing of raw materials, normally used as animal feed or wasted, into human food would be a significant improvement. Recovering proteins suitable for human consumption from oil seeds, green plants and agricultural wastes, are known examples.

Novel components for food from conventional and semiconventional resources will have to be modified to make them acceptable to the general public and this could be achieved by diluting them with familiar foods. In the long term, completely new foods based upon new sources will have to be developed and promoted. The experience gained during the development of foods from conventional sources can then be successfully used to promote the unconventional material.

We conclude that biotechnology could contribute greatly to feeding mankind on an industrial scale. Various micro-organisms and substrates offer a great variety of processes for the production of proteins, fats, carbohydrates and other nutritive compounds. On a small scale the transformation of agricultural wastes into microbial animal feed is also feasible. The high growth rate of any biomass and ease of its process control are advantages which allow production to follow demand closely and thus can avoid any long-term storage with possible spoilage.

Although the world food situation seems to be critical and the protein gap seems to increase, a chance for overcoming world-wide famine exists at least in principle. Scientific and technological solutions offer basic extensions in food production. It is now essential to promote them and use them for the benefit of all human nutrition (Fig. 7).

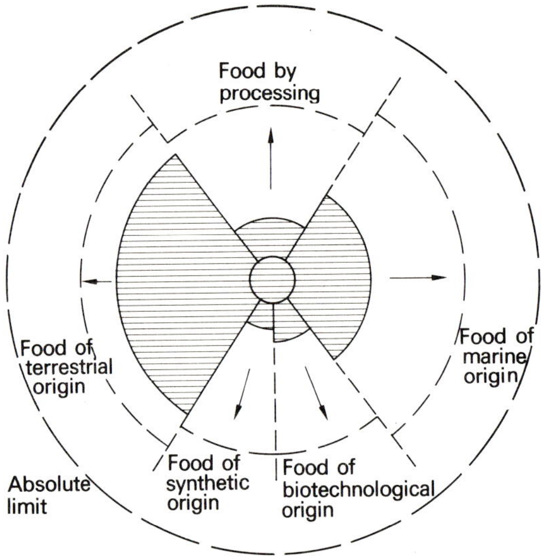

Figure 7. Projections for future status (▤) and potential (→) extension of nutrition for mankind.

LITERATURE CITED

1. H. Kahn, W. Brown and L. Martel, *The next 200 years: A Scenario for America and the world*, W. Morrow, New York (1976).
2. P. Collins, Facing the facts about food. *Nature* (*London*) **256**, 85 (1975).
3. M. Witt, Überfluß an Fleisch und Milch in den Industriestaaten – Tödlicher Eiweißmangel in den Entwicklungsländern – die Schere öffnet sich immer weiter. *Umschau* **77**, (12) 375 (1977).
4. D. Meadows, *Die Grenzen des Wachstums, Bericht des Club of Rome zur Lage der Menschheit*, Deutsche Verlagsanstalt Stuttgart. *Limits to Growth*, Potomac Associates, Washington DC (1972).
5. A. O. Herrera, *Grenzen des Elends, Das Bariloche Modell: So kann die Menschheit überleben*, Gustav Fischer Verlag, Stuttgart (1977).
6. M. Mesarovic and E. Pestel, *Menschheit am Wendepunkt, 2. Bericht an den Club of Rome zur Weltlage*, Deutsche Verlagsanstalt, Stuttgart. *Mankind at the Turning Point*, Dutton Inc., New York (1974).
7. N. S. Scrimshaw, Effect of infection on nutrient requirements. *Am. J. Clin. Nutr.* **30**, 1536–1544 (1977).
8. M. C. Latham, Nutrition and infection in national development. *Science* **188**, 561–565 (1975).
9. M. Ganzin, Present and future world protein supplies. *Folia Vet. Lat.* **VI**, (Suppl. 1) 34–48 (1975).
10. H. Stone, World protein requirements. *Notizario* **14**, 23 (1973).
11. T. T. Poleman, World food: perspectives. *Science* **188**, 510–518 (1975).
12. H. D. Cremer, Fehlernährung als Weltproblem. *Zentralblatt für Bakteriol. Hyg. Abt. 1.* **163**, 96–110 (1976).
13. M. Winnick and J. A. Brasel, Early malnutrition and subsequent brain development. *Ann. N.Y. Acad. Sci.* **300**, 280–282 (1977).
14. G. H. Beaton and J. M. Bengoa, *Nutrition in preventive medicine*, WHO, Geneva (1976).
15. *Energy and Protein Requirements.* FAO Nutrition Meeting Report, Series No. 52 (1973).
16. W. B. Clapham and M. D. Mesarovic, Report of the Club of Rome. *Ann. N.Y. Acad. Sci.* **300**, 228–238 (1977).
17. J. Senez, Single Cell Protein in Food Strategy of the EEC. *Folia Vet. Lat.* **VI**, (Suppl. 1) 97–108 (1975).
18. W. Schug, Agrarüberschüsse und Nahrungsmittelhilfe, *Kraftfutter* **60**, 170–174 (May 1977).
19. H. Grote, Weltmarkt für Eiweißfuttermittel. *Kraftfutter* **61**, 18–24 (January 1978).
20. E. N. Todhunter, Food habits, food faddism and nutrition. *World Rev. Nutr. Diet.* **16**, 286–317 (1973).
21. D. P. Chaudrey, Famine amid surplus: a problem of production, economics or distribution? *Search* **7**, 124–128 (1976).
22. D. Pimentel *et al.*, Land degradation effects on food and energy resources. *Science* **194**, 149–155, (1976).
23. N. E. Borlaug, *Mankind and civilization at another crossroad*, McDougall Memorial Lecture, FAO (1971).
24. FAO Production Yearbook, Rome (1976).
25. K. Weissermel, Von der Rohstoffversorgung bis zum Krebs. *Umschau* **76**, (16) 509 (1976).
26. K. H. Büchel, Thesen zur Podiumsdiskussion: Chemie und Ernährung (100 Jahre V.C.I.) (Verband der Chemischer Industrie) am 27. Oktober 1977.
27. J. R. Gatt, The water potential of the arid zone. *Ann. N.Y. Acad. Sci.* **300**, 17 (1977).
28. S. P. Meyers, Utilization of water resources in agriculture. *Feedstuffs* **50** (16) 19 (1978).
29. H. F. Wellhausen, The agriculture of Mexico. *Sci. Am.* 129 (September 1976).
30. T. J. Cunha, More basic research on animals needed, *Feedstuffs* **50**, (5 June 1978).
31. J. Janick *et al.*, The cycles of plant and animal nutrition. *Sci. Am.* 75–86 (September 1976).
32. R. A. Brink *et al.*, Soil deterioration and the growing world demand for food. *Science* **197**, 625–630 (1977).
33. O. Kinne and M. Rosenthal, Commercial cultivation (aquaculture). *Mar. Ecol.* Vol. III Part 3, 1321–1398 (1977).
34. J. Hildingstam, Fish farming trends. *Vet. Rec.* **100**, 256–259 (1977).
35. G. Luther, Anwendungsmöglichkeiten neuer wissenschaftlicher Erkenntnisse in der industriellen Großproduktion von Süßwasserfischen. *Fischwirt* **20**, 4 (1970).
36. F. H. Sanderson, Das Auf und Ab der Welternährung. *Umschau* **76**, *Heft 2* 35–39 (1976).
37. *Mon. Bull. Agric. Econ. Stat.* **23**, (4) 12–19 (1974).
38. G. Smith, Whey proteins. *World Rev. Nutr. Diet.* **24**, 88–116 (1976).

39. E. R. Pariser, The potential, the problems, the status of using proteins of aquatic origin as human food. *Food Technol.* **25**, (11), 62–74 (1971).
40. J. J. Doesburg and J. Bon, *Concentrated fish proteins – trends in past and future*, in *Dechema-Monographien*, Vol. 70, p. 71–83. Verlag Chemie, Weinheim (1972).
41. R. Finch, Whatever happened to fish protein concentrate? *Food Technol.* **31**, (5), 44–51 (1977).
42. J. E. Bardach and E. R. Pariser, Aquatic proteins, in *Protein resources and technology: Status and research needs*, ed. by M. Milner, N. S. Scrimshaw and D. I. C. Wang, pp. 427–484, AVI Westport (1978).
43. D. A. Halliday, Blood – a source of proteins. *Process Biochem.* **8** (12) 15–17 (1973), **10**, (4) 11–12 (1975).
44. W. J. Wolf and J. C. Cowan, Soybeans as a food source. *CRC Crit. Rev. Food Technol.* **2**, (1) 81–158 (1971).
45. D. M. Cater *et al.*, Oilseed proteins, in ed. by M. Milner, N. S. Scrimshaw, D. I. C. Wang, *Protein Resources and Technology, Status and Research Needs*, pp. 278–301. Westport, AVI (1978).
46. H. A. Woodham, Oil seed protein, present and future. *World Rev. Nutr. Diet.* **10**, 44–76 (1969).
47. M. W. Adams, M. Milner, G. M. Montfort and L. B. Rockland, Food legumes as a protein source, in *Protein Resources and Technology, Status and Research Needs*, ed. by M. Milner, N. S. Scrimshaw, D. I. C. Wang, pp. 302–317. Westport, AVI (1978).
48. G. O. Kohler *et al.*, Leaf protein in relation to forage crop production and utilization, in *Protein Resources and Technology, Status and Research Needs*, ed. by M. Milner, N. S. Scrimshaw, D. I. C. Wang, pp. 543–568. Westport, AVI (1978).
49. D. K. Parrish, M. Kroger and J. C. Weaver, The prospects of leaf protein for human food – and a close look at alfalfa. *CRC Crit. Rev. Food Technol.* **5**, (1) 1–13 (1974).
50. J. E. Kinsella, Functional properties of proteins in foods: A survey. *CRC Crit. Rev. Food Sci. and Nutr.* **7**, 219–280 (1976).
51. Z. Duda, Meat extenders – their technology and application. *World Anim. Rev.* **11**, 12–19 (1974).
52. S. W. Fox *et al.*, in *Protein Resources and Technology, Status and Research Needs*, ed. by M. Milner, N. S. Scrimshaw, D. I. C. Wang, pp. 569–583. Westport, AVI (1978).
53. J. Mauron, Ernährungsphysiologische Beurteilung verarbeiteter Eiweißstoffe. *Dtsch. Lebensm. Rundsch.* **71**, (1) 27–35, (1975).
54. P. Präve, D. A. Sukatsch and U. Faust, Biotechnology: The Industrial Production of Natural Substances. *Interdiscip. Sci. Rev.* **1**, 85 (1976).
55. H. G. Schlegel and J. Barnea (Editors), *Microbial Energy Conversion*, E. Goltze KG, Göttingen (1976).
56. W. J. O'Brien, Lagoons and oxidation ponds. *J. Water Pollut. Control Fed.* **48**, 1107 (1976).
57. F. Zadrazil and M. Scheiderest, Basis for cultivation of a previously uncultivated variety of Pleurotus. *Der Champignon* **135**, 25 (1972).
58. N. L. Söhngen, Wachstum von Bacterium paraffinus. *Zentralble. Bakteriol. Parasitenk. Dept. II* **37**, 595 (1913).
59. H. Dellweg, Aus der Geschichte der Fermentation. *Branntweinwirtschaft* S. 202. (2 Juniheft 1976).
60. A. Champagnat, Proteins from Petroleum. *Sci. Am.* **213**, 13 (1965).
61. J. Filosa, Improvements in the production of yeast. Brit. Pat. 914, 568.
62. P. Präve, Utilisation of microbes – modern developments in biological technology. *Angew. Chem. Int. Ed. Engl.* **16**, (4) 205 (1977).
63. Conventional and non-conventional proteins. 1st Capri Veterinary Workshop 7–8 October 1975. *Folia Vet. Lat.* **VI**, Suppl 1.
64. D. Pimentel *et al.*, Food products and the energy crisis. *Science* **182**, 443 (1973).
65. A. L. Johnson Jr, *Astronaut. Aeronaut.* 64 (November 1974).
66. A. L. Lehninger, *Bioenergetics*, 2d Edn, p. 123. W. A. Benjamin, Paolo Alto, California (1965).
67. S. Wortman, Food and agriculture. *Sci. Am.* **235**, 31–39 (1976).

The manuscript was received 21 September 1978

Aquaculture
A Critical Assessment of its Potential and Future

PROFESSOR DR OTTO KINNE
Director, Biologische Anstalt Helgoland, Hamburg, Federal Republic of Germany

While important for making additional food available, aquaculture is – in the foreseeable future – unlikely to contribute more than a few percent of additional food to the global protein yield for the growing human population. Some 85% of the world's total annual protein yield is at present produced by agriculture and some 14% by fisheries. Aquaculture must focus more on microorganisms, plants and herbivorous animals, and long-term research programmes should test such feasibilities as utilizing the synthetic mechanisms of low-trophic-level organisms for producing essential portions of the human diet from waste materials and providing tailor-made diets for meeting specific physiological-nutritional needs of man.

Never before in history has the production of sufficient human food attained such dramatic urgency as at present. Alarming predictions on logarithmically increasing growth rates of the human population and reports on large numbers of people, already undernourished or starving, have made progress in food production a major, world-wide concern for politicians and administrators. Progress in food production acutely depends on research. It requires interdisciplinary co-operation between biologists, chemists and engineers.

ECOLOGY AND FOOD

For the next 50 years, official estimates predict a tripling of the present human population from about 4×10^9 individuals to some 12 by the year 2030 (Fig. 1). Already an estimated 500×10^6 people receive less protein than they need, and the daily FAO safe allowance for a 70 kg male adult – 41 g of reference protein (egg albumin) – cannot be provided for about 12% of the world's human population. While the present malnutrition is, to a large extent, a problem of food distribution, food production itself is rapidly nearing its inherent limit.

When we consider here techniques for making more food available we must, first of all, realize that our food cannot really be 'industrially' produced. Human food is, and will remain, the product of biological processes, deeply embedded in synthesis, transformation and re-mineralization of organic matter in ecosystems. Man, just as all other components of the ecosystem, depends ultimately on the normal functioning of these processes. There is no technological development in sight which would allow us to fabricate our food without natural biological activities. All we can do is to support and control those natural processes which make material available which we can then utilize as food, that is, either to cultivate our food organisms or to manage free-living organisms and rely on them as a food source.

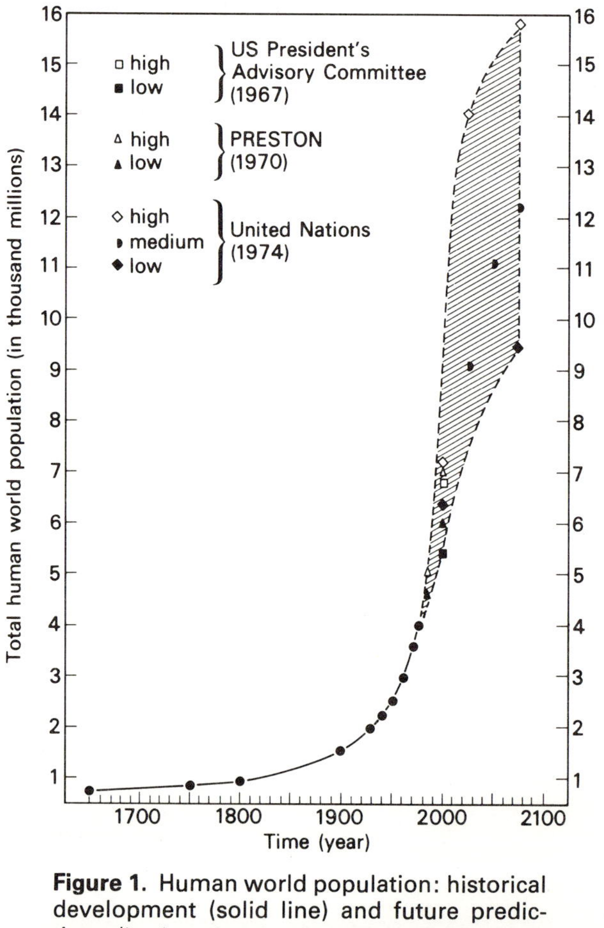

Figure 1. Human world population: historical development (solid line) and future predictions (broken lines). After Kinne and Rosenthal.[1]

CCC-0308-0188/80/0005-0024$0450/1

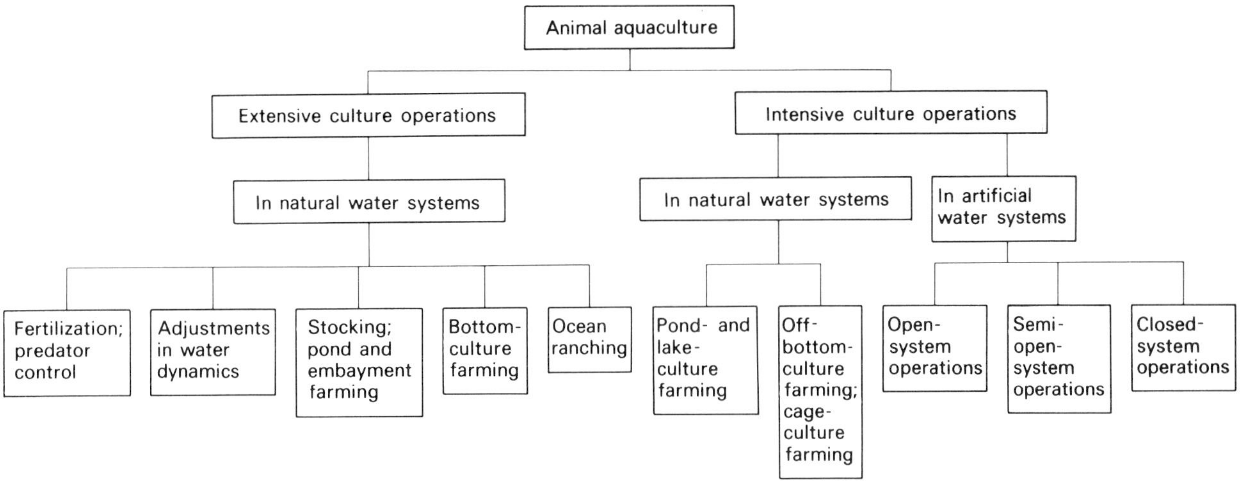

Figure 2. Principal techniques employed in aquaculture. After Kinne[1] and Rosenthal.

These statements illustrate at once that food procurement for thousands of millions of additional human beings involves deep ecological problems. The increasing nutritional needs of a human population, beginning to outgrow its natural endowments, may deform and distort nature's ecological dynamics and – together with man-made pollution and construction – could induce irreversible damage to individual species, to ecosystems and, in fact, to the life-supporting properties of our planet Earth.

There are three major groups of activities which make human food available: agriculture, aquaculture and fisheries. In this review I shall briefly consider some principal techniques of aquaculture; the general potential of aquaculture with special emphasis on mariculture; and the relative importance of fisheries, agriculture and aquaculture. I shall close this article – which is, in part, based on more detailed reviews[1,2] – by offering some conclusions; these include a few suggestions related to future perspectives.

PRINCIPAL TECHNIQUES OF AQUACULTURE

'Aquaculture' may be defined as commercial cultivation of aquatic organisms for nutritional and other human needs. Most aquaculture techniques involve the isolation of the organisms from their natural ecosystem, and subsequent exposure to specific, more or less man-controlled environmental and nutritional conditions. In essence, the aquafarmer strives to support the biological development of the desired product by intensifying selected portions of the energy flow and of material transformation in his culture system. The techniques aim at maximizing quality under conditions which minimize the costs of feed, operation, construction and energy.

Where the organisms are cultured in sea water, we speak of 'mariculture'; where they are kept in fresh water, of 'limniculture'. Most aquaculture farms concentrate on animals, especially fish, crustaceans and molluscs. Aquatic plants and microorganisms still play a very minor role.

Two basic groups of aquacultural techniques can be distinguished: extensive and intensive culture operations (Fig. 2).

Extensive Mariculture

Extensive culture operations are characterized by: a low degree of control; low initial costs, low-level technology and low production efficiency; high dependence on local climate and sea-water quality; use of natural water bodies. Examples of extensive

PROFESSOR OTTO KINNE, a member of our Editorial Board, is the Director of West Germany's largest and oldest marine biological station – the Biologische Anstalt Helgoland (1962). After studying zoology, botany and oceanography at the University of Kiel, he went to Los Angeles (UCLA) as research zoologist. He accepted a position as assistant professor (1958–1960) and later as associate professor (1960–1962) at the University of Toronto. He held special appointments at Hopkins Marine Station, Stanford University, and Marine Biological Laboratory, Woods Hole, and was Canadian Guggenheim Fellow and Senior Foreign Scientist at Arizona State University, Tempe. He returned to Germany in 1962. Other major activities: initiator and organizer of nine international symposia and of two underwater missions with 'BAH I' and the underwater laboratory 'Helgoland'. He has contributed numerous original research papers on marine biological and ecological topics and is editor of several journals and books, including *Marine Biology* and *Marine Ecology*.
Address: Biologische Anstalt Helgoland, Palmaille 9, 2 Hamburg 50, Federal Republic of Germany.

culture activities are: fertilization of a water body used for fish farming, predator control, adjustments in water dynamics, stocking, pond and embayment farming, bottom-culture farming and ocean ranching.

Fertilization involves the addition of either inorganic substances, like phosphates and nitrates, or organic substances, for example suitable sewage or manure, to the culture water in ponds or lakes, thus enhancing productivity. Fertilization can lead to significant harvest increases.

Predator control is achieved by preventing potential predators from entering the culture enclosure, for example by sieves, or by employing chemical or biological methods. For the control of oyster predators, quicklime, salt or biocides have been used with considerable success. Upon contact with the predator, the sea-star, quicklime causes disintegration of the predator's tissue within 24 hours. For maximum efficiency, fine-particulate quicklime is pumped directly on to the oyster bed. Biocides such as chlorinated hydrocarbons and related pesticides are very effective, but these substances are dangerous pollutants and hence should be banned from the aquatic environment. Biological methods of predator control involve organisms which exert detrimental effects on the predator and are collectively referred to as 'biological control'; they rank high for future planning and are more compatible with overall ecological considerations.

Adjustments in water dynamics include such measures as water exchange, aeration and control of flow rates. Stocking of natural water bodies is based on the release of large numbers of laboratory-reared larvae or juveniles. Unfortunately most stocking programmes involving, for example, fish and crustaceans have failed to produce significant harvest increases. However, stocking has been successful for some prawns and turtles.

Pond and embayment farming involves extensive cultivation of shrimps, oysters and fish, in natural water bodies. Bottom-culture farming – largely restricted to molluscs such as oysters, clams or mussels – concentrates on the transfer of spat and juveniles from settling to fattening grounds, as well as thinning and the avoidance of excessive siltation. Ocean ranching, finally, is based on the predictable migratory and homing behaviour of anadromous fish such as salmon. Breeding in the upper reaches of their 'home river', juvenile salmon migrate to the sea with its large reservoir of potential food and return as mature adults to the place of their birth for spawning and death. On their way they can easily be harvested. Salmon hatcheries are relatively inexpensive to operate and the release of recruits in predetermined river areas or in artificial spawning channels can be highly effective. Some salmonids may establish residence in freshwater lakes if prevented from migration; hence, they avail themselves to the sport fishery on a year-round basis. [But see *Interdiscip. Sci. Rev.* **3** 335, (1978) for ecological consequencies of overfishing.]

Intensive Mariculture

Intensive culture operations are characterized by: a high degree of control; high initial costs, high-level technology and high production efficiency; tendency towards increased independence of local climate and sea-water quality; use of man-made culture systems, like tanks, artificial ponds and raceways. Examples of intensive culture operations include pond- and lake-culture, off-bottom and cage-culture farming, as well as culture operations in man-made open, semi-open and closed water systems (Fig. 2).

Pond- and lake-culture farming involves the fattening and, where the culture methods are completely developed, also the breeding (seedling production) of animals such as shrimps or fish. In off-bottom culture farming, oysters or clams are suspended in free water, either on ground-fixed frames or on anchored floats. Pioneered by the ancient Romans, these off-bottom cultures are now used in many countries. Advantages are their high production potential, easy accessibility, low silt sedimentation, reduced dangers from predators and good crowding control. Four types can be distinguished: rack, raft, long-line (Figs. 3–5) and tray methods.

Cage-culture farming is successfully used for fattening fish (Fig. 6). Numerous commercial culture operations are now based on cage-culture techniques, especially in Japan and Norway. The terms open, semi-open and closed water systems refer to different standard techniques of water management. Open sea-water systems can of course only be operated near the shore. The water enters the culture enclosure in a continuous flow and is subsequently discharged. In contrast, in closed water systems the culture medium is constantly recycled. Such recirculation offers increased possibilities for monitoring and controlling essential environmental factors, but the maintenance of adequate water quality does require continuous water management. Pollution from animal activities and from materials used in cultivation procedure must be counteracted by mechanical,

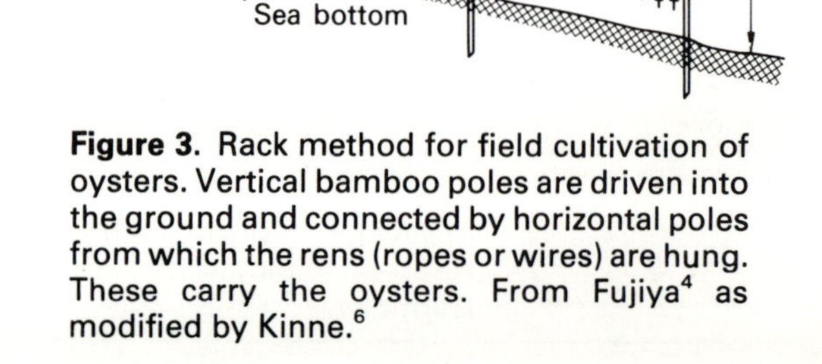

Figure 3. Rack method for field cultivation of oysters. Vertical bamboo poles are driven into the ground and connected by horizontal poles from which the rens (ropes or wires) are hung. These carry the oysters. From Fujiya[4] as modified by Kinne.[6]

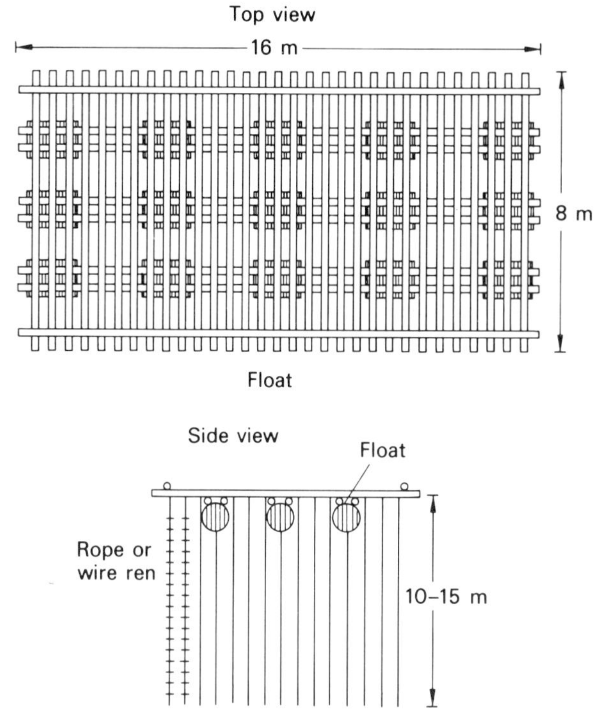

Figure 4. Raft method for field cultivation of oysters. Made from 10–15 cm diameter poles, which are lashed with wire at right-angles, the rafts being buoyed by floats. The oysters are attached to rope or wire rens. After Fujiya[4] as modified by Kinne.[6]

biological or physico-chemical water treatment. It is not possible here to describe the rather complex techniques required for such water management and the interested reader is referred to a recent review by the author.[3]

THE POTENTIAL OF AQUACULTURE

The present world-wide production of aquaculture amounts to about 8 million metric tons of wet weight a year. Some 67% of this are fish, 16.5% molluscs, 16% seaweeds and 0.5% crustaceans. The total amount of protein obtained through this 8 million ton harvest of aquaculture is estimated to be close to 1 million tons per year. In other words, all animal plus plant protein available from aquaculture activities makes up less than 1% of the annual world-wide protein harvest.

For the future, the major potential for increased aquacultural protein production is animal mariculture. In order to assess this, I shall discuss, in very general terms, the requirements, constraints, dangers and possible improvements of aquaculture with special emphasis on animal mariculture.

Requirements

In addition to biological knowledge and technology – which can both be considerably improved – the basic requirements for aquaculture are land, labour and energy.

The land selected for mariculture farms should be close to the sea and have access to clean, that is, unpolluted or subcritically polluted sea water. In general, subtropical and tropical coastal areas are most promising. Where sufficiently high temperatures are not available or where the climate is otherwise unsuitable, the immediate culture environment

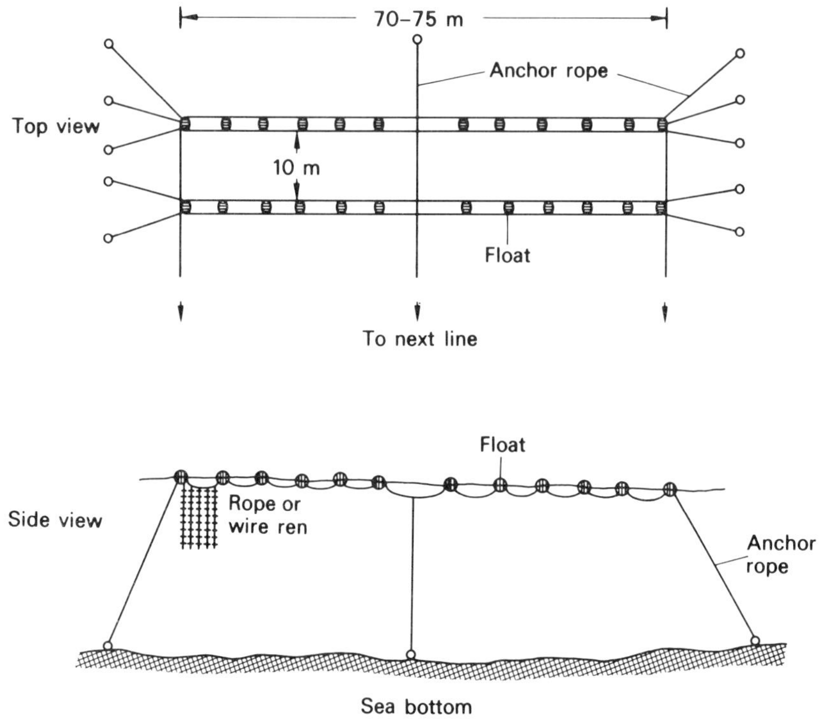

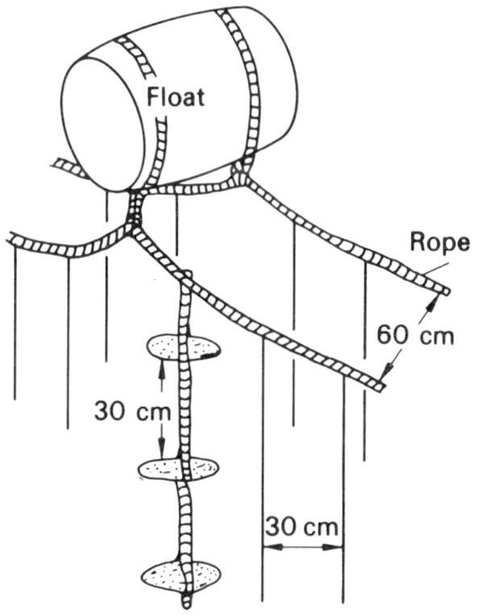

Figure 5. Long-line method for field cultivation of oysters. Long lines are particularly useful in stormy areas and have allowed the extension of oyster cultivation into the open sea. A series of rope-connected floats are anchored and carry the rens (shown here only in the side view for one float interval) to which oysters are attached. From Fujiya[4] as modified by Kinne.[3]

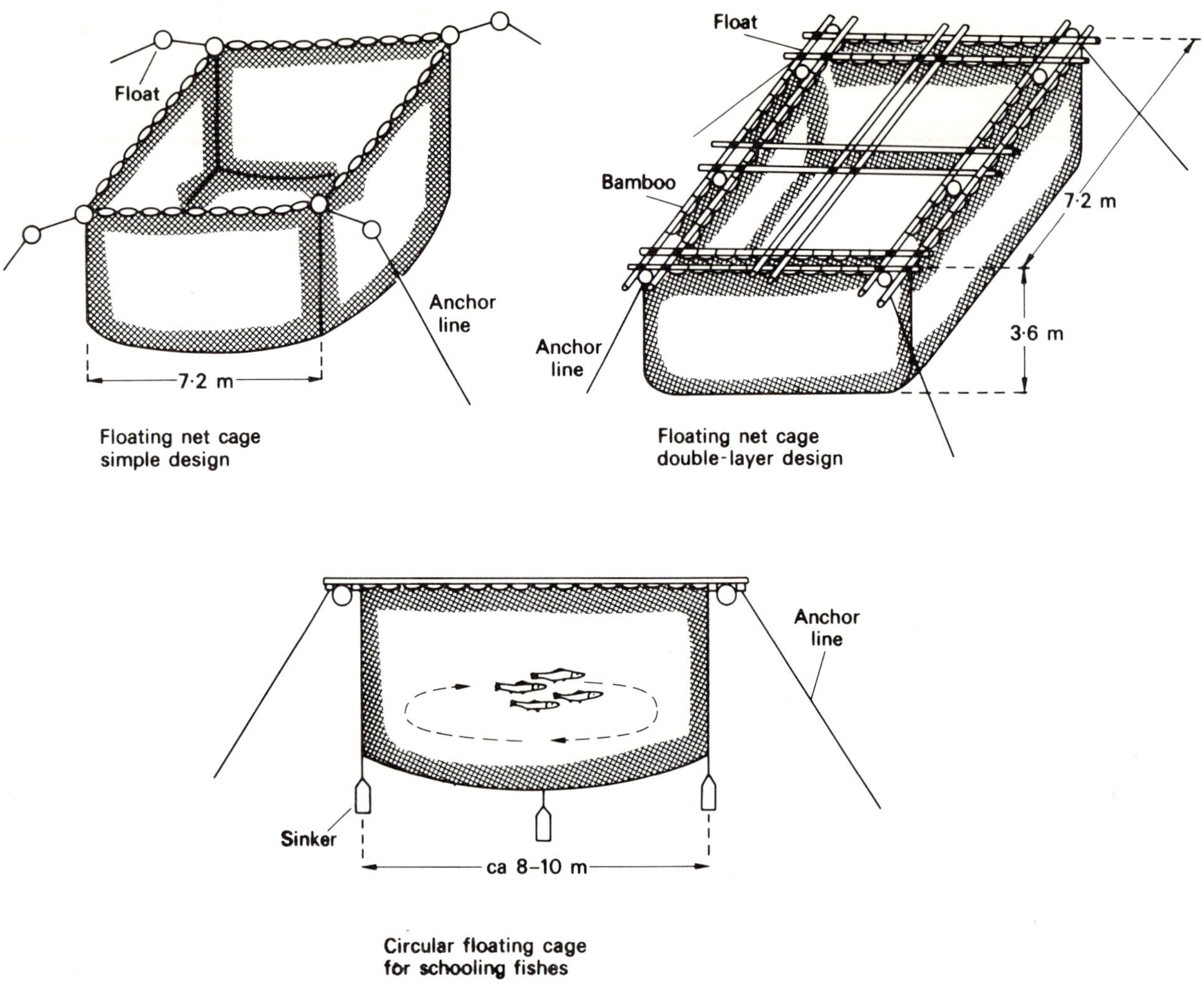

Fig. 6. Cage-culture farming for fattening fish. From Harada[5] as modified by Kinne.[3]

must be controlled – often an expensive proposition. Other requirements related to choice of land are its low-cost labour and energy, adequate transport facilities, availability of seedlings, and acceptable market values compared with competing foods, especially fishery products. For most farming purposes, shallow coastal areas are required – about 1–35 m deep and sufficiently protected from excessive wind, tide and wave effects. Since the total shelf zone on Earth amounts to less than 3% of the global sea-surface area, and since several other commercial ventures compete with the aquafarmer – to name here only industry and recreation – the total area actually available is probably much smaller than optimistic predictions indicate.

In terms of labour, mariculture requires trained or at least adequately trainable people. This may create temporary problems in some subtropical or tropical regions. In addition to a few academically trained scientists and supervisors, a technical staff must be available with sufficient biological and chemical education. In the long run, improvements in labour qualification are possible, and hence labour can hardly be expected to act as a major stumbling block to future developments.

The energy requirements of mariculture projects have so far received little attention. In the light of increasing world-wide energy costs and shortages, it becomes more and more important to consider world-wide food production from the viewpoint of energy expenditures. The fossil energy required for obtaining 1 kcal of animal protein is fairly high: about 35 kcal for catfish protein, 36 kcal for milk, and 78 kcal for beef.[7] For harvesting 1 kcal of gutted sea fish, about 20 kcal of fossil energy are needed.[8] Now, consider that some 5–25% of the sea fish caught is transformed into fish meal, that growing amounts of such fish meal are used as feed for cultured fish, and that the turnover of feed into meat of the cultured fish has an average efficiency coefficient of only 10–30%; consider further that additional energy is required in the culture process and that large amounts of waste are produced – and then one begins to understand that new policies and goals in food protein production must be devised. I shall return to this important point below. In my opinion, only a new approach holds promise for a major breakthrough.

The animals usable for mariculture are limited because of the following essential requirements: they must be resistant to crowding, domestication and disease, have a short grow-out time, a high feed-conversion efficiency, and environmental as well as nutritional needs which can be met by the cultivator without excessive investment. The suitability of an animal for aquaculture increases greatly if it is capable of making major adjustments to the culture environment and if it is amenable to genetic stock improvement. While, in many animals used in aquaculture, commerically desirable properties can presumably be considerably enhanced, very few aquatic animals can as yet be bred and reared under controlled culture conditions. In contrast to agricultural husbandry, animal aquaculture is still largely based on wild populations or on populations in which wild characteristics dominate. In addition, largely unpredictable variations in abundance, catchability and properties of the seed material caught at sea introduce large farming risks.

Constraints

The major constraints facing the aquafarmer are competitive economic interests, legal problems, diseases, predators and climatic restrictions.

Competitive economic interests include construction of industrial plants harbours, roads and seaside homes; dumping of sewage and industrial pollutants; sand and gravel mining; oil drilling; and last but not least recreation. These activities severely limit the total size of the area available for farming.

Legal problems arise from the fact that natural marine waters are either state property or have international status. For the aquafarmer it is difficult to obtain permission to exercise sufficient control over his cultures in these waters. Up to now, ownership problems have remained largely unsolved. Wild animals are usually considered national or international property. Consequently, the ownership of wild seedlings for stocking or the killing of wild competitors or predators may result in legal controversies. Even the release of laboratory-reared larvae or juveniles into their natural environment – be it for purposes of stocking or fattening of migratory 'homers' – entails legal problems of ownership. For proper planning, harvesting procedures require protective legal regulation. While the aquafarmer must insist on a reduction of legal constraints, the multiple utilization of inland waters and seas, and the diverging interests of the users, tend to perpetuate the present legal inadequacies.

Diseases and predators can be effectively controlled only in intensive culture operations. Disease control still poses considerable problems, especially where microbial disease agents such as viruses, bacteria and fungi are concerned,[6] predator exclusion usually constitutes no severe problem in intensive cultures.

Climatic restrictions entail differences between animal requirements and local climate. Where these are significant, environmental control becomes necessary and this may cause considerable additional expense. Particularly dangerous are irregularly recurring extreme conditions such as very low temperatures. Long-term construction stability requirements depend on extreme wind or wave forces which may prevail only once in a decade. Many aquafarmers have learnt to live with this risk. They know that it would be very costly to prevent damage from hurricanes and hence repair their insufficiently stable installations time and again, rather than resort to more rigid but very expensive facilities.

Dangers

Although necessary for increasing our overall food-production potential, large-scale commercial culture operations may endanger the normal functioning of local ecosystems and reduce the aesthetic and recreational value of the areas concerned. Major dangers come from environmental pollution, over-exploitation of natural seed stocks and translocation of organisms.

The release of large amounts of culture-water effluent into rivers or coastal areas can cause eutrophication, odour nuisance or water-colour change. For example, the pollution potential of a farm containing 100 000 catfish corresponds to that of a flock of 150 000 chickens or a herd of 480 steers.[8]

Over-exploitation of seed stocks can endanger the continued existence of the wild population of those mariculture animals which cannot yet be bred under culture conditions. As has been pointed out above, this pertains to the vast majority of animals concerned.

Translocation of organisms due to the transportation of mariculture animals from their home area to the culture site – sometimes a matter of thousands of kilometres – can result in not yet fully known dangers to the local ecosystem. Unless the effluent culture water is carefully sterilized, a continuous flow of larvae, microbial associates and disease agents enters the receiving water body. They may develop there and affect the local ecosystem in unpredictable ways.

Possible Improvements

The potential of aquaculture can be significantly enlarged by a variety of improvements; these in turn depend on a deepening and widening of our present scientific and technological know-how. Major improvements can be expected, for example, from (i) an optimization of culture environment, of culture-system design and nutrition; (ii) a minimization of losses due to disease and stress; (iii) a maximization of desirable genetic properties of the cultured organisms, such as rapid growth, high efficiency of food conversion into body meat, high tolerance to crowding and domestication; (iv) an automation of essential production procedures.

Additional improvements can result from an increased application of polyculture techniques (the accommodation of several organisms with different, usually complementary requirements in one and the same culture enclosure), from an extension and refinement of *in situ* cultures, and from the utilization of local benefits such as heated waste-water effluents or suitable sewage outlets as fertilizer. This list, by no means complete, emphasizes again the need for interdisciplinary research. To the specialist it also demonstrates the rapidly increasing difficulties which must be overcome in order to achieve progress.

A largely untapped, but important source of natural materials which could be converted into human food are the microorganisms: bacteria, fungi and unicells with their enzymes and metabolites; they facilitate the synthesis of a multitude of building blocks (amino acids, fatty acids and peptides) which can be used for building up tailor-made human nutrients. So far, human nutrition has been considered too much in terms of commercial taste. Serving our tongue, rather than our physiological and nutritional needs, we have neglected our health. Numerous human diseases have their roots not in undernourishment but in excessive eating and the wrong foods. Critical investigations of specific nutritional requirements as a function of age, physiological constitution, professional stress and so on should be combined with research on microbial synthesis aiming at the production of large amounts of basic human food components. By adding binders and taste substances these could then be recomposed to healthy palatable dishes for a balanced diet.

Of course, many people will reject such 'synthetic foods'. Tradition and long-term nutritional adaptations have led to strong habits and even stronger convictions. It also may take considerable time before such foods would indeed match man's nutritional requirements and meet his demands for taste and aesthetic satisfaction. However, the goal seems achievable. While effective readjustment in our food habits will be difficult and require much convincing argumentation, in the long run such adjustments appear to be necessary.

In summary, the requirements, constraints, dangers and possible improvements considered above make it quite clear that a significant expansion of aquaculture encounters a variety of problems. Although aquaculture is likely to grow appreciably, for the foreseeable future it seems unlikely that it can make a major contribution to solving the world-wide food crisis.

RELATIVE IMPORTANCE OF FISHERIES, AGRICULTURE AND AQUACULTURE

Fisheries depend largely or entirely on natural productivity in little or non-controlled aquatic habitats. On the other hand, agriculture and aquaculture produce human food under more or less controlled conditions of environment, feed, disease, predation and even genetic constitution of the cultured organisms. However, some two-thirds of the Earth's surface is covered by water and theoretically total marine productivity by far exceeds that of the land. Hence, many look upon the seas as an inexhaustible protein reservoir for the growing human population. This view, however, neglects the important fact that, in contrast to the large and easily harvestable terrestrial plants, the main result of marine photosynthesis manifests itself in the form of microscopic algae, widely distributed and extremely difficult to collect in large quantities. Similarly, the collection of marine food animals requires considerable effort and expense in terms of technology and energy. Only where large predators such as fish or crustaceans aggregate, for example during spawning, migration or on feeding grounds, can great quantities be efficiently collected.

A comparison of aquatic versus terrestrial food production further requires one to note that the quantities of food produced per unit area are not strictly comparable on land and in water. In both cases food production is usually expressed per unit surface area; neither water depth nor water volume are taken into account. With these limitations in mind we shall now attempt a brief comparison of the productive potential of fisheries, agriculture and aquaculture.

The world fisheries provide an appreciable portion of man's food and will most likely dominate the output of aquaculture for many years to come. In 1973 the total annual world fishing catch was close to 66 million metric tons, comprising an average of 14% protein. This amounts to only some 1% of the total amount of food harvested on land and some 5% of the global annual protein yield. From 1974 to 1984 the world fishery is assumed to increase its total annual protein harvest from 9.3 million tons to about 14 million tons. However, information at present available on wild productivity, over-fishing and pollution hardly justifies expectations for an increase much beyond the 1984 estimate without introducing a critical deformaton of the marine ecosystems.

The world-wide annual agricultural protein harvest is expected to increase from the present 130 million tons – about one-third animal protein and two-thirds plant protein – to some 170 million tons in 1984. This predicted increase is based on assumed improvements in irrigation, product conservation, disease control and harvesting techniques. During the same time interval of ten years the annual protein harvest of world fisheries is assumed to increase from about 11 million tons to about 14 million tons a year and the annual aquaculture protein harvest from about 1 million tons to some 1.6 million tons. In spite of the percentage increase expected from aquacultural protein, its total harvest would represent only about 1% of the global protein by 1984.

This comparison demonstrates that agriculture will most likely continue to provide the major amount of human food. Why is agriculture so much more efficient in food production than aquaculture? The major reasons are evident from the following comparison of the essential production conditions.

In agriculture, our domesticated animals are predominantly herbivores: cattle, goats and sheep; only poultry and hogs live on animal as well as plant diets. Plant eaters transform the organic material available efficiently and directly into animal protein. The domesticated terrestrial animals have no absolute requirements for living food. Their dry feeds are readily available, inexpensive, and can easily be stored over appreciable periods of time. The animals reproduce under controlled conditions. This guarantees the availability of offsprings and allows selective breeding for properties desired by farmer and consumer. All major diseases are known and can largely be controlled – as can all predators and most competitors. There is no, or only a limited, need to control the animal's environment. All these points facilitate the efficient production of high-quality protein in large quantities which most people can afford, at least in Western countries.

In aquaculture, most cultured animals are carnivorous. Both aquatic carnivores, as well as herbivores such as molluscs, mullet and milkfish, require more attention from their farmers than most of their terrestrial counterparts. Many of the cultured aquatic animals, especially the marine ones, have absolute requirements for living food – at least during certain stages of their life-cycle. Living feeds are expensive, elaborate to maintain and less convenient to handle than dry feeds.

CONCLUSIONS

Aquaculture opens up an important new avenue for making additional food available. We cannot simply continue to hunt sea food in oceans and coastal waters, exploiting natural productivity in a primitive way which was abandoned on land thousands of years ago. With time, aquaculture is bound to gain in importance. However, the total amount of food produced will remain – at least for the foreseeable future – significantly below that provided by fisheries, and it will not exceed a very small fraction of the food made available by agriculture.

The importance of aquaculture is, to a considerable extent, a function of the area and the habits of the country concerned. The subtropics or tropics qualify as preferred areas and where conditions are suitable – especially with regard to available coastal areas, water supply, climate and nutritional habits – aquaculture can attain high priority as an industry, as for example in Japan, Israel, Mexico and Taiwan.

Improvements in biological and technological know-how can appreciably increase the efficiency of aquacultural food production. Improvements are also possible and urgently needed in the world-wide distribution of the food produced.

In aquaculture, are we proceeding in the right direction? No. Aquaculture is presently focusing too much attention on converting one food source into another for the sole reason of making a commercial profit. An example: with appreciable expense of fossil energy a fish is caught at sea; some 30% of its meat is transferred into fish meal, and this again fed to another fish which tastes better and commands a higher price. In Japan, the economically highly successful yellow-tail cultivation produces about 80 000 tons of cultured fish each year, but must use some 550 000 tons of animal protein, from fish meal, as fish food. In ecological terms this is a waste of energy and material.

At the end of an energy-consuming and waste-producing culture process, only about one-eighth of the originally available fish protein remains. The rest has been sacrificed for transforming perfectly acceptable protein into an expensive luxury product. The conflict between private profit and public needs remains to be solved. Most governments support commercial aquaculture projects with the declared aim of reducing world-wide malnutrition and starvation. They would be well advised to place more emphasis on ecological aspects, on healthy, palatable diets that can be afforded by the poor, and on the cultivation of organisms that occupy trophic positions low in the food pyramid.

Human food produced from microorganisms, plants and herbivorous aquatic animals deserves more attention than it receives at present. Aquafarmers who specialize in the mass production of sea foods from such organisms and for acceptable prices, deserve concerted governmental support – in terms of finance, provision of suitable land and reduction of legal restraints.

Turning to future perspectives, we must first define our own position. While man has developed never-dreamt-of technological means for exploiting nature and for changing the surface of this planet Earth to his own liking, nutritionally he has remained a captive of natural productivity. The logarithmic growth of human populations, both in individual numbers and in impact exerted per capita, creates problems which are unsolvable on the basis of present ecological knowledge. Food production is one of these problems. Others are pollution, translocation of organisms and man-made structural alterations on our planet's surface.

A single member of the ecosystem – *Homo sapiens* – cannot indefinitely change and distort the natural food web to his own advantage without critically deforming the ecosystem itself – a system which has been the prerequisite for his own development and which remains the basis for his future existence. For an ecologist, the potential consequences of food production for 12×10^9 people are a nightmare. There is every reason to predict that the ultimate limitations in food production lie not so much in the scientific and technological know-how or the land and energy available, but in the tolerance of natural ecosystems to man-made deformations.

Looking at the long-term perspectives of food production, the ideal solution would be to utilize the synthesizing abilities of microorganisms, plants and low-trophic-level animals to produce essential portions of the human diet from waste materials. Such recycling copies the natural ecosystem dynamics of synthesis, transformation and re-mineralization of organic matter. It would be in line with ecological principles and considerably reduce man's impact on nature.

The most significant future prerequisite for long-term survival of a civilized human society is a world-wide reduction and control of man's population growth. The present growth rate and the predictions for the foreseeable future dwarf all attempts to combat hunger. Unless we succeed in controlling our own growth we shall continue to swim against the tide of nature.

LITERATURE CITED

1. O. Kinne and H. Rosenthal, Cultivation of animals. Commercial cultivation (aquaculture). In *Marine Ecology*, Vol. III, Part 3, edited by O. Kinne, Chap. 5.2, pp. 1321–1398. Wiley, Chichester (1977).
2. J. E. Bardach, J. H. Ryther and W. O. McLarney, *Aquaculture*, Wiley, New York (1972).
3. O. Kinne, Cultivation of marine organisms: water-quality management and technology. In *Marine Ecology*, Vol. III, Part 1, edited by O. Kinne, Chap. 2, pp. 20–300. Wiley, London (1976).
4. M. Fujiya, Oyster farming in Japan. *Helgol. Wiss. Meeresunters.* **20**, 464–479 (1970).
5. T. Harada, The present status of marine fish cultivation research in Japan. *Helgol. Wiss. Meeresunters.* **20**, 594–601 (1970).
6. O. Kinne, Cultivation of animals. Research cultivation. In *Marine Ecology*, Vol. III, Part 2, edited by O. Kinne, Chap. 5.1, pp. 579–1293. Wiley, Chichester (1977).
7. D. Pimentel, W. Dritschilo, J. Krummel and J. Kutzman, Energy and land constraints in food protein production. *Science* **190**, 754–761 (1975).
8. Y. P. Murphy and R. I. Lipper, BOD-production of channel catfish. *Prog. Fish Cult.* **32**, 195–198 (1970).

The manuscript was received 15 December 1978

Chairman and Cook

Twice as Spicy

PROFESSOR DR HEINZ MAIER-LEIBNITZ

Formerly President, Deutsche Forschungsgemeinschaft, Bonn, Federal Republic of Germany

To achieve desirable decisions from a committee, a novel but dangerous manoeuvre is recommended here: the Chairman to be at the same time also Cook, preparing and serving a superb meal to bring the meeting to a successful conclusion. Theoretical considerations are followed by a detailed description of a meeting-eating.

Anyone who has ever chaired a meeting, knows full well how easily a mistake can be made. But he will also know that there are certain subtle strategems which can achieve his carefully predetermined objectives. Here I shall describe one of these tricks of the trade which, because of its high risk factor, is relatively unknown. It can, however, be most effective. The technique is simple: the Chairman is also a Cook, and whilst the meeting is in progress he prepares the meal which crowns its success.

THE GREAT DUALITY

This dual manoeuvre is full of dangers, and anyone attempting it should be aware of what is entailed.

- ○ Any meal can turn to disaster.
- ○ Millions cook every day. Anyone who claims he can do better than all the others will only antagonize them.
- ○ The dual operation is not appropriate for all committees, as quite a few members will be glad if meetings are short and end quickly. It is also wrong to apply it to industrial tycoons, accustomed as they are to rush for their private planes at the end of a meeting so as to be home for dinner punctually.
- ○ Furthermore, anyone trying to be unconventional in our Capital City (Bonn) is likely to be the target of some insolence. Such an unmanly activity as cooking is an excellent target for social disapproval, and he who attempts to be both cook and chairman can expect only discourtesy.
- ○ For a cooking chairman, the rate of return invitations to other dinners will be very small and his achievements will appear as supernatural to most hostesses.
- ○ Finally, cooking as a hobby demands a great deal of time and effort. Thou shalt have no other hobby: such as golf or party-going.

WHY IS THE DUAL DEED TWICE AS SPICY?

- × He who cooks does something for others, he performs a specific service. Thus, he exposes himself and is seen to be fallible, by no means perfect as it is so often the case for a Chairman and other top executives.
- × Cooking relaxes and brings repose. This is my own experience and is confirmed in the literature. And repose, unconnected with idleness, is balm indeed for any chairman.
- × With regard to sarcasm, it is better to steer it towards such a harmless activity as cooking rather than wait for any inevitable smear campaign. And here comes unexpected help; although envy is an all too human stimulus, cooking will never evoke jealousy in other men.
- × Finally, I know that a meeting can be turned towards success if the chairman is also a cook.

A SPECIFIC EXAMPLE OF A MEETING-EATING

An actual meeting begins at 9.00 and finishes at 13.30. This has to be changed so that the meeting starts at 15.30, thus allowing everyone, from wherever he started out in the Federal Republic to reach Bonn by rail or plane in good time, and to work during his travels. Furthermore, meetings which end

CCC-0308-0188/80/0005-0033$02.00/1

in eatings are quicker, hence 15.30 to 19.00 should be adequate. This has the advantage for the Chairman that he gains an hour for his preparations and he can save a further hour if he has a quick snack lunch in his kitchen.

The menu must be chosen carefully, preferably a day or so in advance, whilst travelling on official business:

Soupe Longueville
Filet Chateaubriand
Gratin Dauphinois
Chicorée Salad
Fromages
Pommes Cardinal

I choose Filet Chateaubriand because it can be fried in five cm thick slices and one frying pan is large enough to cater for eight to ten persons. (German butchers have perfected a method of keeping fillet for a long time under a layer of fat, thus ageing it and always having it ready and available.) Needless to say, the menu has a built-in timetable for the meeting. See below.

The previous evening the dough for the French bread must be prepared, and the next morning there is about one hour's work with about the same again at 14.30. Then, by the time the guests arrive, the kitchen is full of provisions and food, already prepared and laid out, on pans, plates and in pots.

My armchairs are certainly more comfortable than the seats in the Ministry, and hence all Committee members feel that they have been invited to a social occasion; naturally this helps with the affairs of business. The first part of the meeting is without controversy: approval is needed for a new peer-review and grant-approval system. There have been many previous discussions on this subject, so now only minor details need to be settled. But a new expert subcommittee must be chosen. This has also been planned well in advance and a list has been drawn up, taking into account all possible disciplines and geographical regions. But now there is disagreement between the six scientists on the Committee and the two representatives from the Ministry. This has become an almost ritual disagreement, reappearing at each committee meeting, and by now upgraded thus acquiring its own life-style. In principle one name could be replaced by another, but this leads inevitably to a lengthy discussion, as in consequence other names must now be exchanged.

THE CHAIRMAN'S PEREGRINATIONS

The challenger in these discussions is the representative of a small and insignificant discipline who, to convince himself, repeats everything five times and his opponent is an impatient colleague who can raise a rude hullabaloo. At this stage the Chairman gets up and moves towards his kitchen returning innocently after seven minutes, asking the committee what has been decided in his absence. This certainly helps somewhat, and when 40 minutes later the committee must finally vote on their decisions, the Chairman will again absent himself for five minutes.

Now comes the difficult part of the proceedings. Under the new arrangement the large-scale research projects will be reviewed and underwritten by the Ministry, whilst the other smaller ones will be funded by certain Research Associations; for the first time they have sent their representatives to these meetings. This unfamiliar new arrangement does not please our scientists, and they do not understand the largely unscientific reasons for the division between large and small research projects. Further ill-feeling arises from the realization that total financial support for research is decreasing and that their opinions as scientists, are no longer solicited.

Now the Chairman is really in the soup. He knows from his private talks at the Ministry that the new arrangement is inevitable and that the division is pre-ordained. All he can do is to let the arguments rage back and forth, while once again he takes leave for six minutes to review progress in his kitchen. Fortunately the waiter has arrived at this juncture and offers drinks – orange juice, with and without gin, gin and tonic, campari soda and sherry are liked best, whisky no longer being popular. When the Cook-Chairman returns to his meeting, all present realize that they cannot now ask him to go back to the Ministry and make new demands for fundamental changes. The necessary decisions are quickly agreed and after a few further small points have been settled, there is a ten-minute interval before the dinner; naturally the Cook spends these in his kitchen.

During the soup, the representatives from the Research Associations have to listen to some anecdotes about good bureaucratic manners. Thus, the Chairman can throw a last glance at his chef-d'oeuvre in the kitchen. During the main course, the general conversation turns towards the next Nobel prize

PROFESSOR DR HEINZ MAIER–LEIBNITZ – a Member of our Editorial Board – was until a few months ago President of the Deutsche Forschungsgemeinschaft, the organization allocating about DM 750 million annually to pure and applied research in the Federal Republic of Germany. He has been Professor of Physics at the Munich Institute of Technology since 1952 and was Director of the Max von Laue–Paul Langevin Institute at Grenoble 1967–1972, as well as being a Member of the Royal Swedish, Austrian, Bavarian, Finnish, Flemish, French, Heidelberg, and Leopoldina Academies of Science.
Address: 110 Pienzenauer Strasse, Munich 81, Federal Republic of Germany

winners and their new scientific theories. Most committee members remain for coffee after dinner.

At this stage, no further decisions are taken – that would indeed be unfair – but new plans are brought forward, as yet too provisional for any formal agenda. This then is the most important part of the whole meeting, when new ideas are cast about, when everyone can contribute and at the same time learn and receive a new stimulus. It is also the rarest part; it hardly ever takes place, as the Chairman hardly ever cooks.

Now follows in the original German text, the author's detailed recipes for French baguettes, potage Longueville, filet Chateaubriand, sauce Béarnaise, gratin Dauphinois and pommes Cardinal. These have not been translated here as any other Chairman, equally gifted gastronomically and courageous enough to imitate the author, will either know the recipes by heart, studied them in the original text or will have had his reading lessons as a five-year old from Mrs Isabella Beeton.

Französisches Weißbrot
Natürlich können Sie nicht mit einem guten Bäcker konkurrieren. Aber Sie haben einen Vorteil: Sie wissen, wann Ihr Brot gegessen wird, und Sie können es so backen, daß es frisch ist. Brot und französische Butter stehen bei Beginn der Mahlzeit auf dem Tisch.

Am Abend vorher:
Ein Pfund Mehl, das sind vier Tassen, und zwar etwa $\frac{2}{3}$ Weißmehl 405, $\frac{1}{3}$ Manitoba Weizenmehl.
$\frac{3}{8}$ *l Wasser* abmessen, davon eine kleine Tasse Wasser über *40* g zerbröckelte *Hefe* und *1 Teelöffel Zucker* gießen und nach einiger Zeit verrühren. Das Wasser soll lauwarm sein (30°).
Ein Viertel des Mehls (1 Tasse) in eine Schüssel geben und mit der Hefelösung verrühren. Gut zudecken (mit einem Teller). Über Nacht im nicht verschlossenen Wärmefach des Backofens oder sonstwo bei 30 bis 35° stehenlassen. Der Teig geht auf und fällt wieder zusammen.

Morgens noch ein Viertel des Mehls (eine weitere Tasse) zufügen, dazu das restliche Wasser; verrühen und wieder gehen lassen bei 30 bis 35° bis zum Nachmittag. Jetzt das restliche Mehl zugeben, dazu *1 flachen Eßlöffel Salz* (oder weniger) und mit dem Teigrührer einige Minuten kneten. Dann bei nicht zu großer Wärme, also nicht im Ofen, gehen lassen, bis das Volumen sich fast verdoppelt hat, was nach 30 oder 40 Minuten der Fall sein wird. Den Teig mit bemehlten Händen herausnehmen (er soll eine solche Konsistenz haben, daß er daran kaum mehr klebt), auf einer mit Mehl bestäubten Fläche mit einem großen Messer vierteilen und auf ein ganz leicht geöltes Backblech vier lange Stangen auflegen. Wenn möglich, noch zehn Minuten gehen lassen, dann, mit Wasser bestrichen, in den auf 220° vorgeheizten Ofen geben und etwa eine halbe Stunde backen. (Die Zeit muß man probieren, ebenso die Höhe im Ofen; Haushaltsbacköfen sind sehr verschieden.) Das Brot soll oben wenig braun sein, von unten nicht dunkler. Wenn man es von unten anklopft, soll es hohl klingen. Das endgültige Kriterium ist aber der Geschmack. Mir selbst schmeckt es mit etwas längerer Backzeit und kräftiger Kruste. Nach dem Backen herausnehmen und umgekehrt abkühlen lassen.

Potage Longueville
Die äußeren Blätter *eines großen Salatkopfes* (etwa die Hälfte des Salats) und ein kleingeschnittener *Lauchstengel* werden mit *30 g Butter* langsam gedämpft. Dann *6 Tassen tiefgekühlte feine Erbsen* dazugeben, mit $\frac{1}{2}$ *l Weißwein* $\frac{1}{2}$ Stunde kochen, dabei nach 10 Minuten eine Tasse Erbsen herausnehmen, das übrige durch ein feines Sieb passieren.

Vor dem Auftragen mit $1\frac{1}{2}$ *l leichter Bouillon* wieder aufkochen, mit *4 Eigelb* und $\frac{1}{4}$ *l Rahm* abziehen. Die Tasse Erbsen wieder zugeben. Mit *gerösteten Brotwürfeln* servieren.

Brotwürfel. Weißbrotscheiben in Würfel schneiden, mit fast ebensoviel Butter, wie man für Butterbrote verwenden würde, in einer flachen Pfanne im Ofen bei 160 Grad trocknen und leicht bräunen (etwa 30 Minuten). Nach 10 Minuten einmal durchrühren. Die Würfel müssen sehr kroß sein, sonst werden sie in der Suppe sofort weich.

Filet Chateaubriand
Man rechnet *175 g Filet* pro Person, aber nach meiner Erfahrung sind auch 250 g nicht zuviel.

Das Fleisch (aus dem nicht zu schlanken Teil des Filets) in 5 cm dicke Scheiben schneiden, auf ein Gitter legen, damit es nicht naß bleibt.

20 Minuten vor dem Auftragen in einer schweren Pfanne für acht Personen *2 Eßlöffel Öl* und *1 Teelöffel Butter* (5 g) heiß werden lassen, bis die Butter brutzelt (sie dient hier als Temperaturanzeiger). Die Steaks rasch hineinlegen, etwas breitdrücken, daß die Pfanne gut ausgefüllt ist. Auf beiden Seiten scharf anbraten (lieber etwas zu scharf, aber ganz kurz); dann auf Mittelhitze (eventuell auf einer zweiten Herdplatte) bringen, beide Seiten der Steaks sorgfältig mit *Butter* bestreichen, *Salz*, *grober Pfeffer* aus der Mühle, und auf jeder Seite etwa 8 Minuten so braten, daß der Satz in der Pfanne nicht verbrennt, aber vor allem darf sich keine Flüssigkeit ansammeln. Der ganze Bratvorgang soll 20 Minuten dauern.

Die Steaks auf das Tranchierbrett (mit Rinne für austretenden Fleischsaft) legen. Eine *halbe Tasse heißes Wasser* in die Pfanne schütten and unter Kratzen mit der Gabel zum Kochen bringen.

Am Tisch sehr schräg, also quer zur Faser, tranchieren, jedes Steak in 5 Stücke. Die Gäste wählen, ob sie Randstücke oder rohe Stücke vom Inneren wünschen.

Sauce Béarnaise
4 Eßlöffel Schalotten, grob zerkleinert, *4 Zweige Estragon* in Stücken, *20 Blatt Kerbel* in Saucenpfanne mit *2 Eßlöffeln Essig*, $\frac{1}{2}$ *Tasse Weißwein* begießen, *Salz*, *Pfeffer*, auf kleinem Feuer zugedeckt 20 Minuten kochen, absieben; es sollen noch etwa 4 Eßlöffel Flüssigkeit übrig sein. Die Flüssigkeit in die Pfanne zurückgeben, dazu 6 *Eigelb*, *20 g Butter*.

Kurz vor dem Auftragen auf der heißen Herdplatte unter heftigem Rühren mit einem Holzspatel erhitzen. Sobald die Butter geschmolzen ist und beim Rühren etwas verfestigtes Ei vom Boden kommt, ein großes Stück (50 g) Butter zugeben, weiterrühren, bis sie fast geschmolzen ist, wieder Butter zugeben, bis insgesamt *250 g Butter* verbraucht sind und die Sauce recht fest ist, dicker als eine Sauce Hollandaise.

Wenn die Prozedur zu gefährlich aussieht, wenn also das Eigelb in Klumpen gerät, stellt man die Pfanne einen Augenblick auf eine kalte nasse Platte und rührt mit Butterzugabe weiter, bis die Sauce wieder glatt ist.

Zum Schluß kann man einen Eßlöffel kleingehackte *Estragon- und Kerbelblätter* zugeben – fürs Auge.

Gratin Dauphinois
Ein Fürst unter den Kartoffelgerichten und geeignet als Beilage für Gerichte mit wenig Sauce; für eine Person oder für fünfzehn, nur die Zubereitung ist verschieden. Hier ist das Rezept für acht oder mehr.

Nicht zu mehlige *Kartoffeln* (*100 g pro Person*) werden gewaschen, geschält (bei jungen Kartoffeln vielleicht nicht nötig), in 2–3 mm starke Scheiben geschnitten, in Wasser nicht länger als zwei Stunden aufbewahrt. In einer großen Form (etwa oval 40 cm) werden *vier Knoblauchzehen* mit dem Boden eines kleinen Glases reibend zerquetscht und wieder herausgenommen. Dann wird die Form gut *gebuttert*. Der Backofen wird auf 200° vorgeheizt.

Eine Stunde vor dem Auftragen werden die Kartoffeln abgeschüttet und flach in der Form verteilt, darüber *Salz*, *Pfeffer*, *Muskat* und bis zu einem Drittel der Füllhöhe *halb Milch*, *halb Rahm*, darüber *Butterflöckchen*. Beim Backen (1 Stunde bei 200°) muß die Flüssigkeit über den Kartoffeln schäumen und dann allmahlich fast verschwinden, während sich eine hellbraune Kruste bildet. Die Buttermenge muß so sein, daß zum Schluß kein flüssiges Fett bleibt. Hier hilft nur Probieren, denn Kartoffeln sind verschieden.

Pommes Cardinal
Ein Apfel pro Person wird geschält, entkernt, mit *Wein*, *Zucker*, *Zitrone*, *Zitronenschale* gut weichgekocht (der Sirup muß über den Früchten hochsteigen). Anschließend die Äpfel kühlen.

Frische oder gefrorene *Himbeeren* (ein Pfund für 10 Personen) werden durch ein Sieb gedrückt, gezuckert und mit weniger als einem Gläschen *Kirschwasser* versetzt. Jeder Apfel kommt auf einen Teller, darüber zwei Eßlöffel Himbeermus. Dazu werden ein leichter *Schlagrahm*, mit ein wenig *Calvados* abgeschmeckt, und *kleines Gebäck* gereicht.

TIMETABLE FOR THE MEETING–EATING

Night before	Start bread dough with a little flour
Next morning	Add flour and water to dough Prepare soup and raspberry mousse
14.30	Roast croutons, cut meat, prepare salad and place in cool place; cut potatoes and start to prepare sauce
15.30	Committee Meeting begins
16.55	Turn over bread dough
17.35	Shape the dough
17.45	Put bread in oven
18.15	Take bread out of oven Put potatoes into oven to bake
18.50	Finish soup, start the fillet and lay out the cheeses
19.00	End of Committee Meeting Serve soup and turn the fillets
19.10	Take fillets out of pan and prepare gravy with water Finish sauce Béarnaise and turn the salad
19.15	Serve main course
19.40	Serve cheese board
19.55	Finish and serve dessert

The original text appeared in: *Texte + Thesen + Sottisen*, a Jubilee Issue, Edition Interfrom, Zürich. It is reproduced here by kind permission of the Publishers and translated by The Editor.

FRITZ WOLF

Smell: A Challenge to Interdisciplinary Science

LEO NARODNY
Barbados National Council for Science and Technology, Barbados, West Indies

The sense of smell remains an intractable problem and a challenge to interdisciplinary sciences. Not only are there no primary odors or periodic table of scents but there are no satisfactory numerical identifications by visual, auditory or somesthetic methods of measurement. A particle–wave relation may exist to reconcile the correlation of many odors with infrared spectra, but the vibrational theory of smell is currently out of fashion. Structure–activity relationships have been studied in human olfactory processes and have yielded perfume compounds of great value to commerce. Research on pheromones is intense to control insects, and the further study of human pheromones may even reveal an unsuspected communications system. At present, interdisciplinary sciences can well be guided by the artist-perfumer for what seems a millennium of research in molecular biology.

The fundamentals of music, painting and perfumery are similar in that they are the means by which great artists communicate to the world. Scientists have been able to analyze and synthesize the elements of music and painting, but they have not been able to define all the parameters of smell, nor have they agreed on an acceptable hypothesis to account for all the experimental facts. Smell is therefore a continuing challenge to the interdisciplinary sciences of molecular biology, enzyme chemistry, electrophysiology, neurology, psychology and quantum chemistry, to name a few.

PARAMETERS OF SMELL

Perhaps in no other field can the artist guide the scientist as much as the creative perfumer can guide the chemist in analyzing and synthesizing odors. Billot,[1] a perfumer, described an 'attempt at rationalizing the technique of the perfumer: while it is impossible to codify creation in perfumery, since all art contains a share of imagination, it is possible to define to a certain extent the science of mixtures, as is done in painting, and the science of proportions, as in sculpture. Although our understanding of odors and odor phenomena remains limited, certain factors such as tone, volatility, intensity, volume and tenacity can, to some extent, be isolated and intelligently discussed.'

Definitions

By Webster's definition,[2] 'Smell is the most general word, and odor, with which it is often interchangeable, may suggest either a pleasant or unpleasant impression, but smell often conveys the more direct or earthy suggestions. Aroma suggests a distinctive, often delicately spicy odor. Scent often implies less strength than smell or odor. Fragrance is a fresh and pleasant odor. Perfume belongs either to natural or prepared objects and may be either light or heavy.' It might be said that olfaction belongs to science, fragrance to commerce, and scents to the animal world, while as Dr Johnson remarked to a woman, 'Madam, you smell, I stink.'

A German patent claims to have computerized the composition of perfumes, but this seems as premature as computerized music with electronic tones, or computerized painting with colored tiles. However, a computerized sausage aroma may be more interesting than ordinary *knackwurst.*

Information Patterns

Beets[3] points out the analogy of music and painting to the information pattern of smell: 'For a melody, the frequency of its modalities, and for each of the latter, the positions in which it occurs in the sequence of a melody, and for each, the quantity, duration and intensity. . . . A painting can be described by the color of each modality, the positions in which it occurs (any parameter other than quantity and intensity) and for each occurrence, the magnitude of the area, and the intensity of the reflected light.'

Unique Sense

Unlike music, however, where the rhythm of the melody and the modulation of the chords determine the composition, the rapid adaptation, the so-called fatigue, of the olfactory sense prevents such a

CCC-0308-0188/80/0005-0037$06.00/1

symphony of smells or their artistic correlation with motion pictures, 'Aromarama' and 'Smellavision' were not commercially successful partly because of the unpleasant sensory adaptation of the olfactory receptors to a rapid succession of the intense odors, not to speak of the cost of dispensing tons of perfumes through a complex air-conditioning system ducted to each seat. The programmed release of aerosols of such aromas for commercials in television fortunately seems beyond the budgets of most 'soap operas'.

Primary Odors and Anosmias

One of the problems of smell is that there appear to be no primary odors, nor is there a spectrum from which all other smells may be created. Amoore,[4] however, has developed a concept that 'human primary odors may be indicated by the various forms of specific *anosmia*. In this condition a person, of otherwise normal olfactory acuity, finds some particular odor that escapes his attention, while it remains obvious to other people.'

The first primary odor to be studied in detail was the sweat-like smell of isobutyric acid, the second the sperm-like odor of l-pyrroline. Later Amoore included the fishy odor of trimethylamine, the sweet musk odor of pentadecalactone, the malt odor of isobutyraldehyde, and urine (5α-androst-16-enone); he believed that there might be as many as 32 such specific anosmias. Schizophrenic patients,[5] for example, are apparently unaware of their intense smell of *trans*-3-methylhexanoic acid, like apple sauce.

However, Beets[3] points out that although specific anosmias may have wide applications in clinical diagnosis, anosmia may have its limits. 'We do not know whether every informational modality is represented by an experimentally accessible specific anosmia.'

ANATOMY OF ODOR RECEPTORS

The anatomy of the olfactory areas is well known, although the molecular biology involved in the receptor process remains a mystery. In humans the olfactory receptors are presumed to be located in the membrane of the olfactory knobs or 'hands', each with six to eight cilia or 'fingers', flaccidly floating in and on the surface of a 10–60-μm-thick layer of a mucus sol: lipids, phosphatids and polysaccharides. Each cilium has nine pairs of outer filaments and a central pair, surrounded by granules of proteinaceous adsorbents, which in the frog are associated with cyclic AMP.[8,27]

Human Olfactory System

In humans approximately a million of these receptors, each surrounded by supporting cells with microvilli (Fig. 1), are located in the roofs of two dark, silent, humid clefts. These are undisturbed by fresh air except when a purposeful sniff tumbles any odorant molecules up to the roofs of the clefts, covered by hair 'blankets' 2.5 cm^2 each, on both sides of the nasal septum.[11] The air is at atmospheric pressure and each nostril is presumed to have the same concentrations of oxygen and carbon dioxide. Meanwhile oxygen-hungry hemoglobin and proteins and other components circulate in the bloodstream of the nasal septum and passages.

Nervous System

It is not unreasonable to suppose that some adsorption takes place on the granules in the finger-like cilia, like chromatographic columns; the presumably negatively charged membranes of the knob-like hands feel a pattern of smell. Moulton[19] concludes that there exists 'a pattern of activity distributed in space in which the ratio of excitation in a number of parallel channels may define the stimulus'. The analogy with vision is hard to resist: a pattern of visible light focused on 115 million rod cells and 6.5 million cone cells in the retina of the eyes, located only a few centimeters forward of the twin olfactory clefts and serving as an early warning system, to be confirmed by the nose, or perhaps vice versa.

Special Features

Best and Taylor[11] have drawn attention to the fact that 'the end organs of smell, however, differ from those of any other sense in that the cell body of the primary neuron is situated in the peripheral organ itself, and is stimulated directly, without the intervention of a specialized receptor cell. No other sensory mechanism possesses both these features. Though the primary neuron of the visual pathway is

LEO NARODNY, an interdisciplinary physicist, is now Consultant to the Barbados National Council for Science and Technology. He started with a degree in mining from Columbia University, followed by a decade of trading odd lots on the New York Stock Exchange with a statistical formula which he developed. This he followed with a decade of growing and processing of essential oil crops in Dominica. He then turned to optical physics with inventions of rapid photographic processing, the use of synthetic holograms for templates and a fibre optic keyboard; after another decade he ended up as Chief Scientist at the Kollsman Instrument Company.
Address: Martin's Bay, St John, Barbados, West Indies.

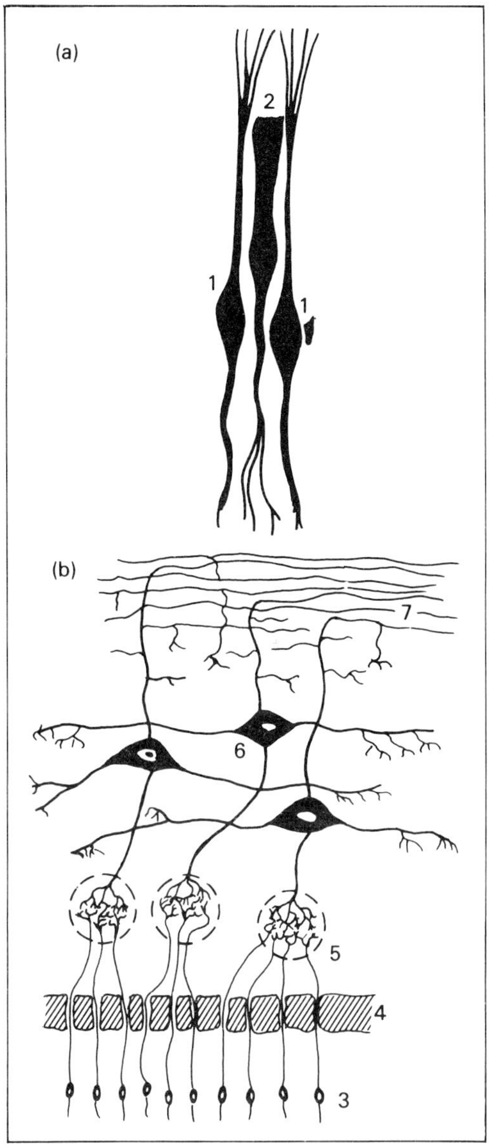

Figure 1. Human olfactory system (Ref. 11). (a) Cells from olfactory mucous membrane separated from one another and highly magnified: 1, olfactory receptors; 2, supporting cells. (b) Cerebral connections of the olfactory nerves: 3, cells of olfactory mucous membrane; 4, olfactory nerve fibers piercing the floor of the skull to reach the olfactory bulb; 5, connections of olfactory nerves with processes of mitral cells in olfactory bulb; 6, mitral cells; 7, fibers passing into the olfactory tract.

situated in the retina, the stimulus is received by the rods and cones; pain sensations are subserved by bare nerve endings, the cell bodies of the pain fibers located in the posterior root ganglia. It is important to remember that here, as nowhere else, the nervous system is in direct contact with the external environment.'

NEUROLOGY OF OLFACTORY BRAIN

Each receptor cell has a small axon which is a thin, 0.2 μm diameter, fiber which transmits information through the skull from the walls and roof of the nasal clefts directly to the olfactory bulb. This acts as a relay station linking the sensory receptors in the nose to the cerebral cortex. The axons from the olfactory cells make synapses in the olfactory bulb with semispherical, root-like formations of the glomeruli which go to layers of dendrites and dendritic tufts of mitral cells, and gemmules of short granule cells which process the signals (Fig. 2a).[18] The mitral cells, shaped like a bishop's mitre, extend laterally and approach synaptically the granule cells. Then follows a deeper layer of granule cells, and finally the output axons of the mitral cells as well as the short axon interneural cells connect to the olfactory cortex.[42,43]

The Olfactory Bulb

Neurobiological investigation[22] of the olfactory bulb has shown that there are inhibitory signals which modulate the pattern received from the olfactory receptor cells. As Shepherd recently reported,[23] 'It is clear from our work so far that inhibition is important in shaping the responses of the mitral cells. . . . It seems safe to say that the ability to distinguish one odor from another depends on interactions between excitatory and inhibitory activity in the dendritic circuits of the olfactory bulb. . . . With the aid of radioactive isotopes we have found that odor stimulation is associated with spatial patterns of activity in the olfactory bulb. The foci of activity are located precisely, in groups of glomeruli which have a high density of dendrites and synapses' (Fig. 2b).

From the olfactory bulb the pattern of smell is transmitted to the cerebral cortex, where the center for smell is situated in the uncus, and the anterior part of the hippocampal gyrus (pyriform area) closely related to the temporal lobe.[44,45]

Memory Banks

A comparison of the odor pattern with the memory of previous odors and the words or gestures used to describe them must therefore relate to each subjective individual. Until we find a universally accepted optical, auditory or somesthetic indicator or numerical measure of the tens of thousands of smells, these 'memory banks' must remain as our smell directories and dictionaries.

THE ODORS OF PEOPLE

Perhaps we can, with practice, extend these memory banks to include recognizable odors of people. Piéron[28] brought fir needles near the nose of a sleeping hunting dog and observed him as he mimicked a chase through the pine woods. Menzel[29] has shown that a dog can identify a piece of wood if his master has held it in his hands for 2 s. Bloodhounds,

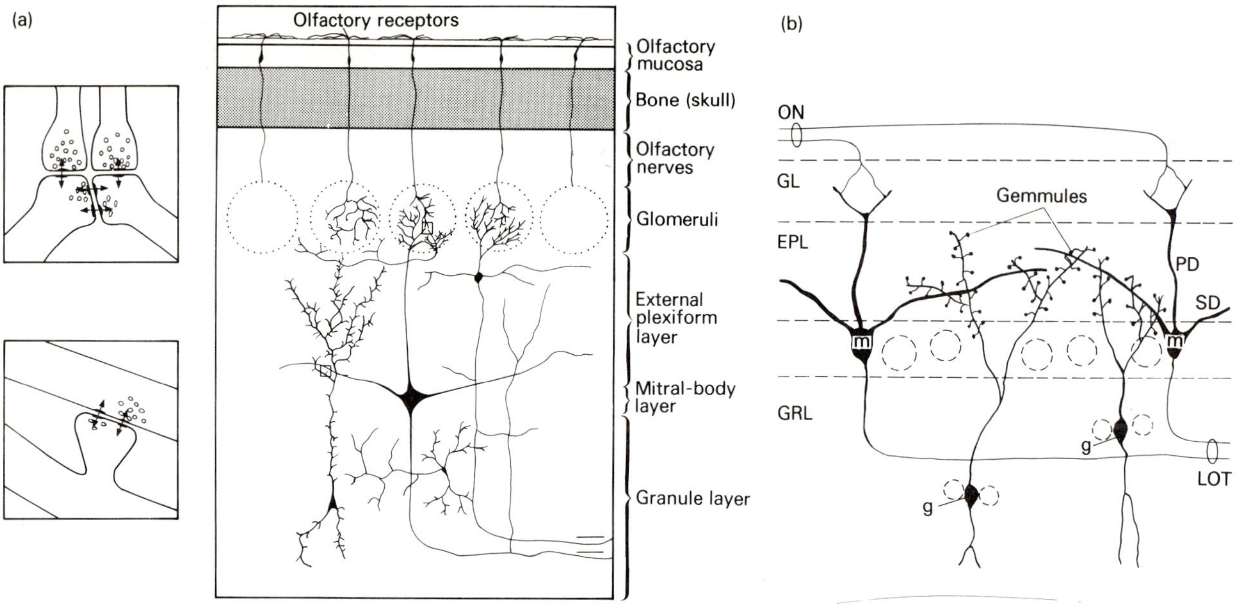

Figure 2. (a) Human olfactory bulb (Ref. 23). The olfactory bulb is a relay station linking the sensory receptors in the nose to the brain, and contains numerous microcircuits that transmit and process olfactory information. (b) Feedback connections in olfactory bulb (Ref. 22): GL, glomerular layer; EPL, external plexiform layer; GRL, granular layer; ON, olfactory nerve; LOT, lateral olfactory tract; g, granule cell; m, mitral cell; PD, primary mitral dendrite; SD, secondary mitral dendrite.

relatively short-sighted, live in a world of odors, and can track footprints many hours old in forests and shops. As Tesla, and later Kirlian, have shown, ions stream from living objects, and are made visible for example by high frequency and high voltage electricity. It is clear that some charged or uncharged particles of chemicals must remain in the air or the surface of the track for the bloodhound to smell.

Olfactory Abilities of Primitive Man

Primitive man undoubtedly depended on his nose as an early warning signal, and as a method of locating food and a mate. Father Labat[6] provided some fascinating observations on the fauna of the West Indies in 1694:

> 'One surprising thing is that barracuda rather attack an Englishman than a Frenchman when they find themselves together in the sea.
>
> It may be that Englishmen have more open pores than the Frenchmen and therefore exhale a greater quantity of attractive scents, but why are they more open? Is there a notable difference between the cells of the body of a Frenchman and an Englishman? I have heard people discuss this fact without arriving at a demonstrative or convincing reason. After having thought about it, it seems to me that it could be by nourishment and temperament of the English race. It is certain they eat a lot of meat little cooked, and almost no bread (hence the Irish proverb that bread is the nourishment of necessity).'
>
> 'Now it is certain that the consumption of a lot of meat produces an odor which is not noticeable in those who eat less meat, at least to those who lead the same lives. That is just the experience we notice among butchers, and I mean the cleanest among them, who are always with meats and eat enough of them for their nourishment: their odor distinguishes them even among the people who do not know them as butchers.
>
> It is not only the opinion of the Caribs, who as we know, have tasted the flesh of all the Europeans who have come to hunt them in their country, and swear that the flesh of the English is the most appetizing and more delicate than the French or the Spanish.'

THRESHOLD OF DETECTION

Seifert[30] found that the odor threshold of 2-methoxy-3-hexylpyrazine was one part in 10^{12} parts of water. The odor component of green Bell peppers has a threshold of two parts in 10^{12} parts of water (2-methoxy-3-isobutylpyrazine). The volume of the whole Earth is just over 10^{12} km^3, which gives some idea of the dilution which is perceptible. However, the potency falls by 10^8 – which we can call 82 deci-'Bells' by analogy with decibels – if we reduce it to its parent molecule pyrazine; this requires 175 million parts per 10^{12} parts of water. This shows the power of the methoxy and alkyl groups on the odor receptors, and should be capable of further interpretation.

Analogy to Vision

Certainly intermolecular forces must be responsible for odor, and a 'chorus line' of a few Bell pepper molecules may be detected like a telephone wire is identified at a distance corresponding to a half second of arc resolution – although the normal limit of resolution of the eye is one minute of arc.[47] As Randebroek speculates, the 'arrangement of the secondary valence forces is responsible for the generation and kind of odor'.[31]

MOLECULAR STRUCTURE OF SMELLS

The relation of the chemical structure of molecules to olfactory activity has evolved recently. Previous concepts imagined specific sites for specific shapes of odorants. Similar to a pin-ball game, molecules acted as spheres or disks, tumbling across the epithelial surfaces floating in or on the mucus layer, and finally fitting into holes of the corresponding shape and size.[26]

The odor of camphor, for example, was hypothesized as the engagement of a nearly spherical molecule of camphor approximately 7 Å in diameter, with a corresponding round hole, thus finally registering the word 'camphor' in the brain.

Some musks were assumed to act like disks or ellipsoids with a cross-section of 40–57 $Å^2$ and a range of length to breadth of 2.8 to 3.3;[3] these fell into slits to yield an information pattern as 'musk-like' in the brain.

Ohloff[9] described the odorants related to ambergris, a secretion of the intestine of the whale *Physeter macrocephalus*, as having a decalin ring sytem with a '1,2,4-triaxial arrangement of the substituents of the ring system, one of the axial group being an oxygen function'. According to this pattern, base compounds of great value to the perfume industry have been synthesized, yielding virtually all the six odor qualities of ambergris: moist earth, tobacco, sandalwood, seaweed, animal musk and a weak fecal note. However, he does not accept the full implications of the stereochemical hypothesis, although he confirmed the importance of the ambra relations. 'The correlation of molecular profile or morphology of odorous substances and the quality of odors constitutes a crude simplification of the olfactory process' (Fig. 3). For example, why is glycerin odorless?

Electrical Activity

Moulton[19] has demonstrated patterns of odor-induced excitation in the frog by electro-olfactograms (EOG); these are electrical responses from several bulbar sites to different odors. He concluded that 'odor recognition is the function of a cell ensemble in which individual receptors possess multiple and overlapping sensitivities. Information reaches the brain as a pattern of activity distributed in space, in which the ratio of excitation in a number of parallel channels may define the stimulus.' In some respects this may be compared to a visual hologram.

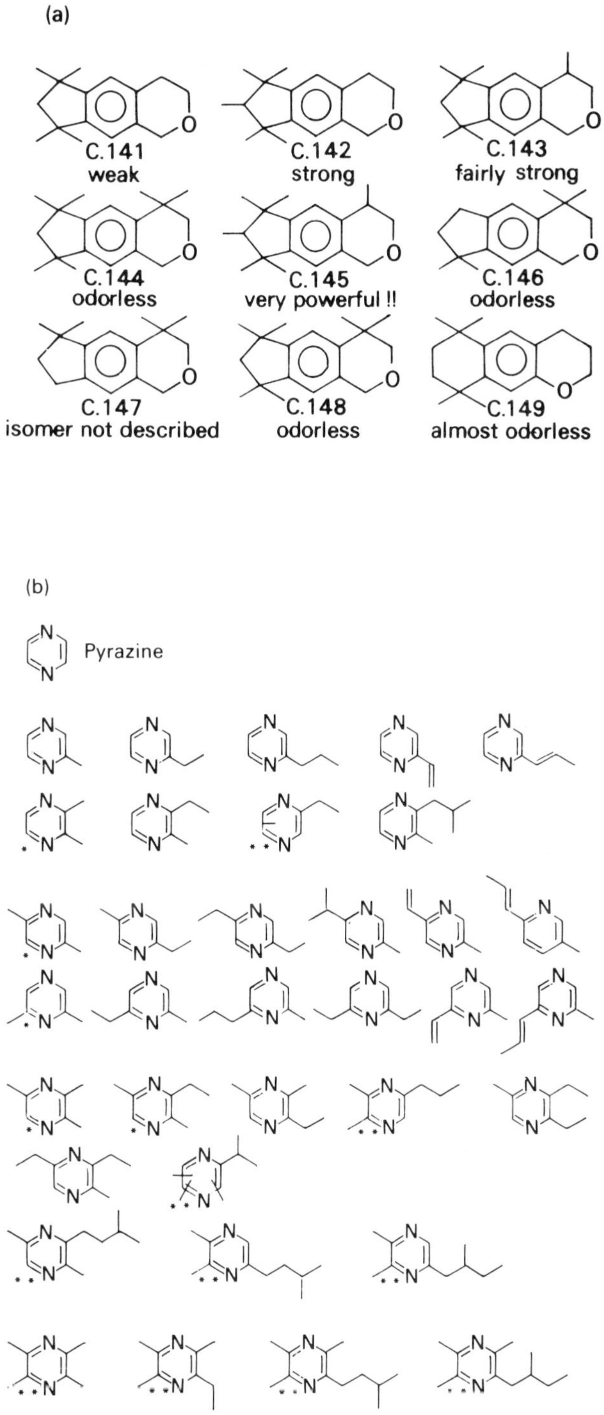

Figure 3. (a) The musk modality (Ref. 3). Variation in structure of musks resulting in weak or strong odors. (b) Some of the 226 odorants of coffee and cocoa related to pyrazine (Ref. 48): *in cocoa also; **only in cocoa. No single one is responsible for the aroma. See p. 40 re threshold of 2-methoxy-3-hexylpyrazine.

Topological Characteristics

Beets[3] has evolved a similar concept in a recent exhaustive monograph on structure–activity relationships in human chemoreception. According to his generalized concept, 'the peripheral membrane of a receptor cell contains many potential receptor sites belonging to many structural types. . . . Biochemical considerations strongly suggest that the composition of the assembly of structural features in an epithelium of adjacent cell membranes varies gradually as we pass from one spot to another. This gradual transition of compositions of site collections from cell membranes indicates the possibilities that the topological characteristics of a chemoreceptory epithelium could be the distinguishing criteria of the various informational modalities.'

Feeling Shapes

This, after all, is similar to the somesthetic sense of touching, like distinguishing a metallic from a wooden shoehorn in the dark, not to speak of the blind learning to feel Braille characters; incidentally this was an accomplishment of Bismarck, who liked to read in bed after dark! We can imagine the nasal epithelium consisting of a million hands with six or eight flaccid fingers on each hand, feeling the basic skeleton and sidechain structures of the molecules as they arrive on the mucus. There they might puncture the membrane of the olfactory epithelium at spacings determined by the location of the active radicals, like Bell pepper mentioned above.

THE VARIABLES OF A PERFUME

Billot[1] emphasized in 1975 the difficulty of comparing equal intensities of completely different character. This is akin to the Purkinje phenomenon[7] when approaching low light levels, where the peak sensitivity of the human eye moves from 555 nm (greenish yellow) toward the blue, peaking at 510 nm (greenish blue). At the threshold of the fully dark-adapted eye every wavelength appears hueless. 'At the present time', Billot continues, 'the perfumer has no practical alternative to prolonged experience based on trial and error.'

Quality and Intensity

In the case of α-ionone, for example, the quality changes significantly with decreasing concentration in air, a fact of utmost importance to perfumers in using traces of powerful odorants or fixatives to create a perfume with base notes, a body and top notes. Appell[15] lists grades of intensity of equal concentrations of odorants with crude nonadienal as the most intense inherent quality down to rosacetol as the least intense quality.

Notes and Volatility

Top notes are usually associated with high volatility, such as lemon and base notes are usually of low volatility, such as oak moss or patchouli.

Mathematically, volatility (Y) may be expressed as a rate of evaporation, proportional to (m) grams per square centimeter of exposed surface to air at saturated vapor pressure (p) in atmospheres, at absolute temperature (t) as $Y = p\ (m/515t)^{0.5}$ as derived by Billot.[1]

Velocity

The atmosphere of a perfume has been defined as its more or less pronounced faculty of dispersion or velocity (v). Billot[1] states that 'In practice it is the perfumed wake which a woman using her perfume leaves behind her. It is to some extent possible to measure its value comparatively as follows: We measure the rate of propagation of the molecules carrying the odor through a long tube of fairly thin diameter. These different times will provide coefficients proportional to the dispersion, thus characterizing the atmosphere, the shortest time, or highest v representing the greatest atmosphere.'

Wavelength

A molecular beam of the odorant may also be projected thermally or supersonically into a vacuum chamber and the velocity (v) of the beam may be obtained directly using a suitable tracer or other detector. By the conventional formula of wave mechanics[14] the wavelength (w) associated with the odorant molecule, with molecular weight (m) divided by Avogadro's number may be represented as $w = h/mv$, where h is Planck's constant.

To find the wavelength derived for a musk-like odorant whose molecular weight is 235, for example, we need only measure the velocity. As a simple numerical example, if the measured velocity is $1\ \text{cm s}^{-1}$, the wavelength $w = 17\ \mu\text{m}$, which is in the infrared. The atmosphere of an odorant with a molecular weight of 235 and velocity of $1\ \text{cm sec}^{-1}$ may then for example be described as a wavelength of $17\ \mu\text{m}$.

Wave Mechanics

As Pressler[13] has suggested, perhaps some combination of wave and particle theory may be developed, comparable with wave mechanics, but a de Broglie is still awaited in olfaction, not to speak of a Schrödinger.

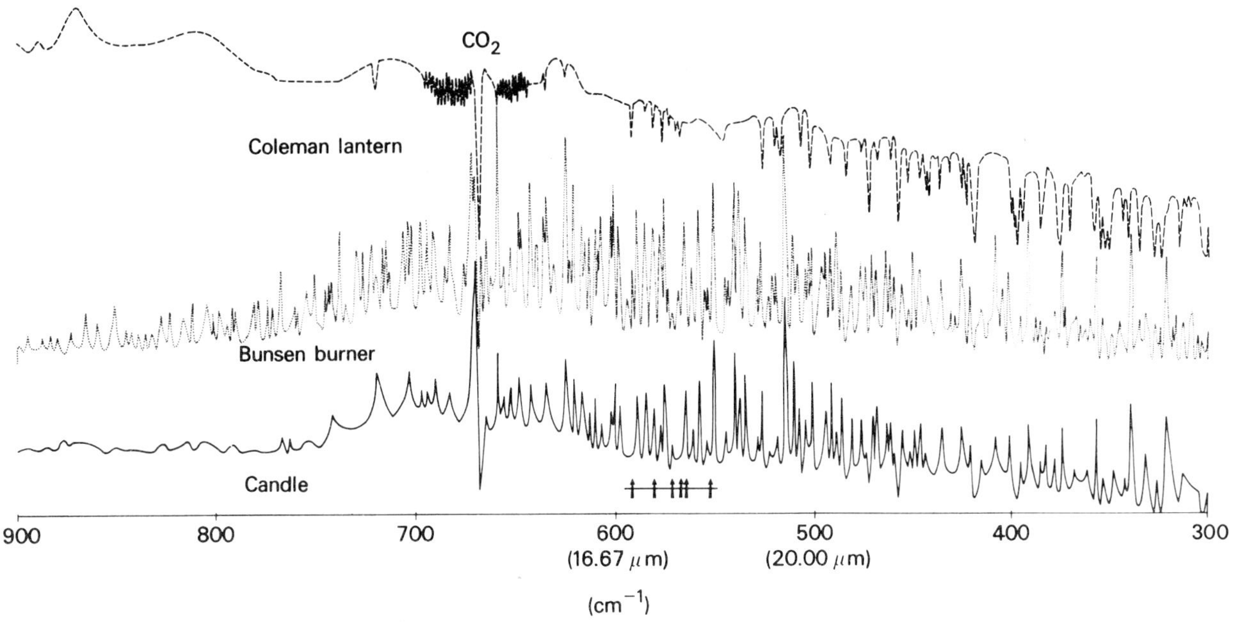

Figure 4. Infrared spectrum of a Coleman lantern, Bunsen burner and candle from 900 cm^{-1} (11.11 μm) to 300 cm^{-1} (33.33 μm) (Ref. 12). Arrows between 600 cm^{-1} (16.67 μm) and 500 cm^{-1} (20.00 μm) represent the pheromone emissions in that region. The male moth is attracted to the candle and Bunsen burner but not to the Coleman lantern which makes a very poor light trap. It is no doubt the processing vortex core (flicker) that self-modulates the candle and Bunsen burner so that the emission occurs. The Coleman lantern not only does not flicker but also is so hot that it burns up the hydrocarbons before emission can even occur.

Callaghan[12] recently reported that the wavelength of the pheromone of the cabbage looper moth corresponded to some of the peaks in a candle flame in the 17 μm region (Fig. 4). Although current literature assumes that smell is a particle phenomenon with the consequent chemoreception resulting in the sensation of smell, Beets[3] admits that 'odor similarity is to a certain extent correlated with similarity in certain regions of the far infrared absorption spectra'.

OLFACTORY THEORIES

Clearly, insects may be able to receive odors both as a wavelength or frequency phenomenon, as well as a particle phenomenon, possibly using the visual sense to confirm the chemoreception or vice versa. Most vertebrates, however, may only be able to receive chemotactic relationships on their olfactory epithelia. However, with the discovery and practical utilization of the first pheromone from the gypsy moth (*Lymantria dispar* L.), biologists excluded the possibility of a wavelength phenomenon from their considerations, especially after the work of Beck[32] was not confirmed for honey bees[33] and the debate between Amoore and Wright[25] on the vibrational theory of odors was never consummated.

For the moment, at least, the vibrational theory of smell is out of fashion. At present it is generally assumed that a single molecule, such as bombykol, may initiate the interaction. The aggregation or arrangement of several molecules or atoms in recognizable lines or shapes is assumed to behave in random fashion in the turbulence caused by sniffing.

Macromolecular Aggregates

In the theory of liquid crystals, a molecular statistical theory[17] has evolved from the swarm theory of Bose which explains the nematic mesophase, or alignment of molecules with the long axes of the molecules parallel and, as a special case, in 'smectic' layers, depending on the potential energy, angular position of the axis and the charge on the surface. As Maitland and Smith observed,[16] 'Despite an intensive investigation over many decades, the quantitative determination of the forces between molecules has proved one of the most intractable problems in physical chemistry. Complex systems involving polyatomic molecules, biological macromolecules and surface have remained unsolved.'

Perhaps the first practical approach to compute the molecular and intermolecular interactions of odorant molecules was described by Claverie at the Third Congress of the European Chemoreception Research Organization at Pavia in 1978. Publication of his methods and applications is awaited from the Institut de Biologie Physico-chimique at the University of Paris.

Cell Puncture Theory

According to the cell puncture theory, large odorant molecules, or an ion-transporting enzyme system, can puncture the lipid part of a cell membrane by nonspecific absorption into the cell membrane, according to Davies.[33] It is known that the normal excess concentration of K^+ inside the nerve cell is able to escape through a puncture in the plasma membrane. Simultaneously, the deficiency of Na^+ is made up locally by an intake of sodium from the aqueous phase surrounding the cell. During its resting period the nerve cell removes Na^+ from the cell and allows in excess K^+ against the concentration gradient.

Neural Impulses

The penetration of a bulky odorant molecule leaves a hole which heals only slowly, so that ions pass through the hole and initiate excitation of the nerve. The neural impulse is presumed to originate where the olfactory receptor narrows to the olfactory fiber. This circuit then triggers the nerve impulses, the frequency depending on the intensity, in turn determined by the number of channels formed in any membrane. Perhaps only one molecule, or a line-up of three molecules, is sufficient to cause ionic leakage, and if three molecules make three simultaneous penetrations of the cell wall, then, by probability theory, the concentration must be greater by several powers of 10, than if only one molecule causes the leakage.

This may explain the multi-million-fold differences in the threshold sensitivity for example of β-ionone and ethane. Davies developed a formula for olfactory threshold whereby the sum of the log of the threshold and the log of the absorption coefficient (for the odorant at the cell–membrane interface) is a function of the critical number of odorant molecules found within a given area, and the thickness of the membrane. This enabled Davies and Taylor[33] to predict the approximate olfactory threshold for 36 odorants.

Chromatography

The work of Patte at the Centre National de la Recherche Scientifique shows that chromatographic measurements may be used to predict olfactory properties in using the solubility factors of dispersion, induction, orientation and pH which can be established from chromatographic measurements. It is hoped that some periodic table can be developed to predict olfactory qualities similar to the table of elements.

OLFACTORY ACTION

Olfactory thresholds may also be related to the structure of the membrane of the olfactory receptor. For instance, certain outer surface membrane proteins can be made to cluster by linking them biochemically.[34] Once clustered they link with protein cables arrayed within the cell, hovering over them with the membrane sandwiched between.

Enzymes are known to face one or the other side, or straddle to face both membrane sides. Many glycoproteins face so that their sugars extend away from the cytoplasm – outward from the outer or limiting membrane, but inward from membranes within the cell and surrounding the organelles. Even the lipids in membranes may be arranged asymmetrically, with more of one kind than another in one of the halves of the bilayer making up the membrane.

Chemotaxis

One clue to olfaction is the behavior of certain bacteria known as chemotaxis.[35] Chemotaxis of *Escherichia coli* and *Salmonella typhimurium*, for example, is usually displayed by counterclockwise propulsion toward nutrients and backstroking by tumbling with clockwise flagellation from noxious chemicals (Fig. 5). These brainless creatures propel themselves by a half-dozen thin (13–17 nm diameter) filaments, or flagellae, which are fastened by four O-rings around the constricted base (7 nm) into the several layers of the bacterial shell. They extend outwards about 2 μm (2000 nm) into the juices of the human stomach or other medium. They respond instantaneously and can propel themselves about 40 body lengths per second in some cases to reach their preferred L-aspartate and less so for L-glutamate.

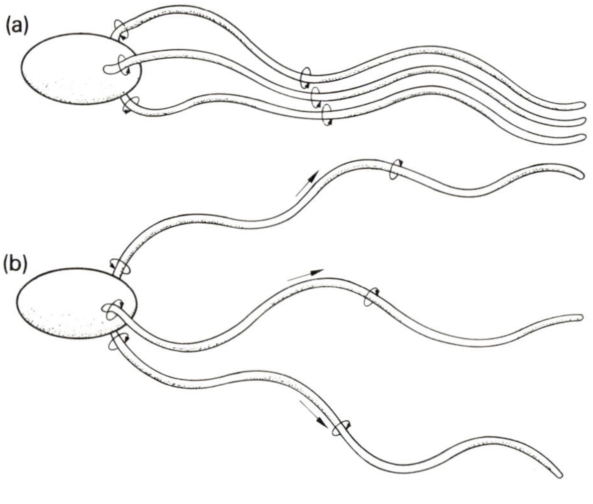

Figure 5. Behavior of chemotactic bacteria (Ref. 49). Swimming and tumbling seem to be accomplished by *E. coli* as shown here. When the flagella (only three of the usual four to eight are shown) rotate counterclockwise (a) the helical filaments form a bundle and function as propellors that push the cell through the medium in runs. When the flagella rotate clockwise (b), on the other hand, the flagellar bundle flies apart. The filaments now tend to pull the cell, but each filament pulls in a different direction depending on its orientation, and so the cell tumbles about.

Negative chemotaxis is greatest for methanol, indole and skatole, also for low pH (below 6.5) and high pH (above 7.5).

Chemotaxis is similarly shown by leukocytes in mammals which pursue specific molecules such as peptides. Their olfactory decision determines their lives, a decision supposed to be made in the cerebral cortex of higher animals, and like the bacteria it is often determined by concentration. For example, with aircraft pilots, as the smell of smoke intensifies, their warning of danger increases.[20,24,36]

Electro-olfactogram and Electro-antennogram

While the interaction of particle or wavelength with neuron remains unsolved as revealed by Callaghan's work on the insect electro-antennogram (EAG), the electrical activity in olfactory receptor cells has been described in detail by electrophysiologists in many countries. Schneider investigated the insect EAG and Ottoson[36] developed the electro-olfactogram (EOG),[10] which has been demonstrated for humans by eliminating the mechanical actions of sniffing by Plattig and Kobal at Erlangen. However, no correlations of odor quality with EOG patterns have been devised to forecast smell from brain waves.

PHEROMONES

Pheromones, according to Birch,[38] 'are substances which are secreted to the outside by an individual and received by a second individual of the same species, in which they release a specific reaction, behavior or development process', literally, a transfer of excitement. These substances must have been known to man since ancient times and Professor Dr D. Schneider of the Max-Planck-Institut für Verhaltungsphysiologie, Seewiesen, Germany, has drawn my attention, by way of a letter to the Editor, to the following story in François Rabelais's classic *Gargantua and Pantagruel*, published in 1532. Rabelais was also an eminent physician, lecturing on Galen and Hippocrates at Paris and Montpellier, and thus well qualified to tell how Panurge, the companion of Gargantua, took his chemical revenge on a Paris lady of high society, who had spurned his amorous advances. Panurge obtained a bitch in heat, killed and dissected out the sexual organs, minced these finely, and during a High Mass at church was able to spread them over her sleeves and clothes (Fig. 6). Rabelais tells how 600 014 randy dogs followed her on her way home, urinated all over the poor lady

Figure 6. Gustave Doré's illustration how Panurge spread pheromones on a Paris lady of high society during Mass. (From J. Bry Ainé's edition of Rabelais's works, Paris, 1854).

and thus created a small rivulet, its contents supposed to be identical to the secret dye used by Gobelin for his tapestries.

Today the biologically best known pheromone system is that of the honey bee. In this species, 9-oxodec-*trans*-2-enoic acid, the female sex attractant, not only lures the drones to the queen during the nuptial flight, but also lures the worker bees to follow the young queen during swarming. Another pheromone, 9-hydroxydecenoic acid, inhibits workers from raising young queens in the hive. As may be imagined, the economic importance of pheromones for insect control has resulted in extensive and accelerated research in many countries interested in agriculture and food production.

Humans, according to Comfort,[38] have a complete set of organs which are traditionally described as nonfunctional, but which in any other mammal would be recognized as part of a pheromone system. They are apocrine glands with hair tufts some of which do not produce sweat, the prepuce and labia with production of smegma, and a conspicuous and apparently unused array, which presupposes an unsuspected communications system.

Insect Pheromones

Butenandt[40] synthesized the first known pheromone of the silkworm moth (*Bombyx*): hexadeca-10-*trans*-12-*cis*-dien-1-ol. The prominent antennae of the *Bombyx* moth (Fig. 7) are presumed to be positioned like a comb trap for molecular particles and chemoreception. Grant's work on insect antennae as receptors for radiation has not been accepted by entomologists except for Callaghan mentioned above.

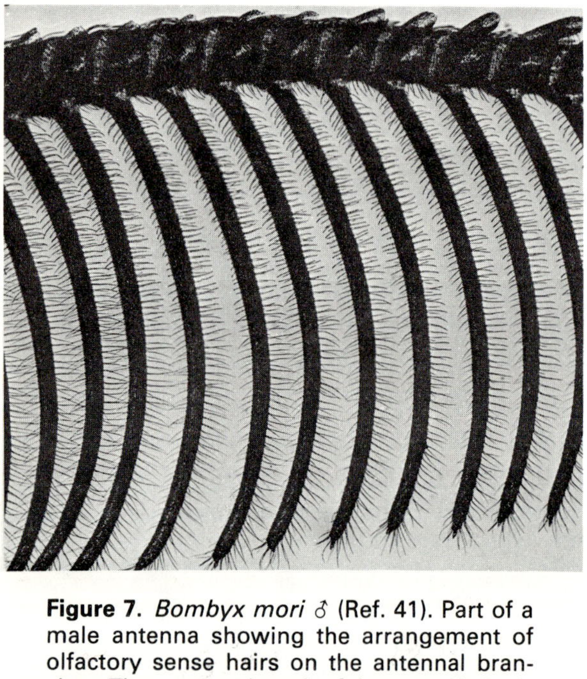

Figure 7. *Bombyx mori* ♂ (Ref. 41). Part of a male antenna showing the arrangement of olfactory sense hairs on the antennal branches. The average length of the sense hairs is 100 μm.

The sensory system responsible for the reception of bombykol is located only on the antennae of the male insect. The threshold of the male moth is reached when each antenna absorbs 500 molecules in 0.86 s, which is the reaction time of the moth. This fast reaction time and the speed of transmission of bombykol are still unexplained. One demonstration by Max Delbrück to the author at Cold Spring Harbor seemed to show that transmission in apparently still air was much faster than the usual diffusion processes.

Species Specific Blends of Pheromones

Roelofs[39] has shown that exact blends of chemicals are used in mating disruption programs for the leaf-roller moth in apple orchards. Exact blends of pheromones are used to obtain species specificity: blends of geometrical or positional isomers which are produced by the female insect abdominal glands, are emitted and perceived in very precise ratios.

Closely related leaf-roller moths use various mixtures of Z-11 and E-11 tetradecenyl acetates or Z-11 and Z-9 tetradecenyl acetates. Knowing the blend for each species gives the timing for the insecticide applications.[37] The red-banded leaf-roller moth has two main scent cells, one type responding to the Z-11 isomer and the other to the E-11 isomer. Signals from these cells presumably are integrated at the central nervous system to activate behavioral responses. Two species may use the same two pheromone components but in different ratios, say 90:10 and 70:30, so that the female moth release rate must be approximated for each species at a given range of concentration.

THE ODOR PROBLEM

Schneider summarizes:[41] 'One may divide the classical odor problem into three sub-problems: 1. The interaction of the odor molecules with the acceptors in the cell membrane. 2. The membrane permeability changes which are induced by the first process and eventually lead to the elicitation of a receptor potential and the following nerve impulses. 3. The integrative function of the nerve centers which deal with the information flux from the peripheral odor cells.'

Interdisciplinary Experimental Design

We can return to the guidance of the artist-perfumer in defining tone, volatility, intensity, volume and tenacity in scientific and mathematical terms, and to the neurologist in defining the spatial patterns of odors and pheromones in the olfactory bulb of mammals or the central nervous system of insects. We may look to the pharmacologist in analyzing the smells associated with mental states like schizophrenia or clinical diagnosis of lesions of the cortex, not to speak of the specific or possibly chiral anosmias similar to color blindness. The forecasting of odors by structural chemistry may become clear if a particle–wave relationship can be established by physicists and biophysicists.[21]

Motivation by Odors in the Next Millennium

Perhaps the psychiatrist can explain the use of a drawer full of dried roses by Richard Wagner and a shelf of over-ripe apples by Friedrich Schiller when they needed musical or poetic inspiration. When explained, odors might be prescribed, not necessarily to sick but to normal people who wish to improve their creative abilities and motivations.[46]

The new 'choose your mood' drugs may at present be dispensed only under medical supervision and are difficult to control, not to speak of addiction and side effects. This may not apply to the dispensation of mood creating and enhancing scents. The motivation to buy an expensive Cadillac car, for example, was enhanced by a scent of vanilla beans on the door interior.[50]

The psychologist may expand some of the 'games people play' to include charades like the contests used in Japan by Lady Murasaki over a thousand years ago. She asked her royal guests to create novel perfumes at her parties. Perhaps the next millennium will evolve new games as well as new chapters in molecular biology.

LITERATURE CITED

As a first guide to the literature, the reader is advised to consult the bibliographies in the six volumes of the International Symposium on Olfaction and Taste (ISOT). The references in the monographs by Beets[3] and Ohloff in *Chemie der Geruchs- und Geschmackstoffe*, published by Fortschritte der Chemischen–Forschung (Springer Verlag), are particularly complete. The wide variety of disciplines makes a complete listing difficult, if necessary.

(It is important to note that 1 nm = 10^{-9} m, 1 Å (Angstrom) = 10^{-8} m, 1 μm (micron) = 10^{-6} m.)

1. M. Billot and F. V. Wells, *Perfumery Technology*, Ellis Horwood, Chichester (1975).
2. *Webster's New International Dictionary*, 2nd Edn, G. & C. Merriam, Springfield, Massachusetts (1944).
3. M. G. J. Beets, *Structure–Activity Relationships in Human Chemoreception*, Applied Science Publishers, London (1978).
4. J. E. Amoore, Specific anosmia and the concept of primary odors. *Chem. Senses Flavor* **2**, 267–281 (1977).
5. A. Comfort, Peculiar odor of schizophrenics. *Science* **166**, 398 (1969).
6. P. Labat, *Nouveau Voyage aux Isles de l'Amerique*, Vol. 1, p. 156. The Hague (1724).
7. Optical Society of America, *The Science of Color*, Thomas Y. Crowell, New York (1953).
8. G. Dodd, A. Menevse and M. Poynder, Cyclic AMP altered by odour molecules. *Biochem. Biophys. Res. Commun.* **77**, 671 (1978).
9. G. Ohloff, Relationship between odor sensation and stereochemistry of decalin ring compounds. In *Gustation and Olfaction*, edited by G. Ohloff and A. F. Thomas, Academic Press, London (1971).
10. S. Takagi, EOG problems, in *Olfaction and Taste*, edited by C. Pfaffmann, Rockefeller University Press, New York (1969).
11. C. H. Best and N. B. Taylor, *The Physiological Basis of Medical Practice*, Williams and Wilkins, Baltimore (1950).
12. P. Callaghan, Moth and candle: the candle flame as a sexual mimic of the coded infrared wavelengths from a moth sex scent (pheromone). *Appl. Opt.* **16**, 3089–3102 (1977).
13. C. Pressler, Theories on odors. *Drug Cosmet. Ind.* **62**, 2 (1948).
14. F. K. Richtmyer, E. H. Kennard and T. Lauritsen, *Introduction to Modern Physics*, p. 185. McGraw-Hill, New York (1955).
15. L. Appell, Physical foundations in perfumery. *Am. Perfum. Cosmet.* (1970).
16. G. C. Maitland and E. B. Smith, The forces between simple molecules. *Chem. Soc. Rev.* **2**, 181–201 (1973).
17. R. Humphries, P. G. James and G. R. Luckhurst, Molecular statistical theory of the nematic mesophase. *J. Chem. Soc. Faraday Trans. 2* **68**, 1031 (1972).
18. A. J. D. de Lorenzo, The olfactory neuron and the blood–brain barrier. In *Taste and Smell in Vertebrates*, edited by G. E. Wolstenholme and J. Knight, J and A. Churchill, London (1970).
19. D. G. Moulton, Spatial patterning of response to odors in the peripheral olfactory system. *Physiol. Rev.* **56**, 578–593 (1976).
20. J. Adler, Chemotaxis for carbohydrates. In *Olfaction and Taste*, Vol. IV, p. 70, edited by D. Schneider. Wissenschaft. Verlagsgesellschaft, Stuttgart (1972).
21. F. N. Jones and M. H. Jones, Modern theories of olfaction: a critical review. *J. Psychol.* **36**, 207–241 (1953).
22. W. Rall, G. M. Shepherd, T. S. Reese and M. W. Brightman, Dendrodendritic synaptic pathway for inhibition in the olfactory bulb. *Exp. Neurol.* **14**, 44–56 (1966).
23. G. M. Shepherd, Microcircuits in the nervous system. *Sci. Am.* **238**, 92–103 (1978).
24. J. Fox, Mutant bacteria. *Chem. Eng. News* **56**, 21–22 (8 May 1978).
25. R. H. Wright, Spectroscopic theory of odor. *Nature (London)* **209**, 571 (1966).
26. E. T. Theimer and M. R. McDaniel, Odor and optical activity. *J. Soc. Cosmet. Chem.* **22**, 15–26 (1971).
27. C. Masson, S. Kouprach, I. Giachetti and P. MacLeod, Relation between intramembranous particle density of frog olfactory cilia and EOG response. In *Proceedings of the Fifth International Symposium on Olfaction and Taste*, edited by Denton. Academic Press, New York (1977).
28. H. Piéron, Observations de psychologie animale. *Rev. Psychiatr. Ser.* **8**, 198 (1904).
29. R. Menzel, Die Verwertung der Riechfähigkeit des Hundes im Dienst der Menschheit. In *Psychical Physics*, p. 102. Elsevier, New York (1949).
30. R. M. Seifert *et al.*, Synthesis of some 2-methoxy-3-alkylpyrazines with strong Bell-pepper-like odors. *J. Agric. Food Chem.* **18**, 246–249 (1970).
31. R. E. Randebroek, Molecular theory of odor with the α-helix as potential receptor. In *Gustation and Olfaction*, edited by G. Ohloff and A. F. Thomas, p. 111. Academic Press, London (1971).
32. L. H. Beck *et al.*, Some theoretical and experimental relationships between infrared absorption and olfaction. *Science* **106**, 511 (1947).
33. J. T. Davies, Olfactory theories. In *Handbook of Sensory Physiology*, IV–I *Olfaction*, edited by L. M. Beidler. Springer–Verlag, Heidelberg (1971).
34. J. L. Fox, Cell surfaces. *Biochem. Chem. Eng. News* **56**, 17 (17 April 1978).
35. J. Adler, Chemotaxis in bacteria. *Annu. Rev. Biochem.* **44**, 341–356 (1975).
36. D. Ottoson, Analysis of the electrical activity of the olfactory epithelium. *Acta Physiol. Scand.* **35**, Suppl. 122, 1–83 (1956).

37. J. Boeckh, Inhibition and excitation of single insect olfactory receptors and their role as a primary sensory code. In *Proceedings of the Second International Symposium on Olfaction and Taste*, edited by T. Hayashi, pp. 721–735. Pergamon Press, Oxford (1967).
38. M. Birch, *Pheromones*, p. 388. Elsevier, New York (1974).
39. W. L. Roelofs, A New hypothesis for sex pheromone perception in moths *Am. Chem. Soc. Div. Pestic. Chem.* **56**, 23 (20 March 1978).
40. A Butenandt, Über den Sexual-Lockstoff des Seidenspinners *Bombyx mori. Z. Naturforsch. Teil B* **14**, 283 (1959).
41. D. Schneider, Specialized odor receptors of insects. In *Gustation and Olfaction*, edited by G. Ohloff and A. F. Thomas, p. 56. Academic Press, London (1971).
42. F. O. Schmitt, P. Dev and B. Smith, Electrotonic processing of information by brain cells. *Science* **193**, 114–120 (1976).
43. E. Dowling and B. Boycott, Organization of the primate retina. *Proc. R. Soc. London Ser. B* **166**, 80–111 (1966).
44. A. C. Allison, The morphology of the olfactory system in the vertebrates. *Biol. Rev.* **28**, 195–244 (1953).
45. M. Okano, A. Weber and S. Frommes, Electron microscopic studies of the distal border of the canine olfactory epithelium. *J. Ultrastruct. Res.* **17**, 487–502 (1967).
46. L. Narodny, The influence of odors on the creative imagination. *Perfum. Essent. Oil Rec.* **48**, 79–82 (1957).
47. S. Hecht and E. Mintz, The resolution of the human eye. *J. Gen. Physiol.* **22**, 593–612 (1939).
48. G. Ohloff, Chemie der Geruchs- und Geschmackstoffe. *Fortschr. Chem. Forsch.* **12**, 227 (1969).
49. J. Adler, The sensing of chemicals by bacteria. *Sci. Am.* **234**, 47 (1976); Recent developments in the field of naturally occurring aroma components. In *Progress in the Chemistry of Natural Organic Products*, edited by G. Ohloff, Vol. 35, pp. 431–527. Springer-Verlag, Vienna–New York (1978).
50. J. Hopson, Scent and human behavior. *Sci. News* **115**, 282 (1979).

The manuscript was received 30 September 1978

Enology

Science and Technology Improve an Ancient Art

PROFESSOR A. DINSMOOR WEBB

Department of Viticulture and Enology, University of California, Davis, California, USA

The emerging science of enology is an interdisciplinary synthesis of chemistry, biochemistry, microbiology, engineering and sensory analytical science. Chemistry analyses grapes, monitors progress of fermentation and assures that the wine is stable and in conformity with legal specifications. Knowledge of the enzymatically mediated reactions of yeasts and bacteria allows control through external conditions, thus yielding better wines. Their production also demands modern engineering methods and materials. Wine is made to be appreciated with other foods on the basis of its smell and taste, hence sensory analytical methods can control its production. Basically grape growing and winemaking must be a very ancient practical art because it would have required centuries to develop the sophisticated state depicted in the earliest written records of the third millennium B.C. Louis Pasteur provided the scientific basis for understanding enology and since his time enology has gradually moved from a practical art towards the science of enology. It is hard to imagine how wines of the future will be better than those we now enjoy, the finest ever produced. Certain it is, however, that they will be still better.

Enology, as it exists today, is a very good example of the application of several sciences to the improvement of an ancient art. Enology is an interdisciplinary synthesis of the applicable principles of chemistry, biochemistry, microbiology, engineering and the emerging science of sensory analysis. Admittedly, wine can be, and is, made in some regions by the traditional techniques of the practical art. Today, however, many of the finer wines and the very great proportion of the wines for everyday consumption, reflect applications of the pertinent scientific and engineering principles.

There can be no question but that development of enology, as a distinct science, is nowhere near as advanced as is the case for chemistry or physics, for example. Even so, one must admit that there is some difficulty defining the dividing line between chemistry and physics or chemistry and biochemistry. The problem is only slightly more difficult in the case of the separation of enology from biochemistry or enology from chemical engineering.

CHEMISTRY

The chemical aspects of enology are relatively easy to discern. Since the winery, in one sense, is a chemical engineering processing plant employing enzymes to convert grapes, the raw material, to a final product, wine, one of the most obvious applications of chemistry will be in analyses of the raw materials and of the product at various stages through processing. Grapes, under ideal conditions, produce enough sugar to yield a stable table wine of approximately 12% alcohol by volume concentration. With the exception of dried figs, dates and honey there are very few fruits, berries, or natural products having such a high sugar concentration.

Sucrose is formed in the grape leaf by the normal photosynthetic process and translocated into the grape berry where it is inverted and stored as a nearly equimolar mixture of glucose and fructose. While the sucrose concentration in the leaves and canes is of the order of only a very few percent, the concentration of invert sugar in the grape berry can reach levels of from 22–26% at full ripeness. Determination of the concentration of invert sugars in the grape berry is thus one of the factors guiding the selection of the harvest date.

The two predominant natural acids of the grape, tartaric and malic, are present at levels of a few percent in the immature, green grape berry. During the ripening of the grape these acids are diluted by expansion of the berry and also metabolized, particularly in the case of malic acid. Acid lost during ripening is a function of ambient temperature with the result that in cool grape growing regions there is likely to be a sufficiency or excess of acid, whereas in the warmer grape growing regions, there may be a deficiency of natural acids at the time sufficient sugar has been synthesized. Thus, acid analysis becomes the second important parameter of the ripening process used in determining the harvest date.

In the colder grape growing regions of the world, or in unfavorable years in other regions, there may simply be insufficient sunshine to ripen the fruit properly. Under these circumstances it is a long-standing tradition that sugar from other sources may be added to the grape juice. In France, this process is

CCC-0308-0188/80/0005-0049$05.50/1

The Vintage in California. Although techniques and equipment have changed over the years, the basics in winemaking remain the same: fresh grapes are crushed carefully to release the rich juices, fermentation occurs, the wine is aged, then it goes to table to cheer men's meals. Note that there was no provision for separating stems from grapes, the workers were Chinese and the pumps were worked by hand. Drawn by P. Frenzeney and published in *Harper's Weekly* (1878). (Courtesy Wine Institute).

called *chaptalisation*, in Germany *Verbesserung*, and in the United States, amelioration.

Each grape growing country and, indeed, region, has its own regulations concerning the conditions under which sugar addition may be practised. Since low sugar concentrations in the grape juice are normally accompanied by excessive acidities, permission is usually granted to dilute some of the excess acid with water at the time the sugar is added. Further reduction of acids can be accomplished by chemical means following fermentation if the resultant wine is still unbalanced with excess acidity. Sugar additions and acid dilution or reduction by neutralization or removal of the acid are all processes requiring careful chemical analysis and control.

Since the time of Pasteur it has been understood that oxygen was a potential enemy of good wines. The *acetobacter* bacteria that spoil wine by changing its alcohol to acetic acid are obligate aerobes. More recently, it has been discovered that even smaller amounts of oxygen can have profound effects in decreasing the quality of fine white wines through reaction with their phenolic materials. In contrast, a small amount of oxygen during the period of wood aging seems to be essential for the proper development of red table wines. Again, these reactions take place between the oxygen and the phenolic materials present. In the case of red wines, however, there is so much more of the phenolics, and they are of different types, that the reaction is favorable to quality rather

PROFESSOR A. DINSMOOR WEBB is Professor of Enology and Chairman of the Department of Viticulture and Enology of the University of California, Davis. Following two vintage seasons in California wineries and the war years in Oak Ridge separating uranium isotopes, he was awarded the first PhD in Chemistry on the Davis Campus. He studied enology at the Universities of Adelaide, Bordeaux and Stellenbosch, South Africa. His research interests are the biosynthesis of aroma and pigment compounds of grapes and wines. Webb is a past-president of the American Society of Enologists.
Address: Department of Viticulture and Enology, University of California, Davis, California 95616, USA.

than being destructive of quality. Careful control of oxygen concentrations in wines is thus of considerable importance. Analytical chemistry is used also to monitor concentrations of the reductant and the bactericide, sulfur dioxide. When used under careful control and with good judgment, it is an important adjunct to the making of fine wines. However, used in excess or at inappropriate times, it can spoil an otherwise very fine wine.

BIOCHEMISTRY

In a broad sense all of enology could be considered applied biochemistry. However, here we shall consider the biochemistry of enology to be only those reactions mediated by enzymes. It is convenient to use this arbitrary separation because the enzymatic reactions, for the most part, occur in the grape juice during the fermentation and immediately after. A few months after completion of fermentation the combination of alcohol and tannins will have denatured and precipitated essentially all of the enzymes. This is particularly true for red wines, but occurs also in whites at a somewhat slower rate and thus over a somewhat longer time period.

Enzymatically mediated browning reactions can be especially troublesome in the white juice from certain varieties. If the grapes are infected with fungi, such as *Botrytis cinerea*, the rate and extent of browning can indeed be severe. Enzymatic browning is not so much a problem in the case of red wines, because the normally intense red color extracted into the wine from the skins during fermentation will usually cover the brown tinges. However, red grapes seriously infected with *Botrytis* may be deficient in overall color because of oxidation and precipitation of the pigment as a result of the oxidative enzymes secreted by the fungus.

Fundamental to the science of enology is a clear understanding of the sequence of reactions of the glycolytic pathway, followed by the decarboxylation of pyruvate to acetaldehyde and its reduction to ethanol. In enology, this series of enzymatic reactions is conducted within the yeast cell suspended in a very sweet medium at low pH and at temperatures relatively lower than normally encountered in biological systems for this series of reactions. Wine, together with the carbon dioxide which usually escapes into the atmosphere, is simply the waste product of the metabolism of the yeast cell.

To simplify, one can say that the yeast would prefer to metabolize sugar under aerobic conditions because the yield of energy, and thus the possibility of growing more cells is greater when oxygen is available. Water and carbon dioxide are the byproducts. Under anerobic conditions, the yield of energy for cell growth is much less, but the by-product of the anerobic reaction is wine plus carbon dioxide. Under the conditions normally obtaining in the winery, the reaction starts out under aerobic conditions but very soon becomes anerobic. Both the enologists and the yeast cell are happy under these circumstances.

Another yeast-enzyme-mediated biochemical reaction in enology occurs during the production of film sherries. Certain species within the yeast genus *Saccharomyces* have the ability to form a film on the surface of the wine when all of the sugar has been used up. The source of carbon for film yeast metabolism is primarily the ethyl alcohol of the wine. Oxygen from the air is used to convert the ethanol to carbon dioxide and water. Fortunately, this reaction is not a very rapid one so that under the normal conditions of a fairly warm and dry cellar, water is evaporated just about as fast as ethanol is metabolized with the result that the wine maintains its normal alcoholic concentration.

Of more importance is the fact that the metabolic sequence is not completed and that some of the by-products are the desired film sherry aroma components. The film of yeast that forms on a newly fermented wine is normally quite thin, but under ideal conditions it can increase to as much as three or four millimeters. Under these circumstances, the upper layer of yeast cells in contact with the atmosphere is metabolizing the alcohol under aerobic conditions. Measurements of redox potential by Fornachon[1] show that at the lower side of the film, where the yeast cells contact the wine, conditions are reducing.

This is confirmed by observation of samples fermented in glass containers, where it is observed that the dark amber color of the new wine becomes progressively lighter as the film thickens and reduction of the brown-colored, oxidized pigments occurs. Further, it has been demonstrated by Augustyn[2] and his co-workers that the ratio of hydroxyethylbutyrolactone to acetylbutyrolactone increases with increasing thickness of film yeast and with increasing time of the wine under the thick film. Changes in aroma and flavor of film wines also can be correlated with the time the wine has been under the complete and active film of yeast.

Bacteria also contribute to biochemical activity in enology. Various *acetobacter* species, in the presence of air, can enzymatically convert ethanol to acetic acid. Certain species also produce ethyl acetate directly. In other cases, it is suspected that the esterases of the yeasts or the bacteria enzymatically mediate the esterification of the bacterially produced acetic acid. In any event, the mixture of acetic acid and ethyl acetate, giving the characteristic odor of wine vinegar, is rapidly produced.

One of the reasons that Louis Pasteur[3] chose to work on wines was that a distressingly high percentage of the wines exported from France at that time were defective because of the presence of acetic acid. Pasteur's genius lay, in part, in his recognizing that, while the acetic acid reaction resulted from exposure of wines to air, it was in fact, enormously speeded up by the presence of bacteria. He realized that the bacteria, and thus the problem, could be controlled by heating the wine to high enough temperatures to kill the bacteria. The potential problem is controlled in the modern winery by two procedures – prevention of exposure of the wine to oxygen by using stainless steel tanks with inert gas blanketing and by judicious addition of sulfur dioxide to the wine. Sulfur dioxide functions in two ways, of course – it reacts with small amounts of oxygen present in the wine and it has a bactericidal effect on the *acetobacter*.

Another biochemical reaction in enology, mediated enzymatically by a number of the lactic bacteria, is the conversion of malic acid to lactic acid and carbon dioxide. This is an extremely interesting reaction and one that has been widely studied by Kunkee[4] and others. The reaction is particularly to be desired in the case of wines containing excessive levels of acid, normally produced in the cooler grape growing regions. Unfortunately, however, the reaction occurs most readily in relatively low acid wines from warmer grape growing regions, and it can constitute a veritable disaster for the quality of such wines.

MICROBIOLOGY IN ENOLOGY

Today it is difficult for us to realize that the science of microbiology dates effectively from Pasteur's work on beers and wines in the 1850s and 1860s. It is particularly appropriate, therefore, that as the science of enology develops, the principles of microbiology continue to be closely interwoven into the fabric of the new science.

Grapes in the vineyard are subject to microbiological attack by numerous fungi. With the exception of *Botrytis cinerea* these organisms can all be considered detrimental and only capable of detracting from the quality of the resultant wines. *Botrytis cinerea*, grown under ideal climatic conditions on the fruit in the vineyard, can lead to a preferential loss of water from the grape berries with the production of the very fine, sweet, low alcohol wines with Botrytis aroma, such as those from the Rheingau and from Sauternes. The high regard with which *botrytis* is held in Germany and France is attested to by the names *Edelfäule* and *pourriture noble* which translate in each case as 'noble mold'. There are, however, problems with *botrytis* infected grapes in that the fungus secretes oxidative enzymes as well as mediating the dehydration of the grape berries. Further, if weather conditions are not correct – a warm moist period followed by a dry cooler period – the *botrytis* will so over-grow the cluster that it produces the condition known as grey mold or bunch rot, and renders the fruit completely inappropriate for fine wine production.

Louis Pasteur not only showed that it was *acetobacter* that were spoiling the wines for French export, but he also finally convinced the French Academy and the scientific community, that wine itself was a product of a living organism, the yeast, *saccharomyces*. At the start of his work Pasteur was young and scientifically unknown, but he had the advantage of his carefully controlled scientific experiments. From his work, documented in *Études sur le Vin*, published first in 1866, the whole specialization of zymology has evolved. Recognized today are a large number of species and strains of yeast comprising not only desirable fermenters for wine production, but a fair range of spoilage yeasts, as well. Among the special types are film formers which give rise to the unique wines of the Jerez district of Spain, the *vins jaunes* of the Jura of France, and the *flor* wines of Australia, South Africa, Russia and California.

Starting with Pasteur's work, and now widely developed as a branch of microbiology, bacteriology is of the gratest importance in enology. Considered mostly from their potential for damage to wine quality, bacteria nevertheless have some good potential for enology. Some strains of the lactics will convert malic to lactic and carbon dioxide, thus reducing the acid taste of wines produced in cold regions. Further, and perhaps of more importance, Pilone[5] has shown that some strains of these organisms add a complexity in aroma and flavor to red wines in which they have grown. This is a field under active investigation at present and one that is likely to be of importance to enology in the future.

ENGINEERING IN ENOLOGY

Most of the steps involved in the production of premium quality grapes in the vineyard and premium quality wines in the winery are amenable to treatment as agricultural or chemical engineering unit operations. Vineyard operations will not be considered beyond pointing out that they are, in general, typical material transfer problems. Similarly, in the winery, where the art, technology and science of enology is practised, a great part of the operations are typical engineering solids, fluids, or gas transfers. Involved also is the engineering separation of these phases. The modern winery almost always commences the production of wines by a smashing of the grape berry and removal of the stems from the mixture of skins, pulp and seeds.

In the case of white wine production, the juice and some of the pulp is normally separated from the skins. The pulp and seeds tend to be distributed between the separated skins and the juice. Red wines, of course, are fermented in the presence of the skin to permit extraction of phenolics, including the red anthocyan pigments from the skins by the alcohol produced during the fermentation. Toward the end of the fermentation, when satisfactory concentrations of pigments, tannins and aroma materials have been extracted, the partially fermented wine is separated from the skins and allowed to finish fermenting.

In many cases of white wine production, before fermentation of the juice, further quantities of

Wine Pressing, Ancient and Modern. A complete tree was used on this German wine press of 1790 to obtain sufficient leverage for wine pressing. (Courtesy Landesmuseum Koblenz, FRG; drawing by G. Stephanow-v. Freydorf). In contrast with this, a modern bladder or concentric-tube press for separation of juice from white skins before fermentation or of wine from red pomace after fermentation. The machine has an inflatable neoprene rubber tube mounted concentrically with the perforated stainless shell. (Courtesy Valley Foundry and Machine works.)

suspended solids will be removed by a settling or centrifugation operation. Recently, the technique of flotation, borrowed from mineralogy, has been applied to the removal of the finely divided, suspended material present in white juices. Here, of course, an inert gas rather than air would be used in the flotation cell. Following fermentation, most of the suspended solids, including the no longer useful yeast cells, will settle to the bottom of the tank and the new wine can be removed by racking – a decantation or siphoning of the clear liquid from the settled solid. Filters, too, are widely used – both for clarification of wines that do not naturally settle bright and for more rapid clarification of wines, where this is economically desirable. Modern enology makes widespread use of membrane filters for rendering slightly sweet low alcohol wines biologically stable in their bottles.

The equipment used in these winery operations encompasses a wide selection of crushers, stemmers, drainers, presses, flotation cells, centrifuges, filters, pumps, liquid transfer lines and solids transfer conveyors. While conditions vary from winery to winery, the ideal is to have each operation conducted with the minimum exposure of the material to oxygen, the minimum amount of shearing forces applied to the liquid, with minimization of the time and costs involved. Bottling systems present special challenges because it is at this point that the engineer must find the best compromises possible for the need to process quickly a large quantity of wine through expensive equipment and, on the other hand, the need to treat the wine as gently as possible.

Of particular concern to the engineer in enology is control of heat transfer processes and temperature control. Fermentation itself generates a relatively large quantity of waste heat – 142 000 joules (34 kilo-calories) per gram-molecule of hexose sugar fermented. In smaller wineries and particularly in older establishments using many small open-topped fermenting tanks, radiation and conduction through the walls provides sufficient surface for heat removal.

In the large modern winery using closed fermentors and particularly in cases where the size of these is large – 100 000 liters or greater – heat must be removed either by having cooling jackets on the outside of the tank or providing cooling tubes within the tank. In most cases, heat removal from the very large tanks is accomplished by running a coolant through a jacket integral with the side walls of the tank.

General consensus and the work of Ough[6] has demonstrated that quality in red wines is enhanced by fermentations at temperatures of 25–30 °C, whereas white wines improve by fermentation at 10–15 °C. It is realized that in the white wine fermentations the temperature within the tank will be essentially uniform, but that in the case of reds, where there is a relatively thick layer of skins floating at the top of the tank, the temperature in this cap may be significantly higher than the bulk temperature in the liquid beneath. Studied only in a very preliminary fashion has been the question as to what the effect of temperature profile with time during fermentation would be on the quality of the wine. Published work has assumed that a constant temperature throughout the fermentation would be ideal, but there seems little theoretical ground for this assumption.

Heat transfer becomes important in enology when one considers the stabilization of the new wine against precipitation of tartrates in the bottle. Potassium acid tartrate is a material of limited solubility (0.37 grams per liter in water at 20 °C, and considerably less in wine). The problem in enology, however, is not the reduced solubility in the alcoholic solution as compared with grape juice, but the fact that the precipitation is very slow and erratic. Thus, it is common to find wines which have rested for as much as two years in the winery without precipitating their excess potassium acid tartrate. Frequently such wines will, upon bottling, promptly deposit beautiful crystals of potassium acid tartrate in the bottle.

In white wines, these crystals may be mistaken for splinters of glass and are thus undesirable. As potassium acid tartrate is less soluble at low temperatures than it is at high temperatures in wine, one method of obtaining stability is to refrigerate the wine, hold it for two or three weeks at the low temperature to induce crystallization, and filter while cold. One hopes that when warmed back to normal temperatures the residual tartrate will not exceed the solubility limit. To accomplish this operation requires cooling equipment and low temperature rooms, or insulated tanks and filters that can be operated at the low temperature. Modern technology would employ a process in which the energy of cooling is partially recaptured by passing the cold wine after filtration against warm wine going into the cold room.

Tartrate stability can be obtained by another technique. As sodium acid tartrate is much more soluble than potassium acid tartrate, if one replaces part of the potassium ions with sodium ions, stability in the wine can be achieved. This ion-exchange process, while widely practised at one time, is no longer greatly favored because it can result in changes of the taste of the wine and it decreases the nutritious benefits of the wine as a high-potassium, low-sodium food.

Wine aging, also, can be considered an engineering unit process. Aging must be understood to occur in two different and distinct phases – wood aging and stainless steel or glass aging. Accomplished during wood aging is a slow diffusion of oxygen into the wine along with absorption of soluble components from the wood into the wine. This combination of effects is highly desirable for red wines, but generally detrimental to quality in white wines. One must except, of course, those few fine varietal whites such as the Chardonnays and certain Rieslings which do benefit from having a touch of wood in their aroma and flavor.

A large stainless steel tank with the ullage space filled with carbon dioxide, nitrogen, or other inert gas functions as a large bottle. Conditions in the tank and in the bottle with a good closure are essentially anerobic in that very little oxygen, if any, can come in contact with the wine. A wholly different set of reactions occurs in the wine under these circumstances. The ultimate quality in wines develops from the proper sequence of the first mildly oxidative phase, followed by the reducing or anerobic phase during aging. While most white wines made today have a very minimum of the oxidative phase, this treatment in wood is essential for development of full quality in fine red wines.

CONSTRUCTION MATERIALS

Technological advances in materials of construction – metals, plastics, glass, oak – have had significant effects on the practice of enology and on the quality of the resultant wines. Perhaps most spectacular has been the elimination of the old problem of heavy metal clouds and haze formation in wines due to corrosion of the iron and copper containing alloys commonly employed in the winery 20 to 30 years ago. Modern stainless steel is essentially inert and thus the problem of the haze development due to the presence of traces of iron or copper has been eliminated.

Although the problem was always soluble through treatment of wines with potassium ferrocyanide, this has never been recognized as a completely safe or desirable process. Use of the large stainless steel tank,

Storage and Transfer of Wine. In the German wine cellar of the 19th century, the cellarman transferred his wine from barrel to barrel by creating air pressure above the level of the liquid. (Courtesy Landesmuseum Koblenz, FRG; drawing by G. Stephanow-v. Freydorf). Similarly, the modern stainless steel wine storage tanks have gas tight closed tops so that when partially filled with wine the upper space may be filled with an inert gas to prevent oxidation of the wine. Similar tanks with cooling jackets incorporated into the walls, and with sloping bottoms and large doors to facilitate solids removal are used for fermentation. Transfer of wine is carried out by small movable electric pumps. (Courtesy Wine Institute, San Francisco.)

in combination with inert gas blanketing, has permitted significant improvements in wine quality as discussed above. The modern winery equipped with this type of storage vessel can always move the wine from the wood at the optimum time, whereas in previous years it was frequently necessary to hold the wine in large wooden containers until it could be scheduled for bottling.

While stainless steel containers are probably ideal, modern plastics of the epoxy type which can be applied to the inner surfaces of mild steel tanks are nearly as inert and considerably cheaper. The plastic coated mild steel tank requires much more caution in cleaning and must be inspected much more rigorously and frequently than the stainless steel tank. Nevertheless, for the very large storage tank it has a very definite place in the present modern winery. Smaller tanks constructed of glass fiber reinforced plastic are also quite useful and completely inert to wines.

Fiber glass reinforced tanks seem to be limited at the present time to sizes of a few thousand liters maximum. With equivalent degrees of inertness to the wine and with roughly equivalent ease of fabrication and installation in the winery, cost per liter of storage tends to become the dominant factor in the choice of material for storage container construction. With present day technology the glass fiber reinforced plastic tank seems to be most economical for sizes up to a few thousand liters, but the stainless or plastic lined mild steel tank becomes most economical in the larger sizes.

Crushed grapes, juices and wines are moved from place to place in the winery through permanently installed transfer lines or through movable rubber hoses. Permanent transfer lines are made of stainless steel or borosilicate glass in the modern winery. Occasionally, even today, one will see a mild steel or bronze transfer line used between the crushers and the fermenting tanks. This practice – while not highly desirable – is acceptable because most of the dissolved metals are precipitated with the yeasts and tannins during the fermentation. Metal transfer lines have the advantage that they are not breakable and that they can be cleaned with relatively strong cleaning agents. On the other hand, the borosilicate lines have the advantage that one can always see what type of material is being transferred and one can verify visually that the lines are clean before filling them with a wine.

SENSORY ANALYSIS

The tasting of wines to determine their quality is inevitably a very ancient practice. Since they are made for consumption with food, wines are naturally bought and sold on the basis of the way they smell and taste. One of the more significant steps toward becoming a science taken by the ancient practical art of enology was the adoption of the principles of sound statistical analysis to control the tasting of wines. Until quite recently, most wines were bought and sold on the basis of the opinion of a single expert. Since this person was usually at the management level of his firm, his decision was accepted without question. Today we realize that even the most experienced taster can have 'off' days and most modern wineries would not make blends or important decisions without the opinion of an experienced taste panel operating under conditions of statistical control.

In the modern winery, most sensory analytical problems fall under one of two categories, consumer acceptance or difference measurements by the expert panel. Consumer acceptance is, of course, of great importance to the winery as it must sell its wines. Filipello[7] presented methods for the statistically valid selection of large panels from which valid estimates of the consumer acceptance of a new product may be judged, but these techniques are very expensive and time consuming. Thus, we find that most wineries will vary the aroma, taste and color only slightly from that of a product already enjoying success in the market. Indeed, in many cases the new label will resemble the label of the successful product in many attributes. An alternative to the expensive large consumer acceptance survey under statistical control is the launching of a new wine type with an expensive and extensive advertising campaign. In the world of the modern winery, a new product launched in this manner may enjoy considerable consumer acceptance as long as the advertising promotion is continued. Very distressing to the producer has been the fact that, in many cases, consumers do not continue their loyalty to the product when the advertising campaign is reduced or terminated.

Once a moderate degree of consumer acceptance of a product has been established, a panel of expert tasters can be used to evaluate the product and arrive as a description of its quality factors. The expert panel then can be used to predict changes in consumer acceptance of small variations in the composition of the particular wine. For instance, variation in the sugar or acid level just barely perceptible to the panel of experts will not be detected by the average consumer as a change in the composition of the wine, but he may very well prefer one variant of the base wine over another. Using these techniques, coupled with very careful monitoring of sales in the region where the test batch of wine is being offered, can eventually lead to maximization of consumer acceptance of a given wine type nation wide.

As Amerine and Roessler[8] point out, the selection of members of panels of expert tasters requires great care. Not only must potential candidates have average to superior sensory acuity, they must have the ability to concentrate on the examination of wines and the drive, determination and persistence to work at tasting for a long enough period to acquire mental olfactory images of many different wine types. Probably the ability to remember what different wines smell and taste like and the ability to concentrate one's senses during a tasting session are of more importance than the possession of a lower sensory threshold to the basic wine constituents. Similarly, the development of a palate in the interested amateur or would-be connoisseur is more dependent on a good memory for smells and tastes and the ability to concentrate during a tasting session than it is on any differences in basic sensory acuity.

Before concluding this review with a brief speculation about the future of viticulture and enology, let me summarize the very first beginnings of winemaking in ancient Egypt and the Middle East and point to the epoch-making discoveries of Louis Pasteur just over a century ago. Only then, could the transition from art to science begin.

PRE-PASTEURIAN ENOLOGY

The antiquity of winemaking and grape growing for this purpose is suggested by the fact that sealed wine jars dating from the reign of King Den of the first Egyptian Dynasty of about 3000 B.C. are well documented. Individual favorite vineyard names are recorded for Kings Jet of the First Dynasty, Khasekhemui of the Second Dynasty, and Djoser of the Third Dynasty. King Djoser was the one who built the first large stone pyramid. By the time of King Tutankhamen, 1352 B.C., wine jars were being labeled with the year of vintage, name of the vineyard, name of enologist and the region of production according to Lesko.[9]

Tomb paintings of this period and somewhat earlier times show a very sophisticated viticulture and enology. Vines were apparently trained on fairly elaborate trellises, carefully irrigated and tended by persons to chase away the birds. Crushing was done by foot and fermentation apparently was allowed to proceed spontaneously in large glazed amphorae. It is interesting that most of the tomb paintings show grapes as red and the wine as red during the third and second millennia B.C. There is a text from Meir and a tomb scene from El Bersagh which shows what appears to be white wine at the time of the Egyptian Little Kingdom, about 2000 B.C.

A 1977 discovery by Matthiae[10] of some 20 000 tablets near the modern city of Aleppo revealed the site to be that of the ancient city of Ebla. These tablets are written in the cuneiform of Sumer, but contain some Eblaite syllabic words. Interpretation was aided by the presence of a number of reference tables correlating this pictographic writing with the Semitic syllabic forms of north-western Semitic language. Dating from the second half of the third millennium B.C. the tablets document a fairly advanced agricultural economy based on grain, cattle and winemaking.

The oldest known writing, on Sumerian tablets dating from 3500 to 2900 B.C., has record keeping as its major purpose according to Schmandt-Besserat.[11] She postulates the following sequence of development. About 8500 B.C. agriculture and village-centered life arose in the Middle East. To aid in record keeping and management of trading operations clay 'tallies' were produced. About 3500 B.C. the technique of enclosing these clay tallies in hollow clay spheres as 'bills of lading' developed. Shortly after, it became the practice to indicate on the outer surface of the clay in pictographs the nature and number of the tallies inside the sphere. From this to the recording of the pictographs on a clay tablet was a relatively simple step. Schmandt-Besserat identifies a tally for beer among the many she has studied, but has not identified any that refer to wine. According to Lesko, the very ancient hieroglyph of a vine with grape bunches supported by forked props is used in words meaning 'vine', 'garden' and 'wine'.

The record we have of tomb paintings, wine jar inscriptions and clay tablets of business records, all indicate a fairly sophisticated and well developed enology by the third and early second millennia B.C. Reasoning that it would have required a long time to develop this varied and sophisticated wine technology, we can estimate that wines have been produced and appreciated at least as early as the middle of the third millennium before Christ.

The Roman Pliny, in his writings on grapes and wines, described the use of three different varieties in Egyptian vineyards. These were the Thasian, Soot and Pine Tree grapes. What they are called now, if they still exist, is not known. Grapes today come in a very wide range of sizes, colors, and physical and chemical composition. Plant breeders have been increasing this number of types by selective hybridizations for only some 150 years.

Two explanations are available to account for the large number of types that existed before deliberate hybridization started. These are, first, bud or other tissue mutations, probably caused by natural radiation which would change the appearance of a cane or shoot of a vine enough to provoke its further propagation by the vineyardist. Second, spontaneous hybridizations would have occurred easily in the early mixed plantings. If seeds from these hybridizations were propagated under selection pressures different from those of the original wild vines' habitat, new types would arise fairly rapidly, especially so in comparison with the rate of production of new types by mutation.

Nevertheless, selection pressures in the earliest hunter-gatherer turned agriculturist's 'vineyard' must not have been too different from the pressures that existed in the vines in the wild. For this reason one would not expect too great a rate of production of new types. This in turn suggests that the first domestication of the wild vinifera grape must have occurred at a very early age in order to have resulted in the many and greatly varied types that exist today.

Early Wine Presses. An ancient German wine press with a wooden screw spindle to exert the pressure necessary to crush the grapes; it is dated 1849. Next to them in the Landesmuseum Koblenz, FRG, two similar presses are preserved which were in 1860 fitted with hydraulic equipment, apparently the oldest of their type. (Drawings by G. Stephanow-v. Freydorf).

POST-PASTEURIAN ENOLOGY

The practical arts of grape growing and winemaking had reached such a fine level of perfection over the thousands of years of their development that it is difficult for us to realize from today's perspective that at the time of Pasteur the biochemistry of photosynthesis and of fermentation were unknown. Indeed, there was a well established school of scientists who maintained that growth of the yeast and bacteria in wines was completely incidental to the changes provoked. What had been developed, and what served as a very firm empirical foundation for the later applications of scientific principles, was the selection of appropriate varieties for the various climatic regions, development of the practices of vine management on an empirical basis, design and construction of winery equipment such as crushers, presses and cooperage to facilitate production, storage and marketing of the wines.

Pasteur's first contribution was to determine that the various species of *acetobacter* were responsible for the conversion of sound wines into vinegar. He also showed that these bacteria could be killed by heat and thus his name became attached to this process for preserving foods in sealed containers. In later, and perhaps more fundamentally important researches, Pasteur demonstrated that it was the enzymes within the yeasts which were responsible for the conversion of the sugars of grapes to carbon dioxide and alcohol. The research of several brilliant scientists was required to finally clarify the many steps involved in this conversion. Pasteur's successors developed the techniques of yeast strain separation and pure culture development. Since Pasteur's time much controversy has revolved around the question of use of single pure cultures in wine fermentations with many of the traditional regions preferring to rely on spontaneous fermentations by the naturally occurring mixture of yeasts.

The last quarter of the 19th century was a period of rapid development of the sciences of chemistry and biochemistry. Particularly striking was the time and effort expended on devising analytical procedures for the constituents of natural products. Following the path blazed by the earlier chemists such as Lavoisier, Priestley and Faraday, we find many researchers working on wine as their material of choice. From these studies came the realization that one could measure the sugar concentration in grapes and set harvest dates based on its accumulation.

In years of deficient sunshine and hence insufficient sugar, one could calculate the required amount of sugar to be added to the juice to secure a reasonably balanced wine – the process of *chaptalisation.* The nature of the various acids of wines was discovered and from this it became possible to detect the spoilage of wines in stages early enough so that preventative measures could be taken. Analytical procedures were also helpful in guiding blending operations, although then, just as now, most blending was done on the basis of sensory analysis rather than chemical laboratory analyses. It was in this latter quarter of the 19th century that the use of sulfur dioxide as an oxidation inhibitor and as a bacteristatic was widely accepted.

Although early winemakers must have discovered very soon that excessive fermentation temperatures would kill the yeast and stop fermentation, man had not been able to do very much about this until the second half of the 20th century when mechanical refrigeration became sufficiently reliable and economical to be used in wineries. The shallow, relatively large diameter fermentation vessel of the latter part of the 19th century gave way to larger and deeper concrete tanks in the 20th century. These were frequently cooled by running water through internal heat exchanging coils. This design has in turn yielded to the very large stainless steel tank of today with the coolant jacket as an integral part of the tank wall (see the illustration on p. 54).

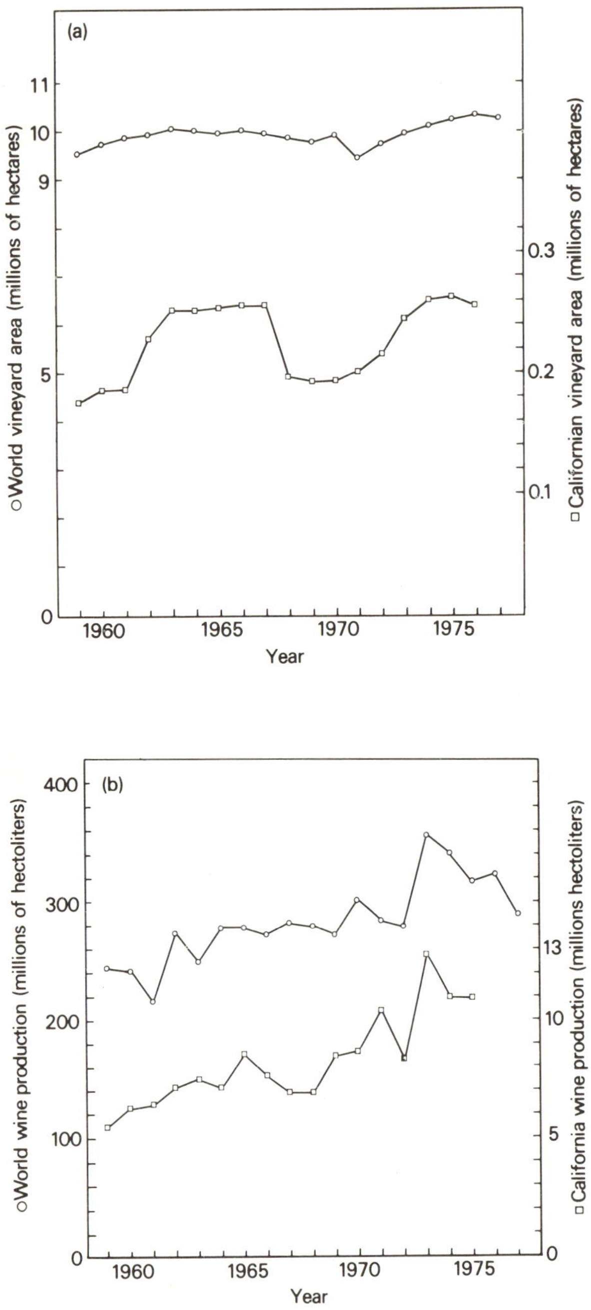

World and California compared in relation to (a) vineyard areas and (b) wine productions, 1959–1976. (Data from O.I.V.V.)

A number of other engineering developments in the fields of pumps, specialized tanks to separate juice from skins and other solids, centrifuges, and large efficient filters, have further changed the winery drastically from its nature at the time of Pasteur and as it existed during the first part of the 20th century.

VITICULTURE AND ENOLOGY OF THE FUTURE

For the first time in some thousands of years, viticulturists and enologists are in a position to control the composition of the fruit from the vineyard and its conversion to wine. Indeed, the vast array of choices accorded these professionals today completely changes the nature of the problems they face. In contrast to the practical artist of the last century who had very little control over his raw material or the production factors in winemaking, today's professional can guide the vineyard management so that fruit of specified composition is delivered to the winery and he can control the winemaking parameters so that the final product reflects his choices of conditions. The problem thus becomes – what are the ideal parameters or conditions?

Choices of the wine type and style will, of course, be dictated by economics but also by tradition established and maintained by interested connoisseurs. I believe that wine styles and types, particularly in the United States, will continue to respond to these two pressures – mass market economics and demands of interested connoisseurs. With the controls made possible by today's winery and vineyard equipment, and with the guidance of current science, the number of styles and types of wine should multiply spectacularly. The maker of wines to be marketed in large quantities will have increasingly sophisticated statis-

Typical of modern and future wine production is this centrifuge for clarifying grape or wine at the La Mont Winery at Di Giorgio, California. (Courtesy Centrico Inc.)

tical survey methods for gauging the true acceptance of his wine by the consuming public of different regions.

Minor adjustments in blends to fine-tune the wine type to individual regions no doubt will occur with greater and greater frequency. The producer of small lots of vintage wines from limited districts is likely to have his feedback from interested consumers in a more direct fashion and in more sophisticated wine terminology. The future seems bright.

LITERATURE CITED

1. J. C. M. Fornachon, *Studies on the Sherry Flor,* Australian Wine Board, Adelaide (1953).
2. O. P. H. Augustyn, C. J. Van Wyk, C. J. Muller, R. E. Kepner and A. D. Webb, The structure of solerone [5-acetyl-dihydro-2(3H)-furanone] a substituted γ-lactone involved in wine aroma. *Agric. Food. Chem.* **19**, 1128–1130 (1971).
3. L. Pasteur, *Études sur le vin, ses maladies, causes qui les provoquent, procédés nouveaux pour les conserver et pour le vieillir,* 2nd Edn. F. Savy, Paris (1873).
4. R. E. Kunkee, Malo-lactic fermentation and winemaking. In *Chemistry of Winemaking,* edited by A. D. Webb, Chapter 7, pp. 151–170. American Chemical Society, Washington, DC (1974).
5. G. J. Pilone, R. E. Kunkee and A. D. Webb. Chemical characterization of wines fermented with various malo-lactic bacteria. *Appl. Microbiol.* **14**, 608–615 (1966).
6. C. S. Ough, and M. A. Amerine, Studies with controlled fermentation. VI. Effects of temperature and handling on rates, composition, and quality of wines. *Am. J. Enol. Vitic.* **12**, 117–128 (1961).
7. F. Filipello, H. W. Berg and A. D. Webb, A sampling method for household surveys. I. Panel recruitment for testing wines. *Food Technol.* **12**, 387–390 (1958); II. Panel characteristics and their relation to usage of wine. *Food Technol.* **12**, 508–510 (1958).
8. M. A. Amerine and E. B. Roessler, *Wines–Their Sensory Evaluation,* Freeman, San Francisco (1976).
9. L. H. Lesko, *King Tut's Wine Cellar,* B. C. Scribe Publications, Berkeley, California (1977).
10. P. Matthiae, *Older Testament* in Science and the Citizen. *Sci. Am.* **237**, 101 (1977).
11. D. Schmandt-Besserat, On the invention of writing. *Syro-Mesopotamian Studies* **1**, (1977).

The manuscript was received 20 January 1979

The Biotechnology of Algaculture

PROFESSOR AMOS RICHMOND and PROFESSOR KENNETH PREISS
The Institute for Desert Research, Sede Boqer and the Department of Mechanical Engineering, Beer Sheva, Ben-Gurion University of the Negev, Israel

Increasing interest in the feasibility of growing algal biomass for commercial purposes derives from the progressively increasing shortage of protein food and animal feed, raw materials and energy resources, and the realization that algae, growing in salt water, could become a useful crop in many of today's deprived, arid lands. Here some basic aspects are reviewed which concern the production of algal biomass and its use. Detailed consideration is given to the possibility of cultivating algae as an alternative source of protein in the arid, protein-poor areas of the world, where brackish or sea water is available. Then the production of major products from marine algae is described and the possibility of producing various chemicals of industrial value from algae is evaluated.

The primary source of all man's food and organic raw materials is solar energy. As humanity expands and its demands persistently increase, man's existence may, in the final analysis, depend on how efficiently he learns to use solar energy to produce the required quantities of food, energy and organic material. As pointed out by Milner[1] and Burlew,[2] present agriculture is very inefficient in this respect, as first, most crop plants utilize only half of one per cent of the sunlight that falls on them, second, most farm crops cover only part of the ground, part of the year and third, only a small portion of the crop plant is edible or commercially useful, since most of the plant body serves for mechanical purposes, roots for anchorage and for drawing minerals and water from the soil, or stems to support leaves and fruit in light and air.

However, even with favourable conditions of light, temperature and mineral nutrients, the maximal rate of biomass production by field grown plants cannot be reached, as it is severely limited by availability of carbon dioxide. Algae, on the other hand, promise important advantages in improving the efficiency of solar energy utilization for valuable biological products. These aquatic plants, which range in size from large oceanic species of over 30 m length down to microscopic unicellular forms, can be grown in a continuous culture; this provides maximal annual production of biomass per unit area, limited only by solar radiation. The algal cell contains relatively little structural material and most of its body may have nutritious or other economic value. In addition, algae such as *Chlorella* can be grown to produce a cell composition with greatly variable major components. Meeting the carbon dioxide requirements of algal cultures is relatively simple compared with field crops, as this gas can be passed readily into the culture medium. Many species of algae have another most important advantage, however, which may hold the key for the development of many arid zones: the capacity to thrive on saline or sea water – resources not suitable for the production of most other plants which man uses today in his conventional agriculture.

ADVANTAGES OF ALGAE IN DESERT AREAS

Many arid and desert areas offer important advantages for cultivation of algae, conditions which usually impose severe limitations on conventional agriculture.

Saline water: Arid zones lack sufficient water. Such water resources as may be available are often of a quality unsuitable either for human consumption or for use in agriculture, since they contain high concentrations of salt or other contaminants. Indeed, saline aquifers have been located in many desert areas all over the world. For example, the saline aquifers of the Negev in Israel, which vary in salinity from 2500 to 6000 ppm TDS (total dissolved solids), could supply 30 million m^3 annually with no depletion, and up to three times that amount with only minor depletion.[3] In addition to brackish water, sea water should be considered for desert algae culture, possibly becoming an important source of water for bioproduction of various useful products. (For comparison, most domestic water supplies in the world have below 500 ppm TDS; sea water varies from 30 000 to 45 000 ppm TDS.)

Temperature: Algal growth increases exponentially with temperature until an optimum temperature is reached. Summer daytime air temperatures in many desert areas approach 40–45 °C on the average, although the water in ponds would usually be a few degrees cooler. The optimal growth temperature for many types of algae is about 35 °C. Hence, normal day temperatures for hot deserts, which would sub-

CCC-0308-0188/80/0005-0060$05.50/1

ject many conventional plant species to severe stress, fall in a range which is suitable for algae production at maximal yields.

Availability of large amounts of solar radiation: The average annual levels of solar radiation are considerably higher in hot desert regions than in other areas of the world, because of their latitudes and the characteristic absence of cloud cover. High radiation levels hold two advantages for algae cultivation. First, the rate of solar radiation available for photosynthesis in the warm regions is the major factor that determines the rate of growth for algae, unlike conventional agricultural crops, where water, soil nutrients and carbon dioxide are usually the major limiting factors for growth. Second, the high radiation levels open up the possibility of using solar energy to process the harvested algae.

Availability of large tracts of empty land: For maximal absorption of solar radiation, algae cultivation must be carried out in shallow, 10–20 cm deep ponds, requiring large flat stretches of vacant land. These are abundant in desert areas.

THE IMPENDING PROTEIN CRISIS

The great increase in world population during the recent decades and man's desire for a better quality of life have made great demands on world food production; some progress has been made in recent years towards closing the gap between increasing population growth with the demand for a better diet and world food production. Significantly, however, most of the progress has taken place in the developed, rain-rich countries, where pressure on the food supply is not usually extreme. The food prospect for the developing, mostly arid countries seems poor. The total yield of today's conventional agriculture is insufficient to feed the present world population of about four thousand million people, and it is estimated that one-eighth of the world's population is malnourished. Williams analysed the present and future world protein production and concluded that by the year 2000 the annual protein deficiency in the world will reach 39 million tons.[4]

Another grim prospect for the future is that where abundant harvests are produced by modern techniques, the energy cost of moving any surplus to areas of want, coupled with the costs of purchase and distribution, prevents one part of the world from feeding another. The established patterns of political and economic dominance are a further hindrance to an equitable distribution of the world's food.

The protein gap is well summarized by a report of the United Nations' Advisory Committee on International Action to Avert an Impending Protein Crisis.[5] It points out that the marked difference in *per capita* food production between the developed and the developing countries is bound to increase as death rates are falling, birth rates remain high and crop yields continue to be low in the developing countries. Moreover, for over one-third of the present population in the developing countries, the protein-calorie balance of the diet is inadequate and the gap between the nutritional requirements and the actual consumption of protein is widening. Indeed, the present nutritional deficiencies in many developing countries are judged to jeopardize the future of many hundreds of millions of the world's people.

In a growing number of developing countries the dependence on food imports is steadily increasing and their improvement of conventional agriculture is not taking place rapidly enough. Hence, the Committee concluded that conventional agriculture should be supplemented by protein-containing foods of unconventional origin, such as oil-seed, fish-protein concentrate and single-cell protein from micro-organisms.

In a recent article, J. A. Bassham[6] suggested covering large areas of the south-west of the United States of America with huge greenhouses made of Sun resistant inflatable plastic where a protein forage legume, such as alfalfa, could be grown the year round, under elevated carbon dioxide levels. He suggested that under these conditions 20 tons per 100 m^2 of proteinaceous material containing some 25% protein could be produced. We propose here that the same objective could be achieved more simply with an algal crop.

PROFESSOR AMOS RICHMOND, born in Tel Aviv, is the Chairman of the Ben Gurion University's Institute for Desert Research at Sede Boqer, Negev, Israel. He received his academic training in the United States, obtaining MSc and PhD degrees in Horticultural Science. In 1975 he became Professor of Biology at the Ben-Gurion University, where he served for three years as Dean of the School of Natural Sciences. His recent research interests include applied algology and he is now heading a group studying the commercial aspects of algal biomass production in saline water.

PROFESSOR KENNETH PREISS, born in Cape Town, South Africa, was a member of Professor Richmond's algae group, where he advised on computer and engineering aspects concerning biomass production of algae. He studied in South Africa where he received his BSc and in London, UK where he obtained a PhD. He is professor of mechanical engineering at the Ben-Gurion University and has been working on the application of nuclear radiation in engineering and in computer-aided design.

Address: Desert Institute, Ben-Gurion University of the Negev, PO Box 2053, Beer Sheva, Israel.

THE NUTRITIONAL POTENTIAL OF ALGAE

The food potential of certain microscopic algae has been studied fairly intensively in recent years. The blue-green alga, *Spirulina*, belonging to the family *Oscillatoriaceae*, is particularly interesting. Rediscovered by the academic world as recently as 1940, *Spirulina platensis* has been collected from the salty lakes and ponds along the northern shores of Lake Tchad since time immemorial, sun-dried and eaten by the Kanembou people, now numbering more than 80 000 persons.[7] *Spirulina geitleri* was similarly collected and prepared by the Aztec Indians at the time of Cortez' arrival in Mexico. *Spirulina* is an easily harvested multicellular filamentous alga of high digestibility and mild flavour which has been found to contain up to 70% protein of excellent quality as shown in Table 1.

The high protein content found in *Spirulina* makes this product a concentrate by itself, having a significant advantage over conventional vegetables with low protein concentration.

To improve the amino acid balance of the *Spirulina* protein additional sulphur-bearing amino acids could be added. This is easily done by including the gluten of wheat or barley in the diet, and in fact, both the Kanembou and the Aztecs used *Spirulina* with whole grain cereals. According to Fox[7] as much as 40 g of dried algae per m^2 per day have been produced. However, a reasonable large-scale goal of only 20 g per m^2 per day would yield 50 000 kg of protein per hectare, annually, compared with beef cattle which average less than 100 kg of protein per hectare per year.[8]

Figure 1. A painting showing the ancient Aztecs straining *Spirulina* from the lake.

Table 1. Essential Amino Acid Composition of Algae – *Spirulina* and *Scenedesmus*, Compared with other Sources of Protein for Feed

Amino acids	*Spirulina* sp.	*Scenedesmus acutus*	Yeast	Soybean meal	Fish meal
Isoleucine	6.0	3.2	5.1	4.6	5.4
Leucine	8.5	8.6	7.4	7.3	7.7
Phenylalanine	5.0	3.9	4.3	4.0	5.1
Tyrosine	4.0	2.8	3.6	2.9	2.7
Threonine	4.6	4.8	4.9	4.2	4.0
Tryptophane	1.4	1.4	1.4	1.2	1.5
Valine	6.5	6.2	5.9	5.2	5.0
Arginine	6.5	5.8	5.1	5.0	7.7
Histidine	1.8	1.7	2.1	2.3	2.4
Lysine	4.6	5.3	7.4	7.0	6.5
Cystine	0.4	1.0	1.1	1.0	1.4
Methionine	1.4	1.4	1.8	2.6	1.4

The nutritive value of *Spirulina* is amplified in that it has a relatively low percentage of nucleic acids, 4%, as compared with the high content of nucleic acid in bacterial protein. The mucoproteic membranes that separate the cells are easy to digest, unlike the cellulose cell wall found in many other nutritional algae; it is completely non-toxic, and its lipids are made up of unsaturated fatty acids that do not form cholesterol. This makes *Spirulina* an interesting food item perhaps for patients suffering from coronary illness and obesity, as was suggested by Durant Chastell.[9]

Algae for Pisciculture

Furthermore, the presence of linolenic acid, indispensible for fish feeding, gives it a great advantage for pisciculture. Durant Chastell reported on the value of *Spirulina* for different aquatic animals during their early life when feeding must be carefully controlled for size of particles and digestibility. *Spirulina* was also used successfully as a complementary feed for molluscs-crustaceans, promoting their growth and sexual maturity, and good results were obtained with larvae, cockles and young shrimps.[9] Feeding *Tilapia rendalli* on a 100% *Spirulina* diet resulted in considerable weight and length increase over the control, sexual maturity and early reproduction being readily observed. Altogether, there seems good evidence for *Spirulina* as an excellent feed for fish due to high food conversion, a well balanced aminogram, a very high content of protein, vitamins, particularly vitamin B_{12}, and essential unsaturated fatty acids.

The quality of protein from the green algae *Scenedesmus obliquus* has been thoroughly investigated by Pabst.[10] Considering any international nutrition parameter *Scenedesmus* compares very well with most common animal feeds, such as soybean extract and fish meal, as shown in Table 2.

Table 2. Nutritional Values of Standard Feeds and Algae

Feed	% Protein	PER[a]	NPU[b]	BV[c]
Spirulina platensis	56	1.80	62	75
Scenedesmus obliquus	54	1.85	61	75
Soybean extract	50	1.82	60	70
Fish meal	67	1.82	66	79
Casein + methionine	90	2.50	86	92

[a] Protein efficiency ratio.
[b] Net protein utilization.
[c] Biological value.

In addition to the high protein content of many microalgae there is a high content of vitamins and minerals which makes algae a promising additive to animal feed. In particular, microalgae contain a high quantity of carotenoids, important for intense colouration of such food as egg yolks, shrimps and fish.

Algae from Sewage

A system for the utilization of sewage for algal biomass production to realize the dual aim of harvesting proteinaceous biomass and water reclamation has been proposed and developed by Oswald[11] and experimentally tested by Shelef.[12] It is claimed that this system has a particularly promising economic future, as the value of the reclaimed water compensates for the overall cost of algal production. Indeed, a recent economic analysis by Pano and his colleagues[13] supports this claim well. Oswald pointed out the prodigious production potential of large-scale algal culture on waste, considering it to be theoretically more economical than most agricultural crops from the standpoint of land usage, water, energy, capital and human resources. He believed that algal culture on wastes will require $\frac{1}{50}$ the land area, $\frac{1}{10}$ the water, $\frac{2}{3}$ the energy, $\frac{1}{5}$ the capital and $\frac{1}{50}$ the human resources required in conventional agriculture to produce an equal quantity of useful organic matter.

Yanai, Mokadi and Berk[14] used several algae species grown on municipal sewage as a substitute for soybean protein in feed for chicks and rats. The content of lead, mercury, cadmium and copper in the algae were determined in relation to the content of these elements in tibia bones, and of mercury in the thigh muscle of the chicks. The levels of aluminium were also checked.

The results indicated that the algae protein gave normal growth and its nutritive value was comparable with that of soybean meal, always provided that the amount of algae protein did not exceed 25% of the total protein in the diet. No abnormalities were noted in the experimental animals, except for the accumulation of a yellow pigment in their skin and fatty tissues. As to metal content, there was no appreciable retention in the chicks, despite the relatively high concentrations of metals in the algae.

Similar, but preliminary conclusions of the carp and *Tilapia rendelli* growth experiments in Dortmund[15] revealed that unicellular green algae were well accepted by fish; up to 12% of the feed by carp and up to 40% of the feed by *Tilapia rendelli.* No increased mortality or malformations were observed in carp receiving diets with 12% algae containing Al^{3+} during an experiment of 42 days. In tank experiments unicellular algae could replace 60% of fish meal or part of the soybean meal of the feed of carp, without ill effects on their growth.

TECHNOLOGICAL ASPECTS

Three major technical aspects are involved in developing commercial systems for the mass cultivation of algae. The first relates to pond construction, the shape, depth, and system of mixing the algae-laden water, and most important, the lining. The second concerns separating the algal mass from the medium and the third relates to the dehydration of the algal mass to facilitate product distribution and storage. A satisfactory solution of these three problems is essential before any large-scale cultivation can be economically pursued.

Pond Construction

Mass production of algae requires an inexpensive method of pond construction. Since the medium must be thoroughly exposed to sunlight shallow race channels a few meters wide and about 20 cm deep are required.

One example of constructing such channels is to line them with black plastic sheeting, 0.25 mm thick. Extending from 50 to 300 m the channels will run back and forth, making pond units of 5000 to 10 000 m^2. The channels should be dug with a small constant slope in the direction of flow, assuring a rate of approximately 30 cm min^{-1}. This should create sufficient turbulence to mix the algae adequately, essential for the distribution of solar radiation to the cells in culture. At the end point of this channel maze, the algal medium will be lifted some 1.0 or 2.0 m, returning it to its highest point.[16] This concept could greatly reduce the cost of pond construction as compared with ponds built of concrete or other durable materials in which the input of material and construction cost is greater by perhaps one order of magnitude. Another method is to construct the raceways 4–6 m wide, with no incline, and provide

Figure 2. Empty pond raceways dug in the ground, covered by 0.2 mm black polyethylene lining.

turbulence with a paddle wheel, which may be the most efficient way of stirring shallow layers of water.

Separation of the Algal Mass from the Medium

Harvesting the algal biomass requires separation of the cells from the aqueous medium. Various methods are possible for this and the size of the algae to be separated governs the choice. Harvesting may be achieved by applying centrifugal force and separating by difference in density between algae and water. Separation can also be performed through flotation of the algal mass by generating a fine stream of air bubbles through the medium. The small bubbles adsorb on the surfaces of the cells and increase their buoyancy, resulting in flotation.

Separation may be greatly facilitated by introducing chemicals into the medium which induce the single cells to flocculate. A commonly used flocculant is aluminium sulphate which promotes separation by flotation and thus the foamed biomass can be easily filtered. However, the algal powder which is obtained after dehydration contains about 20% aluminium and therefore this form of separation may have only limited use as a feed, unless the algae are further treated. But preliminary experiments in Israel have indicated that the aluminium content in the algal powder has no negative influence when it is used for fish feeding. A system designed to extract the aluminium from the dry matter by acidification and washing is feasible, but, of course, increases the cost of harvesting.

Figure 3. Experimental ponds lined with 2.0 mm poly vinyl chloride sheet; note paddle wheel at the far right corner,

Obviously, if available at low cost, an edible flocculant such as 'Rubifloc'[17] obtained from the sugar refining industry, would be very useful in the production of algal single cell biomass for food or feed. Causing the algae to sediment, either by settling on suitable membranes or induced sedimentation by the addition of proper agents is a further technique of separation. The sedimented algal mass may then be collected by various means. Today filtration appears to be the obvious choice for the first stage of separation from the aqueous medium. The type of filter and the process depend on the algae to be separated and on the production rates.

Small micro-algae, such as *Scenedesmus* or *Chlorella* with a diameter of a few microns, pose great difficulty due to their rapid clogging of the filter. However, Dodd and Anderson[18] have devised a micro-algae harvester based on a pre-coated paper belt running over a filtration drum. They propose a process in which a double fabric belt sandwiches the algae-paper mat on both sides, so as to retain the paper fibres during one application of vacuum and water and then to separate the algae from the belt. The pre-coat and residual algae are then removed from the belts by water showers, the dispersed fibres are washed to remove residual algae and the washed fibre recycled to form a new pre-coat. Doubtless, if an economical method for the filtration of microalgae could be developed, mass-algae culture would be greatly advanced.

Filtration is already possible of micro-algae growing in colonies measuring some 25 microns in diameter, such as *Coelastrum proboscideum*. Using a machine developed for the food industry for thickening starch suspension, researchers in Soeder's group at the Kohlenstoffbiologische Forschungsstation, GSF, in Dortmund thickened a suspension of *Coelastrum* some 100-fold, from 0.6 to 60 $g\,l^{-1}$.[19] The energy requirements to collect 1 kg of dry matter from this alga was 0.7 $kWh.^{-1}$, a figure which may very probably be reduced. At present, only the filamentous *Spirulina* is readily filtered on a 300-mesh commercial filter, but precautions must be taken to avoid physical damage to the delicate cells and filaments so as to prevent a leakage of cellular matter.

Dehydration of the Algal Mass

Dehydration forms a problem of major economic importance. Systems differ both in the extent of capital investment and energy requirements, and have a marked effect on the biological value and taste of the product, particularly with regard to green algae with a cellulose encasement. Pabst[10] of Dortmund has shown that dehydrating algal mass with a fine-layer drum-dryer yields an excellent product. He ejects an algal suspension of 6–8%, obtained by centrifugation, onto a rotating, steam heated drum; the algal cells are thus heated for a few seconds and the ensuing dehydration causes the cellulose encasement of the green algae to open up, thereby greatly increasing edibility and biological value of the powdered product. According to Soeder and Mohn,[20] 15.7 kcal

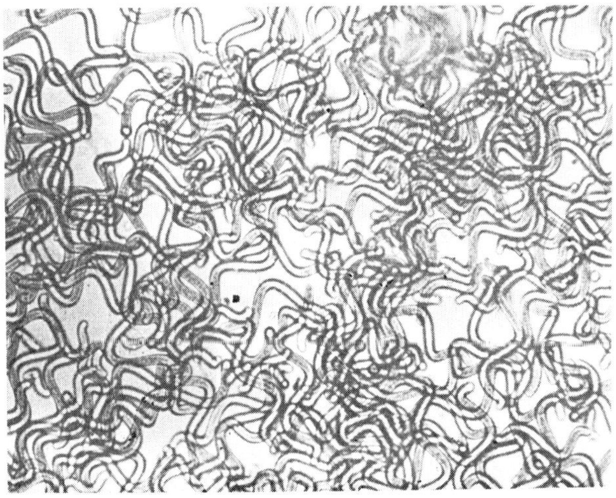

Figure 4. Photomicrograph of *Spirulina platensis* from an outdoor culture ×200 magnification. The culture is practically pure *Spirulina* filaments.

are needed for the evaporation of 18.2 kg water to obtain 1 kg of dry powder with a water content of 4%. This energy input is in addition to the 1.4 kWh which is needed to run the machine itself. It is quite clear that energy expenditure may be saved if other methods of dehydration were tested, such as subjecting the filtered or flocculated algal mass to steam treatment followed by vacuum filtration and concluding the dehydration in a rotary kiln. However, this suggestion has not yet been tested.

Soeder and Mohn also proposed that dehydration may be successfully achieved by mixing algae with dry additives such as straw, sugar beet pulp, meal powder or grains, and thereafter pressing the mixture by extrusion, producing pellets instead of powder. In addition, Mohn showed that dehydration expenditure may be appreciably cut when the biomass is dehydrated to 10% rather than 5%. It is conceivable that certain mixtures could be kept successfully in a more hydrated form than is possible with the pure algal powder, thus reducing the cost of dehydration. Direct dehydration of algal biomass in the Sun is feasible, but it is not recommended as it is harmful to quality and requires large drying areas. Nevertheless, a possibility so far unexplored is the harnessing of solar energy to create a source of heat for dehydration of the biomass.

ECONOMIC ANALYSIS

No detailed economic analysis of large-scale production of algal biomass for protein is available as yet, simply because large-scale production of, say, 100 ha under field conditions and for the sole purpose of harvesting algal biomass, has not yet been attempted. On the basis of our own ideas and experience, a general analysis may be possible, however.

The economic feasibility of algal mass production on brackish water has been examined carefully by Pano and his colleagues from Tahal Consulting Engineers Ltd of Tel Aviv,[13] who have analysed the construction and maintenance costs of a hypothetical 10 ha plant in the Rift Valley in the southern part of Israel.

Assuming an algal yield of 2.5 tons per day during the summer, April to October, and 1 ton per day during the winter, November to March, and taking into account the necessity of changing the entire pond volume eight times per year, they calculated the yearly input for a 10 ha pond as follows:

Total water requirement	480 000 m^3
Major nutrients ($NaHCO_3$, KNO_3, H_3PO_4)	about 1000 tons
CO_2 supplied by locally burnt fuel	280 tons
Giving a yearly output of algal product of	750 tons

Pano assumed a process by which the algae would be separated by vibrating screens to form a 5% solution of algal solids. Thereafter, this mass would be centrifuged, capacity of 35 m^2 per hour, resulting in a 15% solution which would be drum-dried to produce a 90% concentration of dried algal mass containing 55% protein. Their analysis of the cost in rounded up figures shows that investment capital costs would be as follows:

Construction of ponds	US$ 1 300 000
Supplementary nutrients	510 000
Harvesting devices	840 000
Stirring systems	415 000
Drying facilities	980 000
Total	4 045 000

Production costs would be as follows:

Maintenance	US$ 92 000
Fuel and energy	60 000
Nutrients and chemicals	220 000
Labour	94 000
Water	19 000
Total	485 000

Assuming capital recovery at 10% interest and total cost at 10% interest, cost of production would be about $1300 per ton.

In a different analysis, Melamed of the Tushia Engineering Co., Tel Aviv[16] indicated definitely a lower cost of production, based on a hypothetical 100 ha unit. He suggested constructing shallow raceways covered with a 0.25 mm black polyethylene layer on a 5 cm bed of thick sand, would cost $2.7 million for 100 ha, together with central facilities for nutrition, harvesting and processing, as well as circulation devices and pumps. The capital outlay for 100 ha would be some $5.1 million. The annual fixed cost according to Melamed would amount to $1.7 million and the variable cost per ton of produce is $220. According to this analysis, at an annual yield of 80 tons per ha of algal dry product, the cost of production would be only about $450 per ton.

Preliminary analysis by our own group in Sede Boqer indicates that whereas the Melamed estimate may be low, Tahal's estimate is on the high side. It may be correct for small-scale production at the present level of ignorance, but cannot be used as a basis to estimate the future economics of algal biomass production. We suggest that production costs could be sharply curtailed when the following goals have been attained:

◇ Ponds constructed with inexpensive lining resisting breakdown for some 10 years.
◇ Natural energies exploited to produce turbulence and heat for drying.
◇ Yield increased to 10 kg dry algal mass per m^2 per year.
◇ Algal slurry used as local fodder without dehydrating and drying.
◇ The pH in the pond maintained at 11.00 rather than 9.00.
◇ Four, rather than eight, pond refills annually.
◇ Pond system not smaller than 100 ha.

These calculations for the cost of algal biomass indicate that research now becomes worthwhile. In arid zones the cost of protein grown in rain-rich regions is certain to increase with time, and production of protein from algal biomass is one of the very few alternatives which can help to make desert areas significantly more life-supporting.

MARICULTURE FOR THE PRODUCTION OF NATURAL PRODUCTS AND FEED

At present, four major products are derived commercially from marine algae: agar, carrageenin, alginic acid and diatomite. The first three products are extracted from marine *Rhodophyta* and *Phaeophyta*, and the fourth from either marine or freshwater diatom deposits.[21] Agar, carrageenin and alginic acid derivatives are products of different commercial usage, their demand and prices showing a significant upsurge. The following summary gives some basic information.

Alginic Acid Derivatives

These may be extracted from various species, such as *Phaeophyta*, *Laminaria*, *Ascophyllum*, *Macrocystis*, *Ecklonia*, *Lessonai* and *Durvillea*. The alginic acid occurs in the middle lamella and primary walls of the cells, while cellulose is present in the secondary walls. There is considerable variation in the alginic acid content of the cell walls, ranging from 14–40% of the dry weight,[22] high in winter and low in summer. The structure of the insoluble alginic acid and the soluble sodium alginate is very similar to that of cellulose and pectic acid, being essentially a long unbranched chain of mannuronic acid units joined by 1:4 glycosidic linkages.

The use of alginates in industry is based on their chemical and physical properties. They are non-toxic, highly viscous and form gels readily, thus finding use as thickeners in the food industry, filling creams in cosmetics, hand creams in the textile industry, as printing pastes and in the rubber industry for latex production. They are also useful as emulsifiers, and gelling agents in confectionary and meat jellies, as well as dental impression powders, as surface films in the paper industry and as glazes on ceramics. Alginate films may be made into sausage casings and alginate fibres into temporary textile yarns and medical gauze, haemostatic and slowly absorbed into the body.[22]

The harvesting of the weed depends upon the genera used and their habitat. The only one harvested regularly in California is *Macrocystis*. It grows attached to the sea bed by means of huge anchors, bearing stipes and fronds trailing up to the surface, and these may be cut by means of a mechanical harvester attached to the front of a harvesting barge. After the fronds are cut they regenerate and a growth rate of 3.0–4.5 m per week is reported; this is said to represent a greater rate of growth than that of any other land or sea plant so far investigated. The harvested weed is desalted, shredded and digested with soda ash, yielding a crude mass of sodium alginate. Further purification involves filtering, bleaching, conversion to calcium alginate and finally conversion to alginic acid by addition of hydrochloric acid. In this form it may be stored or converted into sodium or other salts.

Carrageenin

This is a cell wall polysaccharide complex of D-galactose-3, 6-anhydro-D-galactose and monoesterified sulphuric acid. Carrageenins are sulphated, linear galactans extracted from marine algae of the *Rhodophyceae sp.*, red seaweeds, and principally from selected species of the genera *Chondrus*, *Gigartina*, *Eucheuma*, *Iridaea* and *Hyphea*.[23] Extraction in hot aqueous alkali is performed for several hours. The products are recovered by filtering the insoluble solids, thereafter obtaining the polymer either by solvent precipitation or by drying on steam heated rolls.[24]

Three major components are recognized; kappa carrageenin (D-galactose-4-sulphate-3, 6-anhydro-D-galactose) and lambda and iota carrageenins, which are also sulphated D-galactose. In the presence of potassium, these compounds gel and are used like alginates to stabilize emulsions and suspend solids in the food, textile, pharmaceutical, leather and brewing industries.[24] According to Doty[23] the modern food industry requires 10 to 20 times more carrageenin than the wild seaweed crop provides. It is thought that up to 36 tons per ha per year represents a practical goal to strive for.

Agar

Agar is the name of a dried or gel-like non-nitrogenous extract from *Rhodophycean* algae. It is used as a medium in the culture of bacteria, fungi and algae and also in numerous industrial processes. Dry agar is insoluble in cold water, but soluble in hot water. A dilute solution of 1–2% remains liquid down to a temperature varying between 35 and 58 °C, according to source and purity. The gel may then be melted at 90–100 °C.[22]

Until 1939 Japan was the largest producer of agar, which was probably manufactured from a species of *Gelidium*. However, *Campylaephora*, *Pterocladia* and *Gracilaria*, which are found in the intertidal zones, were also used.

The dried and washed agarophyte – a term used for plants from which agar may be obtained – is extracted under pressure, clarified, filtered and allowed to gel. After freezing it at −10 °C the gel is then thawed and the water removed together with the soluble impurities. The purified agar is dried into flakes and bleached. It was shown that agar from *Gelidium cartilagineum* is a chain of alternating D-galactose and 3, 6-anhydro-L-galactose residues with a half ester sulphate on about every tenth unit of the galactose.[22]

Agar is used in the food, pharmaceutical, cosmetic, leather, textile and other industries in much the same way as carrageenin and alginates. Here is a worthwhile challenge – to develop the biotechnology for intensive mass cultivation of specific seaweeds.

Algae in Marine Food Chains

Ryther *et al.*[25] and Goldman *et al.*[26] suggested a marine counterpart for the dual purpose system of algae grown on sewage for control of pollution and production of single cell protein. They brought evidence to indicate that algae grown on a mixture of 50% waste water and 50% sea water have a distinct advantage over fresh water algal cultures. The problem of removal of the algal mass was solved by utilizing the algal cells as a first link in marine food chains leading to integrated aquaculture systems. In continuous flow cultures algal yields were relatively constant when grown in dilution rates ranging from 0.5 to 1.0, but there was a distinct difference between the yield obtained in Woods Hole, Massachusetts and Fort Pierce, Florida; 12.7 g dry weight m^{-2} per day in the former and 19.0 g m^{-2} per day in the latter.

Marine Algae for Pharmaceutical Purposes

There appears to be a growing interest in searching for biologically active compounds from marine organisms. In red and brown algae, halogenated sequi- and diterpenes, bromoacetylenes, bromobenzene derivatives and steroids have been described.[27]

In southern China, the *Rhodophycean* alga *Digenia simplex* is made into a drug and is of some importance as an anthelmintic.[27] According to a survey by Mitsui,[28] laminine – which is obtained from the brown alga *Laminaria* – was reported to have antihypertensive activity. Sodium alginate was reported to prevent gastroenteric absorption of radioactive strontium; sodium laminarin sulphate as well as fucidin could be useful also as blood coagulants. Finally, there have been numerous reports on some bacteriocidal products from various algae.[28] Although so far no substance of wide acceptance for commercial use in medicine or veterinary science has been firmly established, further surveys and the intensive study of marine algae may well lead to the discovery of useful chemicals.

SOLAR ENERGY FOR THE PRODUCTION OF BIOFUELS

Hydrogen Production

In recent years the feasibility of using photosynthesis to generate hydrogen as a source of energy has received much attention. It can readily be seen that for energy purposes hydrogen is a much more desirable product than the various organic fuels produced from natural processes.

The idea is based on the observation that under certain light conditions some algae produce small quantities of hydrogen which serve to 'prime the pump' of their normal photosynthesis. Under normal conditions the two products of the process would be hydrogen and oxygen, but as the production of oxygen severely limits hydrogen production it is necessary to prevent oxygen production.

In essence, the idea of harnessing solar energy to produce useful energy through photobiologically produced hydrogen requires that a natural process, which occurs only to a very limited extent, become the major or even sole process of photosynthesis. As photohydrogen production in current biological systems occurs at an extremely low rate, the biosolar conversion of water to hydrogen gas would obviously need to proceed at rates comparable with the normal rates of photosynthesis. The current understanding of the mechanics of photohydrogen production by plants is contrary to the idea that this process might have an economic function. Even when viewed optimistically it is clear that a commercially feasible production of hydrogen through photosynthesis awaits a large and extensive study, and it is even doubtful whether the economic outlook would be significantly brighter by the year 2000.

Methane Production

Another biosolar energy conversion system, originally devised and much explored by Oswald,[11] entails the growth of indigenous algae on waste in outdoor ponds, followed by separation and fermentation of the algal mass to produce methane. The residual nutrients such as nitrogen and phosphorus could be recycled to produce more algae, while methane could be used directly or burned to produce electricity. If the latter process took place, the carbon dioxide produced could be reused and, in addition, the rejected thermal energy could accelerate biological processes in the ponds. However, the electrical energy yields would be only about 20% of the energy originally present in the algae.

In the final analysis, algal dry matter has a combustion heat of 5.5 kcal kg^{-1}, which may yield methane, having an energy value of 1.8 kWh. Today, the process would be economically feasible in the USA if algae could be produced for about \$20 per ton, an objective which seems rather remote. Nevertheless, Keenan[29] studied the bioconversion of solar energy to methane with the algae *Anaboena flosaque* and concluded that such renewable methane may provide a substantial portion of the fuel requirements for power stations, the overall efficiency of the conversion of solar energy to methane being 1%. He estimated that the energy inputs required for the growth and the harvesting of the algal mass would account for only 12% of the fuel value of the methane produced from the algae.

ALGAE AS A SOURCE OF SPECIFIC PRODUCTS

Glycerol Production

Dunaliella sp. are green, unicellular halophilic algae which produce oxygen photosynthetically only in the presence of high osmolar concentrations but without specific requirements for sodium chloride. This alga was found to provide glycerol as a major product of photosynthesis and to accumulate very large amounts of it intracellularly. Ben-Amotz[30] measured the intracellular volume of *Dunaliella parva* and showed that the internal concentration of glycerol was iso-osmotic, around 2.1 M, with medium concentration of 1.5 M NaCl. The extracellular salt concentration of the algal suspension had an immediate effect on the concentration of glycerol in the cells.

These characteristics make *Dunaliella* an attractive subject for biosolar energy conversion in many arid zones, as highly saline water may be readily obtainable thus removing water as a limiting factor for bioproduction. An additional advantage of this alga is that the residue remaining after extraction of glycerol may be used as a proteinaceous feed and for the extraction of valuable chemicals such as β-carotene. An important aspect of *Dunaliella* mass culture is its overall yield when it is grown in a highly saline medium, so conducive to glycerol production. If it could be within an order of magnitude of other algae grown under conditions of lower salinity, glycerol production from *Dunaliella* could have promising commercial prospects.

Starch Production

Many substances of nutritious and industrial value could be produced abundantly by algae. An interesting example was provided in a recent study by Pirt and Pirt[31] with a high temperature strain of *Chlorella*. At low dilution rates starch content reached approximately 50% of the total dry weight of the cells. They found a very high rate of starch production, reaching 50 mg of starch per 1 gh^{-1} of algal biomass, less starch.

Biochemicals and Natural Products

The utilization of blue-green algae for the production of biochemicals and pharmaceuticals offers a special advantage as cell walls are absent. Laceration of the cells, necessary in the course of the extraction process, is easy and effective. Components in the electron transfer system, such as ferredoxin, ferredoxin-NADP reductase, plastocyanine, cytochrome C or components of the mechanism of carbon fixation such as the enzymes RuDP carboxylase or fructose diphosphate aldolase, are only some examples of soluble and easily extractable enzymes.

E. Tel-Or of our Algology Laboratory at the Institute for Desert Research at Sede Boqer summarized the potential for high cost products from non-marine micro-algae, as follows:

Pigments: Carotenes and xanthophylls may be particularly useful. β-carotene, which is also provitamin A, is attractive for the food industry as a food dye. *Chlorella pyrenoidosa* may contain upto $\frac{1}{2}$% of its dry weight in carotenes, while *Scenedesmus obliquus* may contain up to 1% of xanthophylls, which may also have possible uses in the food industry.

Steroids: One interesting type is ergosterol which is pro-vitamin D_2 and another is chondrilasterol, a major steroid in *Scenedesmus* which may contain up to 0.35% of this steroid. Poriferasterol is a major steroid in *Ochromonas danica*, containing the highest amount of steroids known in algae, namely 1% of its dry weight. The separation of the steroids is rather costly and it seems that the most promising have an unsaturated sidechain. However, little information is yet available on the economic feasibility of steroid production from algae.

Lipids: These form another interesting group that seem worthwhile exploring. For example, a few species of algae have a rather high content of arachidonic acid, potentially useful for the synthesis of prostaglandins.

Vitamins: Vitamins, and especially vitamin E, alpha-tocopherol, have a commercial demand. *Euglena sp.* is particularly high in this vitamin – up to 0.1% of their dry weight.

Special organic products: Mannitol or sorbital accumulate as storage material in certain algae.

When considering algae production for the purpose of special products it seems clear that by genetic manipulation many new species could be produced in which the synthesis of particular organic molecules is significantly accentuated, thus widening the potential of algae to produce useful products.

Special Health Foods

According to the Japan Food Research Laboratories,[32] *Chlorella* powder has some remarkable properties as a health food, and it is believed that, as a baby food, *Chlorella* promotes growth in infants and children. For adults, *Chlorella* invigorates and activates exhausted humans in old age, slowing down aging and eliminating chronic fatigue. The same report claims *Chlorella* powder to possess various therapeutic qualities, such as curing duodenal and gastric ulcers, as well as various kinds of wounds; preventing and curing liver necrosis; regeneration of vascular elements, regulating blood pressure and preventing a decrease in leucocytes.

PRODUCTION OF MICROALGAE TODAY

In the commercial production of algae for food and feed, important advances have been made by producers in south-east Asia and at the *Spirulina* production facility of Sosa Texcoco, SA near Mexico City, where naturally grown *Spirulina* is separated from an alkaline lake. Most of the production factories are in Taiwan and are concerned with *Chlorella* which is usually grown mixotrophically, that is, the algae consume organic carbon compounds, usually acetic acid, and also utilize carbon dioxide photosynthetically during daylight.

In Taiwan more than a thousand tons of *Chlorella* were produced under intensively controlled conditions in 1977 and used primarily as a health food. The cost of *Chlorella* production in these factories is about $8–10 per kg, thus restricting the distribution of this type of food and method of production to a relatively small scale. However, there is growing recognition of the potential for mass algal culture as a crop to augment bioproduction in arid zones. One reason obviously relates to the possibility of growing algae in brackish or even sea water. Another reason is the very high ratio of protein produced per unit volume of water, so typical of many algal strains. Up to 2.5 tons of protein a year can be produced per 1000 m^3 of water, more by about one order of magnitude than can be obtained with conventional crops. Thus, interest in algae production is increasing in India, where the immediate objective is to develop a method of algae production and of processing at the village level for cattle and poultry feed. Additionally, in Peru, augmentation of food production in general and of protein in particular is an important national goal.

Both the USA and Belgium have shown an interest in algae as a special feed for larvae of molluscs of commercial importance such as oysters and clams for example, or as food for rotifers which in turn are used for feeding fish larvae. In California, a commercial outdoors plant is being set up to produce *Spirulina* as a health food. In southern European countries, such as France, Spain, Italy and Bulgaria, research groups exist concerned with the production of algal biomass for animal feed.

SUMMARY

We have described here the prospects for exploiting algae for a wide variety of industrial purposes. We believe that although at present uneconomic for protein production, the future potential of this method is very promising for special products such as protein-rich foods. Perhaps this reflects a revolution in plant production using species and systems which differ greatly from those accepted in agriculture today. It seems appropriate to compare the present situation in algaculture with that thousands of years ago at the beginning of agriculture, when the cultivated plant species and the cultivation methods used reflected only a preliminary selection by man. From simple adaptations of natural habitats evolved the new strains and sophisticated production methods now available.

Thus, we believe that from a consistent effort towards optimization of algal biomass production, harvesting and product processing, as well as genetic engineering and basic research for a better understanding of the relevant biosynthetic pathways will come, step by step, biotechnologies of significant economic importance. This would seem particularly useful in arid lands in which populations are malnourished and cultivation of plants by conventional method is severely handicapped. There algaculture in saline water enjoys distinct relative advantages.

LITERATURE CITED

1. H. W. Milner, Algae as food. *Sci. Am.* **189**, 4 (1953).
2. John S. Burlew (Editor), *Algal Culture Publication 600*, Carnegie Institution of Washington DC (1953).
3. A. Issar, The Institute for Desert Research, Sede Boqer, Israel, personal communication.
4. R. Williams. Protein synthesis in the world. *Chem. Technol.* January, 1975.
5. *International Action to Avert the Impending Protein Crisis.* United Nations Publication. Sales No. E 68 XIII 2 (1967).
6. J. A. Bassham. Increasing crop production through more controlled photosynthesis. *Science* **197**, 630–638 (1977).
7. R. D. Fox. A solar energized system for algaculture. Laboratoire de la Roquette, 34 St Bauzille de Putois, France.
8. J. K. Bhattacharjee. Micro-organisms as potential sources of food. *Adv. Appl. Microbiol.* **13**, 139–161 (1970).
9. Durant Chastell and Sanches, Sosa Texcoco SA, Mexico. FAO – Technical Conference on Aquaculture. Kyoto, Japan (1976). The Tecuitlatl (*Spirulina*) and the Aquaculture.
10. W. Pabst, Die Massenkultur von Mikroalgen. *Kraftfutter* **58**, 2 (1975).
11. W. J. Oswald, Productivity of Algae in sewage disposal. *Sol. Energy* **15**, 107–117 (1973).
12. G. Shelef, C. G. Colueke, W. J. Oswald and H. K. Gee, A study of fundamental factors pertinent to microbiological waste conversion in control of isolated environments. SERL report No 66–8. Berkeley Sanitary Engineering Research Laboratory, University of California (1966).
13. A. Pano, J. Berend, A. Simovitch and A. Ollian, Feasibility study of algal protein production, TAHAL Consulting Engineers Ltd, Tel Aviv (1978).
14. S. Yanai, S. Mokady and Z. Berk, Algae grown on waste water as a source of protein: Toxicological and Nutritional aspects. Proceedings of the 11th FEBS Meeting Copenhagen (1977).
15. Gesellschaft für Strahlenforschung Abteilung für Algenforschung und Algentechnologie. Dortmund, (1977).
16. A. Melamed, 'Tushia' Engineering Co., Tel Aviv, (Personal communication.)
17. Einsatz von Abwasser Algensystemen für kombinierte Wasserrückgewinnung und Proteinerzeugung Gesellschaft für Strahlenforschung M. & H. Dortmund. FE Nr. 77661 (1976).
18. Caldwall Connel Engineers, Melbourne. Field testing an algae harvesting process. Report submitted to Bureau of Environmental Studies, Department of Environment, Australian Government, Canberra (June 1975).
19. J. Soeder, Zur Verwendung von Mikroalgen für Ernährungszwecke. *Naturwissenschaften* **63**, 131–138 (1976).
20. J. Soeder and H. Mohn, Technologische Aspecte der Mikroalgenkultur. Symposium Mikrobieller Proteingewinnung (1975).
21. F. E. Round, *The Biology of the Algae*, Edward Arnold, London (1973).
22. B. Guiseley, Seaweed colloids. *Kirk-Othmer Encyclopedia of Chemical Technology*, Vol. 17. No. 2, pp. 763–784.
23. M. S. Doty, Farming the red seaweed, *Eucheuma* for carrageenins. *Micronesica* **9**, 59–73 (1973).
24. P. R. Standel, Some Polymers from Marine Colloids, Inc. Rockland, USA (June 1976).
25. J. H. Ryther, W. M. Dunstan, K. R. Tenore and J. E. Huguenin, Controlled eutrophication—increasing food production from the sea by recycling human wastes. *Bioscience* **22** (1972).
26. J. C. Goldman, J. H. Ryther and L. D. Williams, Mass production of marine algae in outdoor cultures. *Nature* **254**, 594–595 (1975).
27. A. Pletscher, F. Hoffman-La Roche and Co. Ltd, Basel, Switzerland, in FEPA Kongress at Venice (1975).
28. A. Mutsui, The use of photosynthetic marine organisms in food and feed production. The International Conference on Bio-Saline Research. NSF, Washington, DC (1977).
29. J. D. Keenan, Bioconversion of solar energy to methane. *Energy* **2**, 365–373 (1977).
30. A. Ben-Amotz, Photosynthetic and Osmoregulation Mechanisms in the Halophilic Alga *Dunaliella parva*. PhD Thesis, Weizmann Institute of Science, Rehóvot, Israel (1973).
31. M. W. Pirt and S. J. Pirt, Photosynthetic production of biomass and starch by *Chlorella* in a chemostat culture. *J. Appl. Chem. Biotechnol.* **27**, 643–650 (1977).
32. Lab. No. 6020428-1 report issued by Japan Food Research Laboratories (1976).
33. Israel Scientific Research Conferences, Production and use of micro-algal biomass, Acre, (September 1978). National Council for Research and Development, Jerusalem (1978).

The manuscript was received 20 January 1979

Theory and Methods of Environmental Values Research

PROFESSOR RICHARD N. L. ANDREWS AND MARY JO WAITS
School of Natural Resources, University of Michigan, Ann Arbor, Michigan, USA

Environmental value claims have become major issues in recent public decisions, particularly in such cases as large-scale construction projects, natural resource management plans, and even scientific and technological research and development programs. This paper proposes a theoretical framework for the investigation of environmental value claims, and reviews the methods currently available, or under development, that might be used for that purpose.

Environmental value claims have provoked widespread debate and controversy during the past decade. They have been invoked with particular force against actions proposed by government agencies, causing both uncertainty and anguish to administrators who had thought that they knew how to achieve their missions. The question of environmental values has also stimulated activity in the research community, leading to promising lines of inquiry but also to important gaps and unanswered questions. The purpose of this paper is to propose a theoretical framework for investigation of such value claims and to review the methods that might be useful for that purpose.

ENVIRONMENTAL VALUE CLAIMS

Unlike trees or people, values are not mental objects that can be listed and counted; a value is, rather, a statement of a relationship, an estimation of worth of some object to an individual or in a particular situation. The subject may be a human individual, but need not: we speak also of the value of peace to human societies and of sunlight to plants. Likewise the object may be either an empirical or a perceived condition: I value trees in a city but I also value what I perceive to be beauty and freedom. It is the relationships that are valued – of means to an end, of resource to beneficiary, of environmental condition to organism – not some *a priori* list of objects called values. We shall follow common practice in continuing to use the word as a noun, but with the understanding that ultimately it refers back to a relationship rather than to an object.

The value relationship may take three forms, and the distinction of these forms explains the principal differences in usage of the concept. These forms are preference, obligation and function. *Preference* is a relationship of individual desire: I like (prefer, value, . . .) one ecosystem more than another, solitude more than crowds, or fishing more than water skiing. *Obligation* is a relationship of social norms: citizens in a democracy are expected to honor (value) the norms of picking up their own litter, not contaminating water supplies, respecting others' rights, not destroying public or common property, and adhering to legal processes for making decisions, even if these norms conflict with their individual preferences. Some would refer to these sorts of values as moral imperatives or absolutes. *Function* is a relationship of usefulness or service or system maintenance: stratospheric ozone is valuable because it protects biological life from excesses of temperature and radiation and fertile soil is valuable to a society because it provides a source of food supply. Such functions represent value relationships whether or not they are reflected in social norms or individual preferences.

The distinctions among these meanings are fundamental to understanding environmental values, and to developing methods for considering them in administrative decisions.

INDICATORS OF VALUES

A paramount question in environmental values research is how to identify and characterize the environmental values that might be affected by a proposed action. Valid representation and measurement of such values is the first prerequisite of any appropriate consideration of them. It is essential, therefore, that choices of indicators and methods be made consciously and knowledgeably.

If values are relationships rather than objects, they cannot be measured directly but must be characterized through empirical indicators. The choice of indicators and the inferences drawn from them, however, are the source of many of the most important unresolved questions in this field of inquiry. In many studies, environmental values have

CCC-0308-0188/80/0005-0071$04.00/1

been represented by monetary and other quantitative indicators, but there has been little analysis of the validity and appropriateness of these indicators; nor has there been much dispassionate analysis of intangible value claims that have been made in many environmental controversies. It is likely that far more satisfactory representation of environmental values could be developed if more systematic and comparative study were devoted to this task.

All identifiable values can be considered at least at the nominal level, although some of the deepest conflicts arise at this level in deciding how each value should be classified. For instance, swamp and barren desert conjure up oppositely valued images from wetland or arid ecosystem, even as classifications. Other levels introduce both the advantages and the hazards of greater quantitative precision. Ordinal scales permit ranking of environmental values, but only if such values are in fact independent and transitive, and fall along a definable continuum – conditions that may not hold, especially for aesthetic preferences or choices between different types of ecosystems.

Some ecosystems or vistas may simply be different, not orderable as better or worse, or more or less valuable; or their orders may be reversed depending on the criterion of ranking. Interval and ratio scales permit mathematical evaluation of tradeoffs, but only if agreement is assumed concerning units of measurement – size or distance, dollars, or whatever – that validly express the values in question. In general, interval and ratio scales have the advantage of permitting more powerful quantitative analysis, but are invalid and misleading unless those scales are appropriate to the underlying phenomena.

The essential research question, therefore, is to determine what environmental values can validly be expressed in an interval or ratio scale, and if so, which; and what values cannot be carried beyond the nominal or the ordinal, and how best to incorporate these into decisions. The issue is not whether or not they can be quantified at all, but what methods and indicators most appropriately characterize each. It is useful therefore to review the principal approaches that have been used to characterize each of the three kinds of value relationships – preferences, norms and functions – and to discuss the advantages and limitations of each. The purpose of this review is to provide both an introduction to this field of inquiry, and a rudimentary structure for organizing it and identifying further research needs.

PREFERENCES

Preferences are the most extensively researched of our three categories, perhaps because they follow most directly from the central philosophical doctrines that underlie much of modern western thought – rationalism, individualism, majoritarian democracy and utilitarianism. In law and politics as well as economics, it is conventionally assumed that the individual's autonomous preferences should be preeminent, whether measured through the ballot box or in the market place, or, more recently, through public hearings and opinion surveys.

Preferences are subjective, and they are held by individuals. Since they are subjective, they are known initially only to the individual (if even to him), and cannot be directly observed; they must therefore be inferred, either from statements or from other elements of his behavior. Since they are held by individuals, moreover, they can provide criteria for government actions only if they are somehow aggregated; yet the process of aggregation itself introduces serious difficulties of both theory and practice. Preferences are held with differing degrees of intensity, even by the same individual; they depend on the range of alternatives available, and on the order in which alternatives are presented. They lead directly to 'tragedy of the commons' outcomes if the aggregation is not somehow constrained by the capacity of the environment to satisfy them.

VERBAL INDICATORS

Procedures for identifying preferences through verbal expressions normally involve direct solicitation of individual responses. The responses may be in various forms, such as expressions of willingness to pay, direct statements of preference, social and environmental indications, and expressions of judgment.

Willingness to Pay

Two questioning approaches are commonly used to identify willingness to pay. Both assume that price

RICHARD N. L. ANDREWS is Associate Professor and Chairman of the Resource Policy and Management Program at the University of Michigan; he holds a joint appointment in the School of Natural Resources and the Urban and Regional Planning Program. He has written extensively on environmental policy and planning and has been employed as a staff member or consultant by several government agencies.

MARY JO WAITS holds a Master's degree in Public Administration, and has worked on the professional staff of two city and one state governments, two legislators, and an independent research institution. She is now a coastal zone planner for the State of Alaska.

Address: Resource Policy and Management Program, School of Natural Resources, University of Michigan, Ann Arbor, Michigan 48109, USA indicates value and that individuals can be persuaded to reveal their true value preferences. The most direct approach is interviewing, asking each individual the dollar magnitude or marginal change in cost he is willing to pay for goods or a service. Recreational preference studies, for instance, commonly ask individuals 'at what admission price would you stop using this area for recreation?' A second but related technique is to conduct 'contingent games', such as bidding games or auctions, in which individuals are asked to state what their preferences would be in the face of a specified contingency. This approach has been used to identify a demand curve for a recreational area, to determine the demand or preference for public television, and, more recently, to measure aesthetic preferences and to develop dollar values for environmental quality.[1]

Expressions of Preference

Environmental preference studies seek expressions of 'wholly personal, subjective appreciation of (or repugnance for) specific landscapes'.[2] A common method is to present photographic, video or other graphic samples of natural and built environments to individuals, and have them rate the scenes using ordinal scales, paired comparison, or semantic differential scales. Kaplan *et al.*[3] used color slides to elicit respondents' preferences among methods of storm drain landscaping. Daniel *et al.*[4] presented slides of differentially treated forest areas (for example, clear cut, uniform strip cut, conventional log and others) in order to determine respondents' perceived aesthetic value and preferences among the different vegetative treatments. Another method is to ask individuals to identify from memory the features of an environment that they consider significant, desirable, or even simply noticeable.[5]

Yet another approach to preference identification uses survey research to identify people's ordinal preference rankings either among environmental concerns or of environmental concerns among other issues such as health, police protection, welfare and education. Dillman and Christensen, for instance, solicited rankings of public preferences for government expenditures among a series of programmatic areas including pollution control; Miller used a similar methodology to study the ordering of interests in environmental quality among urban leaders; and Frederickson and Magnas solicited people's rankings of the relative importance of water pollution among public policy problems.[6]

Finally, Delbecq and Van de Ven have developed an approach which they call 'nominal group process' for soliciting public concerns and preferences in the context of public hearings and workshops on proposed actions.[7] In this procedure, people are called together in a group as for a public hearing or workshop, but are then asked to record their preferences independently before discussion, both for agency use and to ensure that all concerns are documented, heard and discussed. The results almost certainly reflect the preferences of unrepresentative populations, because of the self-selection bias inherent in public hearings; however, the method is at least a possible improvement on the traditional public hearing.

Social and Environmental Indicators

Preferences vary over time, and may change with such factors as age, income, social class and lifestyle. Likewise, preferences may vary from generation to generation. Some work in the area of social and environmental indicators, therefore, is now directed toward monitoring the dynamics of environmental preferences in human groups. Recent baseline investigations have sought to establish a series of measurements bearing on social well-being and environmental quality, which can be monitored at more or less regular intervals and used to chart changes in social and environmental conditions over the years.[8]

Expressions of Judgment

In contrast to the methods discussed so far, so-called 'judgment capture' methods assume that people generally do not know their preferences in advance and need help clarifying them in specific contexts. Techniques to help with these difficulties have been developed primarily in the areas of decision analysis and futures forecasting. Hammond *et al.*,[9] for instance, asked people a series of questions involving various choices and used regression analysis to analyze the weights individuals assign to each issue in a situation. The individual's judgments were then compared with others' and the results fed back to them as a basis for clarification of issues and better resolution of conflicts.

Decision analysis is also used to aid in the selection of a preferred judgment from a spectrum of alternatives. Its procedures include techniques for systematically describing and organizing problems, techniques for recording preferences, and techniques for estimating likelihood. Common methods for organizing the problem are decision-flow diagrams or decision trees. Preferences are then solicited and modeled, usually by such methods as bidding and asking games,[1] the construction of multi-attribute utility functions[10,11] and 'policy capture' techniques.[9,12] Finally, probability estimates are generated of the magnitude and likelihood for potential outcomes of decision alternatives, and the optimal solution – that is, the alternative with the highest expected utility – is calculated.

Third, opinions of professionals rather than surveys of the general public are often relied upon for judgments about preferences and probabilities. There are techniques in the area of futures forecasting – Delphi methods, cross-impact analysis and relevance trees – that gather, as systematically as possible, judgments and opinions about the future from several experts or groups of people.[13] With these approaches experts are selected, assembled, and queried independently. The responses of the others are then fed back to the participants so that each has an opportunity to revise his predictions on the basis of his knowledge of others' opinions. What emerges is a consensus of expert opinion concerning predictions of preferences or probabilities for future events.

Problems with Verbal Indicators

Preference statements may be inaccurate for many reasons, which can be quite difficult to correct or even to detect. Most obviously, people may be affected by the choice of questioning procedure, or led by value-laden language in the survey instrument, or confused by instructions – all common problems of survey research. More subtle are problems of strategic bias: if a person wants a park, for instance, why should he not exaggerate his stated willingness to pay for it, especially if he knows that a favorable result will be used not to charge him that price, but to tax everyone for it? Finally, the very asking of the questions may create artifact answers – speculative preferences that are expressed only because a question was asked about them when the respondent may not have held or even have thought about them beforehand.

Preference statements may also be based on erroneous assumptions about the implications of these preferences. For instance, people often ignore questions of the reliability of data; they tend to assume that even small samples of data are representative of the populations from which they are drawn; and they tend to be overconfident about the quality of inferences that they can draw from uncertain information.[1] More generally, people's expressed preferences are highly dependent upon their perceptions, information and experience. A real estate developer may perceive farmland quite differently from the farmer and his preferences differ accordingly. A city resident cannot be expected to care much about wilderness if he has not experienced it, nor a rural resident about the city – although people who have such experiences may then learn corresponding changes in their preferences.

Finally, people's verbal expressions of preferences may fail for several reasons to correspond to their behavior in real situations. Verbal expressions are frequently responses to hypothetical questions, which may not capture either the categories or the cues by which people perceive and act in real situations. A person may express a willingness to pay if asked, but in a real situation may not perceive choices in economic terms, let alone decide to act on economic grounds. More generally, preferences are relative rather than absolute and express value orderings only within particular contexts. This circumstance causes some problems with all indicators of preferences, but it causes especially difficult problems with verbal indicators as any questioning procedure must necessarily abstract a limited number of characteristics from reality and convert them into a hypothetical case; it is likely to neglect potentially significant details.

NONVERBAL INDICATORS

The inference of preferences from nonverbal behavior is a second related approach, but not identical to the first. In this approach, people's behavior is treated as the empirical manifestation of their preferences, on the assumption that what people value is reflected in what they do, else why would they do that rather than something else? The strength of nonverbal measures is that they reveal what people are willing to act upon: to do, to use, to pay for, to vote for, to organize against, to relocate near or away from, to spend time on, to pass laws about, and so forth. As empirical observations, these indicators avoid some of the biases of strategy and of hypothetical choice that may color preference statements. The observations may be direct, such as observations of spontaneous behavior; but more often they are indirect through the use of data and trend analysis, prices, social profiles and case studies.[14]

Consumers' Market Behavior

Observation of market behavior is a common method used by economists for identifying preferences. The property value approach to measurement of environmental quality is an example of this general method.[1,15] In this approach it is assumed that an individual, free to choose the site of his home, will try to balance the cost of land or property against the utility – transportation costs, amenities, and in this case environmental quality – of different sites. Therefore, the differentials in costs or property values between the chosen and the nonchosen sites should reflect the individual's marginal willingness to pay for environmental quality improvements. The valuation of environmental quality is thus inferred from individual market decisions.

Two other economic approaches are the use of shadow pricing and of opportunity costs.[16] Shadow pricing involves estimating the demand function for a related private good, and inferring from it an estimate of the demand for the public good. Opportunity costs are dollar values assigned on the basis of opportunities lost because of commitment of resources elsewhere, providing an economic surrogate for the dollar value of the intangible resource.[17]

Time and Effort Budgets

These budgets may be based upon either observation or interviews, each inferring preferences from behavior patterns. Their assumption is that if an individual prefers it he is willing to expend time and effort and sacrifice other alternatives in order to achieve this. For example, if a person devotes considerable time and effort to wilderness backpacking, we may conclude that for some reason he prefers it to other uses of that time and effort. Similar accounting may also indicate preferences among vacation sites, sections or views in a park, or particular discretionary activities.[18]

Observing people's political and organizational memberships is another approach for identifying the direction of people's interests, related but not identical to time and effort budgets. An individual associated with the Sierra Club, for instance, is generally assumed to have an interest in the activities of the club, since he contributes at least money and perhaps time or effort to the club's goals or purposes. Environmental preferences of individuals are thus indicated by their affiliation with organizations or political parties known to respect certain environmental preferences.[19]

Problems with Nonverbal Indicators

An important difficulty in using behavioral choices as indices of values lies in their ambiguity and even inconsistency. No single action or behavior unambiguously indicates anything about an individual's values. In the case of the backpacker, it might be the exploring, the socializing, the risk-taking, or the escape from daily pressures and stress, or it might be some more direct appreciation of a particular physical environment. Conversely, people do many different things in different roles, and rarely achieve consistent orderings of preference among them. They also change their behavior with age, lifestyle, experience, necessity and other circumstances. Only by placing the behaviors in a larger pattern of the individual's life and identifying the larger complexes of lifestyle preferences of which they are elements, can we get any useful information about the individual's preferences.

Second, individual behavior sometimes does not permit the achievement of one's preferences, since action by others would be required – as for instance in a commons situation, where the cumulative effects of many people's actions overwhelm the impact of any one individual's behavior change. Behavior also provides no evidence of what the preferences would be if contextual constraints were changed: if a mass transit system were built, if the person got a raise or a better job offer, or if his company adopted flexible work hours, his behavior might change rapidly from previously observed toward stated preferences.

In addition, any attempt to generalize preferences from an individual to the general public must be, at best, tenuous. If we know that individuals like to backpack to escape pressures, that risk-averse individuals accept low-risk jobs, or that individuals who like to travel may prefer living near airports, it might be possible to identify homogeneous subgroups; but even then it would be problematic to aggregate results from different subgroups to arrive at preferences for the larger population. Different methods of aggregating individuals' preferences yield widely divergent results, depending in part on the criteria used and in part on the order in which preferences are aggregated.

Most important, however, psychological research suggests that behavior shapes preferences as well as the converse, and that people thus develop their preferences retrospectively as they attempt to give meaning and justification to their experiences. If this is the case, the attempt to infer preferences from either verbal or nonverbal indications may have no valid predictive utility for decisions.

Administrators and researchers therefore must choose the behaviors from which they will infer preferences, knowing that any choice may remain open to legitimate dispute. More often, unfortunately, they simply adopt a convenient set of indicators – market behavior, voting behavior, organization membership, recreational use statistics, or whatever – without explicit recognition of their arbitrariness, limitations or biases.

NORMS

The study of norms begins with the society and the group, of which the individual is a member and from which he learns to prefer some things to others. The society is more than the sum of its members; it is a functioning system, longer-lived than its members, shaping their behavior far more than most of them shape it. Its main instrument for shaping them is its body of norms, or rules governing its members' behavior, which are enforced on each member by the collective social and sometimes physical pressure of the other members. Many norms change over time, and few are universally held; most are enforced by smaller groups and subcultures within the society. In all cases, however, value relationships are inferred from aggregate human behavior, in the sense not simply of summing up individual preferences but of identifying long-term needs, constraints and predictable patterns of the experiences of human groups and societies.

Two sorts of norms must be distinguished: norms about what should happen, or normative ideals, and norms that shape what does happen. In a sense this distinction parallels the distinction between preference statements and behavior as indicators of value, in that the former are more likely to be voiced than acted upon, while the latter are inferred from what people actually do.

Normative Ideals

Consider the holy, the good, the beautiful, the moral duty. These have long been considered normative ideals in human societies even though they are rarely fulfilled in action. They are by definition unlikely to be revealed by studies of majority behavior, and probably unlikely to emerge from a summation of individual preference statements as well. Normative ideals are expressed most often by poets, philosophers, artists and professionals rather than by popular majorities or utilitarian administrators.

One way of identifying environmental norms is thus to study the writings, philosophies and principles to which people appeal in advocating particular decisions. Consider Aldo Leopold's poetic accounting of *The Land Ethic*; Ian McHarg's philosophical basis for *Design with Nature*; Barry Commoner's writings on the ecological theory of conservation; or Christopher Stone's case for the existence of legal rights for natural objects.[20] They and others have advocated normative ideals concerning environmental conditions. The norms are not presented as statements of preference but as universal principles, which everyone ought to respect, regardless of what they prefer. Such norms are in fact used as justifications for decisions, and therefore deserve the attention of researchers as well. Such claims must be judged by reference to their outcomes rather than their origins; what is important, for example, is verification that the survival of human societies is really closely allied to the survival and good functioning of the ecosystem, or the verification that the quality of life is really maximized under certain environmental conditions like clean air, open spaces and access to recreation areas.

Norms in Practice

Whatever the ideals to which societies aspire, their actual norms – rules which in fact guide their members' behavior – are by definition observable in behavior. Some are formally recorded as laws, regulations and guidelines and can be studied as such. The Endangered Species Act, for example, prohibits destructive behavior toward threatened and endangered species; the National Historic Preservation Act protects historic sites against destruction by government actions; and the Clean Air Act regulates activities affecting air quality. The statutes may also indicate rewards and penalties for the specified behavior.

Lawsuits under these statutes provide additional insight concerning social norms. For which statutes do people consistently and persistently seek enforcement? Under what circumstances is a given behavior ordinarily approved or disapproved, and what behavior patterns are consistently favored in conflict resolution?

Several researchers have used situations of conflict to identify norms. Tribe *et al.* and Feiveson *et al.* identified norms at stake in the Tocks Island Dam controversy,[21] and Sayre *et al.* found norms in decision-making by the electric power industry.[22] Patton and Cormick have attempted the experimental use of environmental mediation to identify and negotiate environmental conflicts.[23] All these studies prove controversies to be useful in highlighting norms as well as preferences.

It is important, however, to recognize the bias of the analysis. Looking at situations of conflict may bias analysis toward active rather than underlying environmental norms: the focus is on those that are most obvious in people's behavior rather than on those that are taken for granted or are not presently threatened. Rarely do behavior conflicts reveal norms or preferences independent of immediate political issues.

FUNCTIONS

The study of ecological functions requires two distinct but related steps: the understanding of ecological systems, and the evaluation of the conditions under which the ecosystem and its functions will continue to serve human purposes. There are research approaches that include both processes, however, and representative examples are discussed below.

Environmental Planning Approaches

Three well-known approaches to environmental planning and resource analysis were described in a report by Belknap *et al.*[24] The approaches included those developed by G. Angus Hills, Philip H. Lewis Jr and Ian L. McHarg, each of which was intended to provide a systematic inventory and classification of environmental conditions of interest to people. Each also assigned nonmonetary indicators of value to the environmental conditions they described: Hills on the basis of a capability rating, which represented the natural capability of an area to support at the highest intensity possible any given land use or activity, and McHarg on the basis of inherent characteristics of the process, productivity, maintenance and potential hazards to human use.

These approaches dealt primarily with physical landscape characteristics and their inherent potential for development, and tended to underestimate the role of social, economic, technological and behavioral factors.[25] They also underestimated the complexity of environmental interrelationships: McHarg's ecological approach, for instance, emphasized life processes as a total ecological complex, yet his transparent overlays failed to capture ecological interactions (not to mention social and economic interactions), flows and changes over time in the interactions and cycles.

Carrying Capacity Analysis

The determination of carrying capacity is another procedure which considers the environment a support-system for interrelated human and natural activities. Derived from ecological concepts, the determination of carrying capacity requires identifying the limiting conditions and capabilities of the environment to support these activities without unacceptable changes in environmental quality. Measures of carrying capacity used in coastal regions, for instance, include natural characteristics, user perceptions and service system capacities.[26] Measures for wilderness recreational areas have included management objectives, visitor attitudes and recreational impact on physical resources.[27] In both the coastal and recreational studies, capacity was defined in terms of the natural environment, man-made systems such as water and sewer infrastructure, and human experience of the environment.

Functional Relationship Studies

Functional analyses of ecosystems attempt to specify the ways in which components of the ecosystems interact. These interrelationships are called functions or services of ecosystems and include the absorption and breakdown of pollutants, the cycling of nutrients, the binding of soil, the degradation of organic waste, the maintenance of the natural balance of gases in the air, the regulation of radiation balance and climate, as well as the fixation of solar energy.[28] It is these functions that maintain clean air, clean water, a productive Earth and a balance of creatures; these functions are essential to humans in that they provide food, fiber, energy and other material needs for survival. Nevertheless, evaluating the contributions of ecosystem functioning and processes to human welfare is a complex task.

Some studies have calculated replacement costs for ecosystem functions – for instance, estimating the costs of repairing or replacing damaged or cleared vegetation, the costs of dredging sediment from basins or the costs of treating aquatic wastes.[28,29] Another calculated the income-capitalized dollar value of an intact tidal marsh, computed on the basis of its energy flow.[17] Those who oppose putting dollar values on ecosystem services or non-resources have suggested that they be valued on the basis of natural art value, as a reverence for original landscapes, or on the basis of long-standing existence.[30] With this latter approach, the assumption is that long-standing existence in nature carries with it the unimpeachable right to continue in existence.

Several difficulties complicate the valuation of ecosystem functions. It is impossible to survey all the properties or interrelationships of any natural area or system and the dangers of overlooking important relationships are great. This is particularly evident in attempts to calculate replacement costs: rarely do we identify or repair all the damage to ecosystems, and because of the interconnected nature of the systems the evaluation of lost individual services or lost functions is inevitably misleading.[28] Furthermore, the relationships between the effects of human intervention and damage to ecosystems need considerably more research. Much damage is natural in origin, and many changes may be merely adaptations rather than damage as such.[31] It is also true that functions change. If, for example, a technology or new process such as a less expensive tertiary treatment of sewage, is discovered that can replace an ecosystem function, such as the absorption and breakdown of pollutants and the degradation of organic waste, then the functional importance of the ecosystem – at least, of that particular ecosystem in the present context – may decline proportionally.

CONCLUSIONS

This survey has ranged broadly across current areas of research that relate to indicators of environmental values, although we have reviewed here only the publications that bear most directly on our subject and not work in parallel fields that might be brought to bear by analogy. A striking characteristic of this literature is the insularity of much of what we have reviewed. Environmental values research at present is extraordinarily heterogeneous and fragmented. This is not altogether bad, in that it includes many approaches and much activity, but it does suffer from a lack of coherence and cumulation of learning.

Each investigator has looked at what he took to be environmental values through the lens of his own discipline, but it seems that none have stopped either to develop a unified theory or to validate their own methods and procedures against those of others. Future research agendas should therefore be interdisciplinary and consider the clarification of theory and validation as well as the synthesis of existing methodologies. This will be at least as important as the development of new approaches.

Acknowledgements

This review is based on a recently released research report by the authors entitled *Environmental Values in Public Decisions*, supported by the US National Science Foundation. All conclusions are the sole responsibility of the authors.

LITERATURE CITED

1. David S. Brookshire, Berryl Ives and William D. Schulze, The valuation of aesthetic preferences. *J. Environ. Econ.* **3**, 325–346 (1976); Electric Power Research Institute (EPRI), *Proceedings of a Workshop on the Measure of Intangible Environmental Impacts*, Sigma Research, Inc., Richland, Washington (1977); J. A. Sinden and A. C. Worrell, *Unpriced Values for Environmental Policy: Concepts, Measurement and Analysis*, Wiley–Interscience, New York (forthcoming).
2. Kenneth H. Craik and Ervin H. Zube, *Issues in Perceived Environmental Quality Research*, Institute for Man and Environment, University of Massachusetts, Amherst (1975); E. H. Zube, R. O. Brush and J. G. Fabos (eds.), *Landscape Assessment: Values, Perceptions and*

Resources, Dowden, Hutchinson and Ross, Stroudsburg, Pennsylvania (1975); Louise M. Arthur and Ron S. Boster, *Measuring Scenic Beauty: A Selected Annotated Bibliography*, Rocky Mountain Forest and Range Experiment Station, US Department of Agriculture Forest Service, Fort Collins, Colorado (1976).

3. S. Kaplan, R. Kaplan and J. S. Wendt, Rated preference and complexity for natural and urban visual material. *Percept. Psychophys.* **12**, 345–346 (1972); R. Kaplan, Preference and everyday nature: method and application. In *Perspectives on Environment and Behavior*, edited by S. Stokols, pp. 235–250. Plenum Press, New York (1976).
4. Terry C. Daniel, Lawrence Wheeler, Ron S. Boster and Paul R. Best Jr, Quantitative evaluation of landscapes: an application of signal detection analysis to forest management alternatives. *Man–Environ.* **3**, No. 5 (1973).
5. Richard M. Rozelle and James C. Baxter, Meaning and value in conceptualizing the city. *J. Am. Inst. Plann.* **38**, 116–122 (March (1972).
6. Don A. Dillman and James A. Christensen, The public value for pollution control. In *Social Behavior, Natural Resources and the Environment*, edited by William R. Burch, Jr, Neil H. Cheek, Jr, and Lee Taylor, pp. 237–256. Harper and Row, New York (1972); Delbert C. Miller, Power structure studies and environmental management. In *Environmental Quality Analysis: Theory and Method in the Social Sciences*, edited by Allen V. Kneese and Blair T. Bower. Johns Hopkins University Press, Baltimore (1972); H. George Frederickson and Howard Magnas, Comparing attitudes toward water pollution in Syracuse. *Water Resour. Res.* **4**, 877–889 (Fall 1968).
7. Andre L. Delbecq and Andrew H. Van de Ven, The effectiveness of nominal, Delphi, and interacting group decision-making processes. *Acad. of Manage. J.* (December 1974).
8. John P. Robinson and P. R. Shaver, *A Study of Resource Use in Urbanizing Watersheds*, US Army Corps of Engineers, Washington, DC (1970); Angus Campbell, P. E. Converse and W. P. Rodgers, *The Quality of American Life: Perceptions, Evaluations, and Satisfactions*, Russell Sage Foundation, New York (1976); L. W. Milbrath, *Indicators of environmental quality.* Draft prepared for UNESCO, Division for Socio–Economic Analysis, Sector of Social Sciences and their Applications (July 1976); Daniel B. Tunstall, *Developing indicators of environmental quality: the experience of the Council on Environmental Quality.* Draft paper presented at the Conference on Environmental Indicators–State of the Art, Science Center, Berlin (December 1976).
9. Kenneth R. Hammond, John Rohrbaugh, Jeryl Mumpower and Leonard Adelman, *Social Judgment Theory: Applications in Policy Formation*, University of Colorado Institute of Behavioral Science, Boulder, Colorado (1976).
10. Ralph L. Keeney and Howard Raiffa, *Decisions with Multiple Objectives: Preference and Value Tradeoffs*, John Wiley, New York (1976).
11. Howard Raiffa, *Decision Analysis: Introductory Lectures on Choices under Certainty*, Addison-Wesley, Reading, Massachusetts (1970).
12. Jean M. Johnson, Policy capturing in energy. In *Methodology of Social Impact Assessment*, edited by Kurt Finsterbusch and C. P. Wolf, pp. 341–346. Dowden, Hutchinson and Ross, Stroudsburg, Pennsylvania (1977).
13. Joseph F. Coates, Some methods and techniques for comprehensive impact assessment. *Technol. Forecast. Soc. Change* **6**, 341–357 (1974).
14. Kurt Finsterbusch and C. P. Wolf (eds.), *Methodology of Social Impact Assessment*, Dowden, Hutchinson and Ross, Stroudsburg, Pennsylvania (1977).
15. V. Kerry Smith, The treatment of intangibles in benefit–cost analysis. *Environ. Control Newslett.* **4**, No. 5 (7 May 1976).
16. R. N. McKean, The use of shadow prices. In *Problems in Public Expenditure Analysis*, edited by S. B. Chase, Jr. Brookings Institution, Washington, DC (1968); Nicholas H. Coomber and Asit K. Biswas, *Evaluation of Environmental Intangibles*, Genera Press, New York (1973).
17. Anthony C. Fisher and John V. Krutilla, Valuing long run ecological consequences and irreversibilities. *J. Environ. Econ. Manage.* **1**, 96–108 (1974); J. G. Gosselink, E. P. Odum and R. M. Pope, The value of the tidal marsh. Louisiana State University Center for Wetland Resources, Paper No. LSU-SG-74-03, Baton Rouge (1974).
18. Stuart F. Chapin, Jr, Free time activities and the equality of urban life. *J. Am. Inst. Plann.* **28**, 411–417 (November 1971); Gerald A. Gutenschwager, The time budget–activity systems perspective in urban research and planning. *J. Am. Inst. Plann.* **39**, 378–387 (November 1973).
19. Joseph Harry, Richard Gale and John Hendee, Conservation: an upper–middle class social movement. *J. Leisure Res.* **1**, 246–254 (Summer 1969).
20. Aldo Leopold, *A Sand County Almanac*, Sierra Club/Ballantine Books, New York (1966); Ian L. McHarg, *Design with Nature*, Natural History Press, Garden City, New York (1969); Barry Commoner, *The Closing Circle*, Knopf, New York (1972); Christopher D. Stone, *Should Trees Have Standing?* Avon Books, New York (1974); *American Values and Habitat: A Research Agenda*, working paper from the Research Agenda Project on American Values and Models of Human Habitation, American Association for the Advancement of Science, Washington, DC (1976).
21. L. H. Tribe, C. S. Schelling and J. Voss (eds.), *When Values Conflict: Essays on Environmental Analysis, Discourse, and Decision*, Ballinger, Cambridge, Massachusetts (1976); Harold A. Feiveson, Frank W. Sinden and Robert H. Socolow (eds.), *Boundaries of Analysis: An Inquiry into the Tocks Island Dam Controversy*, Ballinger, Cambridge, Massachusetts (1976).
22. Kenneth Sayre (ed.), *Values in the Electric Power Industry*, University of Notre Dame Press, South Bend, Indiana (1977).
23. Leah K. Patton and Gerald W. Cormick, Mediation and the NEPA process: the Interstate 90 experience. Paper presented at the Conference on Environmental Impact Analysis, University of Illinois, Urbana (May 1977).
24. Raymond K. Belknap, John G. Furtado, Richard R. Foster and H. Duane Blossom, *Three Approaches to Environmental Resource Analysis*, Department of Landscape Architecture, Harvard Graduate School of Design, Cambridge, Massachusetts (1967).
25. Carl Steinitz *et al.*, *A Comparative Study of Resource Analysis Methods*, Department of Landscape Architecture Research Office, Graduate School of Design, Harvard University, Cambridge, Massachusetts (1969).
26. David R. Godschalk *et al.*, *Carrying Capacity: A Basis for Coastal Planning?* Department of City and Regional Planning, University of North Carolina, Chapel Hill (1974).
27. George H. Stankey, *Visitor perception of wilderness recreation carrying capacity.* USDA Forest Service Research Paper INT-142 (1973).
28. Walter E. Westman, How much are nature's services worth? *Science* **197**, 960–964 (September 1977).
29. Walter E. Westman and David W. Conn, *Quantifying Benefits of Pollution Control: Benefits of Controlling Air and Water Pollution from Energy Production and Use*, California State Energy Commission, Sacramento, California (1976); J. Cairns Jr, K. L. Dickson and E. E. Herricks, *Recovery and Restoration of Damaged Ecosystems*, University Press of Virginia, Charlottesville (1977).
30. David W. Ehrenfeld, The conservation of non-resources. *Am. Sci.* **64**, 648–656 (December 1976).
31. Rajendra K. Sharma *et al.* (eds.), *The Biological Significance of Environmental Impacts*, Proceedings of the Conference on the Biological Significance of Environmental Impacts, University of Michigan, Ann Arbor, (June 1975).

The manuscript was received 15 August 1978

High Technology Accidents: Unpredictable and Inevitable

In a recently designed high technology system, oxygen gas was required, but, because of certain inherent limitations, the gas had to be stored as liquid in a cryogenic tank. Utmost reliability was specified, and the oxygen subsystem was tested and retested, it was in fact overtested! A thermostat switch inside the tank welded permanently shut through overload cycling during ground tests. When Apollo 13 was on its way to the Moon on 13 April 1970, the oxygen tank exploded in deep space. However, devoted help from a team of 400 scientists and engineers, linked to Mission Control in Houston, Texas, provided emergency instructions to the crippled spacecraft on its way to and from the Moon and, with some luck, these finally led to a safe splashdown in the Pacific Ocean. This accident was totally unpredictable. 'It was a serious oversight in which all parties shared' (Apollo 13 Review Board, NASA, 15 June 1970).

Nothing Perfect

By definition, accidents are unforeseen contingencies, and unpredictable. But that they are also inevitable, is only now slowly being realized. Nothing, but absolutely nothing, made by man is perfect and therefore everything will have inherent faults, be they in design, materials or manufacture. All that the best of engineers can aspire to achieve is to reduce the number of accidents, to accumulate sufficient experience, allow a large enough safety factor and include one or more back-up systems to continue operations when a component fails. And most important of all, where humans are involved, provide for their utmost training by the simulation of all kinds of accidents and supply the operators with instruments of simplicity, clarity and reliability. The most shocking accidents, at a world-wide death rate of several tens of thousands a year, still occur, after 80 years of experience, on our imperfect car–road system, with operator negligence being the most common cause. Familiarity with road deaths has certainly bred contempt.

High Technology Accidents

If today the car–road system is considered as one of very low technology – and this possibly explains its high accident rate – let no one think that high technology is free from its own inevitable accidents. Apollo 13 and the nuclear accident at Three Mile Island, the recent spate of DC-10 aircraft disasters and the Flixborough cyclohexane explosion, or even such earlier examples as the *Titanic*'s sinking or the burning of the *Hindenburg* Zeppelin – both hailed as highest technology in their days – prove that accidents are unpredictable and unavoidable. When compared with the frequency of natural disasters such as earthquakes, floods and circular winds, with their regular annual death rates in the hundred thousands, the comparative rarity of man-made disasters is a great tribute to the engineering profession.

Failure Certain

The components of spacecraft and of nuclear power stations are certainly manufactured to very high standards. Specifications called for 99.99%

The Control Room of the Three Mile Island Nuclear Power Station. (Courtesy of Metropolitan Edison Company, a Member of the General Public Utility System of the USA)

perfection of the Saturn V rocket and the Apollo spacecraft for all their 15 million separate parts. Arguably 1500 items had a failure probability of one! To avoid this dangerous situation, back-up systems, duplication and even triplication were employed. Comparable figures for nuclear power stations are no doubt very similar. Obviously cost and material considerations will always prevent any man-made item from being 100% perfect and even lengthy testing, again provided time, money and facilities can be allocated for this, is no absolute guarantee as Apollo 13 proved. But major equipment failure was not the main cause of the near disaster at Three Mile Island, although the failure of a few minor components started the incident.

Nuclear Accident Forecast

'On Wednesday, 28 March 1979, 36 seconds after the hour of 4.00 a.m. several water pumps stopped working in theUnit 2 nuclear power plant on Three Mile Island, 10 miles southeast of Harrisburg, Pennsylvania, USA. Thus began the accident at Three Mile Island. In the minutes, hours and days that followed, a series of events – compounded by equipment failures, inappropriate procedures and human errors and ignorance – escalated into the worst crisis yet experienced by the nation's nuclear power industry.' (*The President's Commission on the Accident at TMI, John G. Kemeny Chairman*, published Washington, DC, 31 October 1979.) That such an accident was to occur sooner or later, was brilliantly forecast by Thomas N. Scortia and Frank M. Robinson in their book *The Prometheus Crisis*, published in 1976; it was made into the film *The China Syndrome* two years later.

Training Inadequate

One reads in the Kemeny report 'that training did not adequately prepare them (the operators) to cope with the accident at TMI-2' (p.91) and 'the fact that they failed to realize that these conditions resulted from a LOCA (Loss of Coolant Accident) indicates a severe deficiency in their training to identify the symptoms of such an accident' (p.96). Also on page 91: 'Frederick and Faust (two reactor operators) were in the Control Room when the first alarm sounded, followed by a cascade of alarms that numbered 100 within minutes. The operators reacted quickly as trained to counter the turbine trip and reactor scram. Later Faust would recall for the Commission his reaction to the incessant alarms: "I would have liked to have thrown away the alarm panel. It wasn't giving us any useful information" '.

The Control Panel of CERN's 400 GeV Super Proton Synchrotron as used by a single controller from 1978 on. (Courtesy CERN, Geneva)

Mammoth Costs

This must be one of the most damning statements in the whole report. If instruments fail to convey relevant information about unusual conditions, then either operators have been inadequately trained, instrumentation is antiquated and useless, or as in the case of Three Mile Island, both. Modern high technology demands that only the most superbly trained and thoroughly experienced operators control such technology. Nothing below astronaut standard should be acceptable, when so much is at risk. Now, more than a year later, the accident has left behind a mammoth task of decontamination of the reactor and its containment building, as well as the disposal of large amounts of radioactive waste; together with the provision of electricity, equivalent to that lost from the Three Mile Island generator, the costs of the accident to Metropolitan Edison, the Electric Utility owning the reactor, has been estimated at about $1.5 thousand million. Undoubtedly it will be cheaper in future to have better trained and more highly qualified staff attending to modern computer system controls in nuclear power stations.

Salary and Social Status

In Western Society it is today common industrial and military practice to advance salary and social position with increasing responsibility. The captain of a civilian jet aircraft, being responsible for the lives of 500 passengers during his eight hour working day, may earn $100 000 a year. And what were the salaries and social positions of Messrs Frederick and Faust at Three Mile Island? Or for that matter of any other operators at any other nuclear power stations anywhere else, with their responsibility, in case of accident, for the lives of say five million people? If it is objected that it might well prove impossible to keep extremely well qualified and highly paid operators with salaries and positions of jet pilots sufficiently interested in the routine running of power stations, the answers would be that constant training exercises and simulation practice would dispel boredom. Most of the Apollo astronauts' lives were spent in simulation practice, and boredom was never one of their complaints. One could perhaps also learn from the logistics of the Apollo 13 rescue team, spread over the whole of the continental USA and linked by telecommunications to Mission Control. A similar network for advice and rescue of nuclear power stations in difficulties could surely be worked out – it would be infinitely cheaper than another accident like Three Mile Island.

Steam Age Instruments

And finally, one must condemn the steam age instrumentation systems on which Messrs Frederick and Faust had to rely and which served them so badly. Nuclear scientists and engineers at CERN, Geneva, recently faced the task of controlling their new Super Proton Synchrotron; they solved it by adopting a novel hierarchical computer control philosophy and technique. With it the operator in charge, a highly qualified physicist or engineer, faces a single television monitor screen which has projected on to it a series of relevant legends and of image buttons. These can be touched with his finger and thus connect him with any desired component to be operated. Further interaction with the system is achieved at CERN with a rolling ball selector and a knob mounted on an incremental encoder. The whole of the accelerator can thus be started and controlled with only these three operator system interface tools: the touch button screen, the rolling ball and the knob. (See *Interdiscip. Sci. Rev.* **3**, 329–331 (1978).) When I recently spoke with Dr M. C. Crowley-Milling, the designer of the CERN control system, he assured me that it could have wide-scale applications wherever complex systems require monitoring and control, as for example in nuclear and other power stations.

Interdisciplinary Lessons

Therefore these must be the interdisciplinary lessons of the Three Mile Island accident:

* Accidents will inevitably happen at nuclear power stations as anywhere else and they will always be unpredictable.
* To keep accidents to minor inconveniences, it will be essential to have in the control rooms operators trained to astronaut standards, and to accord to the best of them salaries and status equivalent to captains of a jet aircraft.
* Only the most modern electronic computer aided instruments and executive controls can find a place in the high technology operation rooms of nuclear power stations.
* And finally, when the inevitable and unpredictable has happened, as unfortunately it must, a highly efficient aid and rescue service should be available; they will have had frequent rehearsals and thus have become near perfect.

Everything Dangerous

None of the above can be construed as an argument against an expansion of nuclear industries and the building of many more reactors, as the UK and Japan have decided, and France is already achieving. Just as nothing man-made is perfect, so nothing undertaken by man is free from danger. Coal-mining still levies annually a few hundred deaths world-wide instead of the many hundreds when the industrial revolution started; petroleum products are still highly inflammable and explosive as almost each aircraft accident proves anew. The genetic danger to unborn generations from untreated and uncontrolled radioactivity is indeed something neonomian – but there are many historical examples of one generation using up nature's resources without caring for future consequences. There was the deforestation of England's oaks for ship building, the overgrazing and erosion in the Tennessee Valley and the many ruthless wars of colonization, such as the complete eradication of Tasmania's aborigines.

Blessing in Disguise

Therefore risks and dangers are inescapable if Western Society decides to continue living at its present standards. If then the above stated interdisciplinary lessons and the other recommendations of the Kemeny Report on the Three Mile Island accident are learnt and acted upon, the Three Mile Island accident may well be called a blessing in disguise. After all, not a single human being received any excessive radiation. If we act now, swiftly and wisely, the same may hold true for all future accidents at nuclear power stations.

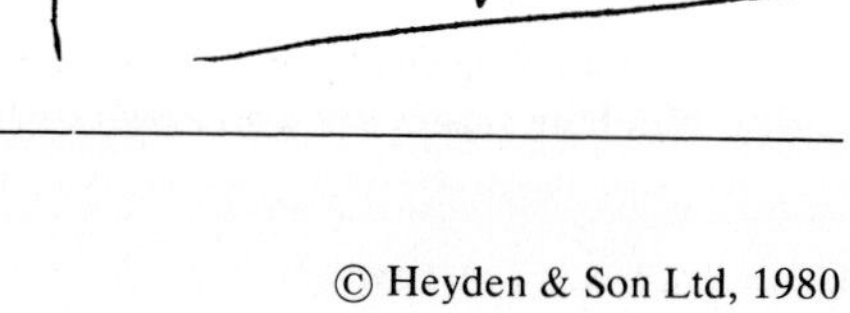

The Law of the Sea

COMMENT

New Technology Prompts New Law

PROFESSOR DR MICHAEL BOTHE

Department of Law, University of Hanover, FRG

Recent years have brought tremendous advances in the technical possibilities of exploiting the seas, their resources, their soil and subsoil. But this progress has also meant an increased threat to their ecological balance and the conservation of their living resources, the menaces of over-exploitation and pollution.

Traditional Law

Traditional international law divides the sea into two areas with different legal regimes: a relatively small strip of coastal waters under national sovereignty and the high seas which all States are free to use. Today, this freedom, if unrestrained, entails two problems: It may lead to over-exploitation and it may also lead to serious factual inequalities among States, as the possibility of using this equal freedom depends very much on one's technological affluence. The recent development of international law of the sea has witnessed two trends: first, considerable extension of the national jurisdiction of coastal States beyond former coastal waters and the continental shelf; second, international regulation of exploitation of resources beyond national jurisdiction.

The Disadvantaged States

Both trends involve serious problems: first, that of the position of those States which do not have the advantage of extended, rich and exploitable waters off their shores; more generally speaking, these land-locked and geographically disadvantaged States have a strong interest in a particular use of the Seas which would now fall into national jurisdiction, such as freedom of passage for military purposes. Second, there is the problem of the beneficiary of an international regime for exploiting resources in areas beyond national jurisdiction. These are the two fundamental issues which have to be solved by the international law-making process.

The Long Conference

In order to meet the challenge of new technology and to find a balance between these diverging interests, the United Nations convened the Third Law of the Sea Conference to produce an international agreement regulating these issues. This conference has now met since 1973; it has reached agreement on many questions, but some hard core issues remain controversial. And it seems doubtful whether the Conference will produce what it is supposed to: a new general international convention on the law of the sea. This Conference is the object of a research project and of publications from the Max Planck Institute of Comparative Public Law and International Law in Heidelberg, analysing and evaluating the developments which have taken place at the Conference and discussing possible solutions of the remaining controversial issues. The following is a summary of the results.

The Exclusive Economic Zone

During the current Law of the Sea Conference, national claims of jurisdiction are much wider and more comprehensive. Although the claim to 200 mile territorial waters put forward by a number of States has not been accepted, the concept of a 200 mile exclusive economic zone, EEZ, is being generally recognized, giving the coastal State an exclusive right to the economic use of the resources of that zone, namely fisheries, resources of the sea-bed, protection of the environment and marine research. The right of navigation, including passage of warships through straits, is not greatly affected by this concept as it stands now. But what is still to a considerable extent controversial, is the right of other States to exploit the resources of the EEZ, in particular the claim of land-locked and geographically disadvantaged states to an equitable share of the wealth accruing to States with long coastlines. Furthermore, there are well-founded fears that marine research may be unduly restricted in the EEZ.

Mineral Resources of the Deep Sea

In areas beyond national jurisdiction, thus defined, the main issue is that of an international regime for the exploitation of the mineral resources of the deep sea. It has been claimed that this wealth constitutes a 'common heritage of mankind', and therefore, that it should be administered by an international agency, the 'Authority', and the benefits deriving from exploitation of these resources should go not only to the technologically advanced countries which alone possess the necessary technology and capital, but also, or even mainly, to the developing countries. While this idea seems to be accepted in principle, its actual chances of being put into practice appear rather remote, due to some fundamental clashes of interest which have emerged between developing and developed countries. In view of the huge investments which have been and will have to be made for deep sea mining, it is in the interests of the technologically advanced countries to have access to exploration and exploitation licences and to be sure of economically sound conditions, namely the safety of their investment and a reasonable expectation of profit. A procedural safeguard of these interests would be an essential part of the votes in the decision-making body of the Authority which is to reserve the grant of the relevant licences for the developed countries.

International Enterprise

The interest of the developing countries is to ensure that the profit of deep sea mining flows to the maximum extent possible towards their own development; and furthermore that raw materials from the sea do not adversely affect the market chances of commodities produced by them on land. Therefore, the developing countries are against free access to licences, and favour exploitation by the international Agency itself, through the international 'Enterprise', for which the advanced countries would have to provide the necessary technology on still-unknown conditions. They also demand production limitations in favour of land-based minerals, and they have a keen interest in adopting a one State-one vote rule for the decision-making process of the Authority.

Parallel System Compromise

A possible compromise, also accepted in principle, lies in the adoption of a 'parallel system': exploitation will be made by both national licensees and the 'Enterprise'. What remains controversial are the duration of this compromise, the licence conditions – duration and technology transfer – the composition of the decision-making bodies and production limitations. These questions are fundamental not only for the deep sea mining issue itself, but also as a precedent for the regulation of production in what is usually called the New International Economic Order.

The Likely Losers

Whether or not a compromise is possible cannot be forecast. But if a compromise is to be reached, it must be reached now. Otherwise, the danger of unilateral action by the technologically advanced countries becomes very real. Legislation to that effect is already being introduced in some States. Once vested rights have been created by national exploitation, it will hardly be possible to bring this exploitation under the roof of an international regime. On the losing side, one would not only find the lofty idea of a common heritage of mankind, but the developing countries themselves, which might end up without a share in these resources that could play such a considerable role in reducing the North–South inequalities.

A Hierarchy of Universities

LORD TODD, OM
President, The Royal Society of London, UK

To remain competitive in the present technological revolution we must recognize that our traditional type of university education is not appropriate to more than a fraction of those entering higher education. Ageing university staff and a sad waste of skill of junior researchers in the United Kingdom, Europe and the United States are the symptoms which should encourage a diversity in universities, some to specialize in teaching, others in research, and thus lead to a hierarchy of universities.

More than ever before, our daily existence is dependent on advances in science-based technology and our future depends more than many people seem to realize upon the use we make of the new technologies which will develop on the basis of today's discoveries in science. Failure to choose wisely among the various choices open to us, or, even worse, to ignore them in the vain hope of continuing to operate antiquated technologies successfully in the competitive arena of world trade spells disaster for any industrialized country. Yet this is what we have been doing in Britain in recent years although the extent of our economic decline is currently hidden from an unthinking public by the fortuitous, but temporary, inflow of wealth from the North Sea oilfields.

Time was when the area of choice open to governments in the formulation of national policy was limited and the factors governing choice comparatively straightforward and simple to understand. But that time has long since gone. The development of science-based technology that followed on the heels of the industrial revolution continues to gather force and there is no way in which it can be halted. Human society cannot escape the consequences of new knowledge which will emerge from science in the future and, as the rate of accretion increases, so too will the complexity of choice and the number of options open to governments whatever their political colour.

In a democracy like ours scientific expertise among politicians is hardly common and today governments are bombarded from all sides with a babel of advice from pressure groups, much of it misinformed or heavily biased. It is not my purpose here to argue in detail the mechanisms by which the need for external and independent advice should be met, but it seems to me that the Royal Society is a body uniquely constituted to organize the provision of that advice. I believe that this is an area in which the Society should be more active than it has been in recent times, and the setting-up and further development of study groups and discussions is an earnest of that belief.

MORTGAGING OUR FUTURE

Future developments in science and technology cannot be predicted; none of us can foresee the discoveries which will be made or the technologies to which they will give rise. All we can say – and that with some certainty – is that they will surprise us. But what we do know from the recent history of our own country is that the survival of a great nation and the standard of living enjoyed by its citizens depend on their ability and readiness to be in the forefront of new technologies as they emerge. And the best way of doing this is to be master of the science on which these technologies rest. In other words those countries will be the most successful which make discoveries in science and then exploit them through technology.

Our record in discovery is good but during this century our performance in the highly competitive area of technological innovation has been, to say the least, disappointing. To recover the economic ground we have lost as a result will demand a greatly enhanced effort, and perhaps also a change of heart, on the part of our people. But any recovery – and we have the opportunity for one now through Nature's gift of North Sea oil – will be of short duration if we cut back on our scientific research for financial or other reasons without considering the effect our economies may have on our future stock of scientists. There is, I fear, a good deal of evidence to suggest that we may even now be mortgaging our future in this respect.

During the past year the Officers and Council of the Royal Society have been much involved in efforts to promote the development of research groups around promising research scientists. Such activity was foreshadowed in remarks I made in my anniversary address in 1977 and I am glad to say that we are now making progress. Although the amount of money available to us is small in relation to the overall need, we hope the contribution we can make will not be wholly insignificant and may have the effect of

CCC–0308–0188/80/0005–0085$02.50/1

encouraging other bodies to promote first-class research and raise the morale of those who are capable of doing it. For morale is still low in our universities and especially so among the younger members of the academic research community. The evidence for this is circumstantial and perhaps to some extent subjective but I believe the decline has now reached a point at which not only urgent consideration but also action is called for if not only our research but our whole university system is not to suffer permanent damage.

MISGUIDED EUPHORIA

It is common to put the blame for this on the stagnation of the British economy and the consequent shortage of money for education and research. Our economic difficulties certainly play a large part but I believe the real root of the trouble lies in the misguided euphoria which in the early sixties caused us – in common with most other industrialized countries – approximately to double the number of our universities and greatly to expand our student numbers. No one would seriously dispute the thesis that higher education – tertiary education would perhaps be a better expression – should be available to all those able to benefit from it, but in those heady days higher education was equated with university education of the traditional pattern.

How wrong this was has been amply demonstrated by subsequent events. The sudden expansion in student numbers involving as it did the entry of many with no real motivation for the type of education provided by our traditional universities (and which was, by and large, adopted by all the new ones) was in my view a material factor in the disturbances which marked the late sixties and early seventies. The universities survived the shock of these student disturbances surprisingly well but the long-term effects of the sudden and prodigious expansion of higher education are now becoming increasingly apparent. They are, in fact, basically demographic although in many respects exaggerated by our nation's economic decline.

The rapid growth of the university system in the 1960s brought about a vast expansion in tenured staff usually by the recruitment of relatively young men and women taken from the normal supply of university graduates and not invariably of the highest quality. Many, if not most, of these remain today at the same institutions and are likely to do so for perhaps another two decades under the tenure system which is almost universal. This phenomenon is, of course, not confined to the United Kingdom and it is giving increasing concern in most other industrialized countries, including the United States.

The secretariat of the European Science Foundation has tried to collect and analyse such statistical material on university staffing as is available for a number of European countries – United Kingdom, West Germany, France, Denmark, Norway and Switzerland – and has published its findings in the Report of the Foundation for 1978. The results are strikingly similar for all the countries examined. Broadly speaking, they indicate that in most European countries the age distribution of teaching staff in universities is now at a level where 50%, or in some of them as much as 60% of the staff has an average age of about 40. The figures show, too, that in the countries examined in more detail (those listed above) the majority of those in post today will remain in office until the mid-1990s, when they will begin to reach retirement age. The demand for replacements during the next fifteen years or so will be exceedingly low and the effect could well be exaggerated if economic difficulties cause universities to economize by the short-sighted policy of suppressing posts that become vacant; this has, indeed, already occurred. The overall result both on research and teaching could be disastrous and persist for many years.

AGEING UNIVERSITY STAFFS

A detailed study of chemistry departments in British universities has been made by Professor Colin Eaborn who summarized his findings thus:

> At present only 7.8% of the staff of chemistry departments are below 35 years of age; the proportion will probably fall to about 4% in five years' time and rise to only about 9% in ten years time. The proportion below 40 years of age, now 26%, will fall to about 12.5% in five years and to about 10% in ten years time. During those ten years the proportion of staff over 50 will rise from the present 28% to about 62%. These figures have serious implications for British chemistry, and thus for the chemical and other science-based industries.

LORD TODD, Order of Merit, has been President of the Royal Society of London since 1975; he was the Master of Christ's College, Cambridge, 1963–1978, and he received the Nobel Prize in Chemistry in 1957. Lord Todd was Professor of Organic Chemistry, University of Cambridge, 1944–1971, and he has been honoured by scientific societies, governments and academies the world over.
Address: The Royal Society, 6 Carlton House Terrace, London SW1Y 5AG, UK.

Indeed they have – and the situation in other physical sciences is unlikely to differ greatly. Viewing Europe as a whole, the European Science Foundation suggests that the chance for a junior research assistant to reach a permanent university appointment has gone down from about 70% during the 1960s to about 15% in this decade. The outlook is indeed bleak, faced as we are with a continued ageing of university staffs over nearly two decades and a denial to departments of the invigoration which young recruits can bring to them. The other side of the coin is that younger scientists who would in previous decades have made a solid contribution to research in an academic environment will now be denied that opportunity and either abandon research or seek a career elsewhere. Surprising though it may seem, the academic research profession has within a few years been transformed from one of the most mobile to one of the most static. This is particularly so in the United Kingdom, where the economic stagnation of recent years has all but staunched the flow of academics in mid-career into other occupations.

Many young scientists are seeking to retain their present precarious positions by hand-to-mouth grants because they fear that a change to another university or research institute might jeopardize their chance of obtaining a permanent appointment. This same attitude is probably responsible for the falling-off in applications received by the Society for fellowships under the European Exchange Scheme. To many, it would seem, the risk of losing a small chance which may exist at home seems too great for all but the most venturesome to follow the path of their predecessors, who normally received some part of their training abroad. In such circumstances it may well be that the diminishing chance of obtaining a university position could lead to a kind of negative selection process among those who stay at home, and in any case it militates against international scientific cooperation and understanding.

DECLINING STANDARDS

To these essentially demographic problems affecting our outlook in research others associated more directly with inflation and economic recession could be added. One only will I mention here; in the past, universities and research institutions spent something like 65% of their budget on staff and 35% on maintenance, including such things as research and library expenditure. As a result of inflation the proportion being spent on salaries and wages has been increasing and this has in many cases been reflected in a shortfall in research budgets, which has had to be met by increasing the demand on Research Council funds. How long such difficulties as these will persist it is impossible to say – they are likely to be with us until we emerge from the economic morass in which we are floundering. In these circumstances it is all the more necessary that the universities and those responsible for sponsoring the university system should grasp – and more courageously than they have been doing – the nettles which now abound, many but not all of them a consequence of underlying demographic problems. Several awkward, even painful, questions arise.

Is it, for example, to continue to be taken as an article of faith that all established academics are capable of first-class research and that all students obtaining first- or upper-second-class honours degrees should be encouraged and provided with the wherewithal to pursue academic research? The answer to both questions must, I fear, be no. I have already indicated my view that the very rapid expansion of university staffs which has occurred was bound to involve some who were not of first-class research calibre. Moreover, we have seen in the past decade or so an approximate doubling of student numbers in our universities, the increase being drawn very largely from the same social classes of our population as in the past. Despite this the proportion of first-class honours degrees awarded each year in science at least has not diminished; it is difficult not to equate this with a lowering of standards.

If these doubts are justified and we continue to believe that each department in every one of our present universities should have a substantial research school the strain on our financial resources may become intolerable. The resulting decline in the standard of our research will in due course extend through to our technology and so militate against the nation's economic recovery. Yet this is the road down which we appear to be travelling owing to over-rapid expansion and the inflexibility of our university employment patterns. The dual support system of the University Grants Committee and the Research Councils would ideally ensure that extra funds for the support of research were concentrated on those most likely to spend them well; as things now stand, however, while there is no reason to believe that really able young scientists with good projects to put forward will be denied temporary funding to initiate them, the prospects of their being given a sufficient measure of permanence to build up a centre of excellence in their institution is small indeed. Since in research only excellence begets excellence it is essential that universities individually or collectively should face up to this problem.

TO ENCOURAGE DIVERSITY

But can all university departments or even all universities really become centres of excellence? This question – and the inevitably disappointing answer to it – has been lurking in the minds of British academic scientists for the past decade but has not been openly faced. In reality, however, it is not possible – and it may not even be desirable – that all departments of, say, chemistry or physiology should stand out for the high quality of their research in some corner (and still less in all corners) of their discipline. Some, for example, have too small a staff adequately to sustain undergraduate teaching and to supervise at the same time programmes of postgraduate work which all departments in all disciplines appear to regard as the essential breath of life. Need this requirement for large postgraduate programmes be universal?

It is a remarkable fact that in proportion to population there are more institutions awarding PhDs in physics in Britain than in the United States. In the United States we find many more institutions devoted to professional education and applied sciences and there are many distinguished universities whose reputation rests substantially on the quality of their basic teaching. Such diversity in our institutions should be encouraged. Why should there not be some differentiation between those who teach and those who pursue research? Even if by proceeding along these lines it should turn out that some of our universities become concerned essentially with the undergraduate education of a more practical and vocational type than they now seek to provide, that would not be the end of the world.

Unfortunately the present machinery for the support of higher education and research in the United Kingdom was not devised for the encouragement of diversity. For all its undoubted virtues the University Grants Committee system does require that universities should compete within what is increasingly a common framework of objectives. These may be laudable enough as far as they go but with grants being geared to undergraduate numbers, individual universities are not necessarily rewarded for doing well what they are best able to do.

The result is that competition between British universities is almost always competition on familiar terms – increasingly and unhappily competition at the margin for the increased tuition fees with which students are endowed by local authorities. Preliminary, and anecdotal, evidence suggests that in this competition Oxbridge and the older civic universities are winning out. If this be the case and the trend continues, then we will end up with a hierarchy in which the universities lower in the pecking order will be trying to do the same things as the others but doing them with students who are not suited to them and would do better and become more useful citizens with a more vocational type of education such as is – or should be – provided by our polytechnics. Rather than let things drift slowly and painfully towards such an arid pattern, why should we not seek boldly to alter patterns and develop diversity in our institutions now? At a time when a major effort is required to remain competitive in the technological revolution that is now occurring, the idea that our traditional type of university education should be universally applied for reasons of social prestige is dangerous as well as foolish.

A SAD WASTE OF SKILL

If we are to change our present university pattern the mechanisms by which we support research in academic institutions may also have to undergo some changes. Project grant applications will no longer be assessed simply on their merits alone without reference to the circumstances under which the research is to be carried out. Research projects in a given field will tend to be concentrated in one, or in only a very few, centres with considerably larger research groups than are usual today. It may well become necessary for the research councils to be more selective in the way in which postgraduate studentships are allocated to university departments; more radically, they may even have to think of making the grants to students of exceptional promise rather than to their potential supervisors. Another convention that may have to go is that every reasonably able PhD graduate can expect as of right to have two or three years of postdoctoral research during which he can establish a claim on a tenured research post.

The frequent frustration of this expectation is one of the saddest of the current symptoms of malaise in our university research. Many postdoctoral fellows, many of them skilled and imaginative people, have discovered that there are more of them than of permanent jobs in what they have come to think of as their own field of research and must turn to something quite different. Nobody will deny that this entails a sad waste of skill; unhappily it is a circumstance that will not naturally go away until the British economy is once again buoyant. And the vigour of our research enterprise would surely suffer if all those concerned were now provided with a formal tenured career structure as many of them are now asking despite the current staff structure in our universities.

And so I come back to the over-rapid expansion of our universities in the 1960s and to the disastrous age-distribution in our university staffs which has resulted from the way in which it was carried out. The unfortunate circumstance that we have since then entered upon a period of acute economic recession has entailed severe restraint on the money available for teaching and research and led to resources being spread too thinly over too many centres. I have indicated where I believe our problems lie and have only posed some of the questions which could be asked about research in our universities and the ways in which it is promoted. But in the last analysis it is difficult to see any real progress being made unless we can do something about our ageing university staffs and the lack of openings for our brightest young academic scientists.

Unfortunately there is no quick or easy answer. From their own resources universities themselves can probably do little and the breakdown of the quinquennial government-grant system makes forward planning well-nigh impossible for them. I believe that the efforts being made by the Royal Society to support future leaders in research and the Science Research Council's scheme of advanced fellowships are valuable but in the aggregate they can provide only a small contribution to the solution of a very large problem. A reduction in the university retiring age to 60 without any reduction in pension would undoubtedly speed up return to a normal age distribution, but whether in present circumstances government would be willing or indeed able to face the very large expenditure which would be necessary is doubtful. Encouragement of voluntary retirement at 55 with generous financial compensation has also been suggested but would probably not be welcomed by more than a few individuals, and even if it were generally acceptable it would be altogether too costly.

HIERARCHY IN UNIVERSITIES

Nevertheless I believe the problem must be tackled and that some of the unpalatable things I have stated here may help to point the way. For example, if we accept that there should be a kind of hierarchy in the universities and that some of them will be much more devoted to vocational teaching and less to research than others, then only in a proportion of our universities need the situation be treated as urgent. These urgent cases could well have a retiring age of 60 with full pension introduced even if only temporarily so that a more normal flow of young academics could be reintroduced in them. This would certainly cost money but a great deal less than any blanket procedure applied to the academic system as a whole. I have not attempted a detailed calculation but I believe the overall cost would be tolerable. But it would involve the introduction of much more diversity into our university system than we now have, and this alone would make it worth while.

To do anything like this with our dual support system in its present form would be difficult and it is not surprising that under present circumstances many academics are beginning to ask whether the present system can continue. The more successful universities are probably right in thinking that they would secure a greater share of the resources available if they were able to compete within a more flexible framework, and suggestions for change will certainly increase if the prediction of a decline in student numbers in the 1980s proves well founded. Perhaps it is not too soon to be thinking of the best form such a change should take.

The above review was part of author's Anniversary Address to the Royal Society, delivered 30 November 1979

The Franklin Institute
Commemorating a Great Interdisciplinary Scientist

DR BOWEN C. DEES
The President, The Franklin Institute, Philadelphia, USA

Continuing with the many interdisciplinary activities of Benjamin Franklin, the Institute named after him more than 150 years ago embraces today two large research centers, one scientific press, a museum and far-flung international activities, including research at the South Pole. The Institute's important pioneering activities to establish American industry and technology are recalled and emphasis is placed on the Institute's present research work.

Since 1824 The Franklin Institute has occupied a unique place in science and technology. Over its history it has played many useful roles regionally and nationally; it has provided research services to American government and industry, carried out basic research on its own, published important scientific and technical materials, served as a broadly based professional society, sponsored important exhibitions and conferences, maintained a major research library, provided several mechanisms for improving the public's understanding of science and technology, and awarded premiums and, in recent years, highly prized medals for outstanding accomplishment in science, technology and industrial management. There is no other institution quite like it anywhere.

The group of Philadelphians who took it on themselves to found an organization 'for the promotion and encouragement of manufactures and the mechanic and useful arts' found it natural to name the new organization after Benjamin Franklin, whose memory was still fresh in the minds of those living in Philadelphia in the 1820s. No man before or since Franklin has left his mark so indelibly on his adopted home town. (Franklin was born in Boston in 1706, but moved to Philadelphia when he was not yet 20.) As a successful printer, typefounder, author, merchant and publisher, Franklin became well known while still a young man. Indeed, starting with literally no capital, he had achieved financial independence by the time he was 42 – and at that age chose to retire from his business pursuits in order to devote more time to his scientific work, which had become an exceedingly important part of his life.

THE INTERDISCIPLINARY BENJAMIN FRANKLIN

Carl van Doren[1], speaking at The Franklin Institute in 1939, said: 'Franklin had, I think, the most eminent mind that has ever existed in America. He appears, somehow, to be a syndicate of men. We study him as a scientist, as a diplomat, as a statesman, as a business man, as an economist, as a printer, as a humorist and wit, as a great writer, as a sage, and as a landmark in the history of human speech about the common ways of life. What had been said before, he so often said better. He was great in friendship, and in his later years was probably the most renowned private citizen on Earth.'

Franklin was truly an interdisciplinarian. The first steps toward a scientific meteorology were contained in his 1743 notes on the origin of northeast storms; in 1751 he wrote a pamphlet which anticipated Malthus; he was the first scholar to study the Gulf Stream; he made the first measurements of heat absorption as a function of color; his were the pioneer investigations of the heat conductivity of different substances; and his work with electricity was of course of first-class importance.

In speaking of Franklin's work as a scientist, Dr Robert A. Millikan[2] once wrote: 'Franklin lives as a physicist because, dilettante though he is sometimes called, mere qualitative interpreter though he actually was, yet it was he who with altogether amazing insight laid the real foundations on which the whole superstructure of electrical theory and interpretation has been erected.' He continued '. . . within two years of the time of his first experiment Franklin had

The Benjamin Franklin National Memorial at The Franklin Institute. Statue sculpted by James Earle Fraser.

CCC–0308–0188/80/0005–0090$06.00/1

acquired a keener insight into the fundamental nature of electrical phenomena, not merely than anyone had acquired up to his time, but even than any of his successors acquired for the next hundred and fifty years, when, about 1900, the scientific world returned essentially to Franklin's views.'[3]

Franklin preferred to think of himself as a scientist. Even at the age of 81, having completed his enormously demanding and crucially important work as a diplomat in Paris, he asked a friend to join in rejoicing with him over the fact that, after 50 years of public service, he could now return to Philadephia and resume his scientific endeavors. Franklin was a freemason and for a time Provincial Grand Master of Pennsylvania.

Dr Franklin founded or helped to establish many organizations: the Library Company, the American Philosophical Society, the University of Pennsylvania, our first fire insurance company, and our first volunteer fire-fighting unit. Yet none of these bears his name. Thus, The Franklin Institute, charged as it is with the duty of honoring Franklin's memory as well as promoting the advancement of science, carries a broad and significant responsibility.

THE INSTITUTE'S ELEMENTS

Sometimes referred to as a 'not-for-profit conglomerate,' The Franklin Institute at present consists of a number of major operating divisions and a central administrative staff – which also carries out certain program activities. The largest division is the Franklin Research Center – an applied research and science information organization employing more than 500 people, with offices in Philadelphia, the District of Columbia, Little Rock (Arkansas), Munich, Luxembourg and Tokyo. [For details see below.] The Institute's first research division, the Board of Managers from 1870 to 1878, provided in his 1924, shortly after Henry Welchman Bartol, a successful Philadelphia businessman, bequeathed his estate to The Franklin Institute. [For details see below.] Mr Bartol, who had served on the Institute's Board of Managers from 1870 to 1878, provided in his will that the Institute should use his legacy to establish a research unit 'along the lines of' the Mellon Institute of Pittsburgh. During its 55 years the Bartol Research Foundation has been continually engaged in cosmic ray research, but has also carried out outstanding basic research work, both theoretical and experimental, in other areas of physical science – notably astronomy, astrophysics, nuclear and particle physics, solid state and surface science.

The third division of The Franklin Institute is its Science Museum and Planetarium, a major science center devoted to the task of making science and technology more understandable to young and old alike. [For details see below.] As one of the leading museums of its kind, the Institute's Science Museum serves an audience of over one million annually, and is recognized as being one of Philadelphia's major tourist attractions, drawing visitors from many other parts of the United States as well as from overseas.

In addition to the activities of its divisions, The Franklin Institute has several other significant elements in its range of responsibilities. In 1973, President Richard M. Nixon signed legislation passed by the United States Congress that year which declared the central rotunda of the Institute's main building to be the Benjamin Franklin National Memorial – an action which simply gave *de jure* recognition to a longstanding fact. This rotunda, completed in 1934, is an impressive marble hall more than 80 feet across and 80 feet high, containing a four times lifesize seated statue of Dr Franklin created from Seravezza marble by the well known American sculptor James Earle Fraser. Arrayed around the statue are four roomettes in which are displayed articles owned by, or associated with, Franklin during his lifetime – the largest collection of Frankliniana known to exist anywhere. This National Memorial is unique, in that it is the only memorial erected in honor of a founding father of the United States which is not governmentally owned and maintained. No federal funds went into the contruction of the Memorial, and only in recent months has consideration been given to the possibility of federal funds being made available to assist with its maintenance and operation.

A continuing series of conferences was initiated in 1974 as one of the elements of the Institute's sesquicentennial celebration; three such Franklin Conferences have now been held. The third in May 1978, focused on *Science and the Congress* – and was jointly sponsored with the Office of Technology Assessment, a service arm of the US Congress. The Fourth Franklin Conference on *Innovation and the American Economy*: *An Issue for Action* (held in November 1979) was concerned with the general question of how innovation flowing from research and develop-

DR BOWEN CAUSEY DEES, President of The Franklin Institute since 1970, is a career scientist and educator. Born in Batesville, Mississippi, Dr Dees graduated from Mississippi College and New York University, PhD, Physics, 1942. He has been awarded honorary degrees by Mississippi College, Lehigh University, and Philadelphia College of Textiles and Science. Prior to joining The Franklin Institute, Dr Dees served as vice-president and provost, University of Arizona. He also held various positions with the National Science Foundation in Washington, DC, and served as physicist and division chief, Scientific and Technical Division, General Headquarters, Supreme Commander for the Allied Powers, Tokyo. He is a Benjamin Franklin Fellow of the Royal Society of Arts, London.
Address: The Franklin Institute, Philadelphia, Pennsylvania 19103, USA.

ment can be made more useful in assuring the economic welfare of the United States. As heretofore, leading national figures from the scientific and technological community, along with economists, businessmen and governmental leaders, discussed in detail the issues arising from consideration of a major topic of national concern. Each Franklin Conference thus focusses on an important topic which, in the minds of those organizing the conference, would have been of interest and significance to Benjamin Franklin.

The Institute's Meetings Program brings to Philadelphia a number of outstanding lecturers who present their ideas on subjects ranging from parapsychology to the relationship between science and art. Now in its 155th year without a break, this lecture series is one of the most significant of its kind to be found anywhere. While most topics discussed fall within the domain of science and technology, other subjects are sometimes treated – especially those that center on Franklin or that deal with the relationship between science and industry.

AWARDS AND JOURNAL

The Institute looks to its Committee on Science and the Arts – which traces its origins to 1824 – for advice and guidance in selecting the recipients of its several scientific medals, awarded each fall. During the course of a typical year, the work of hundreds of scientists and engineers is reviewed by the 72 members of the Committee. Once an important scientific or technological development has been identified for detailed Committee scrutiny, a subcommittee composed of several experts will investigate in detail the discovery or topic to decide who deserves recognition.

Many hundreds of hours are devoted to this task over the course of the year, at the end of which the Committee recommends that the Institute's Board of Managers makes awards of the Franklin, Cresson, Fahrney, Longstreth, Levy and other medals. A detailed outline of the awardee's contributions is prepared which tells why the award of the specific medal is justified. A formal dinner in the Benjamin Franklin National Memorial provides a distingushed setting for an awards ceremony at which the medals are presented. There have been a number of cases where recipients of other notable awards (Nobel prizes, National Medals of Science and the like) have had the nature and importance of their contributions recognized first (or earlier) by the Institute.

Since its inception, the Institute has been a membership organization. Currently some 7000 individuals are members; and a large number of these are family members, members whose dues entitle their entire family to membership privileges.

The *Journal of the Franklin Institute* deserves special mention. As one of America's oldest and most distinguished scientific and technical journal, it continues to be distributed on an international basis. First published in 1826, it has shifted emphasis several times over the past 153 years. Currently, it specializes in publishing technical and scientific articles of particular interest to workers in the engineering disciplines, to applied physicists and to others concerned with the application of mathematics to physical problems. Three or four times each year, however, special issues are published, which feature the proceedings of a symposium or concentrate on some significant topic.

HISTORICAL BACKGROUND

The principal founder of the Institute was Samuel Vaughan Merrick. Sam Merrick was born in Maine, but early in the 19th century came to Philadelphia, where an uncle had already established himself as a prosperous businessman. In 1823, when Merrick was 22, his uncle offered him an opportunity to enter a manufacturing business – a company that made fire engines. At that time there was no engineering school in the entire country. With what proved to be too much self-confidence, he set about trying to set up an institution which would offer appropriate training for individuals such as himself. As he said later: 'I called a meeting at the Philosophical Hall of some fifteen or twenty gentlemen who I supposed would take an interest in so useful a movement. The night of the meeting came, and no one responded to the call. After talking the matter over the next day with several who professed an interest in the project, I called a second meeting with a similar result.'

But young Sam Merrick was persistent. He learned that another young man, William H. Keating, professor of chemistry at the University of Pennsylvania, had developed ideas along comparable lines, and he persuaded Keating to work with him toward setting up the new institution. Working together, they were able to obtain the interest and support of about a dozen others, who helped to move the idea forward. On 5 February 1824 a meeting was held in the County Court House, now called Congress Hall because it was in this building that the US Congress met when it was first set in motion in Philadelphia. This time, Sam Merrick's idea caught on – the hall was full of interested people, most of whom became members of the new organization: The Franklin Institute of the State of Pennsylvania for the Promotion of the Mechanic Arts.

At that meeting a constitution was presented and adopted. In describing the things to be done by the new Institute, this document said:

> The object of this Institute shall be the promotion and encouragement of manufactures and the mechanic and useful arts, by the establishment of Popular Lectures on the sciences connected with them; by the formation of a Cabinet of Models and Minerals, and a Library; by offering premiums on all objects deemed

worthy of encouragement; by examining all new inventions submitted to them; and by such other means as they may judge expedient.

The first Board of Managers included a number of able people – men who were already established or who in later years became well-known for their leadership in Philadelphia. Merrick did not attempt to take over as titular head of the new organization: a prosperous Philadelphia manufacturer, James Ronaldson, was chosen as the first president. Not until the year after Ronaldson's death in 1841 did Merrick assume the presidency – although the record shows that he was heavily involved in virtually all of the activities of the new institution in the years from 1824 to 1841.

PIONEERING ACTIVITIES

In his book on the early history of The Franklin Institute, entitled *Philadelphia's Philosopher Mechanics*, Dr Bruce Sinclair[4] says: '... during the years 1824–1865, ... The Franklin Institute became the most important technical organization in the country. From its varied activities Americans learned the most rational and effective means of advancing their technology. They discovered that progress did not rest on native ingenuity, however free it might be. They came to realize that government had a legitimate interest in technology and the responsibility for control when public welfare was at issue ... and that specialization was just as important in technology as in science ....'

A few of the Institute's early activities that justify this statement were these:

◇ The Institute carried forward the first cooperative industrial research project undertaken in America: a detailed experimental study of the type of waterwheel most suited for use in specified circumstances. This study was conceived in 1829, and received support from mill owners all along the Atlantic seaboard. Commenting on the study, the British expert George Rennie[5] said in 1832 before the British Association for the Advancement of Science that 'the recent [water] wheel experiments of The Franklin Institute ... eclipse everything that has yet been effected on this subject.'

◇ The Franklin Institute initiated and carried out the first research project ever to receive grant support from a department of the US government: an inquiry into the causes of steamboat boiler explosions. This research, conducted between 1830 and 1835, remained for many years the definitive work of its kind. As late as 1872 Secretary of War William W. Belknap noted that: '... the experiments which have led to the most reliable information hitherto possessed by the scientific world on steamboat boiler explosions were instituted and conducted by an American institution (The Franklin Institute of Pennsylvania), under the patronage of our government, over thirty years ago.'[6]

◇ In its first year, 1824, the Institute sponsored the first Exhibition of American Manufactures ever held. This was the first of many such Institute sponsored exhibitions, all of which were specifically designed to encourage more manufacturing of high quality goods in America, thus decreasing the dependence of the new nation on imported goods.

◇ The Institute's *Journal* quickly became a most important contributor to the nation's technological progress, through the publication of information useful to manufacturers and artisans as well as to those concerned with scientific matters. Dr Brooke Hindle[7], one of America's most distinguished historians of science and technology, in an article entitled 'A bibliography of early American technology' has written: 'By all measures, the most important technological journal was the *Journal of The Franklin Institute* .... It has a unique value before 1863 because of its record of patents granted in that period, for the originals, along with all supporting evidence, were lost in the Patent Office fire.'

◇ Beginning in 1824, and continuing until 1923, the Institute conducted a drawing school which offered instruction in draftsmanship and related matters to young men, many of whom were to become architects or engineers as a result of their studies in this school. Later, in 1850, the Institute started a School of Design for Women, in which young ladies were taught textiles, furnishings and furniture design. Spun off several years later, the school still exists under its modern name, Moore College of Art.

These are only some illustrations of a much larger set of important activities of The Franklin Institute in its early years. Obviously, as with any such organization, the goals and thus the program of the Institute shifted over the years. Throughout the 19th century the Exhibitions of American Manufactures remained an important part of the Institute's work; the last of this series was held in 1899. But The Franklin Institute contributions were also important in related areas as well. Thus, in 1869 the Board of Managers began discussions which in due course led to a recommendation to the Select and Common Councils of the City of Philadelphia that there be steps taken to assure Fairmount Park becoming the site of the nation's Centennial Celebration. This and related subsequent actions on the part of the Institute were directly responsible for the highly successful 100th anniversary party held in Philadelphia in 1876.[8]

In 1884 the Institution organized a very successful International Electrical Exposition – the first of its kind in America – which helped in many ways to advance the electrical industry in the United States. Held entirely under Institute auspices and with no state or federal subvention, although made international in character by Act of Congress, a large building was erected especially for this exhibition, which proved to be a great success. Some 300 000 visitors viewed the exhibits, and among the notable spectators or participants were Lord Kelvin,

Thomas A. Edison, Alexander Graham Bell, Sir Oliver Lodge and Sir James Dewar.

Much more could be said of the rich and varied history of The Franklin Institute; indeed, several books have been published on the subject in addition to the Sinclair volume cited above. Hence it is clear there is an important tradition of public service which undergirds the present organizational efforts of the Institute.

RECENT DEVELOPMENTS

In an attempt to reflect modern terminology and thinking in the basic document of the Institute, The Franklin Institute's Charter was modified in 1971. In that document the title of the organization was shortened to: *The Franklin Institute* – leaving off 'of the State of Pennsylvania for the Promotion of the Mechanic Arts.' In addition, the Charter was made explicit as to the goals of the institution, for the Charter purposes are now stated as 'the perpetuation of the memory of Benjamin Franklin and the promotion and encouragement of science, technology and the mechanic and useful arts.'

Staff and Board attention was given in 1972 to the question of making still more specific the role of the Institute, with the result that a 'Statement of Purpose' was considered and ratified by the Board of Managers that year. It reads as follows:

Drawing upon a tradition of dedication to the advancement of both pure and applied science, and of effective service to mankind, The Franklin Institute commits itself anew to maintaining its posture at the frontiers of knowledge, to the search for ways of solving or ameliorating those urgent problems of society that are susceptible of solution by the application of one or more of the disciplines of science or applied science, and to the task of enlightening the public concerning these problems and the role of science in seeking solutions.

Specifically, while continuing to promote inquiry at the cutting edge of basic research, the Institute will devote its resources to the synthesis of the many kinds of expertise in the natural and social sciences and in technology that are required to attack with significant impact:

- our complex and urgent socio-technological difficulties;
- the problem of improving the public understanding of science and technology; and
- fundamental questions in man's quest for a better understanding of nature's laws.

In addition to enhancing the role of science as an integral component of our contemporary culture, The Franklin Institute remains, as always, interested in helping to see science and technology used – and used wisely – in ways that will improve the economic well-being of people everywhere. This was clearly one of Dr Franklin's goals – and the Institute believes it is honoring Franklin in pursuing efforts in these directions. But, in retrospect, it is hard to see how a

The Franklin Institute (main building) with the Franklin Research Center on the left, in Philadelphia, Pennsylvania.

more ambitious role for the Institute can be imagined than the one called for by James Ronaldson, The Franklin Institute's first president, when he said, in 1833:

> The objects of the Franklin Institute are to develop the resources of our country; to throw all the light, possible, on the processes of industry, and to make the bringing of the products into market as simple as the nature of the case will admit; to cultivate whatever is calculated to make a demand for labour, and the employment of all, according to their respective capacities and means, to the end that there should be none who shall not be in the enjoyment of the necessaries and comforts of life.

THE FRANKLIN RESEARCH CENTER, FRC

Dr W. B. Ligett – President

The Franklin Research Center, at its inception known as the Franklin Institute Laboratories for Research and Development, was formally established as a not-for-profit contract research organization in 1946. The charter of FRC was to utilize – in a manner consistent with the purposes of The Franklin Institute – the staff, the facilities and, most importantly, the experience in the management of complex interdisciplinary research and development projects, which had been acquired in the course of contributing to the World War II defense effort. The undertaking of such an activity on an organized basis was particularly appropriate, as the Institute had carried out contract research and development work more than a century earlier on behalf of both government and industry.

The Institute's contributions to the nation's defense efforts during the early 1940s covered a broad range of ordnance research and development, spanning airborne fire control, small arms design and testing, and major weapons development, including the 280 mm atomic cannon. The disciplines assembled to conduct these programs were those of the physical sciences, primarily mechanical and electrical engineering, stress analysis and metallurgy, and not surprisingly, the initial programs of the newly formed contract research organization utilized those existing skills. Since then, however, the nature of the programs has changed to match the changing needs of our society and in many instances successfully anticipating them; consequently it was in a position to respond on call. Today finds the Franklin Research Center engaged primarily in programs involving energy, environment, health and safety.

FRC has conducted research in the nuclear area for two decades, mostly directed toward insuring its safe application. The projects in this field range from serving in a consultative capacity to the regulatory bodies, to computer aided mathematical analysis of the design of reactors, piping systems and containment vessels, to the development and application of procedures for the experimental evaluation of nuclear power plant components – such as motors, pumps, valves, switches, electric cable and coating materials – to determine their performance under adverse conditions.

During the last five years, owing to the increasing concern over the nation's energy supplies, FRC has broadened its programs in energy to encompass energy conservation as well as other forms of energy: oil and gas; coal; geothermal; and the many variants of solar power, including direct solar heating and cooling; as well as wind and hydroelectric energy. These programs have often called upon the historical strengths of the organization, such as knowledge of materials, stress analysis and mechanical design but, increasingly, have involved market and economic analysis; study of the infrastructure of an industry, the relevant environmental factors and the impact of government regulatory agencies; and the need to inform the public about complex technical, social and economic issues.

The solar programs illustrate well the breadth of activities now conducted by FRC: conception and development of a low-cost, flat-plate solar collector; participation in the design of solar homes, schools and commercial buildings; strategic analyses of the commercial potential of solar energy for industrial clients; development of chemical systems for storage of solar energy; economic analysis of solar assisted heat pumps; study of the impact of solar energy on the electric utility industry; and a broad program of collection and dissemination of information to developers, architects, builders, bankers and the public at large by publications, seminars and conferences, trade shows and traveling exhibits.

Wide Range of Activities

The environmental programs of FRC include the determination of the effects of ocean dumping of sewage sludge; the design of a process and construction of a pilot plant for disposal of municipal wastes; the development of methods for the detection of asbestos in air and water; and a great many studies of the hazards of specific substances for both government and industry. There are also environmental aspects to a substantial portion of FRC's health and safety programs.

Health and safety, taken together and under a broad definition, today constitute the largest area of FRC's activities. As was the case with energy, most safety projects in the early years of the organization relied, and many still do so, upon the physical science and engineering capabilities of the staff. Examples of projects based upon a knowledge of materials include development of prosthetic devices, polymeric materials compatible with human blood and tissue, contraceptive devices, and a technique permitting the nonsurgical insertion of a heart patch. Important human tissue studies have been dependent upon the application of advanced analytical instrumentation. Sophisticated engineering knowledge was essential to the construction of a computer model of the human thorax and use of the model to predict the results of collision impacts.

In recent years, the soft sciences have become an increasingly important component of FRC's health and safety programs. Accident investigation, and studies of highway and tunnel lighting, have melded engineering expertise with the behavioral sciences. Electrical engineering and statistical analytical techniques have been combined in a nationwide survey of the safety of a consumer appliance. Analysis of medical records and participation in the development of health planning systems are wholly systems science activities.

To reflect the increasing recognition that advances in technology can originate anywhere in the world, FRC has greatly expanded its activities related to the transfer of technological information in recent years. Staff and facilities provide services from 11 locations in Philadelphia, Washington, Little Rock, Munich, Luxembourg, Bonn and Tokyo. Using state-of-the-art equipment and techniques, FRC responds to the specific needs of its clients or end users. Functions at any of the 11 facilities may thus include collecting, abstracting, indexing and entering published and unpublished scientific and technical material into automated data bases; operating the resulting data bases and information centers or clearinghouses; collecting, extracting, manipulating and validating technical data; and developing reviews, monographs, special reports and handbooks of specialized technology.

Technological information is disseminated through a wide range of techniques, including seminars and conferences, on- and off-site training, stationary and mobile exhibits, and conventional paper publications. FRC staff also conduct applied research in information science, and provide specialized services such as computer assisted foreign language translation. Major United States government sponsors include the Environmental Protection Agency, the Food and Drug Administration, constituent institutes of the National Institutes of Health, especially the National Cancer Institute and the National Institute of Occupational Safety and Health, the Department of Housing and Urban Development, the Department of Energy, the National Science Foundation, and the Department of the Army. Other sponsors include state and local governments, trade associations, and industrial and commercial firms of all kinds and sizes.

The FRC facilities in Europe and in Tokyo are staffed almost exclusively by nationals of the several countries in which they are based, and the European locations are managed through a subsidiary organization, the Franklin Institute GmbH, headquartered in Munich. In all instances, the overseas locations function as extensions of the home base, conducting projects on behalf of United States clients; increasingly however, they are serving both government agencies and commercial clients of their own geographical areas. These overseas sponsors include the Commission of European Communities, the Federal Republic of Germany's Umweltbundesamt, the International Energy Agency and the Japanese Ministry of Construction, as well as private organizations.

Franklin Institute Press

Since its inception in 1976 as an arm of FRC, the Franklin Institute Press has provided an outlet for publishing FRC generated scientific and technical results, as well as manuscripts from a broad spectrum of distinguished specialists, scientific societies and branches of the National Institutes of Health. The constantly expanding Press catalog currently includes more than 50 books and monographs, three primary journals, five secondary journals, and two newsletters; the subject areas relate primarily to medical – biomedical sciences, analytical chemistry and toxicology, alternative energy sources and related public policy and cognitive sciences. Audiences addressed by Press publications range from highly specialized researchers to the general public and junior and senior high school students.

At present there is a staff of approximately 550, including part-time personnel and those who also serve the Franklin Institute Research Laboratory, Inc. The FRC is headquartered at Parkway at 20th, Philadelphia, Pennsylvania 19103.

FRANKLIN INSTITUTE RESEARCH LABORATORY, INC., FIRL

Dr W. B. Ligett – President

The Franklin Institute Research Laboratory, Inc. was established in 1978 as a for-profit contract research organization primarily to serve industry and other private organizations or persons, and particularly to conduct those programs which are of a proprietary nature and thus not of full and immediate benefit to the public at large. In its early years, FIRL will utilize in the main, through contract mechanisms, the staff and facilities of the Franklin Research Center, but will as necessary also call upon a number of other organizations and individual consultants with whom affiliations have been established.

Current areas of emphasis include market and economic analyses; product and process development; mechanical, electrical and electronic design; failure analysis; acoustics; development and application of sophisticated analytical methods; energy audits and energy conservation; design of data processing systems; design of information systems; and library services.

FIRL is a wholly owned subsidiary of The Franklin Institute and has its headquarters at Parkway at 20th, Philadelphia, Pennsylvania 19103.

BARTOL RESEARCH FOUNDATION

Dr Martin A. Pomerantz – Director

Since its inception more than 55 years ago, research on the cutting edge of knowledge has been Bartol's central purpose. Over the years this unusual organization became internationally recognized for its continuous contributions to several important areas on the forefront of contemporary basic science. Functioning autonomously in carrying out its scientific programs under the guidance of only two directors who have served since its founding, Bartol is unique among American institutions devoted to fundamental research in physical science.

In the early stages of its history, staff members were predominantly Fellows to whom Bartol provided a rare opportunity at the completion of their PhDs to further their education and research insulated from the relatively excessive teaching demands that then prevailed in colleges and universities. As the Foundation grew in scope and size, its educational activities expanded, including, for example, arrangements with degree granting universities through which PhD dissertation research was carried out under the supervision of a member of the Bartol staff. Bartol also served as a haven for faculty from nearby colleges, permitting them to conduct research for which facilities were nonexistent at their own institutions. Later, outstanding undergraduates from the United States and abroad, participating in Bartol's research programs during the summer, experienced firsthand what the life of a dedicated research scientist is like.

With growing realization of the fact that advanced teaching and research are inexorably intertwined, a complete graduate program in physics and astronomy was launched in 1968 through an affiliation with Philadelphia's Thomas Jefferson University, a medical institution with a long and illustrious tradition. Bartol, at its home on the Swarthmore College campus, became the University's Department of Physics. Consequently, nine PhD and three MS degrees were awarded. The aggregate experience thus obtained in graduate science education and the changing professional climate in physics formed the basis for the next step – the establishment of an intimate alliance with a department of physics and astronomy at a major university.

An invitation in 1976 to affiliate with the University of Delaware and to occupy quarters that were made available for Bartol in the same building as the Department of Physics, presented the ideal opportunity to effect an appropriate transition to a new and unique joint program of research and education under conditions that will insure the retention of Bartol's independent identity and special character. Since the move to its new quarters in Sharp Laboratory in 1978, this mutually beneficial relationship has flourished. The Joint Program in Physics and Astronomy is enhancing the high quality and strong traditions of the two parent institutions – The Franklin Institute and the University of Delaware. The combined full-time faculty now exceeds 30 professors with a wide variety of scientific interests, in addition to postdoctoral research associations and the graduate student population of 45.

In addition to its campus headquarters in Newark, Delaware, Bartol maintains field stations in the Arctic at Thule, Greenland, and in Antarctica at McMurdo and South Pole. There is also an astronomical observatory at Swarthmore, and an astronomical program has been initiated recently at the geographical South Pole to exploit the unique advantages of that site.

Interdisciplinary programs ranging from esoteric studies of the relationships between high energy physics and cosmology, to fundamental investigations of solar-terrestrial relations, and finally to the application of newly developed nuclear physics techniques to such diverse fields as marine biology, geology, archeology and various branches of engineering, are all included among the projects currently in progress.

THE FRANKLIN INSTITUTE SCIENCE MUSEUM AND PLANETARIUM

Joel N. Bloom – Director

The Franklin Institute Science Museum and Planetarium traces its heritage to the Cabinet of Models mentioned in the 1824 Constitution of the Institute. The present building and form did not, however, open to the public until 1933 and was then modelled on the Deutsche Museum of Munich.

Today it seeks to engender widespread public understanding and education in science and technology, and an appreciation of their broader cultural and social contexts in the belief that such understanding is a fundamental national need.

In keeping with this general goal, the Science Museum operates a large public museum and planetarium and an associated informal educational program which serves over a million people annually. It seeks to speak to: the facts and processes of science and technology; the past, present and future of science and technology; the promises and limits of science and technology; and the place of science and technology in the broader fabric of society.

The Fels Planetarium of the Institute seats 350 persons and employs a Mark IV Zeiss instrument as well as 200 auxiliary projectors and special effects devices to produce highly dramatic multimedia shows on a wide variety of astronomical topics. In recent

Cosmic ray research activities carried out at an Antarctic research station by the Bartol Research Foundation of The Franklin Institute.

Flight simulator in the Hall of Aviation at The Franklin Institute Science Museum, Philadelphia, Pennsylvania.

years, experimental programming has also drawn upon the visual and performing arts.

The Museum encompasses some 100 000 square feet of exhibit space – 70 000 in the English renaissance building constructed in the early 1930s, and 30 000 in an outdoor science park that opened to the public in 1976. Approximately 50 000 square feet of additional space is devoted to the Planetarium, lecture halls, classrooms and support facilities.

Most of the exhibits are designed and fabricated in house and emphasize visitor participation. Major long-term exhibitions include mathematics, physics, marine, railroading, aviation, energy and the graphic arts. A 707 aircraft, a 350 ton Baldwin locomotive, an 80 foot Foucault Pendulum, a walk-in heart model, a man-made lightning demonstration, an observatory equipped with a 10 inch refractor telescope, a collection of ship models, a flight simulator, and a working model Fourdrinier papermaking machine are featured attractions.

In recent years, extensive efforts have been made to deal with the arts and humanities and social issues primarily through special exhibitions. Typical efforts along this line include: Futures – an examination of the Limits to Growth issue and alternative American futures, developed for the US Bicentennial in 1976; Allegheny West Houses – a full-scale replica of an urban renewal site designed to illuminate both the technological and social factors involved in rehabilitating row houses in Philadelphia; Electronic Music – an exhibition dealing with both the esthetics and physics of this new medium of artistic expression. Films, lecture-demonstrations, and special events supplement the exhibition program.

Ancillary educational programs and outreach efforts include: workshops and courses for the general public; special programs for mentally gifted youngsters conducted in concert with local school systems; Travelling Science Shows, lectures and demonstrations, presented in school auditoriums and public places; a satellite Museum-in-the-Mall located in The Gallery, a major shopping mall in the central business district of Philadelphia.

The Museum has responsibility for maintaining and making useful a number of important artifactual, iconographic, bibliographic and archival collections, among them extensive material of the Wright Brothers, the Graff Collection of architectural drawings, maps and pamphlets documenting the Philadelphia waterworks of the last century, and a great quantity of material on the early history of the Institute and the role it played in the development of industrial America. In recent years, much work has been done to catalog, organize and conserve these research holdings. Efforts are underway to establish the Center for the History of American Technology based on The Franklin Institute Library's major holdings in history, technology, industrial development and related fields. The Center will facilitate the studies and investigations of both staff and outside scholars.

As evidence of its commitment to the primary role of the Museum as an educational institution, the Museum includes a Department of Museum Research. The Department is a national model for such activity, and its studies and research on museum learning, exhibit and program effectiveness, visitor behavior, exhibit labeling and the like serve both the Institute and the museum community at large in a major way.

LITERATURE CITED

1. C. van Doren, 'Meet Dr Franklin' in *Meet Dr Franklin* p. 1. The Franklin Institute, Philadelphia (1943).
2. R. A. Millikan, *Scientific Monthly* (January 1939).
3. C. van Doren, 'Meet Dr Franklin' in *Meet Dr Franklin* p. 14. The Franklin Institute, Philadelphia (1943).
4. B. Sinclair, *Philadelphia's Philosopher Mechanics: A History of The Franklin Institute – 1824–1865*, the Johns Hopkins University Press, Baltimore and London (1974).
5. G. Rennie, 'On hydraulics as a branch of engineering.' *J. Franklin Inst.* **19**, 134–135 (February 1835).
6. Quoted in T. Coulson, 'The first hundred years of research at the Franklin Institute.' *J. Franklin Inst.* **256**, 9 (July 1953).
7. B. Hindle, *Technology in Early America*, p. 35, University of North Carolina Press, Chapel Hill (1966).
8. In its February 1873 Report to the Congress, the US Centennial Commission first notes that others had made suggestions regarding a centennial celebration; it then says: 'But the practical and fruitful inception of the present project was a communication from the Franklin Institute of Philadelphia... making application to the municipal authorities of that city for the use of Fairmount Park, for the purpose of commemorating the one hundredth anniversary of American Independence....'

The manuscript was received 25 May 1979

The Influence of the Interstellar Medium on Climate and Life

The Black Cloud Revisited

PROFESSOR RAYMOND J. TALBOT JR

Department of Space Physics and Astronomy, Rice University, Houston, Texas, USA

Recent studies of the gas and dust between the stars, the interstellar medium, reveal a complex chemistry which indicates that prebiotic organic chemistry is ubiquitous. The relationship between this interstellar chemistry and the organic chemistry of the early solar system and the Earth is explored. The interstellar medium is also considered as likely to have a continuing influence upon the climate of the Earth and other planets. Life forms as we know them are not only descendants of the organic evolution begun in the interstellar medium, but their continuing evolution is also molded through occasional interactions between the interstellar medium, the Sun and the climate on Earth.

Of the many types of astronomical influences upon life, it is natural that the greatest emphasis is placed upon those processes which may have immediate effects upon us. In particular, those effects characterized by timescales ranging from days to hundreds of years are the ones which appear to be the most important because we can envisage that they may affect our lives. Examples are solar-terrestrial phenomena such as the 11 and 22 year sunspot cycles and the Earth's orbit and polar axis orientation which governs the weather changes over a year.

This paper discusses phenomena which fall at the other extreme, that is, processes which probably were important millions of years ago and will probably be important millions of years in the future. However, remoteness in time should not be mistaken as signifying that these processes are not important. Indeed, we would not be here if these processes had not occurred to begin life and mold its evolution.

Two aspects of the interstellar medium (ISM) will be described. The first aspect probably should not be categorized as an influence of the ISM; instead I want to suggest that we are part of the ISM. The second aspect might be called astroecology or astroclimatology; specifically, those ways by which the rest of the ISM can have influenced the environment of the Earth and thereby have affected climate and life on Earth.

Many physicists have been attracted towards astrophysics as a research area because it is inherently an interdisciplinary subject; for example, studies of the structure of stars require applications of classical mechanics, electricity and magnetism, plasma physics, atomic physics, nuclear physics, solid state physics and general relativity. In recent years the chemistry of the ISM has become a very active area of discovery and research. (See *Interdiscip. Sci. Rev.* **2**, 124 (1977). The objective of this paper is to point out that astronomy is now casting its interdisciplinary net even further with astrobiology and astroecology, not to mention the potential for astrosociological and astropolitical studies, should radio searches for extraterrestrial intelligence bear fruit.

ISM — ITS CHEMISTRY

Only during this century have astronomers been able to study the ISM, and the last 10 years have provided an explosive growth of knowledge. For example, about half or more of the ISM is in the form of molecular hydrogen and this has been accessible to study only during the last few years. This new knowledge, combined with recent studies of meteorites and planets, has revealed many new details about the formation of the solar system from the ISM.

It is necessary to review briefly the chemical composition and physical state of the ISM. The first part of Table 1 shows the so-called cosmic abundances of the major chemical elements.[1] To within factors of two or three, these figures represent the abundances found in the Sun, other stars, and the ISM. Hydrogen is most abundant, and there are about one-tenth as many atoms of helium. The next step down are the elements carbon, nitrogen and oxygen, the CNO group. One readily recognizes the coincidence so convenient for us, namely that combined with hydrogen the three most abundant heavy elements happen to be the basic constituents of life, at least life as we know it.

CCC–0308–0188/80/0005–0102$05.00/1

Table 1. Chemistry of the Interstellar Medium

Element abundances: Number relative to hydrogen

H	1.0	C	3.7×10^{-4}	Si	3.5×10^{-5}
He	0.1	N	1.2×10^{-4}	S	1.6×10^{-5}
		O	6.8×10^{-4}	Fe	2.5×10^{-5}
				Mg	3.6×10^{-5}

Interstellar Molecules: Observed

2	3	4	5	6	7	8	9	11
H_2	H_2O	NH_3	HC_3N	CH_3OH	CH_3C_2H	$HCOOCH_3$	$(CH_3)_2O$	HC_9N
OH	H_2S	H_2CO	HCOOH	CH_3CN	CH_3CHO		CH_3CH_2OH	
SiO	SO_2	HNCO	CH_2NH	NH_2CHO	NH_2CH_3		CH_3CH_2CN	
SO	HCN	H_2CS	H_2CCO		CH_2CHCN		HC_7N	
SiS	HNC	C_3N	NH_2CN		HC_5N			
NS	OCS							
CH^+	HCO^+							
CH	HCO							
CN	CCH							
CO	N_2H^+							
CS								

One might surmise that CNO atoms form the basis of life simply because they are the most abundant. However, Wald[2] has effectively argued that this is not so, but rather that it is because of their unique properties: for example, because carbon combines with itself, because of the special properties of H_2O, and because CNO atoms readily form multiple bonds and thus enable the construction of polymer chains.

These abundances are found in newly formed stars and in hot regions of the ISM. However, there are regions where the ISM appears to have smaller amounts of atoms heavier than helium. It is known that these regions of space also contain micron size particles, the so-called 'dust'. In some of the regions the dust is so dense that it appears as an opaque cloud blocking most or all of the light from stars behind it.

By studying the way in which the obscuration of light varies with wavelength and polarization, it is possible to estimate the properties of those dust particles. They appear to consist of a matrix of ices – probably water, ammonia and carbon dioxide – and embedded within those ices are small solid grains of silicates, metals and/or graphite. These compound dust particles are referred to as dirty ice. The investigation of this dust component is one of the most intriguing yet difficult and frustrating tasks in astronomy. Intriguing because of its complexity and the fact that dust clouds appear to be the birth sites of stars. Frustrating because one only has the shadow of the dust with which to work.

One of the significant recent advances in radio astronomy is that it is now possible to study the spectrum of molecules in the ISM. What has been revealed is that the dust clouds are also molecular clouds. It appears that these dust grains are efficient catalytic devices for converting atoms into molecules. Table 1 shows the currently known set of molecules, listed by the number of atoms.

Molecules based upon CNO dominate. The reason can be at least partly attributed to their being the most abundant raw materials; however, the unique properties described by Wald are what enable the formation of molecules with large numbers of atoms. One should keep in mind that this is by no means a complete list. New molecules are being discovered. As this Table was prepared the eleven atom molecule HC_9N was discovered. The omissions from this list are dominated by the limitations of the available techniques. For example, the presence of the linear molecule HC_9N almost certainly indicates the presence of HC_9H; however, the latter symmetrical molecule has no accessible spectral features. It is the asymmetry of the nitrogen substitution which produces an accessible radio line.

THE PHYSICAL NATURE OF THE ISM

Table 2 summarizes some of the physical properties of the ISM. For convenience the ISM has been cate-

PROFESSOR RAYMOND J. TALBOT JR studied physics at the Massachusetts Institute of Technology and California Institute of Technology. After receiving a PhD in physics from MIT in 1969 he has been active in astrophysics research at Rice University, the Institute of Astronomy in Cambridge, England, and at the Lawrence Livermore Laboratory. Since 1971 he has been on the faculty at Rice University. His research interests include stellar structure and evolution, nucleosynthesis, galactic structure and evolution, and the astroecology processes discussed in this review.
Address: Department of Space Physics and Astronomy, Rice University, Houston, Texas 77001, USA.

Table 2. Major Components of the Interstellar Medium

	Intercloud	HI clouds	H_2 clouds
Density (cm^{-3})	$\leqslant 0.2$	$1–10^2$	$10^2–10^7$
Temperature (K)	$\geqslant 7{,}000$	100	10
Fraction by mass	20%	40%	40%
Fraction by volume	90%	~10%	$\sim\frac{1}{2}$%

gorized into three groups according to the density of hydrogen. The intercloud medium is the low density neutral and ionized atomic hydrogen which tends to fill the space between the clouds. Its only importance for this paper is that it fills about 90% of the volume of space, but it comprises only about 20% of the mass.

Clouds are found in a wide range of sizes, shapes and densities. To simplify, the various clouds have been divided into two groups according to whether hydrogen is predominantly in the form of neutral atoms – HI in astronomical notation – or in the form of H_2 molecules. The transition from one form to the other occurs gradually with increasing density, but a density of 100 is a convenient criterion for dividing them. Note that clouds are much colder than the intercloud medium. In fact, some very dense clouds appear to be only about 3 K. Note also that these two categories of clouds contain roughly equal amounts of mass. There are far fewer numbers of the denser clouds, hence they fill a much smaller fraction of the volume.

So far the discussion has included only those interstellar molecules for which there is unambiguous spectral evidence. Table 3 lists some of the more complex molecules that have been discussed, but for which the interpretations are not yet firm. It appears that there is the evidence to support the hypothesis that the synthesis of interstellar organic molecules does not stop at 11 atoms. Within the last couple of years Sir Fred Hoyle and Chandra Wickramasinghe[3–9] have discussed observational evidence for biopolymer chains such as polysaccharides (cellulose) in the ISM. The Russian physical chemist Goldanskii[10,11] has discussed the physical processes by which molecules in the dirty ice conglomerates can be polymerized into very long chains through the action of ultraviolet photons and high energy charged particles. The conclusion drawn by Hoyle and Wickramasinghe is that the interstellar grain factories are capable of producing an extremely wide variety of organic chemical components.

The data which support these ideas are subject to controversy and debate, but they are not totally speculative since there exists some evidence. Hoyle and Wickramasinghe have not been bashful about going beyond the data and on into more speculative ideas. For example, they have discussed the possibilities for the interstellar production of self-replicating peptide chains and interstellar organic grain clumps, having the basic functions to entitle them to be classified as living – specifically, consumption of food (that is, eating other clumps), expulsion of waste and reproduction by division.[5]

Table 3. More Complex Interstellar Molecules Proposed

Evidence	Interpretation
2200 Å Band	Complex organic molecules
4430 Å Band	Porphyrins
3 μm, 10 μm Bands	Polysaccharides
11 μm Absorption	Hydrocarbons like propene (C_3H_6)

ARRIVAL OF ORGANIC MATTER ON EARTH

Figure 1 outlines some topics and processes pertinent to a discussion of the origin of life on Earth and the possible roles of the ISM. Most researchers appear to agree that the sequence of prebiotic chemical evolution proceeds through stages shown going down this chart, but the locations of the events are uncertain. Locations pertinent to the discussion go from left to right; they are the ISM, the primitive solar nebula as preserved in some meteorites and comets, and the Earth. The letter O marks locations where direct observation shows the existence of the stages. Question marks show locations where there is some evidence but its interpretation is not firm. Hypothesized locations are indicated with an H.

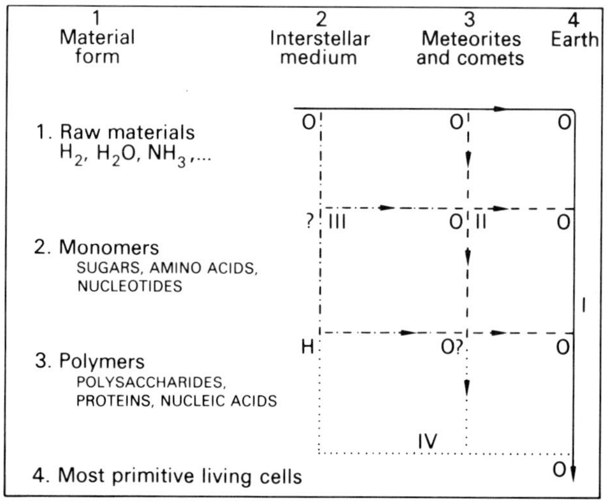

Figure 1. Paths for arrival of organic material and life on Earth. Vertical columns show locations where prebiotic chemical evolution might occur. Horizontal rows show various important stages of that evolution. The path denoted I shows the commonly discussed scenario where the Earth begins with a complement of basic raw materials and all subsequent stages occur in the atmosphere and ocean. Other likely paths are denoted by II and discussed in the text. Less likely and very speculative paths are denoted by III and IV.

The last three columns represent possible parallel sequences; however, it is unclear how independent they are and their relative importance is unknown. The detailed processes which link the stages vertically may vary between columns, but the results at each stage may be similar, simply because they each began with the same raw materials and because the physics of molecular bonds is invariant.

Since the pioneering papers by Oparin[12] and Haldane[13] in the 1920s, laboratory experiments have successfully demonstrated that several types of energetic processes may have produced the conversion of the raw materials in the Earth's primitive atmosphere and ocean into Haldane's primordial soup of amino acids and other biomonomers.[14] These would be processes acting in column 4 of Fig. 1. However, it has not been demonstrated that the monomers were necessarily formed from scratch in the Earth's oceans.

Alternatively, the Earth's monomers may have been acquired ready-made from the organic material in the solar nebula, material similar to that now found in certain meteorites, the carbonaceous chondrites. Chondritic material was produced under near vacuum conditions so its production method was much like our modern methods of freeze-drying. In other words, rather than having been soup made from scratch, Haldane's primordial soup may have been more like instant soup made from a freeze-dried mix.

Support for this comes from studies of the abundances of elements in various classes of meteorites and the planetary bodies such as the Earth, Moon, Venus and Mars. Anders and Owens[15] and other investigators have argued that the Earth's supply of volatiles, like carbon, were not part of the original accumulation of material. Instead, they argue that volatiles came from a late-accreting veneer of carbonaceous chondrite-like material. Although there is no evidence describing the precise form in which the volatiles arrived, since most of the carbon in these meteorites is in the form of organic monomers and polymers, it appears to me that one must seriously consider the prospect that the carbon arrived on Earth in those forms. The extra-terrestrial origin of petroleum has been argued by the late Sir Robert Robinson[16] and Porfir'ev.[17]

Polymerization is the process which links stages in rows two and three of Fig. 1. For the sequence going down column 4, several ideas have been proposed which produce the required dehydration of the dilute monomer soup.[14] Any or all of those special processes may well have occurred, but definite proof is unlikely to be obtained. In contrast, a statement which can definitely be made is that polymerization has occurred in carbonaceous chondrites. About 70% of their carbon is in the form of ill-defined polymers.[18,19]

From what little is known, few if any of those polymers are the highly specialized ones of interest to biology, that is, proteins and nucleic acids. Presumably the material is a collection of all sorts of molecules produced by more-or-less random linkages of whatever monomers were available, a high entropy, random mixture.

Similarly, any prebiotic polymerization process on the Earth would most likely have the same more-or-less random result. The eventual dominance of the highly organized, low entropy, specialized polymers of biological systems may probably be attributed to some form of natural selection. The most hardy polymers – the useful reproducible ones – survived and multiplied in the form of biological systems; the rest were broken down, back into basic components and consumed by the hardier ones.

PANABIOGENESIS

In any theory the materials must have taken a path from left to right in Fig. 1. The commonly discussed theory *assumes* that they arrived by movement across in the top row. Although it is true that the monomers and polymers on the Earth *may* have been produced by processes restricted to the Earth, that is a vertical path in column 4, it is at least equally likely that the material took a path across from column 3 as monomers and/or polymers, i.e. via paths labeled II in Fig. 1.

At present, the existence of monomers in the ISM is not certain and the existence of polymers is even more hypothetical. Nevertheless, if further support is found for them, then one must consider the possibility that they may have contributed to carbonaceous chondritic material and later gone on to the Earth as part of the instant soup mix. This would mean some combination of paths labeled II and III.

These thoughts are summarized by this statement from Hoyle and Wickramasinghe:[8]

> 'We suggest that the origin of life on the Earth involved essentially the assembly of complex chemical components which had already been provided in the last stages of the accumulation of the Earth, and which exist widely and in comparatively great quantities within the interstellar medium.'

This suggestion bears some resemblance to the old idea of panspermia.[20] In that theory small micro-organisms, germs or spores, were cast off from other celestial bodies where life had flourished, and these germs survived a long cold journey through space to fall eventually to Earth. Upon arrival in this hospitable climate, they were resuscitated to flourish again. The modern theory does not carry life from place to place in the universe; instead, it is argued that the production of living organisms takes place throughout the universe – I call it *panabiogenesis*.

The interdisciplinary investigations involved in deciphering the prebiotic chemical evolution stages have been and continue to be very stimulating areas of research. It is an exciting prospect that the merger of efforts in these many areas – astronomy, physics, chemistry, geology and biology – will eventually unravel the ways by which the origin of life on Earth is linked to the chemical properties and processes in the ISM and the early material of the solar system.

One byproduct of these studies is that I believe that we will gradually shift our philosophical perspective to the realization that the solar planetary system and indeed we ourselves are simply one highly complex component of the interstellar medium.

The most far-reaching component of the solar system is its halo of comets which extends from the Sun some 50 000 times the Earth's distance; this is more than one-eighth of the mean distance between stars. Therefore, comets can be called proper interstellar objects, loosely bound to the solar system. Planetary material is that part of the primordial solar nebula which avoided incorporation into the Sun. Hence, one could say that planets are also interstellar objects.

Even if one does not care to stretch the use of the term 'interstellar' in this way, one must concede that the Earth and the rest of the solar system are descendants of the ISM, and hence that we are part of the ISM ourselves. The major difference between ourselves and the interstellar molecules studied by radio astronomers is size and degree of complexity. The essential similarities and differences are displayed in Fig. 2.

The interstellar molecules are made on or in grains about 10 000 Å across. To produce the bonds for large molecules, energy is required; this derives from ultraviolet and visual radiation, as well as high energy cosmic ray particles. Silicate and other solid materials provide a substrate, or even a catalyst, to ease the production of polymers from the icy material.

Earth is not different in its basic features, only size and therefore complexity. It is 10^{13} times larger in linear diameter, 10^{39} times larger in mass and number of atoms. The basic energy mechanisms are similar. The larger number of atoms means that nature has been able to make a larger number of experiments within this single system. Consequently, not only has this system attained and surpassed interstellar grains in the sophistication of its polymers, it has developed sophisticated organized systems of specialized polymers, systems which we call living organisms.

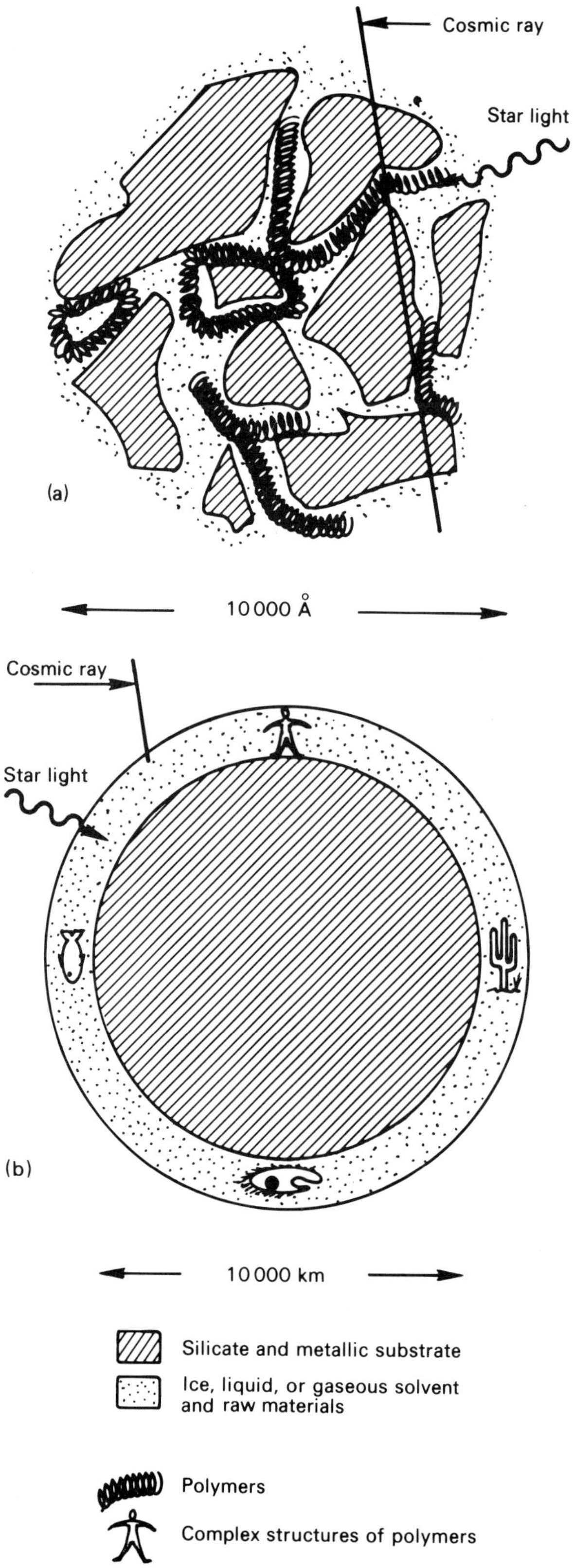

Figure 2. (a) Arrangement of interstellar dust particles believed to be composite mixtures of silicate or metallic or graphite grains and mantles of H_2O, CO_2, or NH_3 ice. (b) Arrangement of the 'interstellar particle' Earth with a silicate and metallic core and an enveloping mantle of solid, liquid and gaseous H_2O, CO_2 and other gases. In both circumstances energy is supplied from the outside by star light and other types of radiation. This energy supply supports the growth of polymer chains.

ASTROECOLOGY: CLOUD DISTRIBUTION

Discussions of the evolution of the solar system generally disconnect from the ISM at the time of formation. This is indeed an extremely good first approximation. The major effect of the ISM is the setting of the initial conditions for the evolution of the solar system.

However, theoretical studies have suggested that the solar system may have been affected by comparatively small continuing influences of the ISM. In particular, the topics of greatest interest are the continuing influence of the rest of the ISM upon our own small piece, the Earth. On the scale of the solar system, the ISM produces a very small perturbation. However, terrestrial lifeforms are fragile components of the solar system, and those small perturbations can have very profound effects.

Recall that about 80% of the interstellar matter is in the form of clouds which I divided into two groups according to their density. Consider the frequency and duration of solar system encounters with these clouds, as shown in Fig. 3.[21]

For a statistical treatment one can regard the clouds as spheres with diameters D and mean spacing L. The solar system moves through the galaxy in a weaving circular path with a varying speed. The present speed is about $20\ km\ s^{-1}$. The system of clouds possesses a distribution of velocity directions and speeds approximately like a Maxwell–Boltzmann distribution, with a mean speed of about $10\ km\ s^{-1}$. Consequently, the relative velocity of the solar system and clouds will vary, but relative velocities of the order of $V = 10\text{–}30\ km\ s^{-1}$ are typical.

The time spent inside a cloud is given by $\Delta t \simeq D/V$. The mean time interval between cloud encounters is given by $\tau = L/V$. The fraction of time spent inside clouds is simply $\Delta t/\tau = D/L$, and this is also equal to the fraction of the volume occupied by the clouds.

Table 4 shows these statistical properties for the interstellar clouds. Shown separately are the lower density HI clouds and two groups of the H_2 molecular clouds.

One sees that denser clouds are farther apart, ranging from about 400 light years (ly) for HI clouds to over 30 000 ly for very dense molecular clouds (30 000 ly is equal to the distance from the Sun to the center of the galaxy). Using a constant characteristic speed of $10\ km\ s^{-1}$, the interval between encounters ranges from about 15 million years to more than a thousand million years. The quantity N_e is computed from the age of the solar system, $4\frac{1}{2}$ thousand million years, divided by τ; that is, it is the number of encounters by the solar system since its formation. N_e ranges from about 400 for the less dense clouds to less than 1 for the densest molecular clouds.

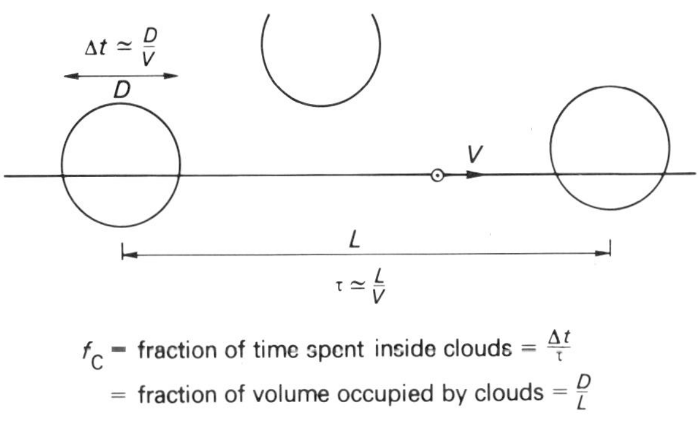

Figure 3. The time spent by the Sun traveling through an interstellar cloud of diameter D is given by $\Delta t \simeq D/V$ if the Sun is moving with speed V with respect to the clouds. The interval between such passages is $\tau \simeq L/V$ where L is the mean distance between the clouds.

Table 4. Statistics of Cloud Encounters[a]

	HI		H_2
Density $n(cm^{-3})$	$10\text{–}10^2$	$10^2\text{–}10^4$	$>10^4$
Separation L (ly)	400	6000	$\gg$30 000
Interval τ (10^6 yr)	15	200	$\gg$1000
N_e	300	20	<1
Diameter D (ly)	30	20	(~2)
Duration Δt (10^6)	1.0	0.7	(~0.07)
Termination radius r_0 (AU)	20	3	<1

[a] Using $V = 10\ km\ s^{-1}$.

Observations show that the mean diameters of clouds do not vary much between these classes, at least for densities less than a few thousand particles per cm^3. Hence, the durations of the encounters are roughly the same for all density categories, about one million years.

CONSEQUENCES OF CLOUD ENCOUNTERS

Now consider the consequences of those encounters. The first question is whether the ISM will even penetrate into the inner solar system. Detailed computations of this problem have not been performed, but simplified calculations suffice to show the primary features of the physical conditions during encounters.

Figure 4 shows the results of these simplified computations for the boundary between the ISM and

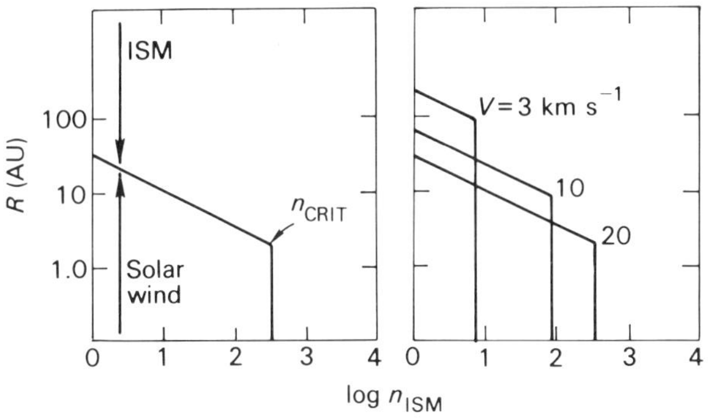

Figure 4. If n_{ISM} is the density of the ISM through which the Sun is passing with speed V, then these figures give the distance R (in astronomical units = AU = distance of the Earth from the Sun) where the solar wind 'terminates' as it reaches pressure equilibrium with the ISM. If $n_{ISM} > n_{crit}$, the attraction of the ISM by the Sun overwhelms the solar wind. The right-hand panel shows how this balancing act varies with V. For very low values of V the solar wind is easily overwhelmed by even the many modest density interstellar clouds with $n \approx 10\ cm^{-3}$.

the solar wind. The radius of this boundary is called the termination radius and its logarithm is plotted as the vertical axis; the units used are astronomical units (AU; 1 AU = the mean distance of the Earth from the Sun = 1.48×10^8 km). The horizontal axis is the logarithm of the density of the hydrogen in the ISM in units of number per cm^3. These figures show how the termination radius varies as we travel through clouds of varying density.

The left-hand panel illustrates the results for an encounter speed of 20 km s^{-1}. The solar wind acts like the internal pressure of a balloon; however, the magnitude of that pressure decreases in proportion to $1/r^2$. The external pressure of the ISM is proportional to the density of the ISM and the square of the encounter speed *V*. The termination radius is the place where these two are equal.

For increasing values of the ISM density, that is, moving towards the right along the horizontal axis, one obtains a decreasing value for the termination radius; that is the line going from the upper left towards the lower right. The solar system is presently in the intercloud medium with density off this scale towards the left; the termination radius for this case is a little over 100 AU.

An important result is that there is a critical radius within which the Sun's gravitational field accelerates or focuses the ISM inward by a significant amount. Of course gravity affects the solar wind also, but the effect is proportionally greater for the ISM because at the beginning it is moving more slowly. The result is that there is an interstellar density, n_{crit}, above which the ISM pressure overwhelms the solar wind at all distances.

When this occurs the solar wind is snuffed out and interstellar material will even begin to fall onto the solar surface; the value of the termination radius is then much less than 1 AU. For $V = 20$ km s^{-1} this occurs for clouds with *n* greater than about 300 cm^{-3}, a moderately common, relatively low density, molecular cloud. This means that for all clouds more dense than 300, the ISM will penetrate to the interior of the Earth's orbit and perhaps even produce accretion onto the solar surface.

The right-hand panel shows what happens for lower values of the encounter velocity *V*. The smaller the value of *V*, the lower the value of n_{crit}. For example, if $V = 3$ km s^{-1}, $n_{crit} = 9$ cm^{-3}; that is a very common HI cloud. Because relative velocities of the solar system and clouds are typically about 10–30 km s^{-1}, a 3 km s^{-1} encounter is expected to be fairly uncommon. However, recall that low density clouds are much more numerous than high density ones. Reasonable estimates of these two competing factors show that there will be a few low speed, low density encounters which will allow penetration of the ISM into the inner solar system.

Table 5 shows the results of making more rigorous estimates of these various effects and the enhancement of the interstellar material density as it is gravitationally pulled towards the Sun. By more careful treatment of the relative speeds of the Sun and the clouds, it is estimated that there have been about 800 encounters with clouds of densities between 10 and 100 cm^{-3}. However, only about two of these will have taken place with the very low relative speeds which permit the material to penetrate to less than 1 AU. For those two events, however, the density at 1 AU will be enhanced by gravitational focusing to over 10^4 cm^{-3}. The luminosity produced if this material falls onto the solar surface is about 10^{-4} of the Sun's present luminosity, not a very significant increase.

Table 5. Conditions During Encounters

	n(ISM) 10–10^2	10^2–10^3	10^3–10^4
Number of encounters: N_E	800	120	16
Number with $n > n_{crit}$	~2	~5	~7
n at Earth's orbit	~4×10^4	6×10^3–10^6	3×10^3–10^7
$L_{ACCRETION}/L_\odot$	~10^{-4}	10^{-5}–3×10^{-3}	10^{-5}–3×10^{-2}

For higher density clouds the number of encounters, N_e, decreases; however, the problem of suppressing the solar wind is made easier so that the net number for which the ISM penetrates to within 1 AU increases. Beyond $n = 10^4$ the number of clouds drops off quickly so that the number of encounters is expected to be less than one.

An interesting behavior of these results is that irrespective of the density of the interstellar cloud, the density of the material penetrating to 1 AU is always about 10^4 cm^{-3} or greater.

Even in the most extreme case the solar luminosity is enhanced by only a few percent. The idea that interstellar clouds might produce significant excursions of the solar luminosity is an old one and it was revived in 1975 by McCrea.[22] He invoked encounters with clouds of densities of 10^5–10^7 cm^{-3} in order to obtain luminosity excursions of the order of 100%. Talbot and Newman[21] investigated the statistics of clouds and showed that there was very little probability of an encounter with clouds of $n > 10^4$. In some other work[23] we discussed the possibility that the lower density clouds may have some effect since their accretion luminosity will appear in the ultraviolet and X-ray passbands. The present solar luminosity in those bands is about 1% of the total.

Consequently, an accretion luminosity of a few percent will be a very high amplitude perturbation in those wavelength intervals even though it is not a large perturbation on the total solar flux. Changes in the ultraviolet flux by 100% will produce large changes in the chemistry of the present upper atmosphere. Under conditions prevailing for the primitive Earth, the ultraviolet flux penetrated to its surface and the thus enhanced flux during that era will have affected lifeforms in the oceans of that time.

During any encounter, the solar wind envelope of the solar system will have been altered. Under present conditions the outward flowing solar wind, together with its embedded magnetic field, provides shielding of the Earth from some galactic cosmic ray charged particles. The received flux of those particles will change during encounters with clouds. The direction of the change cannot be computed with reliability because there are conflicting processes at work: suppression of the solar wind allows cosmic ray penetration to the Earth at 1 AU; however, interstellar clouds themselves act as shields against cosmic rays, so the net flux may decrease. The influence of charged particles upon climate and/or life is not clear yet. Whatever those effects are, passage through an interstellar cloud will alter the situation.

Many investigators have discussed apparent solar-terrestrial connections which show up in the climate.[24] It is not known what mechanism permits the 11 and 22 year solar cycles to influence weather; but it appears that it must be based upon either the solar wind, the solar magnetic field carried out by the solar wind, or charged particle, ultraviolet and X-ray flux variations produced by flares. Whatever the connection mechanism, it will be greatly altered during cloud encounters which suppress the solar wind.

CLOUD ENCOUNTERS: DIRECT EFFECTS ON EARTH

The preceding discussion emphasizes indirect effects upon the Earth modifications of solar processes. There are also direct effects upon the Earth. As noted above, for those encounters where the ISM penetrates to within 1 AU the density at 1 AU is about 10^4 or greater. As the Earth circles the Sun it will sweep up this material. The relative velocity of the Earth and this material will be approximately the vector sum of the Earth's orbital velocity and the free-fall velocity of the material towards the Sun. Simple computation with $n = 10^4$ shows that the Earth will sweep up about 10^{18} hydrogen atoms per cm^2 per year. For comparison, the Earth's atmosphere presently contains approximately 2×10^{25} molecules per cm^2 (N_2 mostly).

Over an encounter duration of 10^6 years the Earth would have an input of about 10^{24} hydrogen atoms per cm^2 (about $1\ g\ cm^{-2}$ or 0.1% of the Earth's atmosphere). Much of that hydrogen would probably escape; however, enough would be present in the upper atmosphere to alter its chemistry significantly. In particular, the combination with oxygen in the upper atmosphere to produce OH radicals and water vapor would greatly alter the transmission properties of the atmosphere.

The ISM contains dust. For length scales comparable with the size of the solar system, the dust is decoupled from the gas so that its density does not experience the same gravitational focusing enhancement; furthermore, solar radiation pressure tends to cancel the gravitational attraction towards the Sun.

On the other hand, the dust is not strongly influenced by the solar wind; however, if it is charged, it is affected by the magnetic field being carried out by the solar wind.[25] If it is uncharged it will penetrate through the solar wind into the Earth at every encounter. About 0.4% of the mass is in the form of silicate and iron grains.

When passing through clouds with densities of 10–10^4 the Earth will sweep up an average of 10^{-9}–$10^{-6}\ g\ cm^{-2}$ of dust per year if they are not charged. The average grain is about 10^{-5} cm in radius, and the opacity of such dust in the visual is $6\times 10^4\ cm^2\ g^{-1}$. The residence time of these grains in the upper atmosphere is a few years, so for a cloud of hydrogen density n the visual optical depth increases by $\Delta\tau = 3\times 10^{-3}\ (n/100)$.

For the large number of encounters with clouds with densities greater than 100 the expected optical depth is greater than 0.3%. Owing to the size of the particles their infrared opacity is some 100 times smaller than the visual opacity, so the infrared transparency is not perceptibly changed, and the cooling of the Earth will not be affected.

However, with an increased albedo in the visual, where most of the input of solar heat occurs, the Earth's surface and lower atmosphere would receive less heat and therefore would cool down. The temperature change of the Earth for this type of process, ΔT, has been computed.[26] For conditions analogous to the infall of interstellar dust over very long periods of time, they find $\Delta T = -35\ \Delta\tau$. Consequently, for passage through clouds with $n = 10$–$10^4\ cm^{-3}$, ΔT is predicted to be of order 1/100–10 K. This would be maintained throughout the duration of the encounter, some 10^6 years. These events will have occurred on the average of once per 10 million years, but the severity will have varied significantly. These changes will be far less if the grains are charged.

The most profound effects occur for clouds in the density range $n = 10^3$–10^4, the value of $\Delta\tau$ (vis) will reach 0.3. For these densities the solar wind and therefore its magnetic field are shut off, so grains will penetrate to the Earth, even if charged. The statistics of these clouds are such that 16 encounters are expected to have occurred, with a mean separation of about 300 million years; the duration of each encounter would be about 1 million years. During these episodes the mean temperature of the Earth should have dropped 1–10 K.

For comparison, the so-called Little Ice Age, from 1450 to about 1850, was about 0.5 K cooler than for the years 1935–1945. The periods of extensive glaciation during the last million years were about 10 K cooler than the interglacial times. Although those particular episodes were probably not caused by interstellar clouds, they provide ample proof that a few degrees of cooling will produce extensive glaciation and attendant severe pressure upon life.

We are presently in a glacial epoch; the polar regions have large ice caps, but during most of the

Earth's history they were ice free. The only two other recorded glacial epochs were the late Precambrian (about 600 million years ago) and the Permo-Carboniferous (about 300). It is tempting to identify this 300 million year quasi-periodicity with the 300 million year average interval between major dust cloud encounters.

It should be emphasized that the dust clouds cannot be the whole story. First, the duration of the individual cloud encounters is about 1 million years or less. Each of the glacial ages mentioned lasted 10–100 million years. One might invoke a dust cloud passage to initiate a glacial age, but then its continued maintainance, or oscillation as the present one is doing, must be attributed to other mechanisms.

The present glacial age is usually said to have begun about 10 million years ago although it is the last phase of a long decline in temperature begun 55 million years ago. If either of these two times were to be attributed to encounters with interstellar clouds, these would be respectively 600 and 3300 light years behind us. It is not easy to verify that prediction by looking for clouds in the path of the Sun. The reason is that the path of the Sun is not a straight line, it wobbles appreciably over times of 10–50 million years. Furthermore, interstellar clouds move with tangential random velocities which are not measureable by us, and in 10–50 million years they will have moved by distances of the order of 30–1500 light years. In spite of my extreme skepticism based upon these uncertainties, it is intriguing to note that there is an interstellar cloud in roughly the right direction and about 300 light years distant. With improved data it may be possible someday to test with precision whether or not we did pass through that cloud about 10 million years ago.

The computed flux of silicate and iron grains is $10^{-10} \times n$ g cm^{-2} yr^{-1}. If each encounter lasts a million years and if we sum over the various categories of cloud densities, we deduce a total influx of 6 g cm^{-2}; this gives an average flux over the age of the solar system of 1.4×10^{-9} g cm^{-2} yr^{-1}. This is about one-half of the average value of 2.4×10^{-9} deduced from lunar soil.[27] In today's intercloud environment there should be an interstellar flux of about 2×10^{-11} g cm^{-2} yr^{-1}.

The peak flux during individual encounters will vary; there will have been many occasions with 10^{-9} g cm^{-2} yr^{-1}, approximately the average, and a few occasions with a flux of 10^3 times that. By studying micrometeor impacts in lunar samples Lindsey and Srnka[28] have found evidence for flux variations which appear to be consistent with this prediction.

When inside an interstellar cloud the environment of the Earth is changed in many ways – the number and energies of cosmic rays, the intensity of solar ultraviolet and X-rays, the interaction between the solar wind and the Earth's magnetic field and the transparency of the Earth's atmosphere. Because the magnitudes of these effects, even their signs, are not yet known, one can say little about their relative significance or their significance compared with other processes which influence climate. However, the rough estimates now available suggest that the magnitudes of these interstellar processes will be large enough to influence climate and life upon Earth.

CONCLUSION

The two basic themes discussed in this review constitute a modern factual version of *The Black Cloud*[29] revisited. These clouds do not communicate their secrets with the same ease as the one in Hoyle's prescient discussion, but deciphering their various influences upon us does provide a similar intellectual fascination.

LITERATURE CITED

1. A. G. W. Cameron, Abundances of elements in the solar system, *Space Sci. Rev.* **15**, 121–146 (1973).
2. G. Wald, Fitness in the universe: choices and necessities. In *Cosmochemical Evolution and the Origins of Life*, edited by J. Oro, S. L. Miller, C. Ponnamperuma and R. S. Young, pp. 7–27. Reidel, Dordrecht (1974).
3. F. Hoyle and N. C. Wickramasinghe, Primitive grain clumps and organic compounds in carbonaceous chondrites, *Nature (London)* **264**, 45–46 (1976).
4. A. Sakata, N. Nakagawa, J. Iguchi, S. Isobe, M. Morimoto, F. Hoyle and N. C. Wickramasinghe, Spectroscopic evidence for interstellar grain clumps in meteoritic inclusions. *Nature (London)*, **266**, 241 (1977).
5. F. Hoyle and N. C. Wickramasinghe, Prebiotic molecules and interstellar grain clumps, *Nature (London)*, **266**, 241–242 (1977).
6. F. Hoyle and N. C. Wickramasinghe, Polysaccharides and infrared spectra of galactic sources, *Nature (London)* **268**, 610–612 (1977).
7. N. C. Wickramsinghe, F. Hoyle, J. Brooks and G. Shaw, Prebiotic polymers and infrared spectra of galactic sources, *Nature (London)*, **269**, 674–676 (1977).
8. F. Hoyle and N. C. Wickramasinghe, Origin and nature of carbonaceous material in the galaxy, *Nature (London)*, **270**, 701–703 (1977).
9. F. Hoyle, A. H. Olavesen and N. C. Wickramasinghe, Identification of interstellar polysaccharides and related hydrocarbons, *Nature* (London) **271**, 229–231 (1978).
10. V. I. Goldanskii, Mechanism for formaldehyde polymer formation in interstellar space, *Nature (London)*, **268**, 612–613 (1977).
11. V. I. Goldanskii, Interstellar grains as possible cold seeds of life, *Nature (London)* , **269**, 583–584 (1977).
12. A. I. Oparin, *Proiskhozhdenie zhizni*, Izd. Moskovkii Rabochii, Moscow (1924); *The Origin of Life*, Macmillan, New York (1938).
13. J. B. S. Haldane, The Origin of Life. *Rationalist Annu.* **148**, 3–10 (1929). Reprinted in J. D. Bernal, *The Origin of Life*, World Publishing, New York (1967).
14. Extensive bibliographies may be found in texts such as D. H. Kenyon and G. Steinman, *Biochemical Predestination*, McGraw-Hill, New York (1969) and S. W. Fox and K. Dose, *Molecular Evolution and the Origin of Life*, W. H. Freeman, San Francisco (1972).
15. E. Anders and T. Owen, Mars and Earth: Origin and abundance of volatiles, *Science* **198**, 453–465 (1977).
16. R. Robinson, The origins of petroleum, *Nature (London)*, **212**, 1291–1295 (1966).

17. V. B. Porfir'ev, *Problem of the Inorganic Origin of Oil*, Naukova Dumka, Kiev (1971).
18. J. M. Hayes, Organic constituents of meteorites: a review, *Geochim. Cosmochim. Acta* **31**, 1395–1440 (1967).
19. R. Hayatsu, S. Matsuoka, R. G. Scott, M. H. Studier and E. Anders. Origin of organic matter in the early solar system—VII. The organic polymer in carbonaceous chondrites. *Geochim. Cosmochim. Acta* **41**, 1325–1339 (1977).
20. S. Arrhenius, *Worlds in the Making*, translated by H. Borns, Harper and Row, New York (1908).
21. R. J. Talbot Jr and M. J. Newman, Encounters between stars and dense interstellar clouds, *Astrophys. J. Suppl.* **34**, 295–308 (1977).
22. W. H. McCrea, Ice ages and the galaxy. *Nature* (*London*), **225**, 607–609 (1975).
23. R. J. Talbot Jr, D. M. Butler and M. J. Newman, Climatic effects during passage of the solar system through interstellar clouds, *Nature* (*London*), **262**, 561–563 (1976).
24. W. R. Bandeen and S. P. Maran (Editors), *Possible Relationships between Solar Activity and Meteorological Phenomena*, NASA publication X-901-74-156, Goddard (1974).
25. E. H. Levy and J. R. Jokipii, Penetration of interstellar dust into the solar system, *Nature* (*London*), **264**, 423–424 (1976).
26. B. Baldwin, J. B. Pollack, A. Summers, O. B. Toon, C. Sagan and W. Van Camp, Stratospheric aerosols and climatic change, *Nature* (*London*), **263**, 551–555 (1976).
27. E. Anders, R. Ganapathy, U. Krabenbuhl and J. W. Morgan, Meteoritic material on the Moon, *Moon*, **8**, 3–24 (1973).
28. J. F. Lindsey and L. J. Srnka, Galactic dust lanes and lunar soil. *Nature* (*London*), **257**, 776–777 (1975).
29. F. Hoyle, *The Black Cloud*, Heinemann, London (1957).

This paper was presented in part at the Symposium on Humans in the Cosmos, Meeting of the American Association for the Advancement of Science, Washington, DC, 15 February 1978.

The manuscript was received 12 August 1978

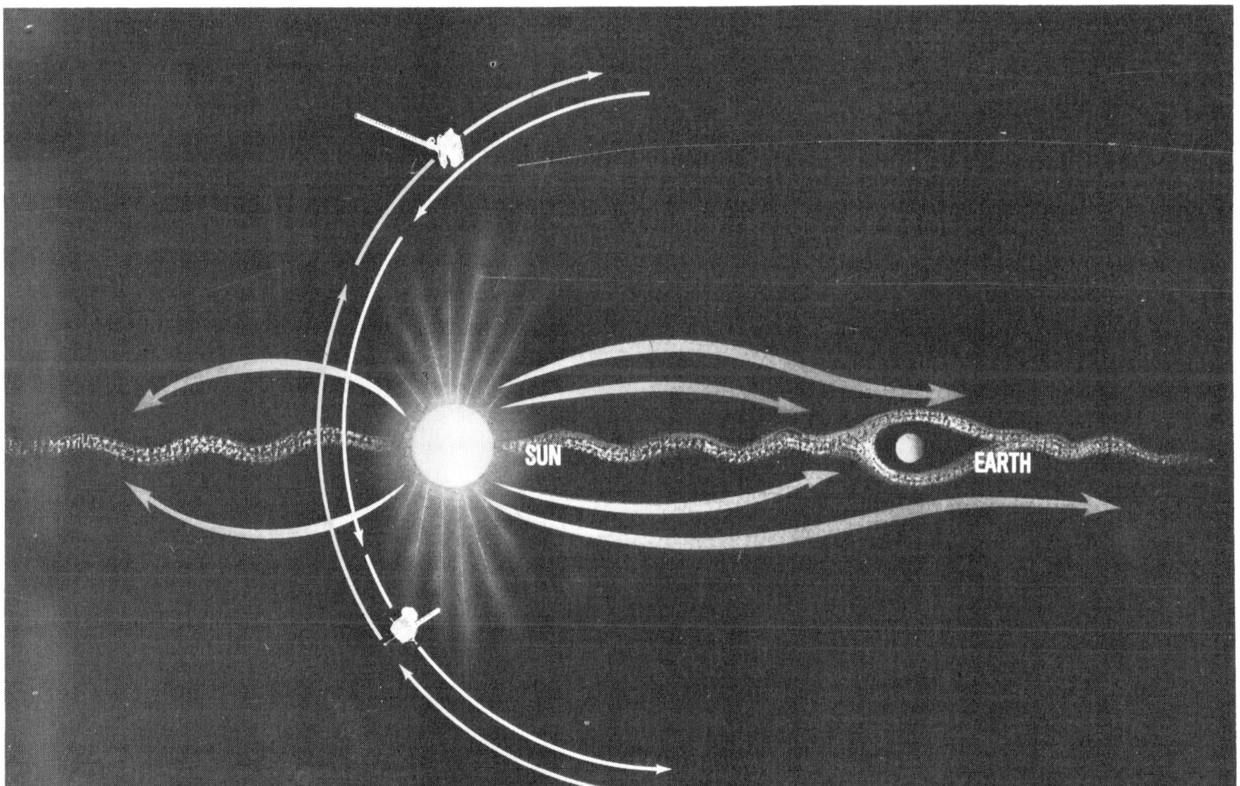

NASA's Proposed Solar Polar Mission. This venture is to explore the three-dimensional space around the Sun by flying two spacecraft out of the ecliptic and over the poles of the Sun itself. The unmanned mission is expected to contribute important new information on many of the solar phenomena – solar wind flow and cosmic rays – that shape and control the space environment of our own Earth. The solar wind is depicted by the snakelike feature in the middle of the picture. The arrows indicate material that comes from the Sun's polar regions and affects the Earth's environment in ways not clearly understood at this time. NASA and the European Space Agency, ESA, would provide one spacecraft each, to be launched in tandem by the Space Shuttle in early 1983. Following a gravitational swing past Jupiter to obtain the energy necessary to travel out of the ecliptic plane, each spacecraft would ultimately pass over both solar poles in a six month time period centering around November 1986. (Courtesy NASA, artist impression)

Australian Archaeology

Implications of Current Interdisciplinary Research

F. W. SHAWCROSS
Department of Prehistory and Anthropology
MAUREEN KAYE
Research School of Chemistry, Australian National University, Canberra, Australia

The last decade has seen a striking upsurge in Australian archaeology. This has been facilitated by the establishment of a reliable chronology and by incorporation of research from other fields such as physics, geology, geomorphology, biogeography, botany, palaeontology and ethnography; the archaeologist in Australia, as elsewhere, finds his work increasingly enmeshed with a variety of scientific disciplines and techniques. What has been the relative fertility of these crossed associations and their benefits or consequences for Australian archaeology? This review will focus on the Lake Mungo locality in western New South Wales to illustrate the interdisciplinary nature of some contemporary Australian archaeological research.

Perhaps the most important result of research at Mungo is that it has shown that 'Australia was not the insular *cul-de-sac* which time and evolution forgot'.[1] Radiocarbon dates from Mungo suggest that Australia was colonized at least 40 000 years ago; for a period of about 30 000 years the shores of the now extinct lake were inhabited. Evidence has been obtained about the physical form of the ancestors of modern Aborigines; something is known about the way of life of these ancient lake shore dwellers – their diet, hunting activities and disposal of their dead. The identification of geological strata has provided horizons in which to order these discoveries; research on palaeoclimates has provided reconstruction of the Mungo environment during the last 50 000 years and in so doing has given environmental reasons for the establishment and abandonment of the settlement.

It was possible to date the geological formations at Mungo only by calling on the extended cooperation of a radiocarbon laboratory. These results then provided appropriate conditions for research in two areas of physical science – thermoluminescence and archaeomagnetism – both of which make use of the alteration by heat of the physical properties of the ancient fire hearths.

The chronology of the archaeological discoveries provided stimulus for some palaeontologists who hoped to discover remains of Pleistocene marsupials accumulated by the ancient hunters. From precise dating of such finds, they anticipated that they could elicit Man's possible rôle in the extinction of the giant marsupials during the Pleistocene.

Mungo re-emphasizes several basic archaeological questions: where had the Australians originated and when did they arrive; how many racial stocks were involved; what were these patterns of dispersion? Current research in anatomy, physical anthropology and genetics is beginning to provide some of the answers, while in turn posing further questions.[2]

AUSTRALIAN ARCHAEOLOGY PRIOR TO THE MUNGO DISCOVERIES

Archaeological investigations in Australia were isolated and sporadic in the 19th century. Natural history[3] had then established the marsupial character of the mammalian fauna, apart from a few easily explained intruders. As Man was the only primate it was difficult, after publication of Darwin's *Origin of Species* in 1859, to argue for Man's independent appearance on this continent. Consequently this raised questions about his origins and time of arrival, questions still apposite to contemporary research.

Late last century most European scholars considered Aboriginal culture a palaeontological survivor from Pleistocene times.[4] Some tens of thousands of years were suggested for its antiquity but, apart from some pioneering work on the possible rate of the formation of shell middens,[5] there was little systematic effort to test this idea.

The discovery of the fossil man *Pithecanthropus* in Java in 1891–1892 was one of the first substantial pieces of evidence in the search for the ancestry of

CCC–0308–0188/80/0005–0112$08.50/1

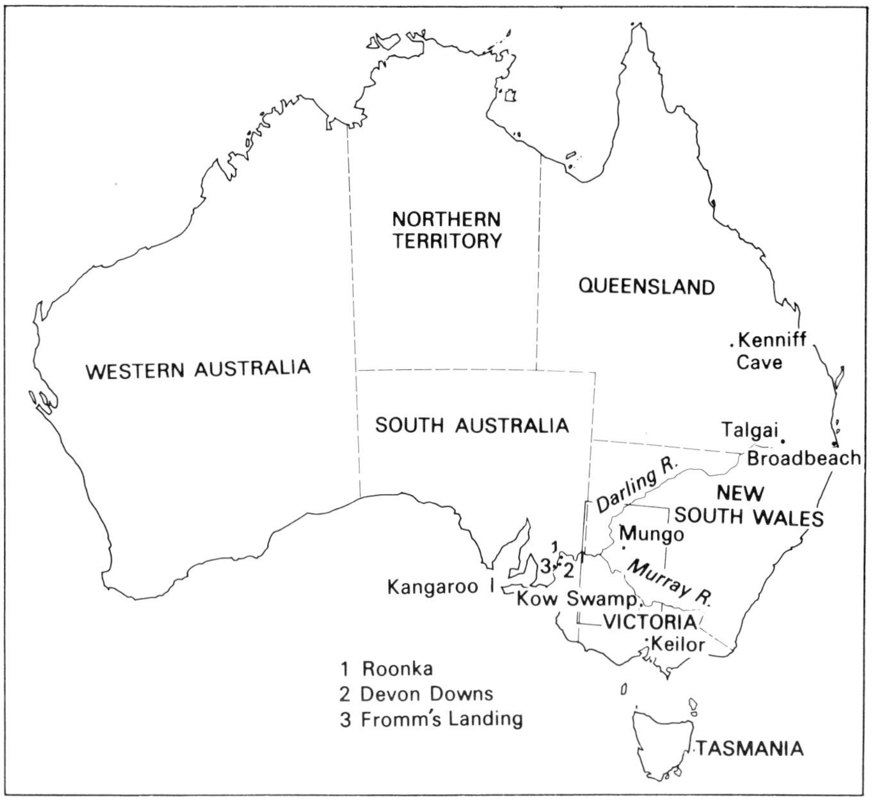

Figure 1. Map of Australia showing Mungo in relation to other archaeological sites.

Aboriginal Man. And this in turn created a more comprehensible background for the subsequent recognition of Australian fossils, such as the first scientific report on the Talgai skull, made in 1914[6,7] (Fig. 1).

In the years between the 1st and 2nd World Wars, the dominant school in Australian anthropology rejected this long chronology and assumed a conservative few thousand years of settlement.[8] This attitude seems to have been influenced by a concern with the sociological present and a reaction against the evolutionist view of Aborigines as survivors of Pleistocene Man.

The most determined supporter of a Pleistocene age was N. B. Tindale[9] who carried out a series of classic excavations in South Australia including the rock shelter of Devon Downs (Fig. 1) which he investigated with Herbert Hale in 1929.[10] Tindale considered that the stratigraphic sequence recorded much of the Holocene. Later he extended this sequence back into the Pleistocene with his recognition of an archaic looking stone tool industry on Kangaroo Island, off the coast of South Australia[11,12] (Fig. 1). Its population had become extinct before the arrival of the Europeans, and he realized it would only have been connected with the mainland during the last glaciation. Tindale's ideas were based on the artefactual and cultural assemblages from his sites and articulated within the broad chronological framework of the Pleistocene and

WILFRED SHAWCROSS is Senior Lecturer in Prehistory (1972) at the Australian National University. He studied his subject at Cambridge, England, and had the good fortune as an undergraduate to be deeply involved in interdisciplinary research into prehistoric animal exploitation in North Africa. Teaching prehistory at the University of Auckland, New Zealand, he pioneered there the study of Marine Biology and Economics in Maori Prehistory. His recent research at Mungo, Australia, has drawn him to semi-arid and lacustrine eco-systems and hunter-gatherer economics.
Address: Department of Prehistory and Anthropology, Australian National University, Canberra, ACT, Australia.

MAUREEN KAYE graduated in chemistry from the University of Cape Town in 1958. She has worked on the analyses of geological, lunar and meteoritic materials at the Universities of Cape Town and Durham, and at the Australian National University. From 1970–1972 she was a co-investigator on the Apollo lunar samples.

Recently she organized a major science exhibition, as part of a national festival of the arts and sciences, *Australia 75*, held in Canberra. The present paper arose from an exhibit, *Glimpses of Early Man in Australia*, which attempted to show how archaeologists use techniques and ideas in the physical and biological sciences, but draw for interpretation on many fields of knowledge in human history and behaviour.
Address: Research School of Chemistry, Australian National University, Canberra, ACT, Australia.

Holocene, a chronology extrapolated from varve dating in the Northern Hemisphere. But his discoveries fell on ground that was barren in congenial ideas; the essential correctness of his pioneer work had to wait until the 1960s for confirmation.

After the 2nd World War there was a quickening of pace in archaeological research, commencing in the museums. F. D. McCarthy of the Australian Museum, Sydney, carried out a series of rock shelter excavations and surveys of rock art sites, and showed that regional differences are a feature of Australian prehistory.[13] An independent but complementary approach was made by the anatomist and physical anthropologist, N. W. G. Macintosh. He systematically investigated Aboriginal skeletal material, in particular the fossil or subfossil finds from the Keilor gravels near Melbourne[14] (Fig. 1).

It was by this time essential for a precise chronology of the Australian finds to be established. The application of radiocarbon dating during the 1950s had provided archaeologists with access to an absolute time scale for the first time. Initially, dating of Australian material was done abroad and then, in 1965, the establishment of a radiocarbon laboratory at the Australian National University (ANU) provided a research facility specifically developed for local conditions.[15]

A New Approach

Until the 1950s Australian archaeology was a minor branch of ethnography applying a fairly rudimentary technology of site excavation and field survey. A new beginning came in 1958 with John Mulvaney's excavation of Fromm's Landing (Fig. 1), a shelter near Devon Downs. Mulvaney, an historian, had been abroad to study the latest archaeological developments which took place after the 2nd World War. The Fromm's Landing excavation introduced new standards of investigation, new theoretical approaches, and was one of the first applications of C-14 dating in Australia.[16,17] During the following decade Mulvaney excavated a series of rock shelters across the continent and achieved progressively earlier dates, which in the central Queensland site of Kenniff cave penetrated the late Pleistocene[18] (Fig. 1). Mulvaney also actively encouraged other archaeological work and interdisciplinary investigations notably with geomorphologists.

During the early 1960s archaeological fieldwork increased substantially, much of it supported by the newly established Australian Institute of Aboriginal Studies. A significant event during this period was the appointment in 1961 of prehistorian Jack Golson to the Institute of Advanced Studies of the ANU, and the subsequent establishment of the postgraduate Department of Prehistory. The new department initiated research programs both within Australia and in the adjacent areas of Papua New Guinea, Indonesia and Island Melanesia. This development marked a shift in the centre of research to the universities. Prehistory was introduced in the Anthropology Department of the University of Sydney which encouraged research on the Australian mainland and Tasmania. By the end of the decade, Mulvaney had produced the first synthesis of Australian prehistory with his book, *The Prehistory of Australia.*[19]

The 1960s witnessed not only an increase in the amount of archaeological research but 'new intellectual cross currents [which] originated in diverse disciplines both within Australia and abroad'[1] significantly altered its emphasis. This new approach crystallized with a series of seminars held in Canberra in late 1968, and published under the title, *Aboriginal Man and Environment in Australia.*[20] It represented, as the editors, Mulvaney and Golson, said, 'the most comprehensive attempt yet made in Australian prehistoric studies to correlate research data in different disciplines, to assess the current state of knowledge and to explore avenues for future interdisciplinary research'. Papers were presented by nine prehistorians, three anthropologists, a historical geographer, a linguist, a physical anthropologist, a geneticist, a biologist, a plant ecologist, and six geologists.

At one of these seminars, the geomorphologist Jim Bowler[21] announced evidence of human occupation between 25 000 and 30 000 years ago at the extinct lake at Mungo (Fig. 1) in western New South Wales, thereby nearly doubling the previous most reliable estimate for the antiquity of Man in Australia.

The Mungo discovery can almost certainly be credited to a working association which had developed some years previously between Mulvaney and Bowler. Without this personal contact it is possible that the formal boundaries of their disciplines might have inhibited the transmission of information and the encouragement that was necessary to establish the discoveries.

With even only a few years for perspective, it seems that these finds came at the right moment for Australian archaeology – they gave impetus, direction and vitality to the informal, if not invisible 'college' which had developed during the 1960s.

MUNGO – PHASE 1

In July 1968, while studying the stratigraphy and Quaternary geology of an extensive system of extinct lakes in the semi-arid region of western New South Wales, Bowler recognized stone artefacts, freshwater mussel shells and fragments of burnt, carbonate encrusted bones in deposits whose age he had already established as late Pleistocene. He realized this material could provide significant evidence for the antiquity of human occupation and that it might assist recent research on the possible rôle of man in the extinction of the so-called marsupial megafauna. He marked the site for further investigation by archaeologists.

Mungo – Its Geological History

Lake Mungo lies in a landscape which can be described as a fossil of the Pleistocene (Fig. 2). It is one of a string of relict lakes, the Willandra Lakes, now a series of oval scars, approximately 25 by 10 km, incised between 10 and 30 m into the surrounding plains. Under present semi-arid conditions, even with exceptional rains, catchment is insufficient to fill the lakes.

Bowler[21] had traced the history of the lakes (Table 1), which correlates with the well defined series of ice ages characteristic of the Pleistocene in the higher latitudes of mountainous regions of the world. Then a single narrow channel joined the Willandra Lakes which were overflow ponds for excess water from the Lachlan River; melt waters from the partially glaciated eastern highlands generated this flow. During periods of high level these lakes would have been fertile oases in a semi-arid region.

Evidence for the detailed history of the lakes comes from the sequence of wind laid sediments of the distinctive crescent shaped dunes or lunettes which mark the eastern margins of the lakes. The dune formation appears to have been cyclic, commencing with high level beaches marking the filling of the lakes. There followed a deposit of quartz sands blown up from the lake shores, just as dunes are formed on exposed sea coasts. The lakes declined when evaporation exceeded floodwater

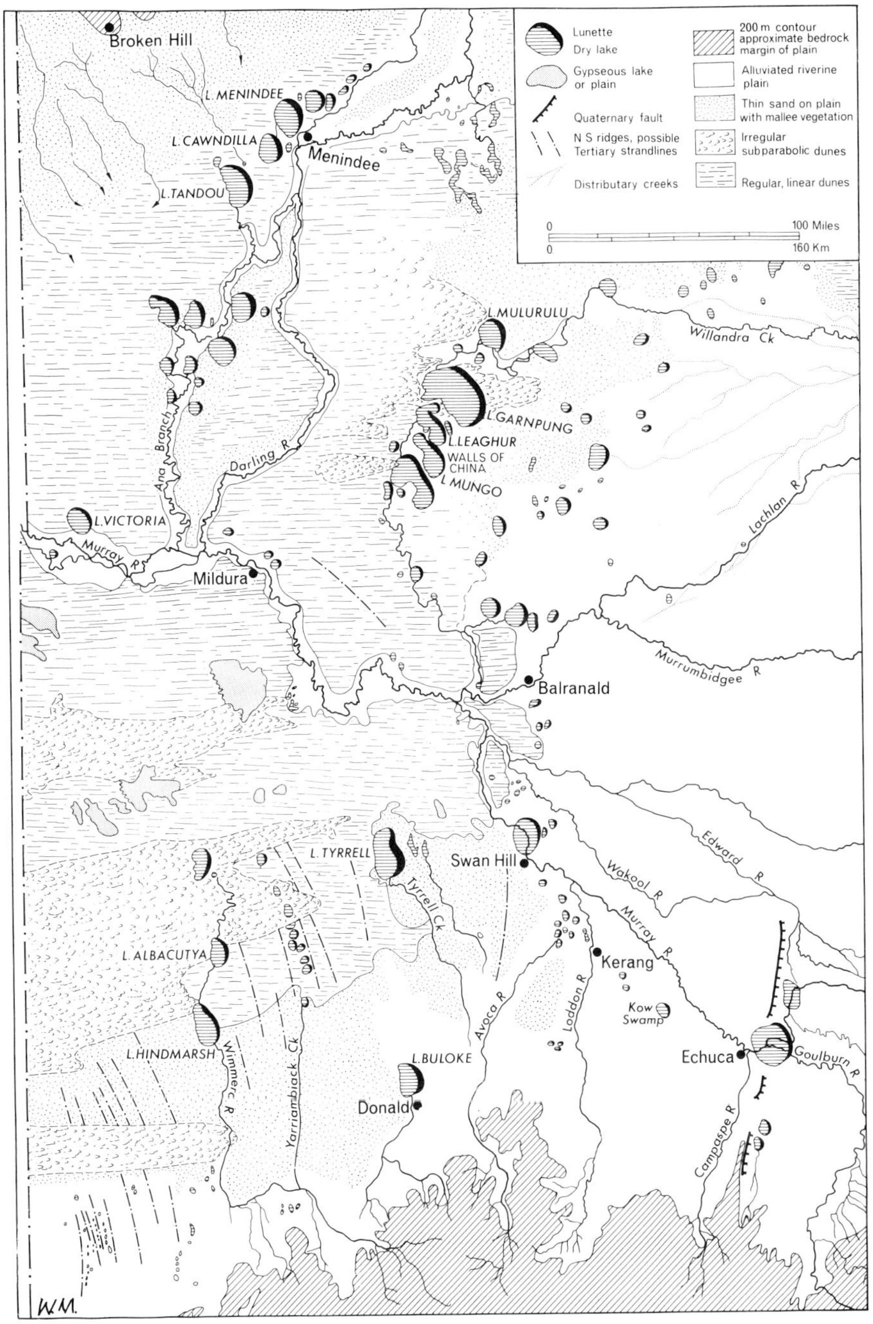

Figure 2. Geomorphic map of north-western Victoria and south-western New South Wales showing the distribution of lunette lakes. (J. M. Bowler.)

Table 1. Environmental changes in south-eastern Australia during the last 50 000 years and their implications for Lake Mungo (Bowler[22b])

SOUTHEASTERN AUSTRALIA	LAKE MUNGO
Before 40 000 years ago	
Climatic conditions are similar to those of today, perhaps slightly colder. The mountains are free of permanent ice, although they bear traces of earlier glaciation in the form of ice-cut basins (cirques). The rivers flow westwards across the plains and by-pass the dry lakes.	The ancient lake basin has been dry for at least 60 000 years. The basin is rimmed on the east by a very old dune ridge which has a dark red soil developed on it. Vegetation covers the lake floor and dune.
40 000–25 000 years ago	
Cold conditions produce permanent snow and glacial ice on the mountains, modifying earlier valley forms. On the plains, the rivers are greatly enlarged and the lakes are filled. The absence of trees permits strong westerly winds to blow sand dunes from stream beds.	The lake has now filled and strong westerly winds produce waves. These winds blow the sand off the beaches so that the red dune is covered. The dune grows to a stable height and shape, and is covered with vegetation, including trees. Men camp, hunt, fish and collect mussels at this lush oasis.
25 000–15 000 years ago	
The snow and ice is reduced on the mountains and trees return to lower slopes. The lakes are reduced in size and rivers change their size and shape.	As the lake water drops and the lake becomes saline, a barren landscape is produced. Strong westerly winds whip up clouds of dust from the drying mud flats, and deposit layers of clayey silt and clay pellets over the dune. The wind-borne salt kills the vegetation. The lake and its associated environments are 'dead'. Men no longer frequent its shores.
15 000 years ago to present	
The disappearance of permanent snow and ice leaves relict glacial landforms near the highest peaks. The treeline rises to its present level. On the plains, the lakes are dry and the rivers are much reduced in size. The rivers change courses leaving ancient abandoned channels such as the fossil Willandra Billabong Creek on which Lake Mungo is located. Sand and clay-rich dunes located along the former lakeshores are legacies of earlier climates.	The lake has dried. Salts are leached down through the lake-floor sediments and the dune. Salt bush grows on the old lake-floor and trees grow on the dunes. Erosion of dune by rain water produces deep gullies. The eroded sands are blown by westerly winds to the east, forming large mobile white sand dunes.

input, and under conditions of increased salinity the sand dunes were capped by much finer, wind blown sediments containing salt crystals. The cycle was completed by the stabilization of the dune and formation of soil.

Bowler[21–23] has identified three such cycles which he correlated with late Pleistocene climates. The latest, named Zanci, commenced with a lake filling about 18 000 BP (years Before Present) and ended its accumulation of sediments about 15 000 BP, coinciding with the final glacial maximum in both Europe and America. The previous cycle, named after Mungo itself, began at an estimated 45 000 BP, just beyond the limits of radiocarbon dating with the available samples, and ended about 22 000 BP.

During the approximately 4000 years between these two depositional cycles a red soil was formed on the surface of the Mungo sediments. Its colour, probably due to oxidized windborne particles from the desert interior, serves as a distinctive marker horizon throughout the lunette. The earliest cycle, the Golgol, is beyond the range of radiocarbon dating. However, its especially deep cap of soil has suggested a correlation with the last Interglacial and an estimate of sediment formation going back beyond 100 000 BP into the penultimate Ice Age.

The Reconnaisance Survey

The Willandra Lakes lay within the field area of the prehistorian Harry Allen who was studying the origins and development of the intensive wild cereal gathering, recorded in the 19th century ethnography of the area. Some 200 km to the south at Kow Swamp (Fig. 1), the physical anthropologist, Alan Thorne, was investigating a remarkable assemblage of human burials of highly distinctive archaic looking individuals.

Another member of the 'college' who participated in the Mungo discoveries was Rhys Jones, then writing up his research on Tasmanian archaeology.[24] This island's long isolation from the continent suggested that its material culture might have preserved elements of that of the mainland during the Pleistocene. Jones was also interested in the geographical and biological factors relevant to Australia's early settlement. One of these factors was the extinction, at an unknown date, of several giant forms of marsupials, including the disappearance of whole genera such as the diprotodonts. He wondered whether they had followed the fate of numbers of Pleistocene mammals in the Americas as a consequence of human settlement, a theory propounded by P. S. Martin.[25]

In March 1969, Bowler returned to Mungo with a party of earth scientists and archaeologists including Allen, Jones and Mulvaney. The archaeologists were attracted by Bowler's report of fossil bones associated with stone tools. (Mungo is not far from Lake Menindee (Fig. 2) where tantalizing but inconclusive evidence had been found for the association of giant marsupials and artefacts.[26])

This group of specialists had an overlapping range of skills basic to their various disciplines which separately would not necessarily have isolated the site's true significance. Bowler had already established a stratigraphic sequence and chronology for the dune sediments, and had delineated the Quaternary environment. The archaeologists realized that many of the scatters of stone tools covering hectares of the lunette certainly belonged to the earliest recognized cultural stage in Australian prehistory, though never previously dated as early as here. Their acquaintance with deposition of food waste on archaeological sites led them to identify cooking areas, hearths and items of diet. From a knowledge of Aboriginal burial practices they suggested that a dense patch of calcined bones was not food remains but a human cremation, an identification later confirmed by John Calaby of CSIRO and Alan Thorne of ANU. The bulk of this cultural evidence, including the cremation was revealed by erosion to be stratified about 20 cm below the surface of the Mungo fossil soil and its age of 24 500–26 500 years was subsequently determined.[27]

This skeleton, Mungo 1, which had been cremated and then pulverized, was that of a young adult female; it fitted within the physical range displayed by modern Australoids.[28] Therefore the discovery filled a palpable void in Australian prehistory – the lack of reliably dated skeletons from earlier stages of the population. The evident deliberation involved in the corpse's disposal implies a great antiquity for Aboriginal death ritual. From the standpoint of our own cultural values there are at least two possible implications in such a rite. First, the cremation is evidence for some form of human self-awareness, perhaps even a concept of personality. Alternatively, the cremation was an attempt at utter transformation and destruction of the corpse, presumably implying a concept of a spirit occupying the body.

The Mungo discovery is possibly the oldest identified cremation, but there are a number of examples of significantly earlier probable death rites associated with the Neandertalers of Eurasia. Perhaps the best known is the Shanidar inhumation, apparently accompanied by a floral tribute.[29] However, no generalization can be made about Neandertal Man's respect for the dead, since other instances are known of an apparent utter disregard for the corpses, for example at the French Grotto of L'Hortus.[30]

Within a few days, the reconnaissance party had perceived the significance of this combination of findings and established the soundness with a fairly high level of probability. Would this be the essential discovery, requiring only follow-up work to increase the level of probability of confidence, or was it an intermediate key to further discovery? Conflictingly, both answers are true. In broad terms the processes responsible for the sediments and the breaks and transformations in sedimentation had been identified at the time of the discovery.

However, cultural processes involved in archaeological deposition are relatively less well understood than physical forces in geomorphology where the principle of uniformitarianism finds models from the contemporary world. But the archaeological models, if based on observation of contemporary human behaviour, are cut across not only by cultural variability but also by the irreversible or linear nature of human history. Admittedly, some historians such as Spengler and Toynbee have held a cyclic view of the broad scheme of their subject (see Collingwood, 1966)[31] but, at the risk of deep entanglement in historiography, it must be emphasized that history does not literally repeat itself.

The prehistorian cannot roll back time and expect to find Aboriginal culture operating in exactly the same manner as it was when it might have been observed in the 19th century. We deliberately use 'might' since observation of cultural phenomena is inevitably partial and highly selective.

'Where the Crow Flies Backwards'[32]

Following the reconnaissance survey, the initial archaeological problem was to identify the cultural factors relevant to the settlement. Harry Allen made surface collections of food waste and artefacts from several areas, totalling 120 m^2, associated with hearths and the cremation site (Fig. 3). He estimated from the level of erosion at which items were exposed that they were approximately contemporaneous with the cremation.[32,33]

The collection of 470 artefacts showed a high proportion of deliberately fashioned tools as against unworked and waste flakes, and was characterized by fairly large pieces. From the high proportion of finished tools to waste material, Allen inferred that they were manufactured elsewhere. Most tools were made from silcrete which outcrops on the western shore of the lake.

The food waste represented small birds, emu eggs, shell fish, several species of marsupials such as wallabies, kangaroos, native cats and rat kangaroos, and significant numbers of fish, particularly golden perch; there was certainly no hint of giant extinct marsupials but rather a preponderance of smaller marsupials and fish. The otoliths of the latter, identified with help from freshwater fishery experts, made it possible to estimate the sizes of the original fish.

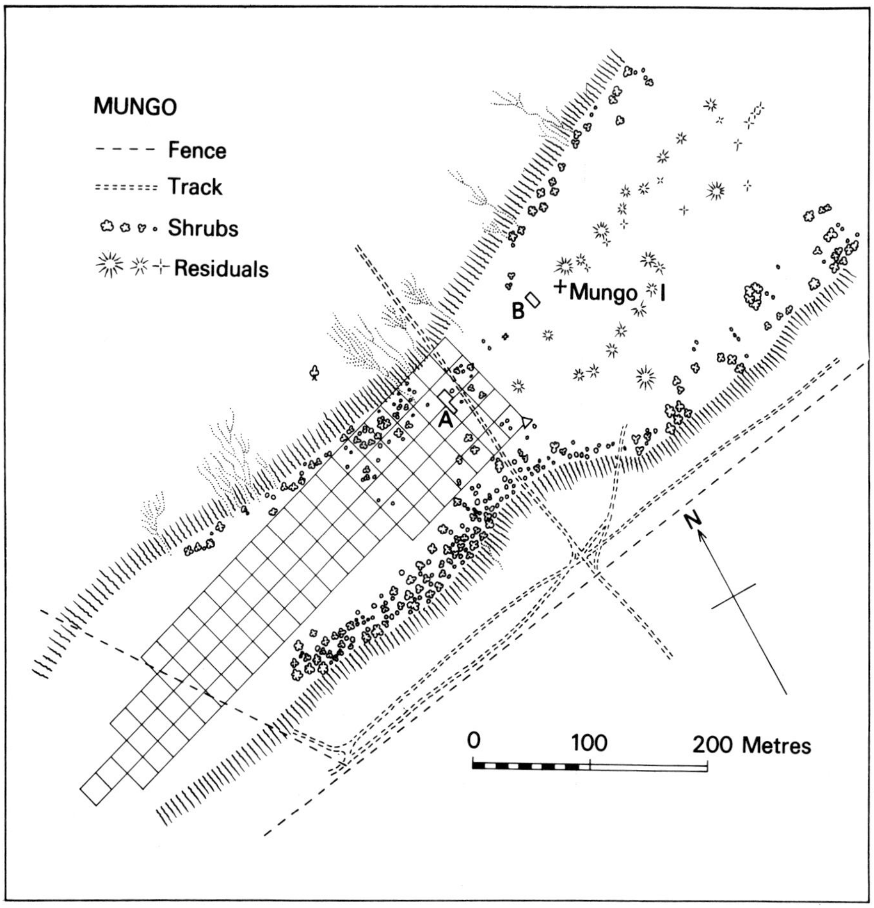

Figure 3. Plan of south-west end of Mungo lunette showing excavation areas A and B and the gridded surface collecting area.

Previously, Allen had studied the 19th century documentary evidence on the Aboriginal people of the Darling River area. He suggested that the former occupants of the Willandra region might have practised a seasonal pattern of migration resembling the riverine people last century, who were attracted to the rivers in time of flood, when fish might be plentiful, and again during spells of low rainfall, but would disperse in small groups during rainy periods. (Rainfall is not markedly seasonal in the region so that this dispersal would have been opportunistic.) Apart from shellfish, the majority of the food animals are still extant in the area and correspond as closely as one would expect to the animal foods eaten last century. The only important segment of the diet missing from the archaeological record, and this is negative evidence, was cereals, whose archaeological evidence would be grinding stones.

Allen's main impression was 'one of a single continuous cultural tradition. Changes took place in this tradition during Man's 32 000 year history in the area, but these were not so great as to destroy the impression of continuity.'[32]

The Giant Marsupials

It had been anticipated that Mungo might clarify Man's rôle in the extinction of the giant marsupials. Genera, such as *Diprotodon*, in some respects a marsupial analogue of the rhinoceros; *Thylacoleo*, a large carnivore; *Zygomaturus*, a large quadrupedal herbivore; and *Procoptodon* which resembled the hopping macropods, are all known from Pleistocene, prehuman deposits[34] and also from a few dubious contacts of association with man. It was resonable to compare the Australian situation with that in other areas, notably America and New Zealand, where the late appearance of Man correlates with a sudden and drastic increase in the rate of extinction of the native fauna.[25]

Presumably, an early site such as Mungo, located broadly within the inferred environment of these extinct marsupials, might well carry evidence in its earliest occupation layers for the destructive impact of the new predators upon the fauna. The prospect of such a discovery was sufficiently promising and, if confirmed, so important that a marsupial palaeontologist was brought into the interdisciplinary team. In this one respect the outcome of field collecting in the first phase was rather unsatisfactory, for no specimens of the extinct giant forms were discovered though, in such a situation, negative evidence is notoriously much weaker than positive.

Archaeometric Studies at Mungo

Following a suggestion by Mulvaney and concurrent with Allen's research, a project investigating the history of the Earth's magnetic field was undertaken at Mungo by two ANU geophysicists, Mike Barbetti and Mike McElhinny.

Throughout the existence of the Earth's magnetic field, neither its direction nor intensity have remained constant. Apart from small-scale secular variations, there have been world-wide reversals of its polarity lasting for periods of about a million years. There is also some evidence for geomagnetic excursions, lasting up to a few thousand years, when the field almost reversed its polarity but then returned to its original configuration.

The history of the geomagnetic field can be traced using iron oxide minerals which have been heated above 700 °C. When these minerals cool down in the Earth's prevailing magnetic field they become stably magnetized in the same direction as that field; the strength of the magnetism they acquire is proportional to the intensity of the causative field. The material is said to have acquired thermoremanent magnetism. The strength and direction of a new field will be imprinted if there is subsequent heating and cooling of the minerals. Both these parameters, the ancient field direction (provided the orientation of the material has remained unchanged) and strength can be measured. If dates can be determined corresponding to the time of the last heating, then a 'spot' value for the ancient geomagnetic field can be determined. (This is necessarily a simplified and condensed account of thermoremanent magnetism, and the reader is referred to *Physics and Archaeology* by Aitken.[35])

When the Mungo Aborigines had used their hearths, which contain iron oxide minerals, they unwittingly preserved a record of the geomagnetic field prevailing at the time. As these hearths were well preserved, had remained undisturbed, and had been reliably dated, they were ideal for archaeomagnetic studies.

And, indeed, the hearths had preserved a record of a geomagnetic excursion which had occurred between at least 31 000 and 28 000 years BP.[36] During this event, now known as the Lake Mungo Geomagnetic Excursion, the field direction departed from the present direction by up to 120°. Associated with it was a five- to six-fold increase in the ancient field strength over the apparent background field intensity. McElhinny and Barbetti tentatively suggest the occurrence of a second excursion associated with low field strengths around 26 000 BP.[37]

Obviously these results, with their implications for processes occurring within the Earth's core, were of particular interest to geophysicists. But there were also archaeological implications: variations in the Earth's magnetic field affect the concentration of C-14 in the atmosphere and hence radiocarbon ages. The production of atmospheric C-14 decreases when the intensity of the geomagnetic field is high since the cosmic-ray flux is reduced; and conversely C-14 production increases when the strength of the geomagnetic field is low.[38]

However, it has not been firmly established whether the Lake Mungo excursions and their associated changes in field strengths were local or global phenomena, and consequently whether they affected the overall cosmic ray flux and, in turn, the C-14 age determinations.

End of Phase One

The initial phase of interdisciplinary research ended with the completion of several PhD programs – Bowler, Allen, and Barbetti. Some members of the original group left Canberra or moved into other lines of research; for those who remained, particularly Bowler and Mulvaney, Mungo represented an uncompleted challenge.

The first phase had followed a natural progression with defined questions, tasks and limits flowing from the original discovery: anatomical reconstruction of the cremation had been made; cultural and economic evidence from the same horizon as the cremation had been collected and studied; palaeomagnetic studies of the hearths had been completed; and stratigraphic analysis of the locality had been concluded. In addition to furthering our knowledge about Australia's early settlement, these discoveries were able to contribute to their respective disciplines.

More could not necessarily be achieved by doing more of the same. A stage was reached when, if new knowledge was to be discovered, new questions would have to be asked. These questions would require application of more specialized aspects of the disciplines originally involved. In retrospect, this resulted in a lower level of interaction between the disciplines even though their results continued to affect each other. At the time it probably appeared that the disciplines would continue their mutual commitment as under the original impetus of discovery.

It would be impossible to sustain a research program on occasional, fortuitous exposures of human skeletons even though the drastic effects of erosion meant that archaeological and anatomical evidence was likely to be exposed and disturbed or destroyed. Furthermore, there was no point in repeating the stratigraphic work in the locality; the logical step to look for the sequence in other areas was taken.

It was in this transitional state of possessing a potential storehouse of discoveries and devising new keys to gain access that the second phase began.

MUNGO – PHASE 2

The archaeological involvement of the ANU with the Mungo area increased further with the creation of an undergraduate teaching department in 1972, as a complement to the existing postgraduate research department. Under Professor Mulvaney's direction, it established new staff appointments, created a demand for field work for practical training, and provided a force of students who could contribute to such work. Consequently it was possible to plan more labour intensive research and, indeed, with a few exceptions, to plan work on a larger scale than previously attempted in Australia.

Planning the Second Phase

In planning phase two there appeared to be three lines of archaeological research. First, it would be desirable to establish the detailed stratigraphic history of occupation. Hitherto ages of finds had been related, where possible, to an apparently densely occupied horizon about 20 cm below the surface of the Mungo soil, or in default, to a later age than this. The usual archaeological approach to stratigraphy is systematic trench excavation, where the relative levels at which all finds are made are observed and related to the visible sedimentological record in the sections. Mungo, with its highly dispersed occupation, offered a new problem by Australian standards, where most work has been in the concentrated deposits of caves, with densities of thousands of artefacts per m^3. It would be necessary to establish the scale of excavation required to achieve reasonable visibility of the archaeological remains. Furthermore, excavation to any depth in the unstable sands of the lunette would necessitate opening up a large surface to reach only modest depths (Fig. 4).

This approach could be combined with an archaeologically more ambitious second research objective: the 'area excavation' of a significant extent of ancient land surface, sealed by subsequent deposition, in order to establish some identifiable pattern of occupation or activities.[39] The objective would be to uncover a Pleistocene campsite, that is, a complete cluster of archaeological features representing the occupation of a human group such as a family or band. This too presented a new problem of scale for Australian archaeology.

With the exception of several cemetery excavations, notably those at Roonka in South Australia, Kow Swamp in Victoria (Thorne[40]) and Broadbeach, Queensland (Haglund[41]), the limits of small team excavations had discouraged the exposure of open site plans, although some tentative observations of living floors have been made within the confines of rock shelters. The model for this research objective comes from the discovery of a number of ancient 'living floors' or 'working floors' from Pleistocene times, particularly those found in East Africa during the past decade or so. The best known are the circular 'hut' at Olduvai[42] and the floors at Olorgesailie.[43] These Pleistocene sites catch the imagination through their immediacy, for one may see on some of them a selective litter of rubbish from human activities of hundreds of thousands of years ago, left as though they had been abandoned for only a short while. Behind the emotional immediacy is the serious challenge of the scientific investigation of these sites as evidence for ancient forms of social organization and indeed, with qualifications, of behaviour itself.[44]

The third objective was to record the density, distribution and probable chronology of the materi-

Figure 4. Completion of excavation area B. The upper, horizontal band is the Mungo soil. The excavation is stepped to avoid the dangerous collapse of the sandy deposits. (F.W.S.)

als eroding at the surface. This required the preparation of a large-scale geological map of the collecting areas and the creation of an extensive, permanent grid to locate finds. To establish the relationship between erosion and exposure it was planned that the surface should be collected annually. Again, the problem of scale arose, for the total area of the lunette is some 540 ha. The sample area of more than 36 000 m^2 that was finally gridded was, by any archaeological standards, an ambitious project (Fig. 3). The third objective could be combined with the second objective of studying the prehistoric settlements on the lake shore and would also help to save evidence from destruction by erosion.

The operation of this research program would ensure the regular presence of competent fieldworkers who could scan the lunette for new exposures of human remains, and their presence would also provide a bridgehead for an expansion of archaeological survey within the region, an area of some three million hectares, that is, about the size of Belgium.

The three archaeological objectives were dependent upon the continued concern of the geomorphologists, including the preparation of the geological map. In addition, Bowler initiated a study of the soil formation processes at Mungo, which was undertaken by a pedologist, Anthony Dare-Edwards. Although the results are of primary concern to the study of soil formation in an arid environment, they have relevance to the archaeological interpretation of the site. The geomorphological study of the area was also extended by Bowler's research assistant, John Magee, into investigations of the sedimentological history of the adjacent lakes. In addition, an independent research project, applying thermoluminescence dating to the burnt clay hearths, was set up.

Thermoluminescence Dating at Mungo

The direction of work on thermoluminescence dating of the prehistoric baked clay hearths under the physicist, Allan Mortlock, arose from a suggestion by Barbetti.

By the late 1960s thermoluminescence dating (TL) had been developed into a usable technique for dating ceramics. (The principles and problems of TL are discussed by Aitken.[35,45]) The method is almost independent of variations in the cosmic ray flux which has a much greater effect on C-14 determinations. TL has been used to 100 000 years and beyond. In contrast to the radioactivity measured in radiocarbon dating, TL emission increases with the age of the sample, although a saturation point may be reached.

Radiations from long-lived radioactive isotopes of potassium, thorium and uranium, present in trace amounts in clay minerals, cause energy to be stored in the quartz crystals. When these are heated, the energy is dissipated in the form of light – thermoluminescence, in addition to, and quite distinct from, normal thermal radiation. The long half-lives of the isotopes provide an almost constant radiation flux so that the TL energy accumulates at a constant rate until removed by heating.

The phenomenon has application to the dating of archaeological materials which contain clay minerals (such as pottery, baked clay from hearths and ovens, and burnt stones) and which, in the past, have been heated above 500 °C. Heating in antiquity removed TL stored over geological time and set to zero a potential archaeological clock. TL energy then accumulated again at a fixed rate; the total stored is proportional to the time which has elapsed since the last heating.

In the laboratory, by heating the sample to 500 °C, the TL emission can be measured quantitatively. In principle, the age of a sample can be calculated using this measurement and by measuring the susceptibility of the sample to acquiring TL, and the concentrations of the radioactive isotopes present in the sample and the soil in which it was buried. In practice the technique is not without its problems.

In 1973, a facility for TL dating was set up in the Department of Physics at ANU. TL dating had originally been applied in archaeology to relatively highly fired pottery. However, it was possible to adapt it to the unusual Mungo fireplaces, which consist of baked clay slabs laid on the sands of the lunette, and which had previously been proven as palaeomagnetic records. Preliminary results on the Mungo fireplaces indicated TL age in general agreement with radiocarbon determinations.[46] Subsequently, using more appropriate techniques, Terry Bell at ANU dated four fireplaces,[47] and Huxtable and Aitken[48] at Oxford dated six fireplaces (Table 2). Recently, Barbetti returned to Australia to establish a TL dating laboratory at the University of Adelaide.

New Fossil Discoveries

As already noted, one of the objectives of the research program was to ensure the regular presence of competent fieldworkers who could scan the lunette for new exposures. This was amply vindicated in 1974 when Bowler found 500 m north east of the cremation[50] the skeleton identified as Mungo 3 (Fig. 5). It was that of a young man, laid out supinely in a shallow grave dug into the lunette at a late stage in the formation of the Mungo unit, estimated at about 28 000 BP.

The full anatomical study of this individual has not yet been reported but preliminary inspection suggests two very interesting features. First, allowing for sexual differences and the range of variation that might be encountered in any living population, the Mungo 1 and 3 individuals could well have belonged to the same population. This tends to confirm that

Table 2. Radiocarbon and thermoluminescence dates of ancient fireplaces at Lake Mungo. Radiocarbon measurements were made on charcoal associated with the fireplaces. (Calculations were made using the 5568 year half-life.) TL measurements at ANU were made on quartz grains from baked sediments from below the fireplaces. TL measurements at Oxford were made on quartz grains obtained from the baked ovenstones from the fireplaces

Fireplace	C-14 date ANU[49] (years BP)	TL date ANU[47] (years)	TL date Oxford[48] (years)
F6	26270±470	31400±2100	21800
F7	30780±520	36400±2500	35300
			29500
			31300
F8	28310±410	32700±2200	37900
F9	27530±340	33500±2300	32000
			32300
F12	28000±410		38600
F14			31400
F16	28140±370		
F17	25570±520		
F20	25310±810		

the oldest known Australians were substantially modern, and cannot be traced in their ancestry through any simple evolutionary scheme in which the earliest settlers of the continent might have been intermediate with the Pithecanthropines of south east Asia.

A further consequence has been to lead to a more cautious approach to the taxonomic position of other Australian human fossils, such as those of Talgai and Kow Swamp. The second feature of interest is in the mode of burial itself, apparently a deliberate rite in which the corpse was covered in red ochre before interment. It might be thought surprising that such varied rites would be practised by what might well be the same population over quite a brief interval of a few thousand years. Of course, this scale is immense in cultural terms, and a major study of Aboriginal death customs in the recent past has shown that groups have in fact practised a variety of such rites at any time.[51]

Excavations

With over 500 ha of lunette offering potential sites it might be wondered what would lead to the selection of particular points for excavation, both in the vertical stratigraphic dimension of time, and in the horizontal one of area and living surfaces. Ideally it would have been desirable to commence with a geophysical survey, such as by proton magnetometer. Apart from practical obstacles, it seemed that this approach could wait until more experience of the deposits and archaeology had been gained. Without this kind of guidance, the selection of excavation area was based on the presence of large quantities of artefacts exposed in the vicinity of the cremation site; for the first excavation a point was chosen within 130 m of it.

This is site A on the plan (Fig. 3) and lies close to the old lake shore, where the surface of the Mungo soil was sealed by later deposits, thus being unquestionably free from later archaeological contamination. This excavation penetrated 2.3 m below the surface of the Mungo unit and stone artefacts were found as deep as 1.5 m. The deepest find was a stone knife or flake, rounded through being rolled in beach

Figure 5. The cranium of Mungo 3 as it was first discovered when exposed by erosion. (F.W.S.)

laid gravels. A radiocarbon sample from above this gave an age in excess of 40 000 years. Practical problems with the removal of overburden and dangers from such deep excavation led to its refilling and transfer to a point B only 25 m from the cremation site, where the surface of the Mungo soil was just below the eroded surface. This site has been reopened, extended and refilled for three seasons, and has exposed some 84 m^2 of the upper part of the Mungo unit, and 12 m^2 at a depth of 2.2 m.

The excavation of such a relatively large area and volume of deposits in an open situation can be approached in terms of solid geometry. Most large, open area sites tend to have fairly shallow cultural deposits lying on the planes of the original land surfaces. Following this reasoning, the 10 cm deep excavation units were adjusted to the 1:50 slope of the distinctive red soil (see section, Fig. 6), a slope which is repeated by the gravel layer below. However, the cultural remains are not in a shallow layer resting on a distinct, former land surface but seem rather to be diffused through the deposits.

Intensity of Human Occupation

If we assume that every artefact and animal carcass on a site is an indicator of human activity, we may use their frequency to measure the intensity of occupation. In practice, the degree to which bone may break up or whether a rat kangaroo bone can be treated as equivalent to that of a red kangaroo of more than ten times its body mass must be considered. Furthermore, human settlement is patchy and there is no reason to expect the same disposal patterns to be superimposed in the same place over thousands of years. Nor is it certain what any arbitrary level in the stratigraphic column means in time and in terms of human settlement. But, with due allowance, a changing pattern of occupation emerges level by level (see schematic section, Fig. 7(a)).

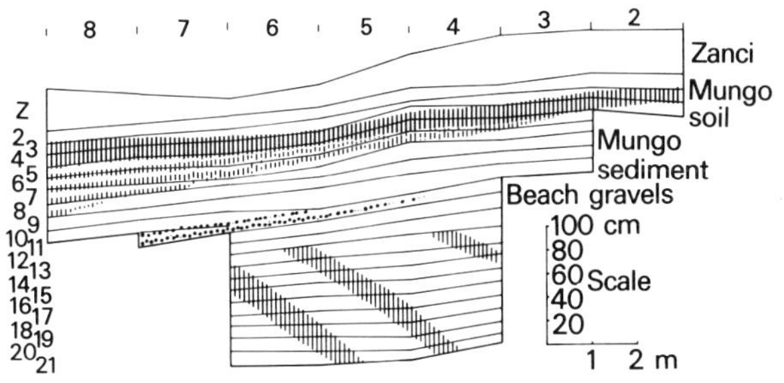

Figure 6. Drawn section of the deposits at Mungo area B. The upper, hachured band is the red, Mungo soil overlying its sandy sediments. The horizontal lines, numbered on the left, are the excavation units conforming with the slope of the soil and beach below. This configuration was not appropriate for the lowermost deposits which slope away in the opposite direction, presumably having been redeposited on the backslope of the dune.

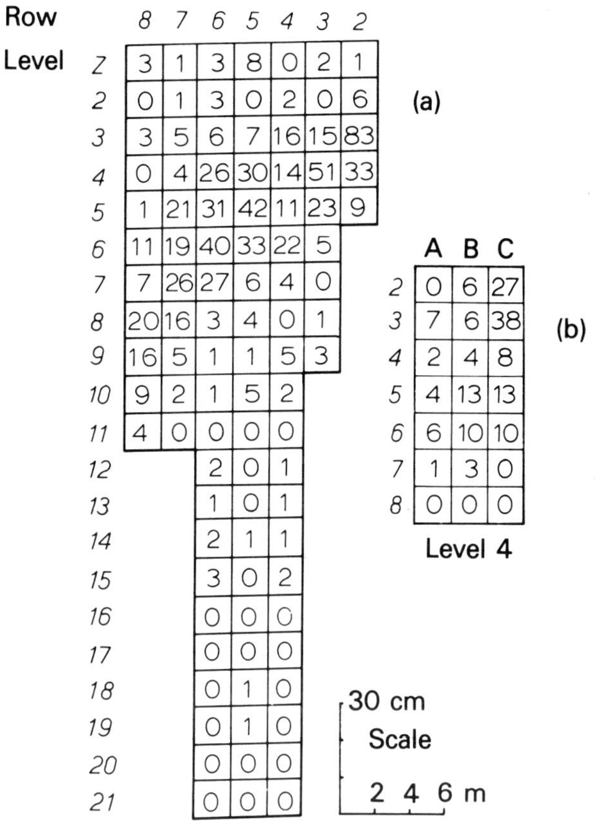

Figure 7. (a) Schematic section Mungo area B: this modification of the drawing in Fig. 6 shows the frequencies of artefacts at successive levels. A plane may be drawn through the units of greatest frequency whose slope does not agree with that of the apparent geological surfaces. (b) Plan of a 'floor', parallel with the profile of the lunette at an arbitrary depth of 40 cm. It is not yet clear whether the patches of different frequencies of artefacts reflect patterns of human deposition of 'noise' from natural causes.

The Zanci stratum has a low frequency of stone tools and virtually no bone at all. This low intensity is also found in the top 10 cm of the Mungo soil and may well be associated with climatic conditions of desiccation and dust storms, unfavourable to Man. Beneath this, however, are some 50 cm of deposits containing high frequencies of artefacts and bone which are preceded by lower intensities, down to and beyond a stratigraphic break which seems to have been caused by a rise in lake levels (Fig. 6). The high waters swept over a previously stable lunette surface, containing some artefacts down to a depth of nearly 2 m. This vertical scale gives a measure of the intensity of occupation through time.

But a closer inspection of the evidence in Fig. 7 reveals a more complex pattern. Within the most intensely occupied half metre of deposits can be discerned a trend with the highest frequency of artefacts deepest in the lakeward, northern end, but progressively retreating south as the deposits increased. This trend is at a slope of 1:27, much steeper than the apparent slope of the lunette. Either this plane of artefact deposition reflects the true geological shape of the former land surface or it may represent a cultural choice in the location of certain activities on the lake's shore and these activities shifted some metres towards the crest of the lunette during the course of several thousand years. Here then is a sign of the horizontal or spatial dimension of the lake shore occupation, though it must be conceded that the emerging pattern is much fuzzier than that of those exceptional African sites which first triggered these research plans.

At Mungo, we might assume that the lunette commenced formation at about 45 000 BP and the sediment increased by steady increments over the following 20 000 years. This would require a century to form 1 cm of deposit; if the settlement of the lake shore was seasonal, an arbitrary 10 cm deep excavation unit would incorporate 1000 successive occupations, should the people have happened to choose *exactly* the same spot for a camp. This likelihood is wholly improbable – with some 50 km of potential shoreline, Lake Mungo can never have resembled a Pleistocene Blackpool. Therefore, there is no reason to divide the artefact densities per 10 cm excavation unit by 1000 and conclude for the most intensely occupied 84 m^2 that 0.16 artefacts had been deposited during any particular season of occupation.

In view of the known low population densities of hunter-gatherers in desert and semi-arid environments, to which Lake Mungo most emphatically belonged, it would be more realistic to start by assuming a single band occupying any 50 m of foreshore as a base during a season. With this population magnitude, a band and its successors could seasonally visit Mungo for 1000 years without ever camping over the same spot. Against this possibility some as yet uncertain qualifications must be raised. First, ethnography suggests that bands tend to be conservative over their camp sites. Geographically, if a place had served well as a camp it is likely that it was chosen for certain advantages. Additionally, hunter-gatherers are not improvident; people may indeed travel light, but do so confident that they can return to an old camp and retrieve old pounding stones, weapons or raw materials when they need them (Peterson, 1968).[52]

The relationships between the intensities of occupations, their changing positions along the lake shore and the rate of formation of the lunette sediments affect our explanation of the scatters of artefacts as camp sites. Certainly the artefacts seem to be clustered in roughly circular patches a few metres in diameter at any arbitrary level (Fig. 7(b)). However, the excavation area is too small to reveal whether such patches form parts of larger, contemporary constellations, or indeed whether they are real at all. If there were such constellations of artefact and food waste patches, and these could be associated with particular baked clay hearths we might begin to feel more confident that we were isolating archaeological evidence for 'camp sites'.

At this stage, however, we cannot be sure that our clusters are not the accidental consequences of excavation by arbitrary units of 10 cm depth, superimposed upon several coincidentally placed successive occupations that have been blended together by a slow rate of sedimentation and the natural and human processes of scuffing and reworking such loose superficial deposits as the sandy lunette.

This kind of problem is unlikely to be resolved by mere refinement of excavation technique, such as by the reduction of depth of excavation units or precise levelling of all finds, which would certainly increase the work load. It is more probable that answers will be found in the pedology and sedimentology of the site and in alternative archaeological approaches. It is unlikely that the Mungo unit had been built up at a steady, slow rate over a period of 20 000 years. Bowler's more recent investigation of the unit shows at least one major break when a thin humic horizon was formed; it is much more probable that the entire unit was built up on a repeated, stop-go fashion, which will become visible when microscopic pedological analyses are applied. It should then be possible to compare the intensity of human occupation with the rate of sediment formation.

One archaeological test of the integrity of patches of scattered artefacts as evidence for contemporaneous occupations, would be to attempt to fit the stone flakes together. It is possible to carry out such a jigsaw reconstruction of flakes and cores when people flake stone experimentally; the method of analysis has been very successfully applied by the French archaeologist André Leroi-Gourhan at the site of Pincevent, where he has revealed the existence of specific working places on the evidence of such reconstructions.[53]

Another source of archaeological evidence for the pattern of occupation may be found in the distribution of artefacts on the eroded lunette surface. There will, however, be some difficulty in making comparisons because of differences of scale of observation, and the confusion caused by superimposition and erosion. In electronic terms the signals may be lost in noise and it will be necessary to develop appropriate 'filters'. It may be possible to develop such filters to emphasize the patterns of particular occupations using the more controlled information available from excavation, and information now beginning to be obtained on the dynamics of modern Aboriginal settlements.[54]

This last point introduces something of a paradox in any attempt to examine such early cultural evidence. If we use the present as a key to read the past, then we are likely to risk distorting that past to fit our conception. This is a point which has been much debated during this last decade and has led to the erection of a variety of guidelines. Some archaeologists argue that it should be possible to establish a scientific archaeology on the basis of rigorous and systematic study of physical evidence, just as physics has developed;[55] taken to its logical limits this would require the entire re-establishment of archaeology, because the present structure is quite certainly not built on these principles.

Another proposal has been that there is a cut-off point beyond which analogies with the recent societies of Man cannot be made.[56] Such a view sees evidence for overall continuities in the behaviour of *Homo sapiens* which would, with qualifications, permit analogies to be drawn between recent Aboriginal culture and that of the ancient people of Mungo. As we have seen previously, continuity, in stone tools and much of the economy, was one of the most important conclusions drawn from the first phase of research. There is one archaeological find made in the second phase, which may add a further hint of continuity. This was a steep sided 'scraper', quite typical of the more completely fashioned tools of Mungo and other early sites, except that its flat underside shows signs of polish as if the tool had been used as a rubber or grindstone (Fig. 8). The microscopic analysis of this wear may show what it was being used for,[57] but in the meantime it is at least a hint of the possibility of tracing cereal grinding technology as far back as this early site so that the sense of continuity, and thus grounds for analogy, becomes stronger still.

THE SETTLEMENT OF AUSTRALIA

As it stands, Mungo is the earliest evidence for the human occupation of Australia. However, its situation, some 450 km from the nearest coastline at the southern end of the continent and deeply embedded in the semi-arid zone, supports arguments, based on studies in biogeography, for the original settlement of Australia having taken place earlier. Inspection of the bathymetric map shows that under the low sea level conditions that existed during the last glaciation of the Pleistocene, New Guinea, Australia and Tasmania would have formed a single land mass, and indeed dry land would have extended out onto the continental shelf.[58] A particularly great extent of land would have existed in the area now submerged under the Arafura and Timor Seas.

This extended coastline would have drawn closer to the correspondingly enlarged islands of Indonesia, though the profound water barriers, known as Wallace's and Weber's lines would still have existed. It is probable that the greater Australian continent would have been most accessible to human migration during the lowered sea level of the last glaciation, that is between approximately 70 000 and 10 000 BP. Thus, if Mungo gives evidence for an age of settlement in excess of 40 000 years there would still be a further 30 000 years during which settlement could have taken place.

The two closest links with the south-east Asian mainland would be across the systems of island stepping stones terminating in Timor and the more isolated Moluccas. If we accept that south-east Asia had contained established human populations for as long as two million years and that, in the course of time, human groups are as expansive and dispersive as any other organism, and if we suppose that the island hopping which had led to the settlement of Indonesia would have been achieved most readily by peoples adapted to water and a littoral environment, we would expect that some forms of water craft, however elementary, had been developed.[59] We do not have to postulate any deliberate migration but a more adventitious sort of movement. Admittedly, the trade winds in this area are easterlies and would tend to blow against any drifting vessels, but, the scale of time is in favour of the occasional reversal of winds which would bring such vessels towards Australia; if anything, one might be more surprised that the continent has not yet earlier evidence for settlement. The relative advantages offered by periods of low sea level are fairly slight.

Pursuing this model further it is a reasonable supposition that the first Australian landfall would have been made close to Timor and that human populations would have settled and adapted to an environment very little different from that which they left. There would then have followed a progressive dispersal round the tropical coastal shelf, in military terms a two-pronged movement. This dispersal could have been rapid.[60,61] It is as well to recall that much of the European settlement of the semi-arid interior of Australia, including such areas as Mungo itself, was achieved within a century and without mechanized transport.

Figure 8. Flat working surface on a steep edged 'scraper'. This stone tool is characteristic of the earliest industries in Australia, but the polished surface, suggesting use as a rubbing stone, is entirely unexpected. (F.W.S.)

The question remains, however, of when the adaptation was made to the semi-arid environment of 500 and more km towards the interior. Slender though the evidence is, it looks as though the discoveries at Mungo may force us to suspect that adaptation had already been made before the deposition of the oldest artefact found. But for a contrary view that sees Mungo as an inland extension of the marine littoral, and thus an example of the initial settlement adaptation, see Bowdler.[62]

INTERDISCIPLINARY CONCLUSIONS

We have shown how one archaeologically significant locality has served as a nexus for interdisciplinary research. The work was organized in a loose federation of disciplines rather than according to a master plan. We have avoided any definition of the term 'discipline', considering that it is adequately covered in this context by appeal to usage. Indeed, we suspect that any overly rigid classification of the disciplines would only reveal inconsistencies derived from the structure of our educational institutions. Another related reason for the nebulous definition of disciplines is the strong element of personal choice in how individuals affiliate themselves.

Researchers tend to view their disciplines as the smallest unit to which a name may be given, and which they might pursue in a variety of academic or government departments. There is a kind of corollary to this, for certainly the research results from Mungo feed back into parts of the educational system which were not involved directly in the field work. For example, Mungo has become incorporated in university courses in geology and geography, and is already becoming dispersed among school curricula. Here then are long-term consequences of such interdisciplinary work which are hard to evaluate.

This issue of what happens to the knowledge derived from the discoveries at Mungo leads us back to a question posed at the beginning of the paper; what has been the relative fertility of these disciplinary crossings and their consequences to Australian archaeology? It would probably be fair to say that the immediate advantages have not been distributed equally. For example, the archaeological evidence has served as a particularly satisfactory vehicle for the physics-related research into archaeomagnetism and thermoluminescence. They have not at this stage contributed to the local archaeology, though the former may well suggest re-evaluations of very early radiocarbon dating, and the latter may become the dating technique for the earliest finds.

Likewise, the archaeology- and physics-based research is relatively incidental to the geomorphology, pedology and palaeoclimatology. These have prospered in the locality because of particularly advantageous sediment formations. Admittedly, the cultural evidence for varying intensities of occupation may indicate degrees of environmental stress, but perhaps what encourages the interdisciplinary cooperation is the site's extreme human antiquity. This emphasizes the locality as a research frontier area in which the problems are more visible and have a human immediacy.

For Australian archaeology, the consequences of interdisciplinary work have, of course, been a head start in the areas of chronology and environmental reconstruction. However, there may well be other less direct consequences discerned. When it is remembered that the foundation work of archaeology was almost invariably on a much smaller scale, much more individualistic and intradisciplinary, labour intensive supported by elementary equipment, and based on applying a simple methodology directly in the field, we may suspect that the interdisciplinary project at Mungo is likely to influence the shape of future research.

This is not to say that individual scale work has been made obsolete, far from it, for this kind of approach is most appropriate to pioneering exploration and the greater part of Australia has never been archaeologically explored. It would be utterly uneconomic, to say the least, to take off a group of specialist scientists on such opportunistic projects.

Mungo's influence may operate at two levels: it serves for Australian archaeology as a test-bed for several disciplines investigating a common area; and, at a more general level, the study of early Man encourages a holistic approach, whereas we are accustomed to see recent, historical and modern Man through the separate and specialized visions of different disciplines. We do not advocate the particular combination of disciplines which have operated at Mungo as the ideal mix and it seems possible that there may be other contributions, but at least we now have sufficient evidence on which to base an evaluation, and the scale and long-term potential of localities such as Mungo provide the kind of conditions under which specialist disciplines can work independently, yet cooperatively.

What do the discoveries at Mungo mean to Australians, not just the participants in the scientific disciplines? The discoveries have certainly received an airing in the news media; as new finds are made they rapidly receive publicity, to the extent that one suspects that the name Mungo now invokes a journalistic conditioned reflex. Most of the consequences are predictable, but there is the very important matter of the effects of such discoveries on the Aboriginal people.

On the one hand, it must be understood that the disturbance of burials presents Aborigines with a sensitive problem. One must possess an unassailable materialist philosophy before one can accept with equanimity that the application of scientific procedures to the remains of one's ancestors is not, in some way, also a violation. Add to this the growing sense of Aboriginal national unity and its consequent extension to ancestors, and we will expect to find some suspicion of, and hostility towards, our work. This has, in some cases, led to the termination of scientifically respectable projects. On the other hand, the sheer antiquity of the Mungo discoveries has also fed back positively into Aboriginal self-awareness. The Mungo dates are quoted in Aboriginal manifestos and letters to the Press. It is, after all, a source of pride that one can point to one's ancestry as descending far into the past of the continent, as against the opinion of only a few years ago, that the Aborigines were relative newcomers.

LITERATURE CITED

1. D. J. Mulvaney, *The Prehistory of Australia*, 2nd Edn, p. 14. Penguin Books, Harmondsworth (1975).
2. R. L. Kirk and A. G. Thorne (editors), *The Origin of the Australians*, Australian Institute of Aboriginal Studies, Canberra (1976).
3. C. Lyell, *Elements of Geology*, 2nd Edn, Vol. 1, p. 312. John Murray, London (1841).
4. Lord Avebury, *Prehistoric Times as Illustrated by Ancient Remains and the Manners and Customs of Modern Savages*, 6th Edn, p. 408. Williams & Norgate, London (1900).
5. E. J. Statham, Observations on shell-heaps and shell-beds. *J. R. Soc. N.S.W.* **26**, 304–314 (1892).
6. T. W. Edgeworth David and J. T. Wilson, Preliminary communication on an Australian cranium of probable Pleistocene age. *Report of the 84th Meeting of the British Association for the Advancement of Science, Australia, Section H.* p. 531 (1914).
7. S. A. Smith, The fossil human skull found at Talgai, Queensland. *Philos. Trans. R. Soc. London, Ser. B*: **208**, 351–385 (1918).
8. A. P. Elkin, *The Australian Aborigines: How to Understand Them*, 1st Edn, p. 8. Angus & Robertson, Sydney (1938). ('One or two thousand years gives scope for many days ... we do not know whether the Aborigines have been in Australia one or many thousands of years.')
9. N. B. Tindale, The antiquity of man in Australia. *Aust. J. Sci.* **3**, 144–147 (1941).
10. H. M. Hale and N. B. Tindale, Notes on some human remains in the Lower Murray Valley, S.A. *Rec. South Aust. Mus. (Adelaide)* **4**, 145–218 (1930).
11. N. B. Tindale and B. G. Maegraith, Traces of an extinct Aboriginal population on Kangaroo Island. *Rec. South Aust. Mus. (Adelaide)* **4**, 275–289 (1931).
12. N. B. Tindale, Relationship of the extinct Kangaroo Island culture with cultures of Australia, Tasmania and Malaya. *Rec. South Aust. Mus. (Adelaide)* **6**, 39–60 (1937).
13. F. D. McCarthy, Culture succession in south eastern Australia. *Mankind* **5**, 177–190 (1958).
14. N. W. G. Macintosh, Fossil man in Australia. *Aust. J. Sci.* **30**, 86–98 (1967).
15. H. A. Polach, Radiocarbon dating as a research tool in archaeology – hopes and limitations. In *Ancient Chinese Bronzes and Southeast Asian Metal and other Archaeological Artifacts*, edited by N. Barnard, pp. 255–298. National Gallery of Victoria, Melbourne (1976).
16. D. J. Mulvaney, Archaeological excavations at Fromm's Landing on the Lower Murray River, S.A. *Proc. R. Soc. Victoria* **72**, 53–85 (1960).
17. D. J. Mulvaney, G. H. Lawton and C. R. Twidale, Archaeological excavation of rock shelter 6, Fromm's Landing, S.A. *Proc. R. Soc. Victoria* **77**, 479–516 (1964).
18. D. J. Mulvaney and E. B. Joyce, Archaeological and geomorphological investigations on Mt Moffatt Station, Queensland, Australia. *Proc. Prehistoric Soc.* **31**, 147–212 (1965).
19. D. J. Mulvaney, *The Prehistory of Australia*, 1st Edn. Thames and Hudson, London (1969).
20. D. J. Mulvaney and J. Golson (editors), *Aboriginal Man and Environment in Australia.* ANU Press, Canberra (1971).
21. J. M. Bowler, Pleistocene salinities and climatic change: evidence from lakes and lunettes in southeastern Australia. In *Aboriginal Man and Environment*, edited by D. J. Mulvaney and J. Golson, pp. 47–65. ANU Press, Canberra (1971).
22. (a) J. M. Bowler, Recent developments in reconstructing late Quaternary environments in Australia. In *The Origin of the Australians*, edited by R. L. Kirk and A. G. Thorne, pp. 55–77. Australian Institute of Aboriginal Affairs, Canberra. (1976); (b) Table 1 was prepared by Dr J. M. Bowler for use in the *Australia 75* Science Exhibition.
23. J. M. Bowler, Aridity in Australia: age, origins and expression in aeolian landforms and sediments. *Earth-Sci. Rev.* **12**, 179–310 (1976).
24. R. Jones, Rocky Cape and the Problem of the Tasmanians, PhD Thesis, University of Sydney (1971).
25. P. S. Martin, Prehistoric overkill. In *Pleistocene Extinctions: the Search for a Cause*, edited by P. S. Martin and H. E. Wright, pp. 75–120. Yale University Press, New Haven (1967).
26. R. H. Tedford, The fossil macropodidae from Lake Menindee, New South Wales. *Univ. Calif. Publ. Geol. Sci.* **64**, 1–165 (1967).
27. J. M. Bowler, A. G. Thorne and H. A. Polach, Pleistocene Man in Australia: age and significance of the Mungo skeleton. *Nature (London)* **240**, 48–50 (1972).
28. J. M. Bowler, R. Jones, H. Allen and A. G. Thorne, Pleistocene human remains from Australia: a living site and human cremation from Lake Mungo, western New South Wales. *World Archaeol.* **2**, 39–59 (1970).
29. R. S. Solecki, *Shanidar the Humanity of Neanderthal Man*, Allen Lane, London (1972).
30. H. de Lumley, *La Grotte de L'Hortus (Valfaunès, Herault)*, Laboratoire de Paléontologie Humaine et de Préhistorie, Marseille (1972).
31. R. G. Collingwood, Oswald Spengler and the theory of historical cycles. In *Essays in the Philosophy of History*, pp. 57–75. McGraw Hill, New York (1966).
32. H. R. Allen, Where the Crow Flies Backwards: Man and Land in the Darling Basin, PhD Thesis, Australian National University, Canberra (1972).
33. H. R. Allen and M. Barbetti, Prehistoric man at Lake Mungo, Australia, by 32 000 BP. *Nature (London)* **240**, 46–48 (1972).
34. A. Keast, Australian mammals: zoogeography and evolution. In *Evolution, Mammals and Southern Continents*, edited by A. Keast, F. C. Erk and B. Glass, pp. 195–246. State University of New York Press, Albany (1972).
35. M. Aitken, *Physics and Archaeology*, Clarendon Press, Oxford (1974).
36. M. Barbetti and M. McElhinny, Evidence of a geomagnetic excursion 30 000 yr BP. *Nature (London)* **239**, 327–330 (1972).
37. M. F. Barbetti and M. W. McElhinny, The Lake Mungo geomagnetic excursion. *Philos. Trans. R. Soc. London, Ser. A*: **281**, 515–542 (1976).
38. M. Barbetti, Geomagnetic field behaviour between 25 000 and 35 000 yr BP and its effect on atmospheric radiocarbon concentration: a current report. *Proceedings of the 8th International Radiocarbon Dating Conference*, Wellington, New Zealand, A104–A113 (1972).
39. I. Hodder and C. Orton, *Spatial Analysis in Archaeology*, Cambridge University Press, Cambridge (1976).
40. A. G. Thorne, Mungo and Kow Swamp: morphological variation in Pleistocene Australians. *Mankind* **8**, 85–89 (1971).

41. L. Haglund, *An Archaeological Analysis of the Broadbeach Aboriginal Burial Ground*, University of Queensland Press, St Lucia (1976).
42. M. D. Leakey, *Olduvai Gorge*, Vol. 3. Cambridge University Press, Cambridge (1971).
43. G. Isaac and B. Isaac, *Olorgesailie*: *Archaeological Studies of a Middle Pleistocene Lake Basin in Kenya*, University of Chicago Press, Chicago (1977).
44. M. B. Schiffer, *Behavioural Archaeology*, Academic Press, New York (1976).
45. M. Aitken, Thermoluminescence and the archaeologist. *Antiquity* **L1**, 11–19 (1977).
46. A. J. Mortlock, Archaeometry at Lake Mungo, New South Wales. *Aust. Phys.* **11**, 213–215 (1974).
47. W. T. Bell, Personal communication of results of PhD Thesis, Australian National University (1978).
48. J. Huxtable and M. J. Aitken, Thermoluminescent dating of Lake Mungo geomagnetic polarity excursion. *Nature* (*London*) **265**, 40–41 (1977).
49. M. Barbetti and H. Polach, A.N.U. Radiocarbon Data List V. *Radiocarbon* **15**, 241–251 (1973).
50. J. Bowler and A. G. Thorne, Human remains from Lake Mungo: discovery and excavation of Lake Mungo III. In *The Origin of the Australians*, edited by R. L. Kirk and A. G. Thorne, pp. 127–138. Australian Institute of Aboriginal Studies, Canberra (1976).
51. B. Hiatt, Cremation in Aboriginal Australia. *Mankind* **7**, 104–119 (1969).
52. N. Peterson, The pestle and mortar: an ethnographic analogy for archaeology in Arnhem Land. *Mankind* **6**, 567–570 (1968).
53. A. Leroi-Gourhan and M. Brézillon, *Fouilles de Pincevent. Essai d'Analyse Ethnographique d'un Habitat Magdalénien*, Centre National de la Recherche Scientifique, Paris (1972).
54. J. F. O'Connell, Room to move: contemporary Alyawara settlement patterns and their implications for Aboriginal housing policy. *Mankind* **11**, 119–131 (1977).
55. R. C. Dunnell, *Systematics in Prehistory*, The Free Press, New York (1971).
56. L. R. Binford, Model building – paradigms and the current state of Paleolithic research. In *An Archaeological Perspective* pp. 244–294. Seminar Press, New York (1972).
57. J. Kamminga, Blood from a Stone: An Enquiry into the Functions of Australian Prehistoric Stone Tools, Doctoral Thesis, University of Sydney (1977).
58. J. N. Jennings, Sea level changes and land links. In *Aboriginal Man and Environment in Australia*, edited by D. J. Mulvaney and J. Golson, pp. 1–13. ANU Press, Canberra (1971).
59. R. Jones, Tasmania: aquatic machines and offshore islands. In *Problems in Economic and Social Archaeology*, edited by G. de G. Sieveking, I. W. Longworth and K. E. Wilson, pp. 235–263. Duckworth, London (1976).
60. J. B. Birdsell, Some population problems involving Pleistocene Man. *Cold Spring Harbor Symp. Quant. Biol.* **22**, 47–70 (1957).
61. F. W. Shawcross, Thirty thousand years and more. *Hemisphere* **19**, 26–31 (1975).
62. S. Bowdler, The coastal colonization of Australia. In *Sunda and Sahul. Prehistoric Studies in Southeast Asia, Melanesia and Australia*, edited by J. Allen, J. Golson and R. Jones, pp. 205–246. Academic Press, New York (1977).

The manuscript was received 1 April 1978.

Physical Science and Archaeology

DR ROBERT E. M. HEDGES
Research Laboratory for Archaeology and the History of Art, Oxford University, Oxford, UK

Archaeometry, the area of co-operation between physical scientist and archaeologist, is at present enjoying a flourishing period. The particular form that this growth has taken is seen to result from the inherently different natures of the two disciplines. Metaphors are developed for these two so that their relationships can be explored, and enabling the special character of archaeometry to be understood. A detailed account of archaeometric work is then given, providing examples of the foregoing abstract considerations.

Physical science and archaeology have, for the most part, developed independently of each other. Furthermore, the two disciplines are quite distinct in the knowledge they embody, in the way they deal with their selective experience of the world, and in their overall aims. It is easy to construct an imaginary boundary separating the one field of human knowledge from the other. Nevertheless, at the present time a particularly fruitful collaboration between scientists and archaeologists is taking place in many laboratories throughout the world; the results of this collaboration are playing an important part in shaping the development of modern archaeology, while the methods employed are leading scientists into new and challenging areas of research.

This activity at the boundary forms the subject of this essay, in which the nature of the interface between the disciplines is examined, the conditions which govern the interchange of information across it articulated, and how these conditions operate in the actuality of archaeological and scientific collaboration demonstrated. Such an intent leads naturally to a description of the major areas of important work in archaeology and science, conveniently referred to as *archaeometry*, where it should become clear how the interdisciplinary character of the subject has influenced its development.

A CONCEPTUAL MODEL

The approach taken here is to interpret 'interface' as a metaphor standing for the complex of relationships that may exist between two evolving systems of knowledge. Each system will be organized according to its unique treatment of information. While detailed analysis of such a situation is not sensible, I hope to provide a conceptual model which will be simple enough to allow useful deductions to be made about the nature of the interface we call archaeometry.

The simplest representation of the physical sciences as an information processing system is as an essentially branching, hierarchic structure. It is a familiar fact to practising scientists that their subject becomes more and more specialized each year. While some cross-links may develop, for example molecular biology, their nature is to continue branching once they are established. This branching implies that a subject may be pursued in relative isolation from the development of similar subjects related by their common level of physical background.

The hierarchical nature of science is also evident in the hierarchy of physical theory, where the major levels of field theory, quantum mechanics, statistical mechanics and classical mechanics provide frameworks for explanations of most physical phenomena. It is evident again in the hierarchical structure inferred for matter: quarks, subatomic particles, atoms, simple molecules, biomolecules, viruses, and so on. It is not surprising that Nature supplies us with a tangible example of such a universal type of structure – that of the *tree* – since both are examples of complex, organized and evolving systems.[1]

Turning to consideration of an analogous structure for humanistic knowledge, a tree-like model is seen to be quite inappropriate. While a degree of specialization is inevitable in reducing the field to a manageable size, humanistic knowledge does not grow in this way. Rather, it has a central concern, that of man in the totality of his cultural experience, to provide the reference about which the details of humanistic studies are organized. To be more specific, and considering archaeology in particular, it is essential for the interpretation of facts from an excavated site to bear in mind not only data from sites nearby in space and time, but also the cultural activities of mankind in quite unrelated circumstances.

CCC–0308–0188/80/0005–0129$04.50/1

Such a structure can be characterized as holistic rather than hierarchic since each part of it must respond to man's culture as a whole. As the structure of knowledge grows, each part becomes modified, and is in turn influenced by the newer additions. Thus, whereas scientific knowledge is liable to be superseded or forgotten by newer research, previous historical or philosophical studies gain a new and different significance in the light of fresh work. Such a holistic form of growth is also to be found in the world of living forms, best exemplified in the spiral growth of seashells.

CONSEQUENCES OF THE MODEL

The opposition of structures, suggested here, is necessarily an over-simplification and an approximation, but it is hoped to show that it constitutes an adequate model to describe the major features of possible relationships between the sciences and the humanities. One can now understand 'interface' to correspond to possible relationships between a branching, hierarchic structure and a spiralling, holistic structure. At this level of simplicity the two structures represent a complete dichotomy, and the inference is that no interaction between the two can take place. If growing in the same environment they would merely inter-penetrate, and the same experience would be comprehended in mutually incompatible ways. In practice the completeness of the dichotomy is not so strict, but nevertheless leads to the split into 'Two Cultures' first widely commented upon by C. P. Snow.[2]

This description, however, is not sufficiently accurate to describe the possibilities of interaction between archaeology and the physical sciences. For, while archaeology has much in common with the humanities, its growth is also influenced strongly by the nature of the evidence with which it deals. Archaeology cannot regard as evidence the surviving atavistic impression our minds may retain of palaeolithic existence, nor even the present day practices of analogous societies, though these may also strongly influence archaeological thinking.

Archaeological evidence has to be confined almost entirely to such material remains that survive from the past, and by and large it is precisely this kind of evidence that is also readily assimilable into the body of scientific knowledge. There is thus a potential dichotomy in the structure of archaeology itself. This can, and does, lead to an actual dichotomy among its practitioners, as is evident for example, in the debate surrounding the emergence of the 'New Archaeology'.[3] What is significant for this argument here, however, is that this special feature of archaeology, which is not found in the majority of humanistic subjects where the evidence or data are far less restricted, allows for a much richer individual interaction with the hierarchical structure of science.

When the archaeometric interface is considered in terms of the 'geometric' picture outlined above, a number of points, basic to its development, can be inferred: (A) the ambivalent nature of archaeology leads to a non-symmetrical relationship between it and science; (B) the interface is likely to involve the branching structure of science at many different points; (C) there is a basic opposition of method inherent in the underlying structure of the two disciplines; (D) information must be communicable across the interface, between structures of opposite type. These points are expanded and their consequences developed below.

The effect of (A) is that archaeology is enabled to formulate questions to which science can respond, but the reverse is not true. The response of science is possible because archaeological evidence nearly always has an objective dimension, for example the material of which a tool is made, rather than its function, which can become appropriate data for scientific information processing. It is usually possible to treat such processing at a single or restricted hierarchic level so that the transformed data can be usefully incorporated into archaeological thinking. (Problems arising where this is not so easy are referred to under (D).) The absence of archaeological response to any scientific question arises from the inevitable holistic transformation of archaeological information, rendering the data inappropriate for science.[4] Thus, point (A) provides a basis for an archaeometric interface to exist (see Fig. 1).

Point (B), the branching structure of science, provides a description of what kind of interface is likely to develop. It follows from a consideration of the possibilities of spatial contact between branching

DR ROBERT E. M. HEDGES—now at the Research Laboratory for Archaeology of the University of Oxford, graduated from Cambridge University where he also received his doctorate for work on the photochemistry of small molecules. He continued research into the chemistry of molecular and atomic collisions at Cambridge and at the Joint Institute for Laboratory Astrophysics at the University of Colorado, Boulder, in the United States. In 1972 he took up an earlier interest in archaeology and his research interests now range over the field of chemical applications to archaeology, with particular interest in the history of technology. During 1977 he was on leave from Oxford at the University of California at Berkeley where he worked on a photochemical project to separate the ^{14}C isotope for a new approach to radiocarbon dating. (See *New Sci.* **77**, 599 (1978).)
Address: Research Laboratory for Archaeology and the History of Art, Oxford University, 6 Keble Road, Oxford OX1 3QJ, UK.

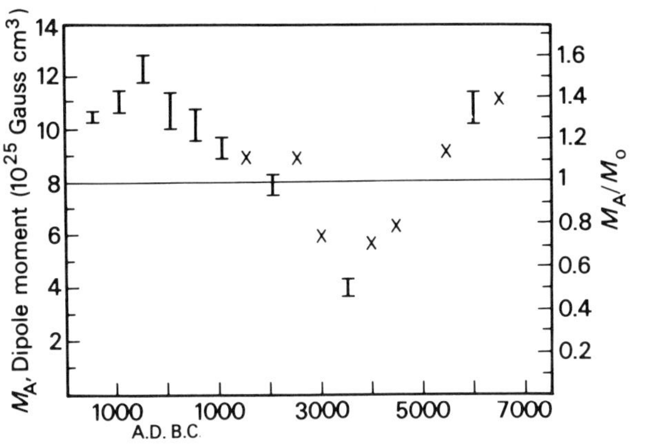

Figure 1. Illustrating point (A) the past variation of the Earth's magnetic field is shown as measured from samples taken from archaeologically dated hearths. It is thus a rare example of archaeology being of service to geophysics. Eventually this curve can be used to help calibrate the radiocarbon curve, and so help refine the application of radiocarbon dating to archaeology.

and spiralling structures and implies that quite diverse specializations within the body of physical science are likely to be involved simultaneously in even a simple archaeological development. These may possibly vary from problem to problem, and the scientific side of the interface must be flexible so that different specializations can be brought to bear and used co-operatively. One of the appealing aspects of archaeometry is in its demand for scientific work at a refreshingly unspecialized level (see Fig. 2).

Point (C), the opposition of methods in the two disciplines, is a reiteration of the basic dichotomy. This may manifest itself in a number of ways, but probably the most prevalent are, on the one hand, the scientist's propensity to isolate the subject of his study as far as possible, and, on the other, the archaeologist's recognition that significance is attached only to an object in its context. Other contrasts in approach include the scientist's search for reproducibility, in opposition to the archaeologist's concern with the uniqueness of human history; or the scientists's concern with possible error arising from a probabilistic distribution in the measurement process, which further mathematical transformation passively transmits, to be compared with the archaeologist's concern with error arising from the probabilistic distribution in the set of interpretations of the data, rather than in the less significant errors in the data themselves (see Fig. 3).

Point (D), information must be communicable across the interface, can be developed in the light of the foregoing. It implies, rather obviously, constraints both to the information and to the interface. Thus, the knowledge structure on each side of the interface must be able to supply a context for the interpretation of the information. If an analysis of an excavated object is made, should the result be given in elemental percentages, as a molecular formula, or in terms of geological or biological structures? Note that the decision here involves a choice of the level of hierarchic processing on the scientific side. Such a choice could have widespread repercussions, perhaps involving quite different people in the collaboration.

Furthermore, connections across the interface must be subtle enough to allow the communication of much more elaborate exchanges of information than 'fact'; particular goals of the evolving disciplines must be understood on both sides. Should an archaeologist be as general as possible in his questions? For example, 'What is this?'; – or as precise as possible? For example, 'What is the chemical composition of this tool?' – when a metallurgical study may be more relevant to an understanding of its functioning.

Ideally, discontinuities at the interface should be avoided and those on either side must discard any fixed notion of where the boundary of their subject lies, while each must develop an awareness of, and some insight into, the thinking prevailing in the other discipline. For example, a scientist confronted with atypical data may be legitimately suspicious of it on non-scientific grounds. Or the archaeologist, when presented with, say, a metal analysis, may have to take into account his perception of the particular chemical environment in which it was found. Naturally good work is best done by a kind of feedback between the protagonists, each with understanding of the structure on the other side.

One example is the determination of the temperature to which a sample of pottery was originally fired. In many cases it is possible to measure this fairly accurately, without being too strongly dependent upon assumptions about the length of time such a temperature was maintained, or the chemistry of the atmosphere prevailing in the kiln. However, the significance of such a result will depend very much on such factors as the uniformity of temperature maintained in the kiln, for some parts could be hundreds of degrees hotter, which in turn depends on the structure of the kiln, on the particular firing practices, and so on.

Therefore, the point in question is not so much the accuracy of the particular temperature measurement of the sample, but the degree to which the sample is representative of the particular class of cultural product, which requires a complex of scientific and archaeological information for its estimation. This case shows how even a simple measurement requires a surprising degree of cross-interaction through the interface.

These points, inherent in the interdisciplinary nature of archaeometry, provide a convenient basis from which to understand how archaeometric knowledge is brought into being. The rest of this article describes current archaeometric research in three of its most important areas. These are the measurement of the age of an artefact, the determination of the place of fabrication, or source of the

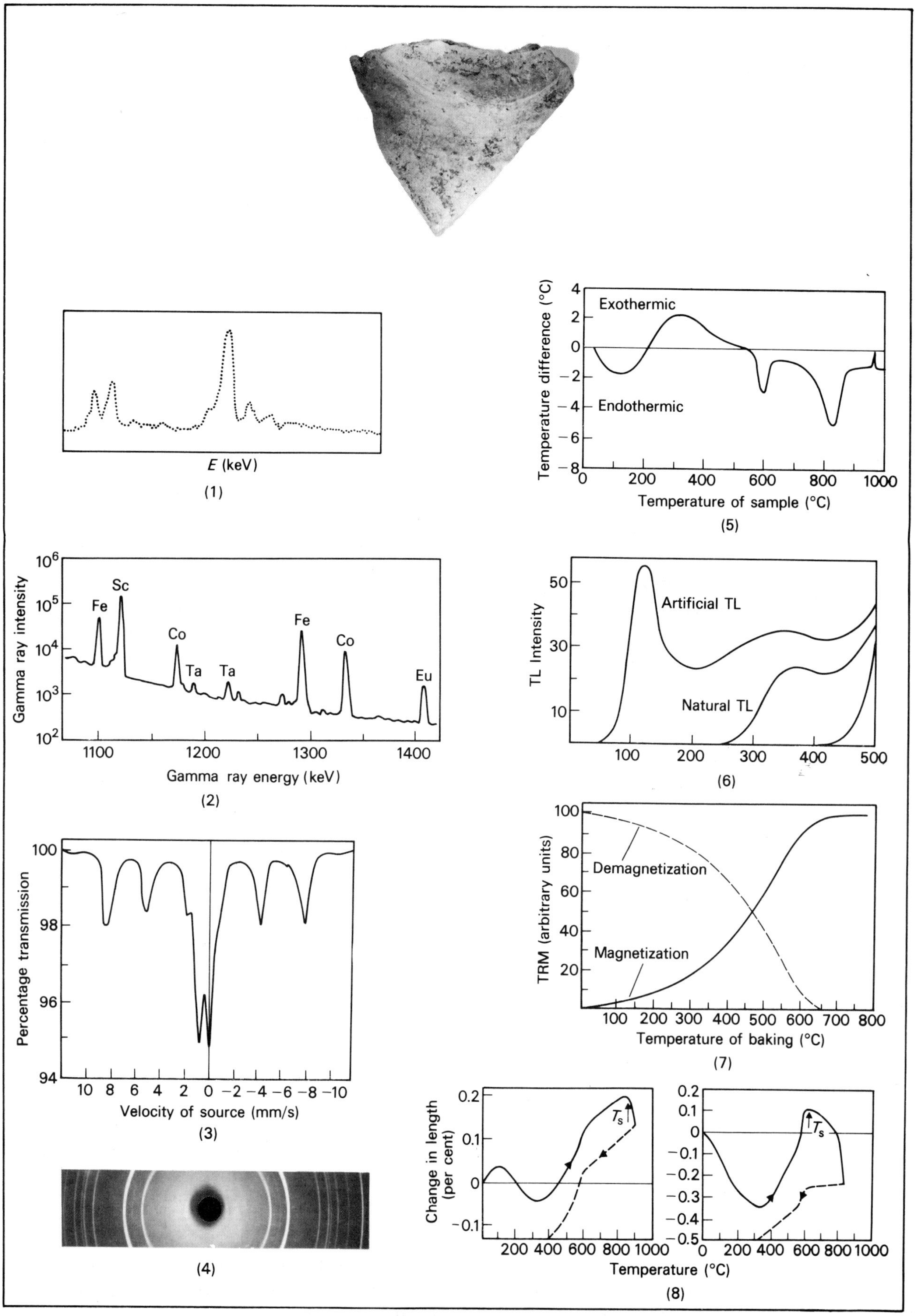

Figure 2. As an illustration of point (B) one pottery sherd may be subjected to a range of scientific studies. (1) X-ray fluorescence spectrum (for major elements); (2) Neutron activation analysis (for trace elements); (3) Mössbauer effect spectrum (for the chemistry of iron present); (4) X-ray diffraction (crystalline components); (5) Differential thermal analysis (thermal history); (6) Thermoluminescent curves (for dating by TL); (7) Remanent magnetization (ancient field strength); (8) Thermal expansion (firing temperature).

Figure 3. An illustration of point (C). The archaeologist studies the Chinese Kuei as a whole. The scientist studies the drillings.

raw material, and the delineation of the technology involved. For each of these three areas the difficulties encountered that are due to the interdisciplinary nature of the work, rather than technical difficulties in the science or archaeology, emphasize in different ways the various points I have derived from the branching spiral model.

AGE MEASUREMENT

The dating of archaeological material by scientific methods is probably the most successful accomplishment of archaeometry. The dimension of Time is well defined in both disciplines; one significant difference is that, for archaeologists, time is measured differentially, that is as a difference between cultural events, while for a scientific measurement the time elapsed to the present is the quantity measured. Thus, a small error in the scientific determination may represent a large error on the archaeological scale, and this places stringent requirements on scientific accuracy. A second difference is that an archaeological date is related to some cultural event, which may not have any significance for scientific measurement; while a scientifically defined event may carry no meaning for the archaeologist.

In dating flints or, more commonly, obsidian, it is not the age of the rock that is important, although there are well developed scientific techniques to establish this, but the time when it was made into a tool. Fortunately this can also be the start of the scientifically determinable process of weathering of the newly formed surface, where the depth of penetration of the hydrated layer in the obsidian becomes a measure of the time of exposure to weathering agencies.

In dating organic remains, for which wood, or charcoal, is the most suitable and abundant, the relevant date is the time when the timber was incorporated into a man-made structure. Here science is given less help, there being no change made to the physical nature of the material by man. In this case the scientifically determined date at which the material ceased to be metabolically active, by becoming heartwood or by death, is all that is possible, and judicious guesswork is necessary to relate this date to that for the archaeological event. This relation will depend on circumstances, for example being quite different for remains of a light wattle structure than for massive oak beams, which might even have been reused.

A third illustration is provided by the dating of pottery. Here again it is scientifically possible to determine the age of the clay, but the archaeologist is interested in when it was first fabricated, or when it was last used. The recently developed scientific method of thermoluminescence (see below) necessarily measures from the time when the clay or, more precisely, the quartz and felspar constituents of the pottery, was last heated past 400 °C or so, and this fortunately corresponds in a useful way to the archaeologically relevant date.

The three methods referred to above do not encompass the whole range of archaeometric dating methods, but are the most significant and widely applied. All three cases illustrate how the scientist's approach to dating is to find some characteristic change in the artefact which runs its course as independently as possible from its environment. This independence is most marked in the case of the steady radioactive disintegration of ^{14}C nuclei which are present at a minute but constant level in all living organisms. At death this constant level is no longer maintained as the organism no longer interacts chemically with the atmosphere, and so the steady rundown in ^{14}C content allows a radiocarbon date to be made from the measurement of the remaining ^{14}C.

While the rate of loss of ^{14}C is as perfectly independent of its environment as any scientist could wish, establishing a date also requires assumptions about the original atmosphere ^{14}C content. Scientifically, this is extremely difficult, involving such complex processes as the circulation throughout the atmosphere, the changing distribution of atmospheric carbon dioxide between air, sea and ice during the Ice Ages, variations in the cosmic ray flux to the Earth and the effect of changes of the Earth's magnetic field (see Fig. 1), the behaviour of the Sun, and so forth. Fortunately, it is possible to calibrate radiocarbon levels against tree-ring dates, and so to show that atmospheric levels have varied by a definite though small amount about the present value. As all contemporaneous radiocarbon dates are affected in the same way, the uncertainties due to the difficulty of validating the necessary scientific assumptions show up primarily when comparing radiocarbon dates with those obtained by other methods.

The method of thermoluminescence is more dependent upon environmental influences than radiocarbon, but nevertheless is becoming an increasingly valuable technique, especially since it refers to material so central to archaeological study. In this case the physical change that is studied is the accumulation of energy in crystalline inclusions in pottery, particularly quartz, resulting from absorption of the very weak radioactivity of the immediate surroundings. Principally this is due to the natural trace levels of uranium and thorium and the rare radioactive isotope of potassium.

The energy, which is stored in the form of radiation damage by the crystal, is released in the form of light on heating the crystal. Dating is possible when the rate of accumulation of damage is known as well as the total stored. This situation is much more complicated than for the decay of ^{14}C. The environment is directly involved in a number of ways; to a first approximation the environmental radioactivity can be measured; but many effects must be estimated. External gamma radiation, partly from cosmic rays and dependent upon the thickness and type of burial soil, contributes to the total radiation; the water content of the surrounding soil has an important effect on the absorption of radiation, and also on the escape of radiating daughter products. Additionally, the accuracy of storage and reproducibility of measurement are sensitive to subtle variations in the physical and chemical integrity and history of the material.

Thus, attempts to meet the demanding accuracy soon encounter an ever-expanding list of interactions to take into account. If even stronger isolation from the environment is attempted, as in measuring the thermoluminescence of zircon inclusions for which radiation sources external to the inclusion itself are negligible, then the date is more sensitive to the remaining inaccuracies, such as variation in the distribution of uranium within the zircon inclusion. Essentially a limit is reached where further isolation is counteracted by the accumulation of additional assumptions.

Dating by obsidian hydration is even more strongly limited by the dependency of the weathering process on the environment. That it is possible at all is due to the fact that the weathering process proceeds by reaction with a layer of water adsorbed on the fresh surface in such a way as to be relatively independent of the water content in the surrounding atmosphere. Nor is the rate of reaction, by which the boundary of hydrated obsidian and the natural rock is maintained as a sharp, slowly moving front which can be easily measured by microscopic examination of a thin section, especially sensitive to such other parameters as the obsidian composition and external temperature. In fact, while chemical weathering takes place on virtually all archaeological materials, this particular process is the only one that has been applied in a quantitative way. Nevertheless, the uncertainties of accuracy of the method restrict its use mainly to relative dating at a single site, or else by comparison with specimens calibrated from other evidence.

DETERMINATION OF PROVENANCE

The second area of archaeometric work, provenance determination, shows some essential differences in the logic of communication across the interface. While the final result, the geographical location of a source of raw material, has meaning equally in the frameworks of both disciplines, the impossibility for the scientist to have complete knowledge of the site, as it was when it provided the raw material, must restrict the certainty of his answer. Inevitably the scientific result must be interpreted in a probabilistic way, although such an interpretation may be incorporated into archaeological thought only with difficulty.

An example is best given for work on the provenance of pottery. It has been discovered that an analysis of the trace elements to be found in surviving pottery sherds, while showing a great deal of random variation, also produces characteristic clustering of compositional types which correspond to pottery believed to be made in the same geographical region. Apparently the variation in clay trace element composition between different clay beds is, in general, larger than the variation within a clay bed, and can still be noticed after the clay has been through all the treatment given it by man, namely selection, weathering, levigation, addition of filler or temper, firing, use and subsequent burial for millenia.

Sometimes this inference can be confirmed by an analysis of modern clay, or brick, obtained from the suspected source, but it is more likely that the superficial high quality deposits were worked out in ancient times for the making of export wares. Consequently an analysis of a representative group of archaeological sherds has to be interpreted in terms of the self-consistent structure of analyses and archaeological inference already built up.

The problem is simplified if the products from a known kiln can be analysed, for example. Even then it can only be given in terms of the probability distributions estimated from previous samples.

If a single, and possibly archaeologically important, sherd is now analysed, and its provenance is determined on the basis of known composition patterns, it is often not helpful for the scientist to report that 'there is a 75% probability that it was made in North Crete, and a 25% chance that it was made around Athens. However, if you include the Cyclades Islands as possible sources, the probabilities change to 60% for Crete, 20% for Athens and 20% for the Cyclades'. Clearly the archaeological question has to be formulated carefully so that results of scientific work can act effectively in the structure of archaeological knowledge. This should not give the impression that the determination of pottery provenance is of little value; a great deal has already been done towards the discovery of trade routes, once the relationship between source and destination became known, while the implementation of large scale trace element analysis, for example by neutron activation analysis, together with computer processing of the data, has made this a particular active area. It should be clear, though, that success in this work depends as much on the ability of the archaeometrist to be effective within the communicational difficulties imposed by the interface.

As a rule, more natural materials can be treated in a more clear-cut way. Pottery represents a middle case, with the analysis of sources of such materials as obsidian providing more definite results, while the analytical study of trace elements in smelted metals presents several additional problems. In particular, a great deal of analytical information is now available concerning the trace element composition of European bronzes throughout the Bronze Age. While the geochemical circumstances of the ore are reflected in the final composition, these may be hidden under much more far-reaching changes brought about by human processes.

Different smelting methods can be estimated, but it is harder to allow for the deliberate addition of small quantities of alloying elements – which may not have had an apparently rational basis – and virtually impossible to take account of the compositional changes involved when scrap bronze from a variety of sources is remelted. Consequently, any interpretation of the analytical information, even at the level where the objectivity of the scientific approach can be maintained, has to be greatly modified by consideration of a whole set of archaeological possibilities, each with a different interaction with the scientific analysis. This is one important reason why progress in the provenance determination of bronzes has been slow despite the wealth of analytical information.

TECHNOLOGICAL EVOLUTION

We see, therefore, that it is in questions involving ancient technology that the interaction across the science-archaeology interface is most dynamic; that is to say, the precise formulation of a question is only possible through considerable feedback between the two disciplines, a process which influences and is in turn influenced by, the type of evidence that is considered relevant. While questions of date, and to a lesser extent, of provenance, can be isolated and treated one at a time, this is hardly possible in the study of the development of technology. The context in which to ask the appropriate question is generally only established after some partial pattern has emerged from the investigation of archaeological materials, and much of this investigation is pursued without preconceived hypotheses of the technical background of the object. Thus, the body of analytical data on European bronzes, referred to above, becomes particularly useful in this regard, as will be made apparent below.

For an unprejudiced scientific examination of archaeological material it is necessary to develop some specialized methods. In particular, techniques of analysis must be non-destructive as far as it is possible. While this is common sense on aesthetic grounds – the sacrifice of museum specimens to reveal analytical information betrays an inconsistent set of values – it also reflects that scientific work can validly address a part of the object, but that it is the whole that is of value in an archaeological setting and which should not be destroyed. Some techniques can be completely non-destructive, such as neutron activation and X-ray fluorescence analysis, while others have been developed to require only a minute sample which can be extracted from the whole with negligible effect.

In the study of archaeological materials many methods may be needed to bring to bear on an object, each yielding complementary information. For example, X-ray fluorescence provides information on the abundance of most major and minor elements. But this only refers to the immediate surface. If the spatial distribution of composition is of interest, perhaps due to changes from corrosion, or inhomogeneities of inclusions deriving from the manufacturing process, then a small sample can be examined on a microscopic scale for analytical variation in an electron-probe micro-analyser. If, however, it is the chemical nature of the substance that is of interest, for example, the material used to opacify an enamel, then X-ray diffraction can be used to study the crystalling structure of the material and identify its constitution with the help of an elemental analysis.

To understand how a metal tool was made, and how well it performed, it is necessary to know not only its alloy composition, but also to interpret its metallographic structure in order to reconstruct the metallurgical processes involved in its formation. The same sample of pottery has been studied for its its elemental composition, including trace elements, its mineralogical composition, its petrographic composition, by thin section microscopy and the analysis of heavy mineral inclusions, for its thermal properties, for its magnetic properties, which relate to the grain size and to the chemistry of iron containing compounds, by Mössbauer spectroscopy, which also relates to the chemistry of iron compounds, and thereby also to the oxidation and temperature of the kiln atmosphere, for the distribution in pore size, by microscopic examination for the degree of sintering and vitreous phase formation, and for its thermoluminescent properties. The results of any one of these tests is immediately and closely related to that for any of the others.

The study of the development of glazing, which owes much to the above techniques, exemplifies the continuous interplay of scientific and archaeological knowledge. Glazing, presumably for decorative effect, occurs at least as early as pre-Dynastic Egypt, say, 3500 B.C., where crystalline quartz and steatite can be found covered with a glassy layer coloured green by the addition of copper or its ore. Glass in the form of small beads is encountered much later, while in the form of shaped vessels is not known for another 2500 years. The formation of vitreous material in a fire from such widespread substance as sand and plant ash and chalk or lime must have been a common occurrence, but for it to stick to a material it is necessary that the expansion of the two materials, or rather the contraction, as the glazed object cools and the glass solidifies, closely match. Since the expansion coefficient depends upon the chemical composition, there is considerable opportunity for technical control of the desired product.

It so happens that both quartz and steatite have expansion coefficients that closely match the range normally produced in glass, and it is easy to understand how such objects became the first to be glazed. Natural development leads to the glazing of small objects made of powdered quartz which are held together and fired with a small quantity of binding ash or soda, and sometimes a little clay. Such material, 'quartz frit' or 'faience', is one of the most ubiquitous and characteristic products of Egypt, and to a lesser extent of Mesopotamia. Faience is difficult to work into larger shapes, especially for containers, and never supplanted pottery.

Control over the production of glaze colours increased steadily with time; a rich deep blue, 'Egyptian Blue', containing mainly copper, calcium and silica, was developed as one of the earliest synthetic compounds, and was in abundant use presumably in imitation of the deep blue and very rare mineral lapis lazuli, which incidentally owes its colour not to copper, but to an unusual structural chemistry of sulphur. However, while glazed faience of increasing variety and sophistication was being made, the potentially much more important ability to glaze pottery had not been achieved. This is understandable, as fired clay has a much lower expansion coefficient, so that a cooling glaze skin on it would contract too much and peel off. It is possible to improve the match by changing the glaze composition, or by carefully working in as much quartz as possible into the body; in modern Afghanistan the unglazed body is covered with a thin layer of quartz prior to glaze application.

We first find glazed pottery in Phoenicia, at about the same time as functional glass artefacts appear. The range of glass composition which can adhere to a clay body is very restricted, and it is to be expected that an ability successfully to glaze pottery would be consequent on advances in glass working. In fact, it is found that once proper glazing is achieved the change of glazing formulation varies very little with time for well over 1000 years. The process was very conservative, and shows little evolution. However, a major technical advance was made during the later Roman Republic, when it was discovered that the addition of 25–40% of lead to a typical glass composition enabled the expansion coefficients to be much more closely matched, as well as producing a glaze with lower melting point and more glossy appearance.[5]

After rapidly spreading throughout the Roman Empire, though the majority of fine wares were still unglazed, Roman lead glazing disappears equally rapidly during the Dark Ages, to be revived during the Islamic conquest and finally became a familiar feature of mediaeval pottery practice. It is interesting that in a lead glaze the addition of copper produces a green rather than a blue colour, and blue glazes were not known during Roman times. However, the resurgence of lead glazing in Islamic times took place alongside the discovery of cobalt ores as a source of blue in glazes, though the occasional use of cobalt in glass and enamel was known to the Romans. This led to the development of the masterfully decorated blue and white ware of the Middle East which in turn influenced the creation of the marvellous blue-and-white porcelain of China.

Enamelling is another aspect of glazing, though here the problems are somewhat changed. The use of lower melting metals, especially bronze, as a substrate, requires that the glass frit shall fuse and flow freely at fairly low temperatures, while the much greater expansion of the metal requires enclosure of the glass in small cells where it is held in compression. Very little early enamel has been scientifically studied; curiously enough, the Egyptians, for all their expertise in metals and faience production, made no attempts to glaze pottery or enamel metal, though they did work at the inlay of cut precious stone into metal on a large scale. Nothing is known about Mycenaean enamels, but by Roman times we find that enamel has a specialized composition, containing enough lead to promote a free-flowing quality, and that the whole range of colourants, stable at the high temperatures of fabrication, and also of opacifiers, is fully developed.

While this overview of a single technical development is but a sketch, and completely omits the relevant archaeological knowledge of the availability of raw materials, of patterns of trade, the level of local technical expertise and of the means of transmitting technical skills, it demonstrates the interplay of science and archaeology in lending significance to each successive discovery.

For example the adoption of iron tools and weapons did not come about by recognition of the superior metallurgical possibilities of the metal. Pure iron is not so hard as a good bronze, but it seems likely that the collapse in the organization of the tin trade brought back a reliance on local resources. Iron ore is widely abundant, though requiring a more advanced technology for its smelting. The spread of the knowledge of iron smelting, and the specialized smithing techniques, believed to originate in East Turkey, can be followed from examination of the construction of early iron artefacts, such as the iron daggers of Luristan. Metallographic analysis reveals a continuing improvement in understanding of the complexities in carburization and heat treatment only fully understood within the last hundred years, which finally led to a mastery in the art of steel making that was achieved not until Viking and Islamic times. Here again, science and archaeology can be seen to make totally interdependent contributions in unravelling one of the major strands in the evolution of human culture.

CONCLUSION

In this essay I have developed two apparently incompatible analogues: the branch and the spiral – in order to understand better what is implied in the word interface. It is not too fanciful, I hope, to project essentially the same dichotomy, and see it as a limitation, into other systems of human thought. It underlies the differentiation into transcendence and immanence in theological ideas of the world, in the opposition of reason to intuition in human discourse and argument, and has even been identified with the different operating processes in the left and right hemispheres of the brain.[6]

With this viewpoint, human culture, particularly in the West, can be seen as tending towards more vigorous growth in the branching, transcendent, rational system of knowledge; such growth has informed our system of values. However, the widespread and recent sense of loss of faith in traditional values, and loss of confidence in the continued development of Western intellectual methods, especially as applied to the broader activities of human culture, is by now a well documented phenomenon, from newspaper to serious sociological study. One may hope that the inherent stress of such a situation will bring about a shift to greater harmony in the relationship symbolized as that between the tree and the seashell.

The newly acquired importance of the study and application of ecology is an especially relevant and significant sign that this might be so. In the same spirit, archaeometry can be seen to be engaged in a practical exploration of where such limits to the growth of thought may lie.

NOTES AND LITERATURE CITED

1. It is worth pointing out that the tree structure applies to science as viewed as an information processing system; that is, once scientific knowledge has become incorporated into the growing whole. If we consider the creative aspect of science, the point where knowledge is born from the unknown, an entirely different model would be required. The 'act of creation' is far less susceptible to shared understanding than is scientific knowledge and at the least must involve the bringing together of branches, rather than their multiplication. See A. Koestler, *The Act of Creation*, Hutchinson, London (1964).
2. C. P. Snow, *The Two Cultures and the Scientific Revolution*, Cambridge University Press, Cambridge (1964).
3. This term refers to a school of archaeological thinking in which a strictly objective and quantitative approach, using the methods of General Systems Theory and Behavioural Analysis, is ideologically given pre-eminence. The debate surrounding its emergence can be followed with entertainment in the pages of the journal *Antiquity* (Cambridge) during the late 1960s and early 1970s. The important and relevant books are: L. R. Binford and S. R. Binford (Eds.) *New Perspectives in Archaeology*, Aldino, Chicago (1969); D. L. Clarke, *Analytical Archaeology*, Methuen, London (1968).
4. The exceptions are informative, however; early scientific work on ^{14}C decay was validated using archaeological specimens – following an intial analysis of New York sewage – but Time is a relatively easy element to isolate from a structure showing holistic growth. Another example is illustrated in Fig. 1. The study of the evolution of domestic animals might provide another example, since human cultural behaviour becomes an important element in the selection pressure. Scientific understanding of the phenomenon would thus require information from the whole sphere of human activity. But such examples are rare.
5. The origin of this discovery is still not known; the earliest examples so far known are from China, and from Turkey. Lead was a component of the common yellow pigment lead antimonate used in faience glazes, and it was also used in helping to colour the reduced copper containing ruby red glass found as inlay in late Iron Age jewellery.
6. R. E. Ornstein, *The Psychology of Consciousness*, W. H. Freeman, New York (1974).

For further reading in archaeometry: M. J. Aitken, *Physics and Archaeology*, Oxford University Press, Oxford (1974), and M. Tite, *Methods of Physical Examination in Archaeology*, London Seminar Press, (1975).

For specialized reports: *Archaeometry*, Oxford (1958–) and *Journal of Archaeological Science*, London (1974–).

The manuscript was received 20 January 1978.

Cell Division Problems and Solutions

PROFESSOR DR N. PAWELETZ and DR M. LITTLE
Institute for Cell Research, German Cancer Research Centre, Heidelberg, FRG

Various aspects of cell division in the eukaryotic cell, the cell with a nucleus, are discussed. A brief history of research in mitosis is followed by an account of mitosis as part of the cell cycle and its characteristic events. The control of cell division in unicellular organisms is compared to that in the differentiated cells of higher organisms. Essential differences between the two types of cells are illustrated by a short description of cellular differentiation. Mitosis in multicellular organisms is initiated by external signals. Several examples of mitosis-inducing substances, especially those which interact with cell surface proteins, are presented.

An analysis of the components of the mitotic spindle is accompanied by a discussion of the current theories of chromosome movement and the assembly of microtubules, the main structural components of the mitotic spindle. Unlike the random polymerization of microtubules *in vitro*, the assembly of spindle microtubules is directed by microtubule-orientating centres. Attempts to study the mitotic process *in vitro*, by isolating mitotic spindles which are still capable of chromosome transport, are also described.

Cell division is one of the most fundamental processes in the life of an organism. The generation of new cells, the replacement of lost or damaged tissue by regeneration, and the growth of any part of an organism is only made possible through cell division. Whereas in the prokaryotic cell, the cell without an apparent nucleus, membrane growth between the duplicated sets of genetic material pushes them apart, in the eukaryotic cell, a cell with a nucleus, a highly complicated spindle-shaped structure has evolved for the separation of the daughter chromosomes. The various aspects of this latter process, mitosis in the eukaryotic cell, are the subject of this article.

Research on mitosis began over a century ago. A great deal of data has been gathered since then, mainly from light and electron microscopic observations, but many important questions remain open. In recent times, valuable expertise has been brought to bear on these problems by the co-operation of scientists from several disciplines. For example, electron microscopists, physicists, chemists, biologists and biochemists have all contributed to our understanding of the assembly of microtubules, the main structural components of the mitotic spindle. Furthermore, the development of new techniques such as immunofluorescence, whereby specific proteins can be located in the cell after incubation with their fluorescently labelled antibodies, has brought fresh impetus to the study of mitosis.

Cell division among unicellular organisms serves to increase the number of individuals. In the multicellular organism, however, cell division is necessary for the replacement of lost cells and for the growth of the organism or of particular organs. The life of a higher organism begins with a fertilized egg cell, which becomes an embryo after a large number of divisions. During the course of embryonal development, some cell divisions give rise to daughter cells which are specialized or 'differentiated' to fulfil definite functions as part of a group of like cells. The growth of any group of cells remains under the control of the whole organism.

Proliferative tissue such as skin, the reproductive organs and the mucosa of the intestines is stimulated to undergo cell division at the same rate as it suffers cell loss. Similarly, wounded tissue undergoes rapid cell division until the wound is closed. A cell which for some reason no longer responds to normal mechanisms of regulation can be fatal. Uncontrolled proliferation giving rise to a malignant tumour, results in the eventual death of the organism.

HISTORY OF RESEARCH ON CELL DIVISION

In 1873 the German anatomist Schneider[1] gave the first detailed description of the various stages of cell division. While investigating the development of the egg and sperm of the flatworm, he observed drastic changes in the shape of the nucleus, which he correctly interpreted as distinct stages in the process of cell division. Shortly afterwards, in 1875, the German botanist Strassburger[2] published the first monograph on cell formation and cell division (*Über Zellbildung und Zelltheilung*). Not long afterwards, in 1882,

CCC–0308–0188/80/0005–0138$05·50/1

Flemming[3] wrote the monograph *Zellsubstanz, Kern und Zelltheilung* ('Cell Substance, Nucleus and Cell Division'). He described each stage of cell division in detail and gave the name 'Karyomitosis' (= thread-like metamorphosis of the nucleus, mitos = thread) to the complete process of cell division, and the name 'mitoses' to the individual figures characterizing the various stages. The expression 'mitosis' is now used to signify those events in the nucleus and on the chromosomes associated with division of the nucleus. Waldeyer[4] introduced the expression 'chromosome' in 1888 to distinguish these structures, which can be well stained, from the achromatic figure (= spindle = mitotic apparatus).

The first investigations of the mitotic process were essentially of a descriptive nature. These were followed by an era of experiments conceived to provide some insights into the mechanism of the mitotic process. Many hypotheses were put forward only to be rejected. Even today there is no generally accepted hypothesis of chromosome movement, and there are some doubts as to whether a universally applicable mechanism will be found.[5]

Mitosis was originally regarded as an autonomous cellular process. We now know that it is very much dependent on events which take place in other parts of the cell cycle.

After cell division the daughter cells undergo a period of growth (G_1-phase = first growth phase) to replenish cell material. This is followed by a period of DNA synthesis (S-phase) in which the genetic material is duplicated. In a second period of growth (G_2-phase) the cell makes its final preparations before going into division. After mitosis the two daughter cells can enter the G_1-phase and run once again through the cell cycle. This series of cell cycle events is typical for growing tissues. When growth is no longer necessary, however, cells can stop cycling and remain permanently, or for a given time, in a resting phase (G_0-phase). Not all the above cell phases are necessarily present in all cells. For instance, in egg cells the G_1-phase is missing and DNA synthesis (S-phase) takes place immediately after mitosis.[6]

THE EVENTS IN MITOSIS

At the end of the G_2-phase a change occurs in the distribution of nuclear chromatin, a complex of DNA and associated proteins. The fine, in part very homogeneous consistency of the chromatin becomes inhomogeneous, and the chromatin condenses to thread-like structures which are visible in the light microscope as individual chromosomes. The nuclear envelope is still intact. During chromatin condensation the cell has already begun to assume the typical ball-like shape of cells in mitosis. At the same time a star-shaped fibrous corona becomes visible around the centrioles, the central body, in the cytoplasm. These fibres can be identified as very fine tubules, the microtubules, under the electron microscope. This phase of mitosis, prophase, is followed by prometaphase in which a system of microtubules is built up between the centrioles as they move apart towards the poles. The growth of the mitotic apparatus, which gradually assumes its typical spindle shape, is accompanied by a disintegration of the nuclear envelope until, eventually, it seems to have completely disappeared. The chromosomes, now lying within the framework of the spindle, are transported to the so-called equatorial plate, the metaphase plate, which lies equidistant between the poles at the cell equator. The failure of any chromosome to arrive at its correct destination can have serious consequences for the organism.

While the chromosomes are being aligned at the metaphase plate, they split longitudinally. Each chromosome, consisting of the two split halves, is connected with the two poles by spindle fibres, bundles of microtubules, which are attached to definite anchorage points on the chromosomes, namely the kinetochores. When the chromosomes are finally aligned at the equatorial plate (metaphase) they are ready for separation and transport to the poles (anaphase).

In contrast to the other phase changes of mitosis, which are difficult to pinpoint, the change from metaphase to anaphase is sudden and simultaneous.

PROFESSOR DR NEIDHARD PAWELETZ, born 1935 in Ossen, Silesia, Germany, studied biology, chemistry and physics at the University of Göttingen. He worked as a research assistant in the Institute for Experimental Cancer Research of the University of Heidelberg before this institute was incorporated into the German Cancer Research Centre, Heidelberg, as the Institute for Cell Research. Here, since 1972, he is head of the department "Cell Growth and Cell Division". His main interests are the ultrastructure of the mitotic apparatus and experimental approaches to the problems of the formation of metastases.

DR MELVYN LITTLE, born 1945 in Manchester, England, was from 1970 to 1975 a research assistant at the Max Planck Institute for Cell Biology in Wilhelmshaven, West Germany, involved in investigations on the mechanism of action of the female sex hormone, estradiol. Since 1975 he has been a member of the scientific staff of the Institute for Cell Research at the German Cancer Research Centre, Heidelberg. He is currently working on the primary structure of tubulin, which is the main structural protein of the mitotic spindle.
Address: Institut für Zellforschung, Deutsches Krebsforschungszentrum, D 6900 Heidelberg 1, FRG.

Here again, the correct and orderly transport of the chromosomes is very important. Any disturbance in the separation of the chromosomes during anaphase can lead to pathological mitoses. As the chromosomes move apart, a region of electron-dense osmiophilic material, characterized by overlapping non-kinetochore microtubules, is formed in the middle of the spindle (the midbody).

The end of anaphase, when the chromosomes have reached the poles, is followed by the telophase. During this last phase of the mitotic process the chromosomes begin to decondense, a nuclear membrane is reformed, and the mitotic apparatus disappears. Already, however, in late anaphase a small ring-shaped furrow has appeared on the cell surface over the original equatorial plane. This furrow becomes increasingly deeper during telophase until only a small plasma bridge containing the rest of the spindle, the Flemming body, remains. The bridge is eventually discarded by one of the daughter cells and is then reabsorbed by the other. The two daughter cells can now participate in another cell cycle. These events are illustrated by Figs. 1–5.

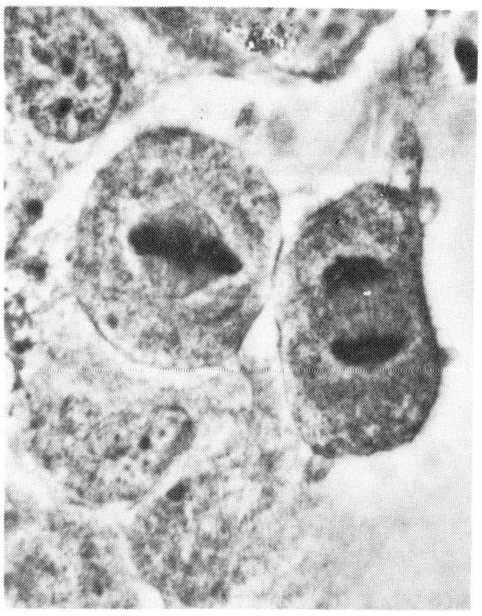

Figure 2. Light micrograph of human tumour cells in mitosis. (*Left*) Metaphase, the spindle-shaped mitotic apparatus is clearly visible. (*Right*) Anaphase, the midbody, central part of the anaphase spindle, can be seen between the separated chromosomes. (1200×)

REGULATION OF MITOSIS IN UNICELLULAR ORGANISMS

Detailed investigations have shown that the individual events of mitosis are dissociable from one another. Similarly, the various phases of the cell cycle are flexible and variable. While the search for signals which trigger the various events has not been very successful, some important aspects of cell division regulation have come to light from a study of cell size in the fission yeast.[7]

Although the size of the individual yeast cells at cell division varies, the mean value remains relatively constant, indicating a homeostatic mechanism of cell size regulation. This could be achieved by two means. In the first case the cell cycle time is kept constant and the growth rate varies, that is, large daughter cells would grow slower and approach the average size towards the end of the cell cycle. In the second case

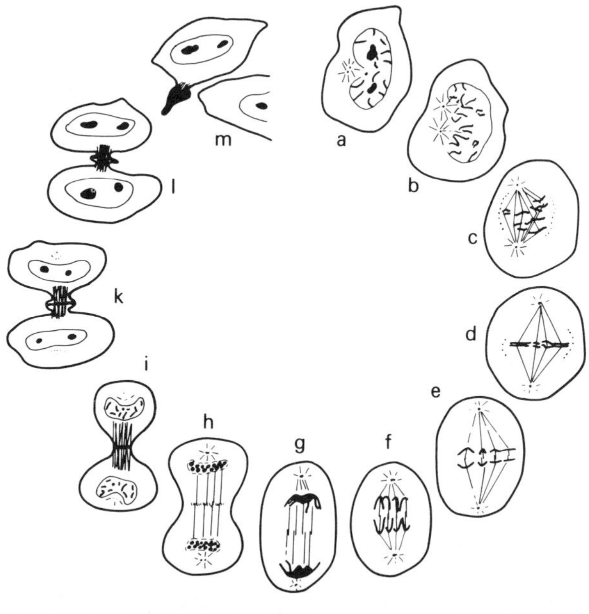

Figure 1. A schematic drawing of the mitotic cycle, reconstructed from electron micrographs. (a) Early prophase. (b) Late prophase. (c) Premetaphase. (d) Metaphase. (e) Early anaphase. (f) Middle anaphase. (g) Late anaphase. (h) Early telophase. (i) Late telophase. (k,l) Reconstruction phase. (m) Interphase.

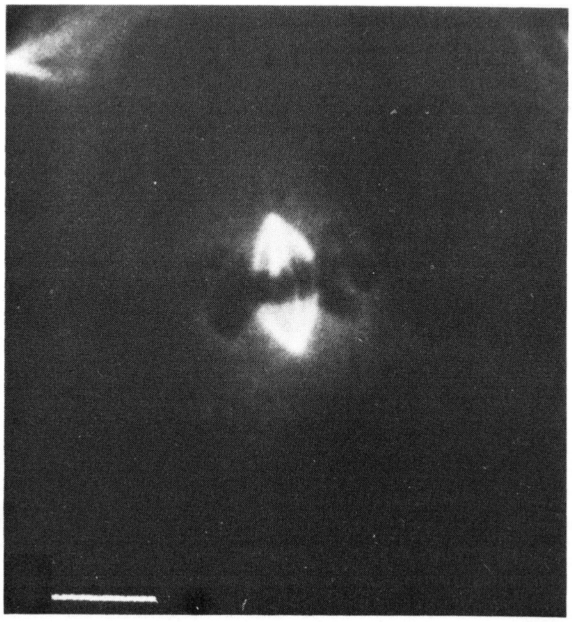

Figure 3. Light micrograph of a rat kangaroo cell in metaphase. Double antibody immunofluorescence technique. The cell was treated with an antibody against tubulin, and then with a fluorescently labelled antibody to that antibody. Against the dark background the bright spindle containing most of the tubulin stands out clearly. (1200×)
(By courtesy of Dr M. Osborn,[53] Max Planck Institute for Biophysical Chemistry, Göttingen.)

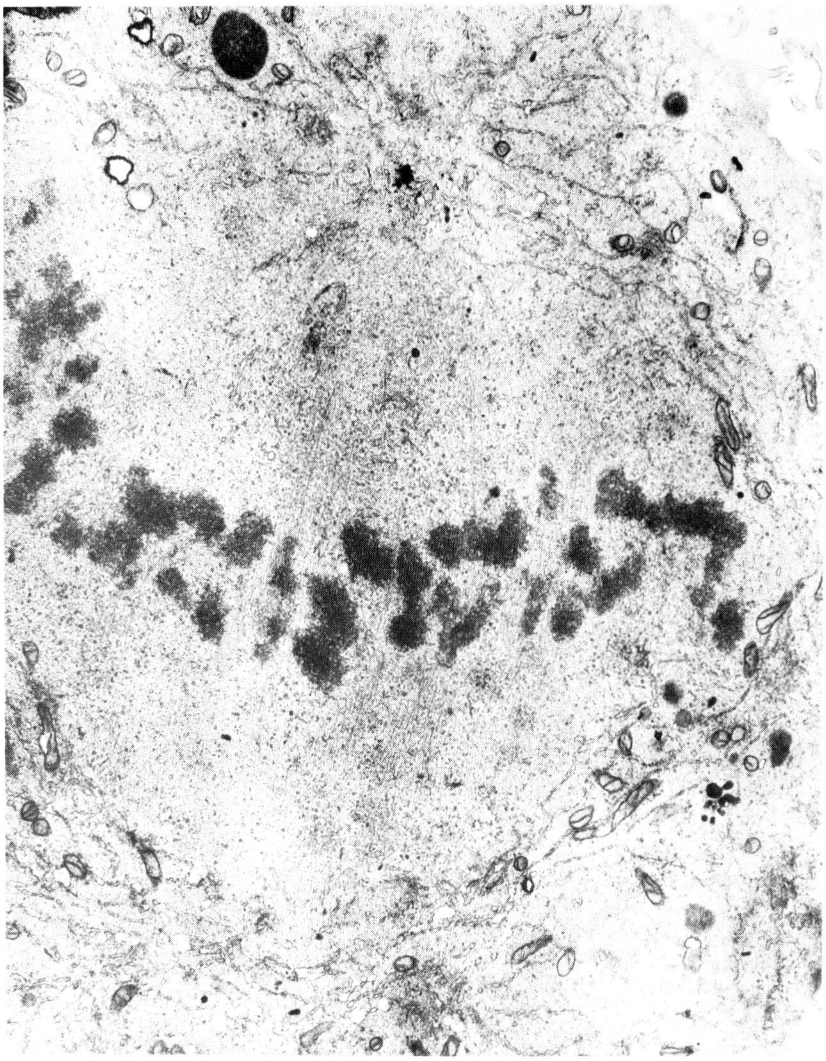

Figure 4. Electron micrograph of a human tumour cell in metaphase. The chromosomes are aligned in the central part of the mitotic apparatus, which is surrounded by membranes. The poles are indicated by a star-like arrangement of short membraneous profiles. (9600 ×)

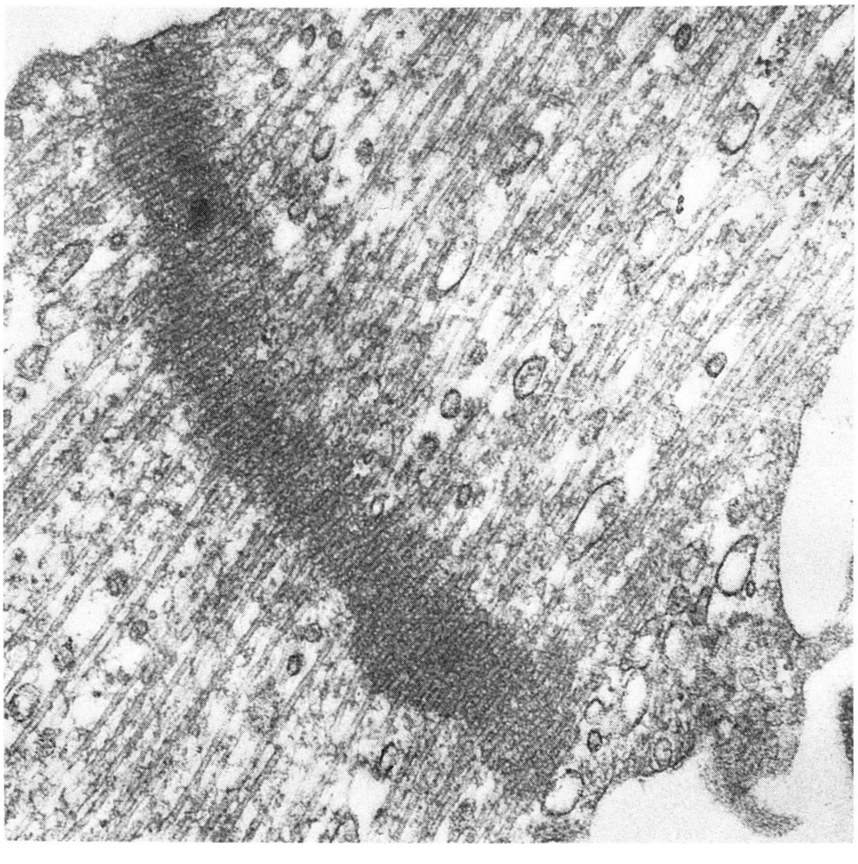

Figure 5. Electron micrograph of the central part of the spindle in late anaphase. The tubular structure of the spindle elements, the microtubules, is clearly visible. In the osmiophilic region the microtubules overlap from both half-spindles. (60 000 ×)

the growth rate would be constant and the cell cycle time variable; in fact, large daughter cells would have a shorter cell cycle and small daughter cells a longer one. The second case has been found to be true for the fission yeast. The mean value for cell size is held constant by regulating the time of entry into mitosis.

The point in the cell cycle at which a size control mechanism acts has been established by experiments with cell size mutants. These are temperature-sensitive mutants which at a permissive temperature have a similar size to that of the wild type of yeast, but a much smaller size at a restrictive temperature. When the temperature is shifted up from the permissive to the restrictive temperature, a rapid decrease in the size of dividing cells and an acceleration of the larger cells through G_2 and into division is observed. This supports the concept of a 'sizer' operating shortly before mitosis.

Extensive research on bacteria has shown that DNA replication is initiated at a constant size irrespective of growth rate. The final size of bacteria at division, however, is dependent on the rate of growth. This has been explained by the concept of a fixed time interval, a 'timer,' following the sizer, in which the final preparations for cell division are made. The constant period of time between the action of the sizer and cell division gives rise to small and large cells in poor and rich nutrient media respectively. Similar experiments with yeast cells, however, give results which are not consistent with the bacterial model. For example, when yeast cells are moved from a rich medium to a poor one, they are accelerated through G_2 and into mitosis, and size at division starts to fall abruptly shortly after the change. The available evidence now suggests that a sizer operates just before nuclear division which is reset by a nutrient change to operate at a smaller or larger size. A size control mechanism which initiates DNA synthesis may also be present, but it is not followed by a timer.

Cell size as a controlling factor for cell division was first proposed fifty years ago, and it has long been known that division in *Amoeba* could be stopped for many days by periodic amputations of cytoplasm.[8] It may be that the concentration of a size control factor, functioning as an activator of mitosis, is synthesized proportionately to cell growth. At a critical concentration of sizer the cell would be induced to start the events leading up to mitosis. Alternatively, the sizer could be an inhibitor of mitosis which is diluted by cell growth until it reaches a concentration at which mitosis can proceed. This last possibility seems unlikely, however, with regard to the above findings on *Amoeba*.

The results with unicellular organisms seem, at first sight, incompatible with some aspects of cell division in multicellular organisms. For example, the daughter cells of fertilized egg cells do not grow to the original size of the parent cell. In the course of further embryonal development, many of the mitoses result in the formation of differentiated cells.

MITOSIS AND DIFFERENTIATION

As mentioned above, the embryonic development of an organism is dependent upon the differentiation of a group of cells to fulfil a particular function. These cells are obviously qualitatively different from the mother cells, although the genetic material remains quantitatively unchanged. A cell cycle in which the daughter cells can carry out functions which the parent cannot, is termed a 'quantal' cell cycle.[9] The characteristic feature of mitosis, however, is that the daughter cells are provided with the same genetic material as the mother cell.

What is the explanation for this 'unequal' mitosis? Cellular processes are regulated by nuclear DNA. A given gene programme, however, must not run indefinitely once it has been started. Specific gene products, synthesized during interphase, the time between two cell divisions, can influence the programme through re-programming. If by chance these products become unevenly distributed in the parent cell, they would also be unevenly distributed in the daughter cells after division. Qualitatively different cells would then arise after the process of re-programming. Although this offers an explanation for the quantal cell cycle, the element of chance seems hard to reconcile with such a precisely regulated process as embryonic development.

The significance of mitosis in differentiation is clear. Changes in the gene programme, in gene regulation, can only be manifested after cell division. Although much data is now available on the differentiation process, research at the molecular biological level is still a very open field.[10]

REGULATION OF MITOSIS IN HIGHER ORGANISMS

In contrast to unicellular organisms, cell division in the multicellular higher organism is mainly regulated by external signals. The organism must maintain coordination between its differently functioning parts. These parts must also be renewed when lost or damaged. For example, when tissue is wounded, the cells in the vicinity of the wound are stimulated to proliferate. It has been postulated that a wound hormone is released when certain cells are destroyed, which induces cell division.

A regularly occurring cycle of cell proliferation and cell loss is represented by the estrous cycle in mammals. An increase in the concentration of the female steroid sex hormones estradiol and progesterone stimulates proliferation of the epithelial cells of the uterus. These hormones are withdrawn if the mature egg is not fertilized and thus fails to imbed itself into the wall of the uterus. The subsequent degeneration and loss of the epithelial cells is followed by the next estrous cycle.

In the last few years, substances which inhibit cell proliferation have been purified from tissues such as liver and skin. These so-called chalones, tissue hormones, are specific for a particular tissue but species unspecific.[11] The equilibrium between hormonal activators and inhibitors has been postulated to be responsible for the maintenance of the *status quo* among the various organs.

Another group of substances, the lectins, are also able to induce mitosis in some animal cells. They are synthesized by plant cells and cause a rounding-up and aggregation of red blood cells. Phytohemagglutin and concanavalin A are particularly well-known examples of the mitogenic lectins.[12] Mitogens are substances stimulating certain cells to enter mitosis. Some proteases may also be classified as mitogens. Trypsin, for example, can induce densely packed cells in tissue culture to divide.[13] The inhibition of mitosis caused by the close contact of neighbouring cells, appears to be overcome after the digestion of certain surface proteins.

An analysis of the cell surface after the administration of certain mitogens in interphase, has shown that the pattern of protein distribution resembles that of cells in mitosis. Transformed cells, that is, cells which continually cycle and no longer respond to the usual regulation mechanisms, show a similar pattern of protein distribution. This strengthens the evidence that specific proteins at the cells surface play a role in the regulation of cell division.

REGULATION OF SIGNALS AT THE CELL SURFACE

Our knowledge of the cell membrane has increased considerably in the last few years. We now know that the membrane is a mobile rather than a static structure.[14] Mitogenic receptors, in the form of cell surface proteins, can be easily transported within the lipid bilayer. Antibodies prepared to purified cell surface proteins can be coupled to a fluorescent dye and incubated with the cells of interest. After washing away the excess antibody, the protein antigens can be localized with a fluorescence microscope.[12]

With the help of this technique, it has now been shown that an originally irregularly distributed receptor protein aggregates after cross-linking with a divalent antibody. Small fluorescent patches become visible, which can move together and form a fluorescent cap. If the cells are treated with non-saturating amounts of concanavalin A prior to incubation with antibody, no movement of receptor proteins takes place. Concanavalin A binds strongly to the carbohydrate part of glycoproteins and appears to inhibit the mobility of receptor protein in the lipid bilayer. This phenomenon has been termed anchorage modulation.

The mitogenic (mitosis-inducing) activity of concanavalin A has already been mentioned. For example, it can induce lymphocytes to start DNA synthesis and form lymphoblasts. If, however, the cells have been incubated with colchicine, which destroys the microtubule system, mitogenesis is blocked.[15] This suggests that the microfilament–microtubule system may be involved in the mediation of external signals to the cell nucleus. Data from various experiments indicate that this system interacts with proteins in the cell surface membrane. Electron microscopic observation of patch and cap formation has shown them to be accompanied by characteristic changes in the orientation of the underlying microfilaments and microtubules.[16]

In the last few years cell surface proteins have been shown to be extremely important for cell–cell interactions. The cells of the immune system, for example, distinguish between the surface structures of self and non-self cells. Contact with non-self cells can initiate a chain of events, including cell proliferation and antibody production, which ends in the destruction of the non-self cell. Although many of the proteins involved in cell surface interactions have been characterized and some of the genes which code them have been mapped, almost nothing is known about the flow of information to the cell nucleus.

SPINDLE STRUCTURES

The fibres of the mitotic apparatus have been shown to be composed of bundles of fine tubules, 250 Å in diameter (Fig. 5). This epoch-making discovery of Porter[17] and of de Harven and Bernhard[18] was followed by electron microscopic observations of the mitotic apparatus of numerous kinds of cell. Although many different spindle types were described,[19] they could be classified into two main groups, namely, spindles which were constructed inside the nucleus (intranuclear spindles) and those in which the chromosomes were integrated into the mitotic apparatus after disintegration of the nuclear membrane (extranuclear spindles, Figs. 2 and 4). The first type of spindle is found mainly in lower organisms, while the second type of spindle is present in all the higher plants and animals. Present data indicates that the mitotic apparatus, like the organisms themselves, has gone through a process of evolution.[20]

All the different types of spindles have a common function, the transport of duplicated sets of chromosomes to the poles. Electron microscopic pictures of the spindle may give the impression that it is a rigid structure which is built up at the beginning of mitosis and then broken down after fulfilling its purpose. On the contrary, the spindle is a very dynamic structure, whose parts are constantly in motion. Small particles, when placed inside the spindle by micro-injection, are carried towards the poles at the speed of chromosomes in anaphase.[21]

In vivo observations of the spindle can take advantage of the birefringence of the spindle fibres under polarized light.[22] Thus, it could be shown that the spindle is a very labile structure which can be disintegrated under the action of low temperature, high pressure and various mitotic poisons. The changes in intensity of birefringence during mitotis indicate a constant change in the gain and loss of microtubules. Recent experiments of Margolis and Wilson[23] show that the subunits of microtubules *in vitro* are constantly added to one end and lost at the other, so that they appear to flow along the microtubules.

MICROTUBULE THEORIES OF CHROMOSOME MOVEMENT

Essentially three theories of chromosome movement have been put forward in the last three years in which microtubules play an active part. The assembly–disassembly hypothesis[24] postulates that chromosomes are moved by an increase or decrease in the length of microtubules attached to the kinetochores. Changes in the length of the microtubules are the result of additions or subtractions of the protein tubulin, which is the basic unit of microtubules. According to the assembly–disassembly hypothesis, this should occur at the kinetochores and in the vicinity of the poles.

Another hypothesis assumes that the microtubules slide past one another as happens during the movement of cilia.[25] The presence of small cross-bridges between some of the spindle structures can sometimes be seen under the electron microscope. Detachment of the bridges at one point followed by attachment at a point a little further along the microtubules, analogous to the actin-myosin system in muscle, could then cause sliding of the microtubules and transport of the attached chromosomes.

The third hypothesis[26] is based on the observation that not all spindle microtubules appear to run alongside one another. Some microtubules lie across the paths of others and appear to form contacts. From the point of contact, the microtubules are assumed to move together in the manner of a zip. If the chromosome microtubules are connected to any of these zipping microtubules, they will be transported for a short distance in the spindle. Repeated detachment and attachment to other zipping microtubules would then bring the chromosomes to the poles.

The assembly–disassembly hypothesis and the sliding microtubule hypothesis have been combined in another mitosis model by Margolis *et al.*,[27] but none of the model systems so far put forward is compatible with all the available experimental data. In the last few years, mechanisms of chromosome transport involving other proteins, besides tubulin, have been postulated.

NON-TUBULIN PROTEINS IN THE SPINDLE

A number of proteins have now been found in the mitotic apparatus which are known to be involved in muscle movement.[28] This raises the question as to whether they also participate in the transport of chromosomes.

The contractile protein with the highest concentration in non-muscle cells is actin. Myosin has also been found, but in much smaller quantities. Microfilaments of actin can be clearly distinguished from other intracellular fibrous structures after decoration with heavy meromyosin, HMM, a fragment of myosin. HMM forms arrowhead-like structures along the length of the microfilaments, which can be seen under the electron microscope.[29] Extremely thin sections of material are employed in electron microscopy, however, so that only very small regions of the cell can be observed. This problem has now been solved by employing fluorescently labelled antibodies. Although the cell membrane must be partially lysed to permit entry of antibodies into the cell, the internal structures are surprisingly well retained.

Incubation of cells with fluorescently labelled antibodies to actin or myosin reveals a random distribution of fibrillar elements in the cytoplasm during interphase. In mitosis, however, a fluorescent fibrous spindle shape can be observed. Its resemblance to the spindle shape under polarized light suggests that the actin and myosin filaments are in close proximity to the spindle fibres. Furthermore, the presence of fine fibrils between the spindle microtubules has been demonstrated with the electron microscope. Other fibrous proteins such as desmin,[30] a 55 K protein[31] and α-actinin[32] have also been recently found in non-muscle cells.

Points of anchorage for the contractile proteins could be provided by the microtubular skeleton of the spindle which would also determine the direction of movement.[33] Alternatively, the microtubules may actively co-operate with the contractile proteins. Surprisingly, substances which inhibit the function of the microfilament system, such as cytochalasin B, have no effect on the transport of chromosomes, whereas the division of the cytoplasm at the end of mitosis is completely inhibited.[34] Thus, the question of a possible participation of contractile proteins in the mitotic process is still open.

IN VITRO POLYMERIZATION OF MICROTUBULES

Investigations of mitosis at the molecular biological level is a relatively new area of research. In contrast to morphological studies where the whole cell is the object of observation, individual molecular events are analysed from which the highly complicated mitotic process can be reconstructed.

A starting point for many experiments was the observation that colchicine destroyed spindle fibres.[35] Colchicine is a poisonous substance of the meadow saffron which inhibits mitosis. By using radioactively labelled colchicine, Taylor[36] could show that it was bound to a protein in the cell which he named 'colchicine binding protein'. This protein was identified as tubulin, the basic unit of the microtubules. Every cell appears to have a pool of unpolymerized tubulin. It is particularly abundant in nerve cells, hence the mammalian brain is the principal starting material for its isolation.

Tubulin is now known to be composed of two subunits, each having a molecular weight of about 55 000. In the test tube, the dimer can form ring-shaped structures which are only stable under definite conditions such as low temperature. On warming to 37 °C, the rings polymerize to long thread-like filaments, protofilaments, which attach themselves to one another in long, thin sheets (Fig. 6). When the sheets are wide enough they begin to bend until, finally, a hollow cylinder is formed. With a diameter of 250 Å, these microtubules appear similar, if not identical, to those of the mitotic apparatus.[37] Although these intermediate structures have not been found *in vivo*, it is assumed that a similar process takes place.

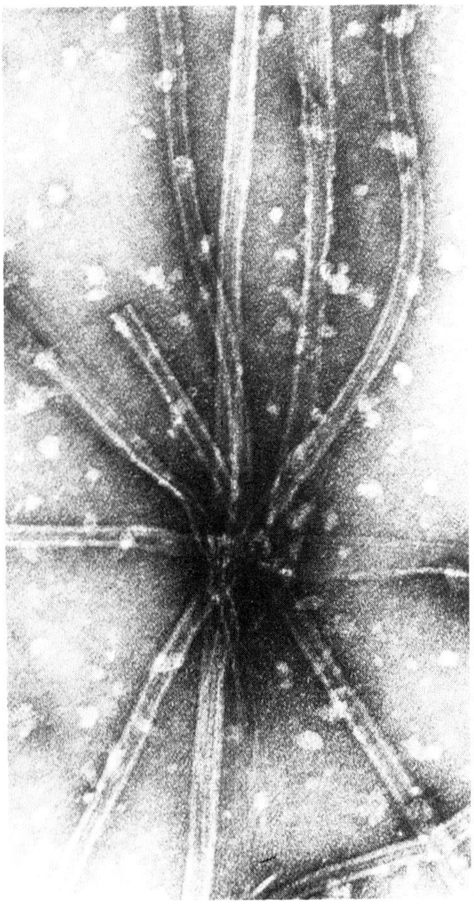

Figure 6. Electron micrograph of *in vitro* polymerized microtubules negatively stained. A longitudinal substructure, the protofilament, can easily be seen. Some microtubules open to form sheets of protofilaments. (120 000 ×)
(Preparation by courtesy of Dr K. H. Doenges, Institute for Cell Research, German Cancer Research Centre, Heidelberg.)

For a long time the polymerization of microtubules was only successful if brain tubulin was used, and it was thought that spindle microtubules might polymerize differently. Recently, however, it has been shown that tubulin from non-neuronal sources can be similarly polymerized.[38] Small quantities of other proteins have also been shown to co-polymerize with tubulin. If these microtubule-associated proteins, MAPs, are lost in the course of tubulin purification, it becomes more difficult to polymerize tubulin. Some of these polymerization proteins have now been characterized, and their influence on the microtubule polymerization process is under investigation.

A protein which does not co-polymerize with microtubules, but makes them especially sensitive to calcium ions, is the subject of much current interest.[39] In the presence of this protein, microtubules are destroyed at much lower concentrations than in its absence.

The sensitivity of microtubules to calcium ions *in vitro* immediately suggests a regulation mechanism similar to that in muscle. *In vivo* experiments with ionophores,[40] molecules which bind ions and promote their transport into the cell, also support the possibility of a regulation mechanism involving calcium. A calcium compartmentation system similar to that of the muscle sarcoplasmic reticulum would necessarily involve a membrane system in the mitotic apparatus. The search for such membranes has indeed proved to be successful for some cell species.[41–43] These findings also give new impetus to the contractile protein hypothesis of chromosome movement.

When microtubules polymerize in the test tube, they do not have any recognizable order or orientation. In the mitotic apparatus, however, they are always directed towards definite centres. The shape of the mitotic apparatus, a spindle, is a direct result of the microtubules converging on the poles. Star-shaped asters are also formed by the orientation of microtubules on to a definite point. The function of such mitotic centres is of fundamental significance for mitosis.

MITOTIC CENTRES

The growth of microtubules has been demonstrated to occur at only one end by following the uptake of radioactively labelled tubulin.[44] It seems likely, therefore, that the growth of orientated microtubules begins with the attachment of tubulin subunits on to centres of polymerization.

At the point of convergence of spindle and aster microtubules in the cells of higher animals, well-defined organelles, the centrioles, can be observed. At interphase, they occur as a pair lying perpendicular to one another. Spindle assembly begins at the two centrioles, also called centrosomes, and is built up between them as they migrate to the poles. Some of the spindle microtubules become attached to the chromosomes at a well-defined site, the kinetochore. Both centrioles and kinetochores are now accepted as being microtubule-orientating centres, MTOCs.[44]

More detailed investigations of the centriolar space have shown that the centrioles are surrounded by a diffuse region of osmiophilic, fibrous material which appears to contain very fine fibrils. Although isolated centrioles can serve as microtubule-orientating centres, recent data suggests that it is rather the diffuse pericentriolar material which is responsible for microtubule organization.[45] Furthermore, some spindles, particularly those of higher plants and lower organisms such as yeasts, contain no centrioles.

The growth of microtubules from the kinetochores can be observed under the electron microscope.[46] When chromosomes with intact kinetochores, but with no attached microtubules, are incubated with a polymerizable tubulin solution, microtubules start to assemble at the kinetochores. Observations on whole cells in prophase, however, show that the kinetochores do not participate in microtubule polymerization, at least not in the early stages of spindle assembly.[47] Moreover, recent experiments on induced mitotic apparatus-like structures in the unfertilized sea urchin egg seem to indicate that the activity of kinetochores may be limited to microtubule capture and attachment.[48] These artificial mitotic apparatuses contain no centrioles but do have many microtubules, often associated with a small amount of osmiophilic material. This suggests that many potential mitotic centres are present besides the 'classical' centres already described.

A large amount of data is now available on the tubulin molecule and its *in vitro* polymerization. There could be some difference, however, between the random and ordered assembly of microtubules and our present knowledge may be only a very pale reflection of the spindle assembly process.

MITOSIS *IN VITRO*

The data from *in vitro* investigations on single events in mitosis is sometimes confusing and difficult to relate to an *in vivo* process. Important pieces of information to link the events are often missing. On the other hand, the whole cell is such a complicated mixture of events that it is extremely difficult to study how a single process functions. Attempts to reduce this complexity have been made in studies of the mitotic process *in vitro*.[49] The goal is to eliminate as many unnecessary parts as possible, while leaving a functioning mitotic apparatus intact. The ideal system would be a purified spindle in a given medium which is able to move chromosomes at a physiological rate towards the poles. This goal, however, is still very far from being realized.

In 1952, Mazia and Dan[50] were the first to isolate a mitotic apparatus by treatment of fertilized sea urchin eggs with cold ethanol, followed by detergent solubilization of as much unwanted cytoplasm as possible. Marine animals' eggs are excellent sources of mitotic spindles for two reasons: they can be obtained in large quantity, and the spindle assembly process can be easily synchronized by simultaneous fertilization. The early spindle isolates, however, were mainly dead ones. Stabilization of the mitotic apparatus, largely due to protein denaturation, was achieved at the expense of its biological function. More recent experiments employ reagents which stabilize the mitotic apparatus, while preserving its lability and the activity of any spindle-associated enzymes. Very life-like spindles have thus been isolated by Sakai[51] which, after the addition of $0.5\ \mathrm{mg\,ml^{-1}}$ of polymerizable tubulin, are reported to move the chromosomes towards the poles at a very small but measurable rate. Further experiments appear to be necessary, however, to decide whether this represents a true anaphase movement.

Electron microscopic examinations of all the mitotic spindles which have so far been isolated, reveal varying amounts of cytoplasmic material. To what extent this represents a necessary part of the spindle, if at all, is not known. Removal of cytoplasmic contamination remains a perplexing problem for all investigators.

Another approach to the *in vitro* study of chromosome movement involves lysing the cells with a minimal concentration of detergent in the presence of polymerizable tubulin to preserve the spindle.[52] Unfortunately, the difference between the amount of detergent necessary for an effective cell lysis and that which inhibits chromosome movement is not very large. A partial lysis of the cell, on the other hand, with low concentrations of detergent, offers a model system for the study of factors influencing chromosome movement. Enzyme or protein antibodies, for example, have direct access to a still functioning mitotic apparatus. Until now, however, not many positive results have been reported using this technique.

All of the mitosis *in vitro* systems suffer from the drawback that the amount of lysis is sometimes so small that the cell is still more or less intact, or so large that the spindle is extracted too extensively.

CONCLUDING REMARKS

Mitosis is dependent on events which take place in other parts of the cell cycle. A resting cell must first begin to cycle before it can enter mitosis.

Preparations for cell division include the synthesis of specific proteins and replication of the genetic material. The signal to start mitosis appears to be determined by a size control factor in unicellular organisms. A more complicated situation exists in higher organisms, where the cell cycle is initiated by external signals. Investigations on the mediation of these signals to the cell nucleus are of central importance in research on the regulation of mitosis.

An understanding of this process would be of considerable value in the fight against cancer. The growth of malignant tumours could be held in check by modulating or blocking mitosis-initiating signals. This approach would seem to be much more promising than the search for a selective therapy based on differences between normal and malignant cell structures, since the latest data indicate that these differences are minimal.

A major problem of mitosis remains the mechanism of chromosome movement. None of the present theories explains satisfactorily how the spindle functions. It is possible that other fibrous proteins, besides the microtubules, are also involved in chromosome transport. An orderly assembly of spindle microtubules is dependent on mitotic centres. The nature of such centres, and the conditions under which they operate, are now the subject of intensive investigation.

An ideal system for the study of the mitotic process would be an isolated mitotic apparatus which still functions. Methods to isolate only those elements necessary for chromosome movement are constantly improving, and significant advances can be expected in the near future.

Acknowledgments

We gratefully acknowledge the fruitful discussion with many colleagues in our Institute and thank Dr M. Osborn and Dr K. H. Doenges for providing us with photographs.

LITERATURE CITED

1. A. Schneider, Untersuchungen über Plathelminthen. *Jahrb. Oberhess. Ges. Natur-Heilk.* **14** (1873).
2. E. Strassburger, *Zellbildung und Zelltheilung*, 1. Aufl. Fischer-Verlag, Jena (1875).
3. W. Flemming, *Zellsubstanz, Kern und Zelltheilung*, Vogel-Verlag, Leipzig (1882).
4. W. Waldeyer, Über Karyokinese und ihre Beziehungen zu den Befruchtungsvorgängen. *Arch. Mikrosk. Anat.* **32,** 1–22 (1888).
5. N. Paweletz, Hundert Jahre Mitoseforschung. *Naturwiss. Rundsch.* **27,** 359–370 (1974).
6. J. M. Mitchison, *The Biology of the Cell Cycle*, Cambridge University Press, London (1971).
7. J. M. Mitchison, The timing of cell cycle events. In *Mitosis, Facts and Questions*, edited by M. Little, N. Paweletz, C. Petzelt, H. Ponstingl, D. Schroeter and H.-P. Zimmermann, Session 1, pp. 1–13. Springer-Verlag, Berlin–Heidelberg–New York (1977).
8. M. Hartmann, Über experimentelle Unsterblichkeit von Protozoen-Individuen. Ersatz der Fortpflanzung von *Amoeba proteus* durch fortgesetzte Regeneration. *Zool. Jahrb.* **45,** 973–987 (1928).
9. H. Holtzer, H. Weintraub, B. Mayne and B. Mochan, The cell cycle, cell lineages and cell differentiation. In *Current Topics in Developmental Biology*, edited by A. A. Moscona and A. Monroy, pp. 229–256. Academic Press, New York (1972).
10. F. Duspiva, Mitosis in differentiation, morphogenesis, and cancer. In *Mitosis, Facts and Questions*, edited by M. Little, N. Paweletz, C. Petzelt, H. Ponstingl, D. Schroeter and H.-P. Zimmermann, Session 6, pp. 121–145. Springer-Verlag, Berlin–Heidelberg–New York (1977).
11. W. S. Bullough, E. B. Laurence, O. H. Iversen and K. Elgjo, The vertebrate epidermal chalone. *Nature (London)* **214,** 578 (1967).
12. G. M. Edelman, P. D'Eustachio, D. A. McClain and S. M. Jazwinski, Surface signals and cellular regulation of growth. In *Mitosis, Facts and Questions*, edited by M. Little, N. Paweletz, C. Petzelt, H. Ponstingl, D. Schroeter and H.-P. Zimmermann. Session 2, pp. 20–42. Springer-Verlag, Berlin–Heidelberg–New York (1977).
13. M. M. Burger, Proteolytic enzymes initiating cell division and escape from contact inhibition of growth. *Nature (London)* **227,** 170–171 (1970).
14. D. F. Hoelzl Wallach, *The Plasma Membrane: Dynamic Perspectives, Genetics and Pathology*, Heidelberg Science Library, Vol. 18. English Universities Press, London; Springer-Verlag, New York–Heidelberg–Berlin (1972).
15. M. E. Baker, Colchicine inhibits mitogenesis in C1300 neuroblastoma cells that have been arrested in G_0. *Nature (London)* **262,** 785–786 (1976).
16. R. D. Berlin, J. M. Oliver, T. E. Ukena and H. H. Yin, Control of cell surface topography. *Nature (London)* **247,** 45–46 (1975).
17. K. R. Porter, Fine structure of cells. *Proc. Symp. Int. Congr. Cell Biol., London*, pp. 236–250 (1955).
18. E. de Harven and W. Bernhard, Etude au microscope électronique de l'ultrastructure du centriole chèz les vertébrés. *Z. Zellforsch. Mikrosk. Anat.* **45,** 387–398 (1956).
19. H. Fuge, Ultrastructure of mitotic cells. In *Mitosis, Facts and Questions*, edited by M. Little, N. Paweletz, C. Petzelt, H. Ponstingl, D. Schroeter and H.-P. Zimmermann, Session 3, pp. 51–68. Springer-Verlag, Berlin–Heidelberg–New York (1977).
20. J. Pickett-Heaps, The evolution of mitosis and the eukaryotic condition. *BioSystems* **6,** 37–48 (1974).
21. R. B. Nicklas and C. A. Koch, Chromosome micromanipulation. IV. Polarised motions within the spindle and models for mitosis. *Chromosoma* **39,** 1–26 (1972).
22. S. Inoué, J. Fuseler, E. D. Salmon and G. W. Ellis Functional organization of mitotic microtubules. *Biophys. J.* **15,** 725–744 (1975).
23. R. L. Margolis and L. Wilson, Opposite end assembly and disassembly of microtubules at steady state *in vitro*. *Cell* **13,** 1–8 (1978).
24. R. Dietz, Bau und Funktion des Spindelapparates. *Naturwissenschaften* **56,** 237–248 (1969).
25. J. R. McIntosh, P. K. Hepler and D. G. van Wie, Model for mitosis. *Nature (London)* **224,** 659–663 (1969).
26. A. S. Bajer, Interaction of microtubules and the mechanism of chromosome movement (zipper hypothesis). I. General principle. *Cytobios* **8,** 139–160 (1973).
27. R. L. Margolis, L. Wilson and B. J. Kiefer, Mitotic mechanism based on intrinsic microtubule behaviour. *Nature (London)* **272,** 450–452 (1978).
28. J. W. Sanger, Nontubulin molecules in the spindle. In *Mitosis, Facts and Questions*, edited by M. Little, N. Paweletz, C. Petzelt, H. Ponstingl, D. Schroeter and H. P. Zimmermann, Session 5, pp. 98–113. Springer-Verlag, Berlin–Heidelberg–New York (1977).

29. H. Ishikawa, R. Bischoff and H. Holtzer, Formation of arrowhead complexes with heavy meromyosin in a variety of cell types. *J. Cell Biol.* **43,** 312–328 (1969).
30. E. Lazarides, The distribution of desmin (100 Å) filaments in primary cultures of embryonic chick cardiac cells. *Exp. Cell Res.* **112,** 265–273 (1978).
31. M. Osborn W. W. Franke and K. Weber, Visualization of a system of filaments 7–10 nm thick in cultured cells of an epitheloid line (PtK2) by immunofluorescence microscopy. *Proc. Natl. Acad. Sci. USA* **74,** 1490–2494 (1977).
32. E. Lazarides, Actin, alpha-actinin and tropomyosin interaction in structural organization of actin filaments in non-muscle cells. *J. Cell Biol.* **68,** 202–219 (1976).
33. W. Z. Cande, E. Lazarides and J. R. McIntosh, A comparison of the distribution of actin and tubulin in the mammalian mitotic spindle as seen by indirect immunofluorescence. *J. Cell Biol.* **72,** 552–567 (1977).
34. T. E. Schroeder, The contractile ring. I. Fine structure of dividing mammalian (HeLa) cells and the effects of cytochalasin B. *Z. Zellforsch.* **109,** 431–449 (1970).
35. B. Pernice, *Sicil. Med.* **1,** 265 (1889), cited from O. J. Eigsti and P. Dustin, Jr, *Colchicine in Agriculture, Medicine, Biology, and Chemistry,* Iowa State College Press, Ames, Iowa (1955).
36. E. W. Taylor, The mechanism of colchicine inhibition of mitosis. I. Kinetics of inhibition and the binding of ^{3}H-colchicine. *J. Cell Biol.* **25,** 145–160 (1965).
37. J. Bryan, A quantitative analysis of microtubule elongation. *J. Cell Biol.* **71,** 749–767 (1976).
38. K. H. Doenges, B. W. Nagle, A. Uhlmann and J. Bryan, Assembly of tubulin from nonneural cells (Ehrlich ascites tumor cells). *Biochemistry* **16,** 3455–3459 (1977).
39. J. M. Marcum, J. R. Dedman, B. R. Brinkley and A. R. Means, Control of microtubule assembly-disassembly by calcium-dependent regulator protein. *Proc. Natl. Acad. Sci. USA* **75,** 3771–3775 (1978).
40. M. Schliwa, The role of divalent cations in the regulation of microtubule asembly. *In vivo* studies on microtubules of the heliozoan axopodium using the ionophore A 23187. *J. Cell Biol.* **70,** 527–540 (1976).
41. P. K. Hepler, Membranes in the spindle apparatus: their possible role in the control of microtubule assembly. In *Mechanism and Control of Cell Division,* edited by T. L. Rost and E. M. Gifford, pp. 212–232. Dowden, Hutchinson and Ross, Stroudsburg, Pennsylvania (1977).
42. P. Harris, The role of membranes in the organization of the mitotic apparatus. *Exp. Cell Res.* **94,** 409–425 (1975).
43. N. Paweletz, Membranes in the mitotic apparatus of mammalian cells. In *Cell Reproduction,* ICN-UCLA Symposia on Molecular and Cellular Biology. Academic Press, New York (1978).
44. G. G. Borisy and R. R. Gould, Microtubule-organizing centers of the mitotic spindle. In *Mitosis, Facts and Questions,* edited by M. Little, N. Paweletz, C. Petzelt, H. Ponstingl, D. Schroeter and H.-P. Zimmermann, Session 4, pp. 78–87. Springer-Verlag, Berlin–Heidelberg–New York (1977).
45. M. W. Berns and S. M. Richardson, Continuation of mitosis after selective laser microbeam destruction of the centriolar region. *J. Cell Biol.* **75,** 977–982 (1977).
46. R. R. Gould and G. G. Borisy, Quantitative initiation of microtubule assembly by chromosomes from Chinese hamster ovary cells. *Exp. Cell Res.* **113,** 369–374 (1978).
47. N. Paweletz, Elektronenmikroskopische Untersuchungen an frühen Stadien der Mitose bei HeLa-Zellen. *Cytobiologie* **9,** 368–390 (1974).
48. N. Paweletz and D. Mazia, the 'nuclear mitotic apparatus' in sea urchin eggs. In *Cell Reproduction,* ICN-UCLA Symposia on Molecular and Cellular Biology. Academic Press, New York (1978).
49. J. R. McIntosh, Mitosis *in vitro*: isolates and models of the mitotic apparatus. In *Mitosis, Facts and Questions,* edited by M. Little, N. Paweletz, C. Petzelt, H. Ponstingl, D. Schroeter and H.-P. Zimmermann, Session 8, pp. 167–184, Springer-Verlag, Berlin–Heidelberg–New York (1977).
50. D. Mazia and K. Dan, Isolation of the mitotic apparatus of sea urchin eggs. *Proc. Nat. Acad. Sci. USA* **38,** 826 (1952).
51. H. Sakai, Y. Hiramoto and R. Kuriyama, The glycerol isolated mitotic apparatus: a response to porcine brain tubulin and induction of chromosome motion. *Dev. Growth Differ.* **17,** 265–274 (1975).
52. J. R. McIntosh, W. Z. Cande and J. A. Snyder, Structure and physiology of the mammalian mitotic spindle. In *Molecules and Cell Movement,* edited by S. Inoué and R. E. Stephens, pp. 31–76. Raven Press, New York (1975).
53. K. Weber, Th. Bibring and M. Osborn, Specific visualization of tubulin-containing structures in tissue culture cells by immunofluorescence. *Exp. Cell Res.* **95,** 111–120 (1975).

The manuscript was received 10 November 1978

The Breguet Dynasty

Two Centuries of Interdisciplinary Scientists and Engineers

CLAUDE A. J. BREGUET
Biarritz and Paris, France

Five generations of the Breguet Dynasty are described from 1747 to 1967. All of them were either illustrious scientists, many of them Members of the Academy of Sciences, or brilliant engineers, more often interested in perfecting their innovations than in commercial success. The founder, Abraham Louis Breguet, the most famous horologist of his century, is widely known amongst clock and watch experts; Louis Breguet, the pioneer aircraft designer and manufacturer, is equally well known amongst aviation historians – but there can be only very few who know the whole continuous story of this remarkable and highly distinguished French family, here described continuously for the first time by a member of the sixth generation of the Breguet Dynasty.

True Democracy allows all to make their maximum contribution to the World. The Breguets belonged to that Democracy and made noble use of this Freedom.
Obituary notice for Louis F.-C. Breguet, 1885, quoting Louis Pasteur.

What has distinguished the Breguet Dynasty from other similar families was the unique place they occupied between the natural sciences and industry. The Breguets were at once scientists and industrialists. On the one hand they had to face the problems of research innovation and production; on the other, the economic problems of industrial production. This held less attraction for them but was essential for their survival and that of their firm. In spite of the diversity of their inventions, they had few opportunities to employ true mass production; it could have enlarged their markets through reduction in costs and would have yielded the essential profits for investment and for new researches (Fig. 1).

To keep going, the Breguets relied more on their imagination and their practical judgment, helped by a rare flexibility of mind, than on external help, whether scientific or financial. Partly because of their Swiss Protestant background, they deeply treasured their freedom and independence. It would appear, as a relic of the craftsman's tradition, that they even viewed the inevitable growth of their firm with trepidation, fearing that little by little it might pass out of their hands, as it finally did in 1967.

To the extent that personality, creative imagination and sincerity are needed for every work of art, the Breguets must be regarded as artists – as artists in engineering and fine mechanics. Posterity has indeed regarded them so, forgetting or neglecting their second *métier* as heads of an industrial enterprise where immediate interests sometimes conflicted with their ideals of human progress. The family lineage of the Breguet Dynasty, from 1747, the birth of the first Breguet, until 1967, the end of any independent Breguet industrial company, gives a graphic survey of my family history and its various endeavours (Fig. 2).

ABRAHAM-LOUIS BREGUET (1747–1823)

One might say that the scientific dynasty of the Breguets began in 1774 when Abraham-Louis Breguet (Fig. 3) set up business as a clock and watchmaker at the Quai de l'Horloge in Paris. He had come to Paris from his native Neuchâtel in Switzerland twelve years earlier and today, 218 years later, the building on the Quai is still occupied by his descendants (Fig. 4). Within a few years of starting his business, although many talented watch and clockmakers flourished in France as in England, Breguet had become without question the most ingenious horologist in Europe. His inventions were of such importance and were carried out with such rare perfection that Sir David Salomons[1] wrote, 'for another century, there was no further progress of note in watches or chronometers'.

To appreciate his contribution to horology and chronometry, one must think back to the middle and the end of the 18th century, full of political upheavals in France, but a period when science and technology had as yet made little, if any, impact on everyday life. Down to the end of the 18th century, all sources of mechanical energy had remained the same as they had been in ancient times: the energy of wind and water, occurring naturally but unreliably. For land

CCC–0308–0188/80/0005–0149$08.00/1

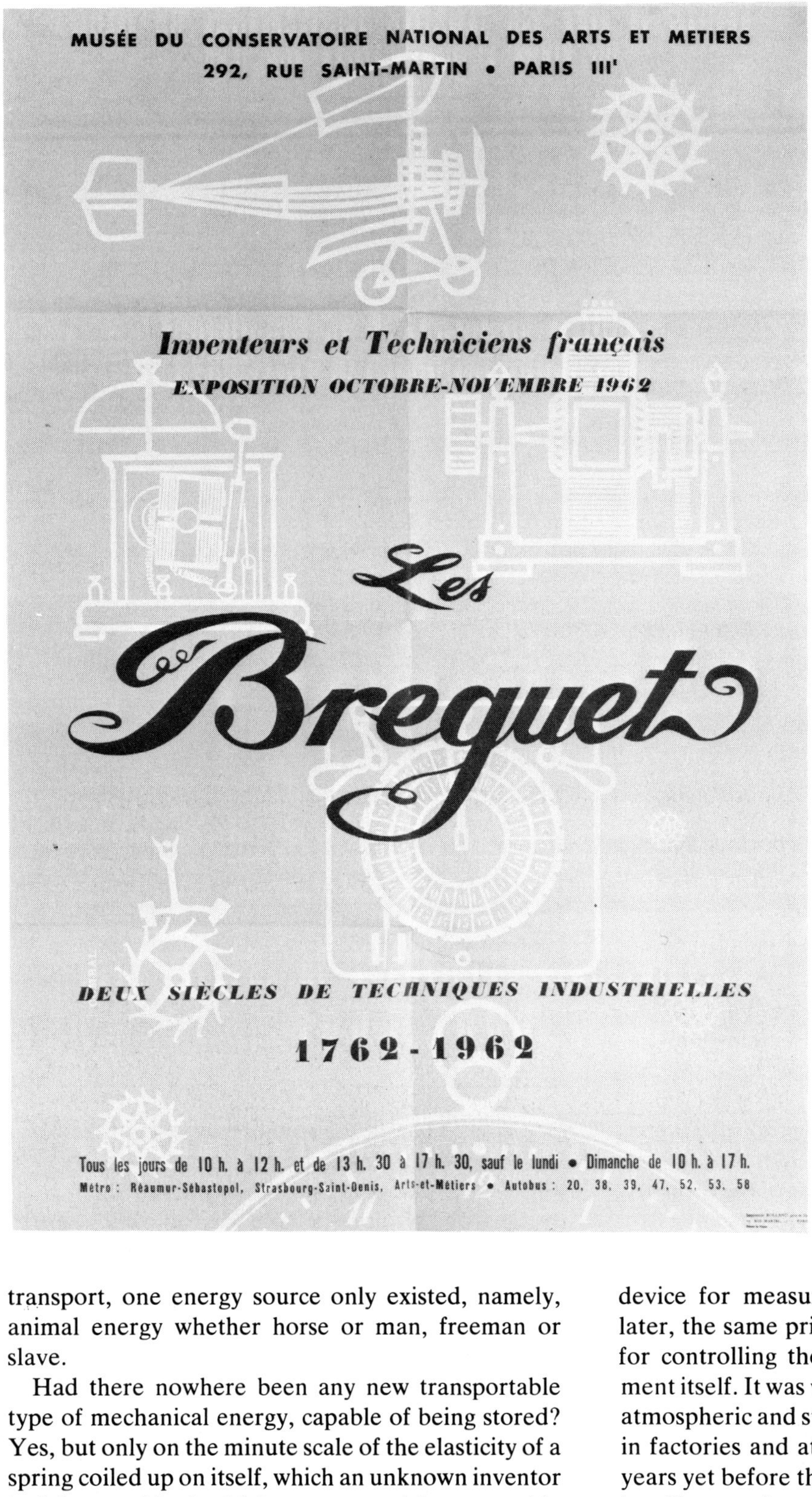

Figure 1. Two centuries of the Breguet Dynasty were honoured by a public exhibition in Paris in 1962.

transport, one energy source only existed, namely, animal energy whether horse or man, freeman or slave.

Had there nowhere been any new transportable type of mechanical energy, capable of being stored? Yes, but only on the minute scale of the elasticity of a spring coiled up on itself, which an unknown inventor had first used in the 16th century, to drive a portable device for measuring time – the watch. A century later, the same principle was to be used by Huygens for controlling the oscillations of the watch movement itself. It was very much later that the first clumsy atmospheric and steam engines began noisily to clank in factories and atop mine shafts; it would be many years yet before the first steam engine would find use in a Breguet factory.

CLAUDE A. J. BREGUET, born 1910, the son of Jacques and the nephew of Louis Breguet, belongs to the sixth generation of Breguets who achieved distinction as scientists and engineers. Claude was attached to the President of the Council, Poincaré; he obtained his Diploma of Political Sciences and his Licentiate in Law by 1936. He entered the aviation works of his family at Biarritz in 1941 and there, until retirement in 1975, worked his way up from draughtsman to *Chef de Service* of the Design Office which included quality control and strength calculations. He has written about horology and aviation and was awarded the Médaille de l'Aéronautique.
Address: La Cle' des Champs, 38 Avenue de Coulaon, 64200 Biarritz, France.

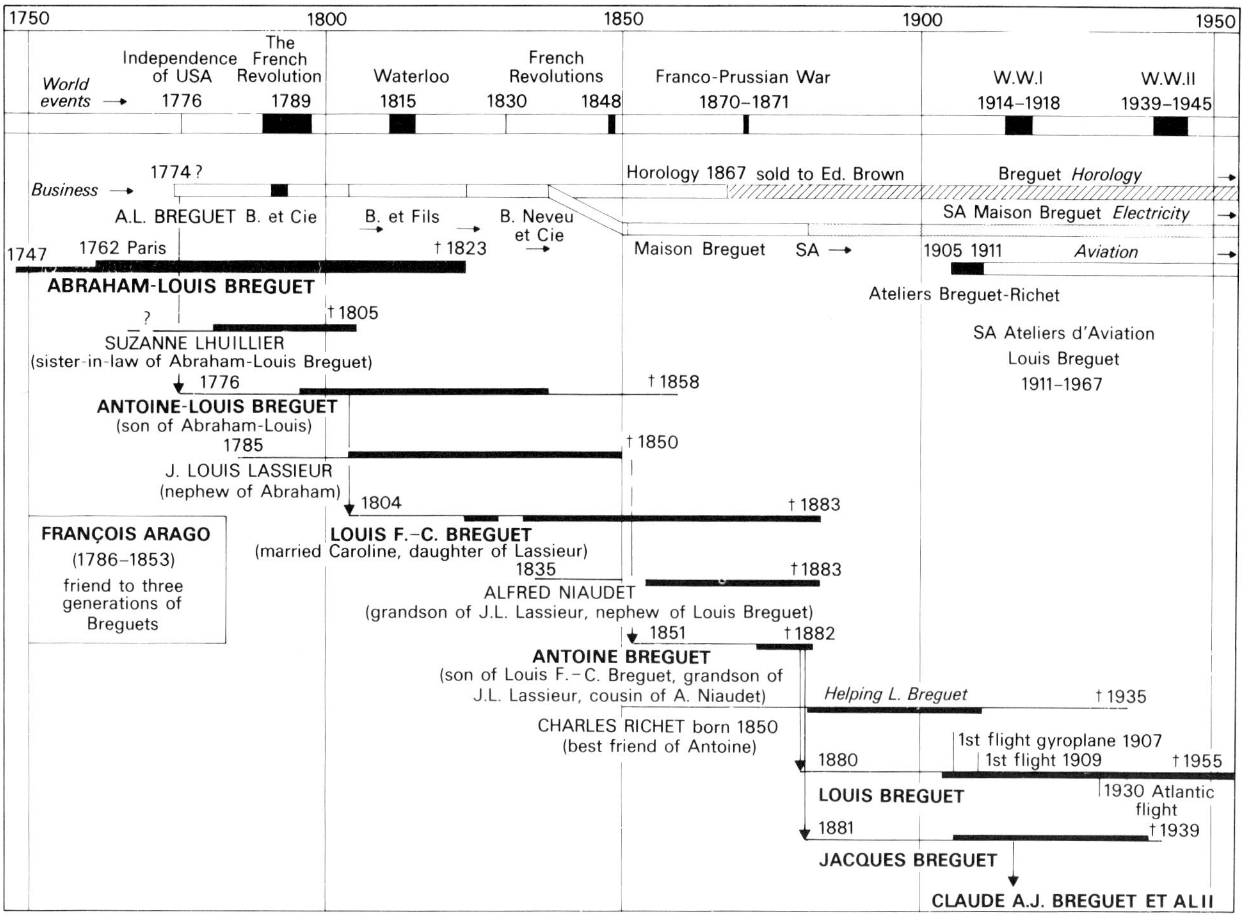

Figure 2. The Breguet Dynasty, 1747–1967.

Horology at the End of the 18th Century

At the end of the 18th century, when technology was just about to start its exponential growth, the making of clocks and watches was the growing point of many

Figure 3. Abraham-Louis Breguet, the Founder of the Dynasty (1747–1823).

Figure 4. The Breguet's family house and workshops, 39 Quai de l'Horloge, Paris in 1815.

new techniques. It took only a little material and little energy to make a watch; certainly, it needed tools and above all ingenuity, work and patience, but there were many skilled workmen and human energy was plentiful.

At this time the demand for clocks and watches entered a period of extraordinary growth as a great need for accurate timepieces had arisen everywhere. Astronomers were becoming ever more exacting in their demands and so were merchant shippers, who had long been seeking a means of measuring longitude with the precision already possible for latitude. Thanks to living on an island, the British had had much more need to find this solution than had their Continental neighbours. The shipwreck of Vice-Admiral Sir Cloudesley Shovell's squadron off the Scilly Isles in 1707 was much in mind when in 1714 an Act of Queen Anne and Parliament in London offered the then very considerable sum of £20 000 to anyone who could make a chronometer of sufficient accuracy to determine longitude at sea. It was John Harrison[2] who eventually won this prize after twenty years' hard work and many disappointments.

Sir David Lionel Salomons[1], the greatest collector ever of Breguet watches, has written: 'England owes its greatness in part to such men as Tompion, Graham and Harrison, for it was through their efforts that the great oceans were surveyed and merchant ships enabled to navigate them with ease and safety.' By the end of the 18th century the power of the Spanish and the Dutch had declined and the French had become the rivals of the English on the high seas and therefore also their competitors in the race for progress in accuracy of time measurement. Talented men like the Leroys, and Berthoud a little later, served the Royal French Navy, French physicists and astronomers. Before the Revolution of 1789 many clever watch and clockmakers competed for the custom of the Court at Versailles; courtiers wanted fashionable but inexpensive watches, watches which had to be chased, worked, enamelled, or even decorated with pearls and diamonds.

Abraham-Louis Breguet's Innovations

The automatic watch of 1780. Breguet had earned European renown as early as 1780 with his automatic watch, 'which runs as long as it is worn'. We must admire his audacity to be the first to put such a mechanism on the market; it was crowned with success but Breguet never claimed to be its inventor, crediting a German Jesuit from Nuremberg of the previous century. Other models made at the time, by his compatriot Perrelet for example, were never offered for sale.

Breguet himself analysed the improvements he brought to horology under the following three headings:

◇ Innovations *in form*: arrangements increasing attractiveness or convenience. Such inventions led only to commercial advantages, gave temporary employment to workshops, enriched some artists, but did not increase their reputation.

◇ Inventions leading to *robustness*, making the product last longer or making it cheaper to repair after an accident. Such inventions remained to enrich the art of horology.

◇ Inventions increasing *precision* of movement, enabling it to withstand irregularities of temperature changes, variations in friction, vibration and changes in position. These inventions were the real riches of horology. They achieved its most important objects – the exact measurement of time (Fig. 5).

A more detailed analysis of Abraham-Louis Breguet's innovations is given in Table 1; their order

Table 1. Abraham-Louis Breguet's most important innovations in horology and physics

1. The perpetual watch	1780
The tone springs	1787
Improvement of repeater at minute intervals	
The sympathetic pendulum, permitting the automatic resetting and adjustment of a watch each evening from a clock	1793–1805
Single-stroke seconds	
Improvement of perpetual calendar	1795
Spring-unwind indicator	
Watch read by touch only	1796
Improvements of hourly day of the month	
Improvements of striking clock	
Engine-turned dials	1798
Natural anchor-lift escapements	1800
Breguet design of lettering and of hands	
Flat watches	
2. The parachute, a very important invention of shock-protected elastic suspension of pivots	1806?
New design of clockwork	
Breguet key	1789
Improvement to jewel bearings	1790
Improvement in the formation of springs	
Improvement (new tool) in the working of hard stones [His early jewels were made in England by Arnold][3]	
Improvement in repeaters	1790
Single-hand watch made from 14 parts for batch production, the so-called 'subscription' watch	1795
3. The Breguet spiral, with its outer end brought back to the centre	
Compensating balance	
Escapement with constant force (1798 patent)	1795
Balance wheel with a revolving-carriage escapement, the *tourbillon*, to reduce positional error as much as possible	1787–1795
Independent seconds	1796
Chronometer with daily rate of 0.75 s variation	
Eyepiece of transit instrument, permitting evaluation of angle to 0.1 s	1802
Counter for recording *pointage* for astronomical use	1823
Marine chronometers with several spring barrels	

Figure 5. The Master's Masterpiece: Breguet's Marie Antoinette Watch. Described by G. Daniels in *The Art of Breguet* as 'the most famous complicated watch made by Breguet, it is said to have been ordered in 1783 by an officer of the Queen's guard and that the order stipulated it was to have every possible complication then known. No limit on the time of manufacture or cost was imposed and all parts normally of brass were to be made in gold. The exact date of starting the watch is not known. Although it has been said that the watch was completed in 1802, a study of the work book shows quite clearly that it was not finished until 1820 by which time it had cost 16 864 francs. The gifted pupil Michel Weber made most of the mechanism for which he was paid 7250 francs for 725 hours work between 1812 and 1815'.

A technical description of this watch was given by Daniels: '*Perpetuelle*, minute repeating, perpetual calendar, equation of time, *secondes d'un coup*, equal lift lever escapement with two-armed compensation balance with recessed screws, helical spring with terminal curves free sprung, winding weight arbors supported in a parachute, every friction surface jewelled, the plates and wheels of gold, rock crystal dial with gilt numerals and subsidiary dial for seconds with inset day of the week dial, at the top left sectors for the state of winding and the equation of time and at the top right the date of the month and thermometer, blued steel hands, plain gold case with rock crystal back to display the movement, repeating piston in the pendant, 63 mm diameter.'

Salomons[1] states that the watch was sold for £600 in 1887 to Sir Spencer Brunton. By 1917 the *Marie Antoinette* watch had found its way into a shop in London's Regent Street where it was purchased by Sir David Salomons; after Sir David's death his collection of Breguet watches, including the *Marie Antoinette*, went to the Mayer Memorial Foundation in Jerusalem, Israel. (*Pictures Courtesy*: *George Daniels, London.* See General Bibliography.)

does not imply any relative importance. In addition to his horological innovations – they include some devices known previously, as French patents regulations were only introduced at the time of the Revolution – we must mention other original inventions which might well be described as those of a physicist.

The metallic thermometer. First of all there was the metallic thermometer (Fig. 6), a logical outcome of the compensation systems for balance wheels. It was presented at the 1819 Exhibition with a note from the physicist Prony, and made use of strips of the three metals gold, silver and platinum. Breguet produced some miniature versions of this thermometer and inserted it in certain watches and also into the setting of a ring, of which the Duke of Wellington owned one.

The semaphore telegraph. Breguet, who, it appears, had also conceived the principle of the mechanism of Chappe's semaphore telegraph,[4] proposed to the French government with his friend Agustin de Betancourt, a famous Spanish engineer, another optical telegraph system. Its advantages did not appear sufficiently convincing to replace the Chappe system.

Metallurgy of steel. In the field of metallurgy, Breguet, with his engraver friend and compatriot Jean-Pierre Droz, who was director of the Paris Mint, invented a method for casting steel which long remained secret.

Creation and development of the Maison Breguet

How large was the firm of Breguet and how was it organized? All we know for certain is that, aided initially by his wife's dowry, Breguet employed only a

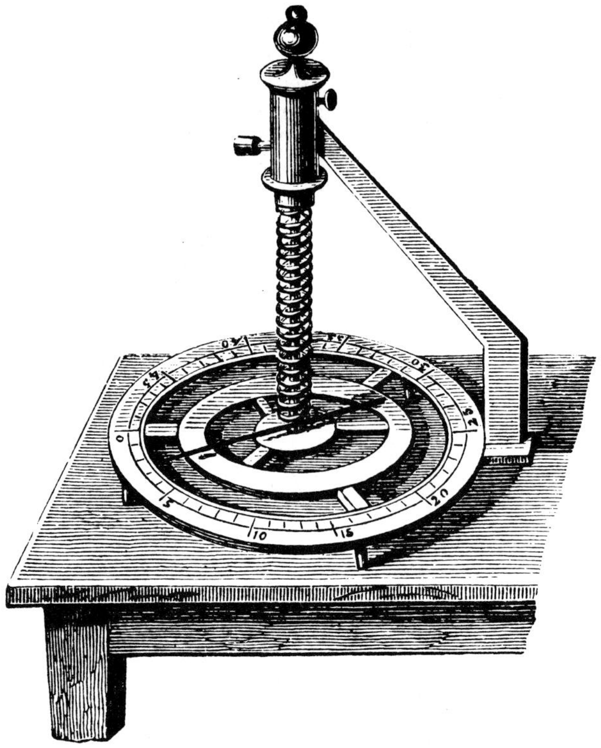

Figure 6. Breguet's metallic thermometer, using the unequal expansion of platinum, gold and silver strips formed into a tight coil to move a pointer over a circular scale. (*Courtesy*: *Ann Ronan Picture Library.*)

small number of workmen – about ten in all, including pupils who came from all over Europe. He also had a large number of people working for him at their homes, specializing in the production of watch-cases, hands, rough blanks and the like. This provided him with a very flexible means of meeting the troubles of the French Revolution.

The team at the Quai de l'Horloge consisted of what we should call a design group and a control group, linked by an assembly and finishing shop. One document informs us that the firm employed over 100 outside workers and that the annual turnover reached Fr. 400 000, five-sixths of which went on labour. By 1823 the assets had risen to Fr. 800 000, including Fr. 450 000 for work in progress.

But Breguet could not do everything himself. We know that he concerned himself only with the technical side and with his outside workers. From 1787 to 1791 he had a certain Xavier Gide looking after the business side, including sales, but this association soon proved disappointing. We should know nothing about the subsequent organization but for the light thrown on it by a letter written by his son Antoine-Louis as late as 18 July 1854:

> 'Until he was fifty years old, my father had been unable to make a penny until my aunt Suzanne Lhuillier came to manage his household. She was a mother to me and she brought in Fr. 30 000 which she was sensible enough not to lend my father but with which she bought his best products. Though hardly able to read and write and having only the bare rules of arithmetic at her command, she took charge of the firm. Well, my friend, it was with this simple knowledge that she shaped our splendid firm.'

This only shows once more that the success of an enterprise does not rest on innovation alone, but on the daily attention to detail, devotion and perseverance and on a family unity.

On the death of Mlle Lhuillier, Louis Lassieur, a nephew of Abraham-Louis Breguet and like him a Swiss, came to take charge of the commercial side of the firm until his own death in 1850. We must recognize, however, that the broad commercial ideas of the firm were entirely Abraham-Louis Breguet's; they were so original that they deserve a fuller account.

The Commercial Methods of Abraham-Louis Breguet

On returning from Switzerland after sheltering there from the troubles of the Revolution, Breguet had to re-equip his workshop and replace the destroyed tools. It was his intention (see letter of 4 April 1795) to concentrate on three high quality models:

◇ a simple watch with a single hand; this was to become the famous 'subscription watch', the first 'batch' production in horology;
◇ a repeater watch, with a standard face or with a decimal face as recommended by the Revolution. (Breguet described this as a fancy watch, *une montre de fantaisie*);
◇ a marine watch.

Breguet described the simple watch as being made of fourteen parts only, though it could contain variants: 'if greater exactitude is required one has only to open the bottom of the case to find a small dial with a hand turning every ten minutes. A repeater model is also available'; and he added, in a short two-page note addressed to the public: 'To achieve the uniformity and perfection I desire I must be able to make several watches at the same time. But for that I shall need financing. To obtain this by a loan would entail enormous interest charges which, as things are, no honest industry could bear.'

The price announced for the silver Subscription watch was Fr. 600, and a cash payment of no more than one–quarter of this sum was called for on placing an order. This was not a mass-produced watch in the modern sense: the stock sheets show no tools for serial production. But by producing watches one at a time, or in small lots, the firm could grow only slowly because Breguet always refused to call on outside capital or take on partners.

One important innovation made its appearance: in the new 'marine watch' Breguet visualized making it 'in such a way that a given part in each watch could be used in all the others simply by tightening two screws; and that if any of them suffered an accident, the damaged part could be replaced in less than five minutes'. Did not this foreshadow the principle of interchangeability, which did not appear in the industry until many years later?

If the so-called subscription method of sale of his watch was not so original as some would have us think, Breguet did adopt in his 1822 catalogue, 'Clocks and watches for common use', a method the originality of which has not been given sufficient attention. What he there offered to his customers was a personal watch:

> 'Here we offer several different arrangements of dials. The illustrations show only a few of the combinations available in watches for common use, but they could be multiplied and varied as required in such details as the diameter of one figure with the thickness of another, or intermediate dimensions, or the degree of squareness of the dial or the arrangement of the hands, some of which could be brought together or separated. Thus, depending on your wishes, we can make watches with a new or special combination of features.'

Apart from these two published notices, and a third on marine watches (1817) recently discovered, Breguet left a small work on a philosophical rather than a scientific subject, entitled *Essai sur la force animale et le Principe de Mouvement Volontaire*. This appeared, unsigned, in 1811 when the public was expecting from him a work on watchmaking. It cannot be said that this essay did much to enhance his reputation. Modern science has rejected most of his ideas, though they do contain in a confused form the concept of the conservation of energy.

Thanks to all the innovations he introduced, Breguet was able to overcome his difficulties and to weather France's Revolutionary periods of misfortune. From the Restoration until his death was his firm's most prosperous time.

He himself received the greatest honours. Supreme among these was his nomination and election, simple workman as he had begun, to the Académie Royale des Sciences. He died rather suddenly in 1823. At his funeral, it was said of him in the name of the Académie that 'his example showed how admirable was the peaceful glory of useful work'.

ANTOINE-LOUIS BREGUET (1776–1858)

Figure 7. Antoine-Louis Breguet (1776–1858).

Antoine-Louis Breguet (Fig. 7) was 47 years old when his father died. He had always worked with him except during the profitable years spent in John Arnold's workshop in London. No one could doubt his expertise and his taste after seeing watches and clocks signed 'Breguet et fils', in which it appears impossible to detect the individual contributions.

Antoine-Louis was a very different person from his father. His boyhood had been marked by the premature death of his mother, by his sojourns in Switzerland and England and by the French Revolution. In London he had acquired a liking for the British way of life and it is therefore not surprising that he twice considered leaving France. In 1811 he thought of setting up a business in the USA, but was discouraged by news and views he received in a letter dated 6 May 1811 from his friend Poletica in Washington:

> 'This country is a paradise for the mass of the people, who multiply like ants, but I do not think it is ready for the fine products of the mechanical arts. There are as yet no connoisseurs nor of course any purchasers. It will be different in another 50 or 100 years, but by then neither you nor I will be here.'

He had a second impulse to leave France, this time for England, around 1813, following the death of his wife. He had not reckoned with the return of the Bourbons in 1815. He then decided to train his only son to succeed him in the business, while he himself would retire to the country some 50 km from Paris. From there he would keep an eye on the staff at the Quai de l'Horloge.

Barely four years after the death of his father, he left Paris and 'managed his company', as he called it up to 1833, from a distance. He died in 1858, after passing the last 30 years of his life in the style of an English gentleman farmer.

LOUIS-F.-C. BREGUET (1804–1883)

Louis-François-Clément Breguet (Fig. 8) was born in 1804, the natural son of Antoine-Louis and Cairo-born Jeanne-Françoise Venture de Paradis whose father was of Marseillaise and Greek origin. Louis

Figure 8. Louis-François-Clément Breguet (1804–1883).

F.-C. was legally acknowledged in 1809 but his parents married only in 1811. Two years later, after his mother's death, he was sent with his younger sister to Neuchâtel in Switzerland to be brought up with the numerous children of his father's best friend, Jean-Jacques Huguenin. Probably his father wanted his two children to be away from the dangers of invasion then threatening France.

He stayed eight years in Switzerland. As he had not shown any special talent, his father subjected him to a very hard apprenticeship ending with a period at Versailles under the clever mechanic A. L. Perrelet, the inventor of differential gears. Following this, he worked as a simple workman at Geneva.

On his return to the Quai, his father put him in charge of marine watches, the production of which had been neglected to such an extent that the firm of Breguet had lost both the title of Clockmaker to the Royal French Navy and the pension going with it. The Navy bought his marine chronometers only in competition with those of Berthoud and Motel; but Louis, by improving his methods of fabrication, succeeded in producing chronometers at moderate prices and selling them to the merchant navy.

During the years preceding the 1830 Revolution, the performance of the firm became progressively worse. Vigorous competition had developed, largely from Abraham-Louis Breguet's old pupils. Industrial companies were being set up for mass production and Breguet's demanding clientèle from the Courts of Europe was gradually replaced by wealthy business men, bankers and industrialists, more interested in the price of a watch than in the elegance of a personal timepiece with costly gadgets.

In a letter of 7 May 1829 to his father, then retired to his country mansion 'Le Buisson', we find Louis F.-C. writing: 'Our firm must advertise in the academic world and above all we must prove that, so far from what some people want to say, we still exist', and again, on 25 September: 'We must break new ground . . . but alas! we no longer have grandfather's talents! . . .'

To do something new was the watchword of the day; and in the years that followed the firm developed by 'doing something new' – but not in the field of horology. Not that Louis F.-C. Breguet did not try to introduce improvements in this field: the firm could hardly turn its back on an art in which it had won a supreme reputation. He did in fact perfect one of his grandfather's timepieces and also simplified the sympathetic pendulum, enabling it to be sold for Fr. 600; but this was not enough.

François Arago (1786–1853)

The direction in which the firm of Breguet developed was prompted by the friendship of François Arago (Fig. 9), one of the greatest intellects of the 19th century. The two families became friendly from the time when Abraham-Louis was elected to the Académie des Sciences at the age of 70; Arago as a prodigy had been elected when only 23. Arago admired the grandfather, was friendly with the son and had an affection for the grandson Louis F.-C., nearly 20 years his junior, who was working desperately to make watches more and more accurate. Noting Louis's ability, Arago enabled him to attend the courses at the Ecole Polytechnique which Arago was giving, in order to improve the theoretical knowledge of Louis F.-C.

We now find Louis taking part to an increasing extent in all Arago's experimental work. In a letter dated 7 March 1832, Louis described to his father an experiment conducted by Arago 'to show the identity

Figure 9. François Arago (1786–1853), a close friend of three generations of Breguets. (*Courtesy*: *Ann Ronan Picture Library*.)

of the magnetic fluid and the electric fluid', adding 'try this out and let me know'.

Why should not the future of the firm lie in electricity? It would mean work, innovation, a search for useful applications and this is exactly what father and son set out to do. But the firm was first reorganized. Antoine the father made it over to his son and to his cousin Lassieur – Louis F.-C. had married Lassieur's daughter Caroline a little while earlier – as well as to his faithful colleague Pierre Tredos. The new name of the firm was 'Breguet, Neveu et Cie' and, to mark the change, the serial numbers of the watches made were started again at the figure 0.

The Work of Louis F.-C. Breguet

The beginnings of applied electricity appear to have started slowly in the Breguet firm. We find electricity timidly appearing in the form of a communication to the Académie des Sciences,[5] 'On a modification to Abraham Breguet's metallic thermometer to enable it to show the effects of small electric currents' (Fig. 10). Then, in 1839, jointly with M. Masson, Professor of Physics at the University of Caen, Breguet deposited with the Académie a sealed packet relating to a new electric telegraph.[6] Two years later they sent to the Académie a memoir on induction – a paper of great interest since the Académie was later to recognize in it the origin of the Rühmkorff coil.[7]

Thanks to the recording thermometer of 1835, the University of Kazan in Russia could measure very low temperatures, far below the freezing point of mercury. On the strength of this success, Louis F.-C. Breguet was appointed a corresponding Member of the University, the Rector of which at that time was Nikolai Lobachevsky, one of the century's leading mathematicians.

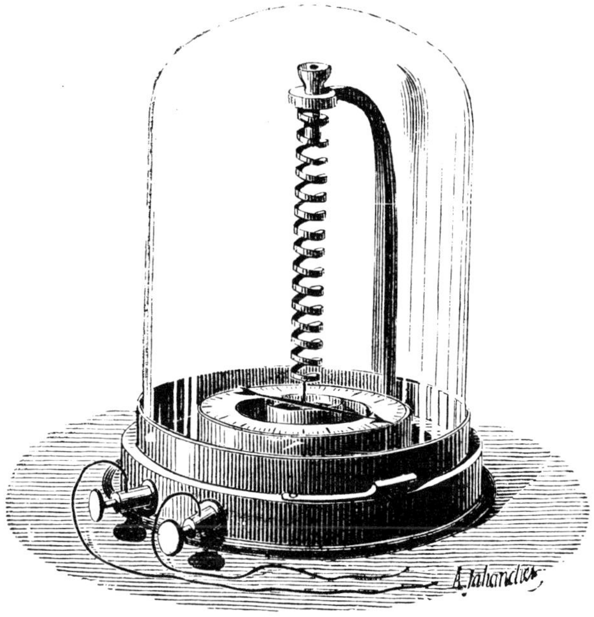

Figure 10. Breguet's modified metallic thermometer, to show the effects of small electric currents; it was the first step towards electrical equipment by the firm of Breguet. (*Courtesy*: *Ann Ronan Picture Library.*)

50 000 rpm in 1843. In 1843 Louis F.-C. Breguet made for Arago a scientific instrument which, like his watches, was driven by two spring drums. The springs drove minute mirrors at high rotational speeds, first at 2000 rps and later at 9000 rps (54 000 rpm). This astonishing result was due to the first application of White's helicoidal gears, which had been machined by tools designed by Louis F.-C. Breguet himself; the instrument is now preserved in the Conservatoire des Arts et Métiers. Using an improved instrument, A. H. L. Fizeau and, later, Paul Cornu measured the speed of light within specified limits of accuracy. One document reports that Breguet succeeded in resolving 1/240 000 second.

In 1845, following a communication to the Académie[8] in which Breguet reported results obtained with an instrument to measure the speed of a projectile at different points in its trajectory for a Colonel Konstantinov of the Russian Army, Wheatstone submitted to the Académie a claim to priority for this invention. Arago, adjudicating for the Académie, concluded that the priority in this invention should be settled between Wheatstone and the Russian, since Breguet had claimed in this very complex work only 'the means of execution and the varied and delicate mechanisms of the mechanical and electrical devices which he had had to use to arrive at the result'.

Disputes and Patents. Other disputes had arisen earlier in connection with the electric telegraph. For this reason, Breguet gave up the use of sealed packets and communications to the Académie as a basis for patent applications after 1847. Académie records show that a claim had been made at that time by Samuel Morse for a 'relay'. It is quite evident here as in many similar examples that it is possible for an invention to be made by different people at practically the same time without the inventors knowing about one another's work. They arise almost spontaneously from the state of knowledge and the economic needs of the time.

The number of patent applications filed by Louis Breguet increased rapidly following 1845, when he installed the first French electric telegraph alongside the Paris–Rouen railway (Figs. 11 and 12). Here he met the complex problems which had to be solved for the control and safety of train movements. The 80 patents and additions granted to Breguet during the half-century 1830–1880 were distributed as follows:

1830–1845	3	1860–1865	11
1845–1850	3	1865–1870	10
1850–1855	10	1870–1875	6
1855–1860	29	1875–1880	8

Only three of these concerned horology, two being on electric horology, Breguet having entered the field later than Wheatstone and Bain, and only one on mechanical horology, namely, the French Patent

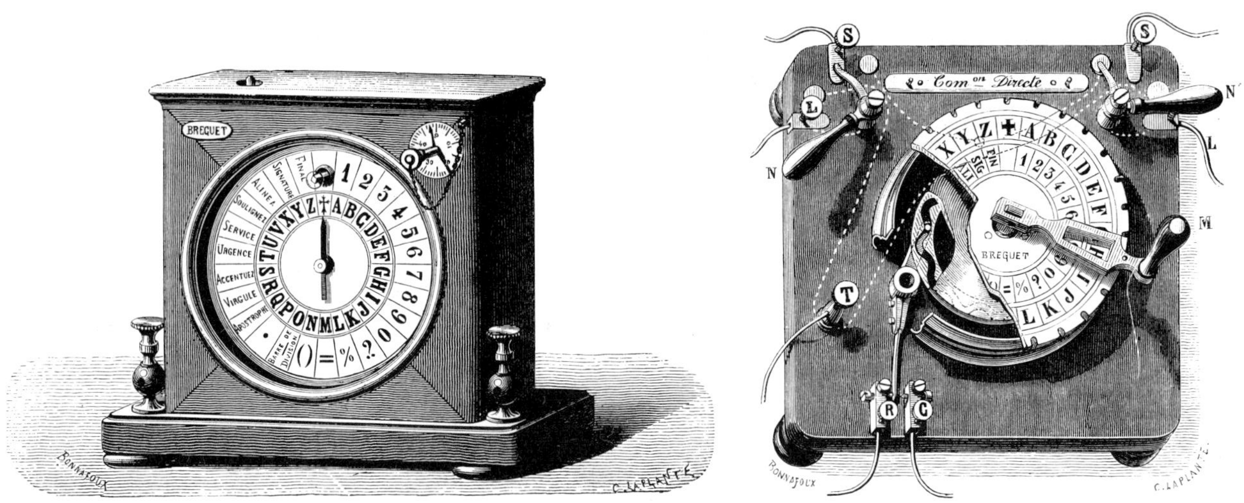

Figure 11. Breguet's telegraphs, left the receiver, right the transmitter. These were for many years standard equipment throughout Europe; with special modifications they were exported to Japan,[4] where in 1870 they inaugurated the public telegraph service. (*Courtesy*: *Ann Ronan Picture Library*.)

39411, 'On an improvement in the arrangement and construction of chronometers'. Of his two patents on electric horology, No. 73414 of 1866 deserves special mention. Long forgotten, since it was not further developed, its importance has only recently been appreciated: it was entitled 'Tuning fork for the regulation of speed: a process applicable to horology and to the adjustment of machines'. It was the basic patent of the electric tuning-fork watch which appeared in the USA at the end of World War II.

The Electric Fuse. Louis F.-C. Breguet, relying on a communication to the Académie, omitted to secure patent protection for what is perhaps the best-known and most widely used electrical device in the world, the 'fuse', originally designed to protect telegraphic equipment from attack by lightning.

Because production expanded to meet orders for telegraphic and railway apparatus and miscellaneous industrial counters, Louis-F.-C. Breguet added an additional storey to his house at the Quai in 1851 in order to provide a special department for telegraphy. The house then assumed its present appearance, which breaks the regular aesthetic line dating from the time of Henri IV.

In 1854 Louis-F.-C. Breguet brought in his nephew Alfred Niaudet (1835–1883) as his principal assistant. The following year he took on many more workers and increased the size of the firm by renting a workshop in the Montparnasse quarter. This was done against the advice of his aged father and in spite of financial difficulties due to the successive bankruptcy of two bankers and a breach of confidence by one of his principal colleagues.

The firm's later non-telegraphic achievements included the regulation and distribution of time, accurate to a second, to master clocks distributed around Paris (Fig. 13); the so-called 'Breguet punch' for exploding mines; and aneroid barometers, made by a patented production method. Louis-F.-C. Breguet helped many contemporary scientists with his mechanical skills during the years following 1860. Thus, he made Marey's sphygmograph, Yvon Villarceau's regulator, Bouquet de la Grye's seismograph, Captain Fleuriais's chronograph and others. We must also mention Pénaud's little helicopter model for which Breguet chose aluminium, a metal of the future but at that time horribly expensive.[11]

Figure 12. Breguet's telegraphs had become so famous that when the first International Telegraph Union Conference, held in Paris in 1865, was to be commemorated by a medal from the Paris Mint, the choice of model was quite obvious. (The Latin text was in 1866 somewhat optimistic, as then neither voice nor telegram could as yet span the planet's orb.) (*Courtesy*: *Ann Ronan Picture Library*.)

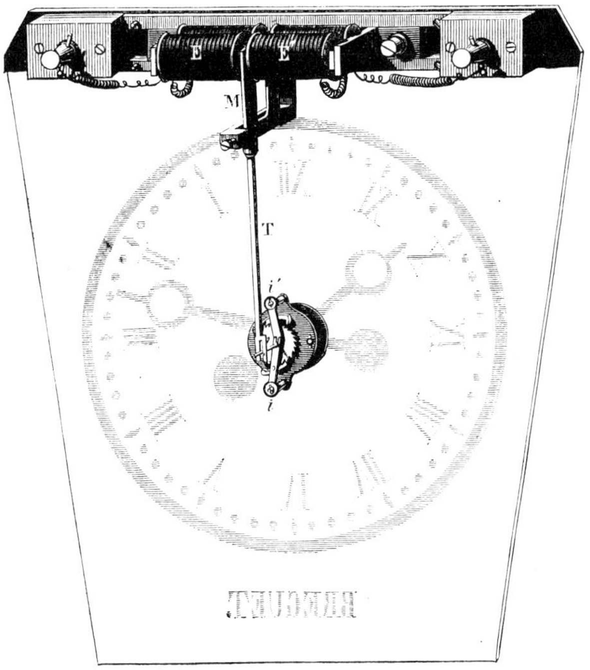

Figure 13. The mechanism of the Breguet electric clock. (*Courtesy*: *Ann Ronan Picture Library.*)

ANTOINE BREGUET (1851–1882)

Louis-F.-C. Breguet's only son, Antoine (Fig. 14), joined his father and his cousin in 1874, by which time the firm of Breguet was already the oldest of the many French electrical companies. In that same year Louis-F.-C. Breguet was elected to the Académie des Sciences.

Figure 14. Antoine Breguet (1851–1882).

Antoine, when he left the Ecole Polytechnic after serving in the war of 1870 as a volunteer in the Corps of Telegraphists, was an engineer of the type found in the novels of Jules Verne. While anxious to help the firm forward in scientific research, he decided that it should abandon its traditional 'craft' aspect and adapt to modern economic ideas. He had only another eight years to live, but these were years of intense activity ending in his death at the early age of 31.

In 1874 the firm's products were much too varied, but they no longer included clocks and watches. That side of the business had been made over in 1870 to Edward Brown; it retained the universally respected trade name of 'Breguet' and followed an independent course in de luxe and precision timepieces.

Electric timing devices, which the Breguet firm had retained, did not develop. In spite of ingenious advances in tooling, the firm was making too many products for its size and employing too many people on a craft basis. One has only to consider the following examples: many types of electric batteries; dial telegraphs, printers and mobile sets; arc lamps invented by various people; bell-alarm systems; safety systems for the railways; laboratory equipment (coils, rheostats, resistances, condensers); lightning arresters; aneroid barometers; and magneto-electric machinery. On top of all this, the firm was working on a dynamo of the future, which could produce the energy the world was waiting for – the machine invented by the Belgian, Zénobe Gramme. Antoine Breguet, as a young engineer, studied this new Gramme dynamo and was the first to establish its theory. The firm acquired a manufacturing licence and began to make several low-power models in new workshops leased in 1872.

Alexander Graham Bell

Bell, who came with his telephone to Paris, quickly formed a friendship with the Breguets, father and son. The industrial production of telephones was studied and then put in hand. The first telephone in France was the one installed between two floors of the Breguet house at the Quai de l'Horloge, between the laboratory and the workshop. We would now call these telephones prototypes, remembering that by that time the telephone had already become highly developed in the USA.

The patents issued in 1878 to Antoine Breguet included five on the telephone, four of which were taken out jointly with Cornelius Roosevelt, Graham Bell's Paris representative. The fifth, issued to Antoine alone, dealt with an original telephone system based on the capillarity of mercury; it was never developed. On his side, Alfred Niaudet worked on batteries and on magneto-electric machinery and wrote numerous scientific papers.

Antoine loved to have many irons in the fire at the same time: the scientific researches, particularly those in collaboration with Graham Bell for the Photophone and the Topophone; the conversion of the firm into a *Société Anonyme* in 1881 under the name 'Maison Breguet'; the General Secretariat of the 1881 Electricity Exhibition; the joint management of the *Revue Scientifique* with his friend Charles Richet, who was to receive the Nobel Prize for medicine in 1913 for his discovery of anaphylaxis, and the foundation of the Société Française des Electriciens.

The Maison Breguet without a Breguet

The Breguet family then suffered a succession of lamentable losses. Antoine Breguet died in 1882, Alfred Niaudet on 11 October 1883 and Louis-F.-C. Breguet, then nearly 80 and unable to bear these misfortunes, barely a fortnight later. There was no longer a Breguet in the firm. The name survived only in Antoine's two sons, Louis and Jacques, then aged only two and three years respectively.

It was fortunate for the company that it had become a *Société Anonyme*: continuity was assured. With its new financial, industrial and commercial structure, it could on 1 September 1882, only two months after Antoine's death, open modern spacious workshops behind the Gare Montparnasse. The steel frames of the machine shops had in fact come from the 1878 Palais de l'Exposition Universelle.

The activities of the new company moved now towards the manufacture of more powerful, larger equipment: telephones gave way to dynamos and electric motors and these in turn led to steam-driven generators (Fig. 15) – and, when the Swede De Laval invented the reaction turbine, Breguet acquired a licence to manufacture.

Low-power arc lamps had been replaced by Edison's filament lamps, and therefore the company fell back on projectors with parabolic mirrors, production of which was transferred later to a small works near Rouen. Industrial development in northern France had started some years earlier with the installation at Douai of some special workshops for the manufacture of electric motors, but the three plants of the Maison Breguet never became large. The great diversity of products, often made to order, the few machine tools or original technical devices, and a staff which was getting progressively older, provided a poor defence against vigorous competition equipped with more up-to-date and efficient machinery.

By the time the two young Breguets, who had decided to become electricians, entered it on completion of their studies, the company had slipped downhill. To their father's mixture of Swiss, Parisian, Marseillais and Greek blood, their mother had brought a new contribution from Normandy and Ireland. She was the daughter of Camille O'Meara, one of Chopin's last pupils, and a niece of Barry Edward O'Meara, Napoleon's doctor on St Helena, whose *A Voice from St Helena* became in its day a best-seller.

Charles Richet (1850–1935)

What had happened around 1825 between Louis-F.-C. Breguet and Arago was now to be repeated. The two young Breguets had the good fortune to find in Charles Richet, their father's best friend, a guide and mentor who was to lead them from electricity into a more elevated and adventurous field.

Charles Richet, assuredly one of the most remarkable men of his time, had a wide-ranging mind open to anything new or useful for mankind. Doctor, physiologist, historian and writer, he was the most marvellous man the youngsters could have desired. Furthermore, for young people with a passion for

Figure 15. A Breguet electricity generator specially designed for marine use. A vertical compound steam engine (right) drives through a elastic coupling, Raffard design, a multipolar dynamo on the left, as patented by Desroziers.

Jules Verne, he owned something to stimulate their zest for adventure: a small island, the smallest of the Iles d'Hyères some miles from Toulon. From his property at Carqueiranne – and to it – he took his own children for their holidays and with them the two little Breguets. And it so happened that in 1897 Richet was experimenting with the model of a steam-driven aeroplane which he had invented jointly with his friend Victor Tatin. The machine flew a distance of only 132 m but what enthusiasm it aroused in young Louis and Jacques!

LOUIS BREGUET (1880–1955)

In 1903, after spending one year at the Ecole Supérieure d'Electricité and following his military service, Louis (shown with his brother in Fig. 16) entered the Douai works to engage in the calculation of electric motors. He applied himself seriously to this work, yet dreamed of taking a personal part in the conquest of the air, following with passion its every progress, every attempt and every failure.

He published a note on the diagrams of asynchronous motors which created a stir at the time and contributed to their progress. It has only recently come to light that he calculated and made in 1907 the electric propulsion engines of the French submarine *Mariotte* which, by reaching an underwater speed of 11.66 knots, set up a French speed record, not to be beaten until after World War II, by submarines of the 'Narval' type.

In the free moments left to him by his work and his family, Louis Breguet started his own researches and trials. He designed and made an electrically driven aerodynamic balance with which he measured the lift and drag forces on various wing sections. With the help of his brother and of Professor Richet, he built his gyroplane No. 1 which rose to 60 cm on 24 August 1907 with its engine and pilot. Up till that time, no one had done better.

Gyroplanes Nos. 2 and 2bis, with sloping rotors and fitted with wings and controls, flew a distance of 50 m in a straight line; they convinced Louis Breguet of the complexity of the helicopter's problems. He decided therefore to follow the formulae of his competitors and do what he could to make rapid progress.

He left his job as engineer to the Maison Breguet and learned to fly by himself, making his first solo flight on 14 July 1909, Blériot having only just flown across the English Channel. By 1911 Louis Breguet had caught up with his competitors and, piloting himself, had beaten some important records. With his friend Robert Esnault-Peltérie – with whom at one time there seemed a possibility of a business link – he realized from the start that the aeroplanes of the future would need more and more powerful motors, that they would have to be made of metal, fitted with variable-pitch propellors, with tricycle undercarriages equipped with brakes amd oil-hydraulic shock absorbers.

Aviation Louis Breguet

The 'Ateliers Breguet-Richet' gave way to the 'Société Anonyme des Ateliers d'Aviation Louis Breguet', which received orders for military aircraft

Figure 16. Jacques (1881–1939) and Louis (1880–1955) Breguet from left to right, sitting in their office in Villacoublay in 1918. Note on their wall chart a world system of airlines.

from all over the world. In spite of these successes, delays in delivery and long production times had put Breguet into serious financial difficulties by the outbreak of World War I. The company was then only saved by accepting a sub-contract for 60 Voisin aircraft and by the gift to the French Army by the Michelin brothers of 100 Breguet aircraft with 200 h.p. pusher motors, machines considered by them to be the best bombers of the day, though the General Staff did not want them.

It was Louis Breguet's great achievement that at the height of World War I he designed freely and alone the 'Breguet Type 14' (Fig. 17), outclassing enemy aircraft, and then organized its mass production when the Army ordered it. Production reached 14 planes a day at the Villacoublay plant and at those of Breguet's seven sub-contractors, including Michelin and Renault. Up until 1920, Louis Breguet made the first test flights of every aeroplane he designed.

After World War I we enter present-day history and the story of Louis Breguet is known to many. Let us recall that a Breguet Type 19 machine was the first to cross the South Atlantic in 1927 in one stage. The 'Question Mark' which made the first direct flight from Paris to New York, with Costes and Bellonte as crew, was also a Breguet-19 Grand Raid TR.

Founder of Air France

In order to find markets for aircraft, Louis Breguet created in 1919 the first French air-transport company. Later, he was among the founders of Air France. In 1934, assisted by René Dorand, Louis Breguet renewed his interest in helicopters and his new 'laboratory gyroplane' beat by a fair margin many records of that day, modest as they were. Among the reasons which prevented their further development were his company's lack of capital and American competition. It must also be remembered that French industry was gravely affected by political and social events from 1936 onwards.

Louis Breguet showed as much of the competitive spirit and mechanical ingenuity in his leisure activities as he did at the work he loved. He died in 1955. His factories, rebuilt after World War II destruction, produced various types of aircraft with varied fortunes, but they did not bring in enough profitable business to enable the company to stay independent. In 1967 the French Government arranged a merger with 'Les Avions Marcel Dassault'.

As for the Maison Breguet, which was being called more and more 'Maison Breguet, Electricité' and of which Louis Breguet had become Chairman and his brother Jacques Breguet General Manager between the two wars, it declined progressively under the pressure of competition, of economic and social crises and finally of World War II. It disappeared about 1966, absorbed by a great French industrial group.

CONCLUSIONS

The history of the Breguet Dynasty and of its interdisciplinary scientists and engineers has illuminated some little-known aspects of the scientific and technological events of the 19th century. It was a period of progress and it might not be difficult to find families other than the Breguets that have made comparable contributions – doctors, actors, musicians, astronomers, industrialists and so on – but these families had almost always worked in the same discipline. The originality of the Breguets might appear

Figure 17. One of the first flight tests of the Breguet-14 is here recorded on 26 September 1916. Louis Breguet is in the pilot's seat with Jacques standing below on the left.

to lie in diversity were it not for one common aspect, fine mechanics, so that the move from one field to another occurred quite naturally.

If, twice in their history, the Breguets changed from one scientific discipline to another, they did so in order to save their enterprise through the affirmation of their personality. They were led by the circumstances of the time and as if in obedience to the voice which Peer Gynt heard on his walk through the forest, calling out repeatedly, 'Be yourself!' On the first occasion the voice was that of Arago and on the second that of Charles Richet.

The products of the mechanical arts are not always the works of individuals but are often collective works such as, for example, architectural monuments. They result from the organized efforts of an interdisciplinary team, the head of which directs, like the conductor of an orchestra. Mutual confidence cannot prosper in the absence of the fundamental freedoms – freedom of the individual, freedom of enterprise, freedom of association – and these have been the conditions in which progress in science and industry flourished during the past two centuries. The Breguets played their part by assuming their full responsibilities as free men. It was no mere coincidence perhaps that their first enterprise was born in 1774, two years before democracy came to America, the first modern democracy.

GENERAL BIBLIOGRAPHY

The Breguets

Archives Breguet (in the possession of the author, Claude A. J. Breguet).

E. Ferret, *Les Breguet*, Paris (1884).

Une industrie parisienne – conférences de Claude et Francçois Breguet. *Bulletin de la Société Historique du 14 Arrondissement* (1962).

Abraham L. Breguet

Charles Fourier, *Eloge de Breguet*, Institut de France (1823).

Baron Ernouf, *Trois Ouvriers français, Paris* (1879).

F. J. Britten, *Old Clocks and Watches and Their Makers*, London (1904).

R. T. Gould, *The Marine Chronometer*, London (1923).

Paul Marmottan, *Une grande marque d'horlogerie sous Napoléon – Abraham-Louis Breguet*, Paris (1923).

Centenaire d'Abraham-Louis Breguet:

Journal Suisse d'Horlogerie et de Bijouterie, La Chaux-de-Fonds (1923).

Revue, *l'Horloger*, Paris (1923).

Catalogues et Brochures sur l'Exposition au Musée Galliera (1923).

Trois montres historiques (August 1953).

L'attrait de l'Amérique – Abraham Breguet et les Etats-Unis (October 1953).

L'attrait de l'Amérique – Antoine L. Breguet et Jean-Jacques Huguenin. *La Suisse Horlogère (Ed. Internationale)* (December 1953).

A. L. Breguet pendant la Révolution Française. *La Suisse Horlogère* (*Ed. Internationale*) (1952).

Claude Breguet, Abraham L. Breguet et sa maison de Russie. *Revue de l'Institut Napoléon*, No. 56 (1955).

Claude Breguet, *Breguet, Horologer*, London (1961).

Louis Armand, Le style de Breguet. *Connaissance des Arts* (March 1952).

Paul M. Chamberlain, *Its About Time*, London (1964).

Cecil Clutton and George Daniels, *Watches*, London (1965).

George Daniels, *The Art of Breguet*, Sotheby Park Bennett, London (1975).

Charles Dupin, *Discours aux Funérailles d'Abraham-Louis Breguet*, Paris (1823).

Writings of Abraham L. Breguet

Essai sur la force animale et le principe de mouvement volontaire.

Note sur les pendules doubles.

Carnets. (Unpublished).

Publications of the Maison Breguet (1800–1823)

Notice: *Souscriptions de montres d'une nouvelle construction.*

Catalogue de l'Exposition de 1819 (1819).

Description de l'horloge marine du Comte de Sommariva.

Horlogerie pour l'usage civil (1817).

Notice sur les montres marines, Brest (1817).

Louis-F.-C. Breguet

Pr. Gaulmier, Venture de paradis. *Revue de Paris* (August 1959).

Exposition de 1834 – Produits de la Maison Breguet, Neveu et Cie. A. Masson et Breguet fils, *Mémoire sur l'induction* (1841).

Exposition de 1844 – Produits de la Maison Breguet, Neveu et Cie. Notice sur les travaux de L. Breguet (1847).

Notice sur les télégraphes de Louis P. C. Breguet (1847).

L. Breguet et V. de Séré, La télégraphie électrique – son avenir, poste aux lettres electrique. *Journaux électriques* (1849).

Manuel de télégraphie à l'usage des employés de chemin de fer à la ligne. Lettre de L. Breguet à C. Erckmann au Corps législatif pour l'établissement de lignes télégraphiques sousmarines (1853).

Notice sur les travaux de L. Breguet (1857).

Catalogue illustré de la Maison Breguet (1873).

Eclairage par la lumière électrique – Instruction pour l'emploi de la pile Bunsen et du régulateur Victor Serrin (1876).

Instruction pour l'emploi de la machine de Gramme et du régulateur Victor Serrin (1876).

Inauguration de la statue de F. Arago à Perpignan Discours (21 September 1879).

Alfred Niaudet and Antoine Breguet

La Science Populaire (30 July 1882).

La Revue Scientifique (15 July 1882).

Collection du Journal des Sciences, *La Nature Carnets d'Antoine Breguet* (1873–1883).

Louis Breguet

Pierre Faure, *Louis Breguet*, Ed. Blondel La Rougery (1938).

Henri Michelet, *Breguet*, Ed. France Empire (1962).

Revue Pionniers (15 April 1965).

Revue Aéro France (February 1956).

Notice sur les travaux scientifiques et industriels de Louis Breguet, Ed. Gauthier Villars (1930).

LITERATURE CITED

1. Sir David Lionel Salomons, *Breguet (1747–1823)*, printed for the author, London (1921).
2. Rupert T. Gould, John Harrison and his timekeepers. *Mariners Mirror* **XXI**, No. 2 (April 1935); Anonymous, *Four Steps to Longitude*, National Maritime Museum Greenwich, 1962. (Catalogue of Bicentenary Exhibition of first successful trial in 1762 of Harrison's Fourth Marine Timekeeper.)
3. George Daniels, private communication to the Editor (22 November 1978).
4. See, for example, A. R. Michaelis, *From Semaphore to Satellite: The Centenary History of the International Telecommunication Union*, ITU, Geneva (1965).
5. *C.R. Acad. Sci., Paris* **V**, 720.
6. *C.R. Acad. Sci., Paris* **VII**, 710 (8 October 1839).
7. *C.R. Acad. Sci., Paris* **XIII**, 426 (23 August 1841).
8. *C.R. Acad. Sci., Paris* **XX**, 157 (1845).
9. *C.R. Acad. Sci., Paris* **XXIII**, 987.
10. *C.R. Acad. Sci., Paris* **XXIV**, 980 (1845).
11. *Revue de l'Aéronautique*, No. 1, p. 29 (January 1876).

The French manuscript was received 8 May 1978; it was translated by L. V. Chilton

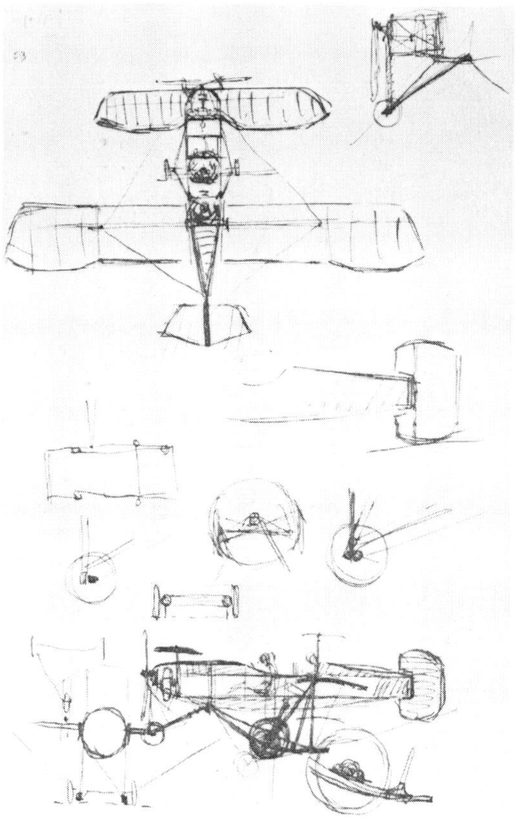

Early aircraft design sketches by Louis Brequet.

Letters to the Editor

COLLISIONS OF METEORITES WITH PLANETS

By Professor Dr Wolfgang Gentner, Max-Planck-Institut für Kernphysik, Heidelberg, FRG

Interdisciplinary Science Reviews **3**, 121 (1978)

Dear Sir

Established facts about meteorites seem to be as follows:

A 1000 tonne object of a type not uncommon in our Solar System, if making impact on Earth, would possibly cause great destruction to life and property. Professor Dr W. Gentner points out that a body of about this mass is so little slowed by the atmosphere that, on impact, the whole projectile vaporizes explosively.

Professor Gentner takes note of the 'Tunguska event' of A.D. 1908, in Siberia. Presumably it was caused by a large meteorite, or comet head, trees being blasted radially for 30 km around.

Arthur C. Clarke, in his *Report on Planet Three* (Chapter 3, 'Meteors' pp. 32–41. Corgi, London (1973)) takes cognizance of both this Siberian event and another in 1947. At Sikhote-Alin, Eastern Siberia, 23 tonnes of material were collected and over 100 craters recorded after the fall of the meteorite on 12 February 1947 (Macropaedia Vol. 12, p. 47. (*Encyclopaedia Britannica* (1974)). No doubt the original mass was greater, and had this iron meteorite not behaved atypically and disintegrated before impact, the surface effect might have been more dramatic. Although Professor Gentner suggests that terrestrial disasters due to meteorite impact have 'become most improbable' I do not see this century's record clearly supporting that point of view.

Professor Gentner observes in his Conclusion that if a large meteorite hit a certain capital city then decision-makers there might well react by sending nuclear missiles somewhere else, with dire results. His estimate of the sagacity of some 'decision-makers' is now shared by many; and it applies no doubt to those in places other than the one he specifies. There is the converse possibility that the responsible decision-makers, through under surprise nuclear assault, might attempt to check that observed effects were *not* due to a meteorite before they *did* react. What data would they use?

A system for giving advance warning of a great meteorite's possible arrival is outlined, in the form of a 'spaceguard' project, in Clarke's *Rendezvous with Rama* (Chapters 1 and 2). We have to accommodate the fact that a 1000 tonne object, composed of even such a low density material as water ice, would be no more than 13 m in diameter. If of iron, the same mass would have but half this diameter. At the low velocity relative to Earth of $10\ \mathrm{km\ s^{-1}}$ it would be over 860 000 km away only 24 hours before impact. It would be hard to see. Denser or faster objects would be still harder to detect 24 hours before their impact with Earth.

Do these facts suggest it is worth assessing the risks and options anew? In just over three centuries (1650–1985) there will have been a tenfold increase in the total population of the Earth, so history's comment on the risk to humans is a little dated. Moreover, history has never before seen the vast urban areas of today, like those of Los Angeles some 25 000 $\mathrm{km^2}$ in extent. We are assured that their number will increase.

It is suggested, then, that consideration be given to the following:

The identification of irregular heavenly bodies such as comets (usually much greater than 13 m in diameter) appears to be done currently by amateur astronomers. Dedicated as they are – is this good enough?

Is it not reasonable to take early steps to devise a meteorite tracking system? It would seem beyond the resources of terrestrial astronomy, since a 10 m diameter object at 1 000 000 km subtends only 1/100 of the angle subtended by Pluto. And it would be desirable to detect bodies at that distance down to 4 m diameter – about 250 tonnes if of iron – with total coverage of the celestial sphere about once an hour. A feasibility study at least for the space-based detection that was used in Clarke's fictional 'spaceguard' should be within our scope. Has one been made? If not, should your Journal suggest that it be done? UNESCO would seem to be the right co-ordinating body if no other is working on this question already. No doubt the many variables affecting atmospheric flight path would prevent even a short-term forecast of the exact impact site. But some forewarning to the probable risk *area* could save much life, reduce injuries and provide alerted rescue resources.

Additionally, a 'spaceguard' facility could be of great help to observational meteoriticists; they might be able to gain useful warning of the arrival even of small bodies, down to 10 cm diameter. This should permit more sophisticated observation of phenomena hitherto chiefly subjects of chance sighting. Any fuller information about the meteorites, and the atmosphere where they make early contact in the important ozone layer, would be of value.

Yours truly,
GEOFFREY BELL MA Cantab.
Bangkok
22 September 2522/A.D. 1979

Professor Dr W. Gentner, Heidelberg, Replies:

I do not consider Mr Bell's proposal to be realistic as it appears to me almost impossible to detect such small heavenly bodies before they strike the Earth's atmosphere. Considerably larger bodies, as for example Icarus (diameter 800 m), are only discovered by pure chance. Countermeasures against such a minor planet hitting the Earth were discussed in the American scientific literature when in June 1968 Icarus came within 6.4×10^6 km of the Earth's orbit. One of the suggested measures was to destroy it in space by an atomic bomb, fired against it by a rocket launched from Earth – an original but somewhat fantastic suggestion! Perhaps Mr Bell knows the relevant literature.

Furthermore, it should be noted that in my review I wrote about a meteorite, considerably heavier than 1000 tonnes. I should perhaps have increased this figure to 100 000 tonnes as only bodies of this mass could cause real catastrophes on Earth.

Mr Bell neglects to consider the possibility of a large meteorite falling into the sea, for example near the shallow coasts of England and France. Such an event would produce a tsunami with even more disastrous consequences than an impact on land.

But these are all calamities of extremely low probability and I cannot imagine that at present anyone would be prepared to pay for such a warning system to negate a very unlikely danger. There are at present many more serious dangers on Earth resulting in the death of many people; we can again see wars and future wars only too clearly.

Editor's Note: On 22 September 1979 an American Vela satellite reported an event within a 4500 km wide circle, covering the South Atlantic and Indian Oceans south and east of South Africa which was variously interpreted as a nuclear bomb test in the atmosphere, a bolt of lightning, a meteorite striking the atmosphere, a freak combination of these two events, or finally a satellite malfunction. No final conclusions, if indeed any, were reached by the American panel of experts called to the White House, or were ever published. See also: *Science* **207**, 1185 (1980).

Book Reviews

PROFESSOR KENNETH V. THIMANN, For. Mem. R.S.

Reviews

Otto Warburg

Professor Dr Hans Krebs, Oxford. Unter Mitarbeit von Dr Roswitha Schmid, München/Stuttgart: Otto Warburg, Zellphysiologe, Biochemiker, Mediziner 1883–1970. Wissenschaftliche Verlagsgesellschaft mbH Stuttgart, 1979. 167 pp. DM 29.00. ISBN 3 8047 0569 3

Who could write the life of Zeus? Apollo? Hera? Surely no one of lesser stature among the gods. So we are fortunate indeed that Professor Sir Hans Krebs has accepted the task of writing on the life and works of Warburg. Not only did Krebs know Warburg for most of his life, but few scientists of our generation have come as near to equalling Warburg's tremendous reputation as has Krebs. The Warburg manometer, the Krebs cycle, are words on the lips of almost every physiologist and biochemist, and perhaps in a few years the two men will receive the ultimate accolade, like ampere, joule and einstein, of having their names spelled with a small letter.

The book is a labor of love. It describes Warburg's life, a great many of his works, his foibles and his many honors. The author's modesty precludes his telling us much about his own contributions, which interrelated with Warburg's in many ways. For Warburg was indeed a giant. He gave us not only major advances in knowledge but also new ways of obtaining it, that is, new and effective laboratory methods. Three of them have come into worldwide use in research in almost every aspect of cell biology, biochemistry and medicine. The method of manometry, adapted and greatly improved from the pioneer apparatus of Barcroft and Haldane, enabled the direct measurement of reaction rates, and thus condensed into a few minutes what had formerly taken many hours.

What is more, it proved in his hands an instrument of the greatest flexibility, adaptable to any biological process in which a gas is evolved or absorbed, either directly or by a coupled reaction, or as a result of illumination. The method of spectrophotometry, in which changes in the absorption of visible or ultraviolet light are followed, has proven of almost equal adaptability, and grew directly out of the young Warburg's acquaintance with several distinguished physicists and physical chemists. (His father, Emil, was Professor of Physics at Berlin.) Finally, the method of using tissue slices for studies of metabolism superseded the old clumsy studies with whole organs, bedeviled with multiple complexities of diffusion, incomplete penetration and variability. Indeed, it was the tissue slice procedure, we learn, that led Warburg directly to his first major discovery, namely that cancer tissue ferments sugar to lactic acid in air, 'aerobic glycolysis', while normal tissues do this only in the absence of oxygen. But Krebs does not hesitate to temper his enthusiam for Warburg's insight with frank criticism, for while he slurs lightly over his own later explanation for this reaction, he points out how Warburg's claim that the glycolysis is the *cause* of cancer was premature; it was rather a symptom of the change in the cells, and the cause lies much deeper – indeed is still beyond us.

Perhaps the discovery best known to the older generation of biologists is that of the 'Atmungsferment' or respiration enzyme. Krebs describes carefully how the concept developed; the role of iron in respiration, the charcoal model of oxidations, the inhibitions by cyanide and carbon monoxide, the reversal of the latter by light (suggested originally by A. V. Hill on a visit) and the ingenious way in which this reversal was made to yield the absorption spectrum of something present in traces far too small to be actually seen. The extensive technical requirements of this experiment – a familiar procedure to us now – had then all to be developed from scratch. This work gained for Warburg the Nobel prize in 1931. Krebs obtained his twenty-two years later. But the work was followed by a long controversy with Cambridge's David Keilin, who had detected iron-containing catalysts of respiration by other methods. It was 10 years later before the two discoveries could be brought into harmony and the controversy ended. This was perhaps the first major example of Warburg's extreme self-confidence and unwillingness to move from a position he had taken on experimental grounds.

Another long controversy, which the book sets out in considerable detail, concerned the role of hydrogen-transferring enzymes in the respiration process. Essentially, hydrogen is successively removed from organic compounds and combined with oxygen, till finally only carbon dioxide is left. Warburg would not accept the biological significance of Thunberg's and Wieland's demonstration that cells and tissue cause the decoloration of methylene blue, namely its hydrogenation, by means of their special hydrogen-transferring enzymes. Krebs tells us that it was not until Warburg paid a visit to Johns Hopkins University in Baltimore, where Barron personally showed him this reaction, and also showed that the reduced dye could be recolorized in air, that Warburg showed an interest in the process.

But then his great experimental genius took off in earnest, and within three years he had brought to light the first of a new group of enzymes, the 'yellow enzymes,' compounds of riboflavin and protein, that bring about the hydrogen transfer. He found that the riboflavin was combined with phosphate and adenine in a 'Flavin-Adenine Dinucleotide,' FAD. But more than that, an additional, colorless substance, a coenzyme, was a necessary participant, and the very next year he had purified this and identified it as containing nicotinamide, also known as niacin, one of the B vitamins. (As it turned out, there are two coenzymes containing niacin.) The final identification was lucky, we are now told, for actually Warburg had the anlytical data but not enough of the material for a regular organic chemical structure determination, and it was Schoeller, the director of Schering–Kahlbaum's chemical firm, who identified it from the very precise data by looking it up in the great *Beilstein* catalog of known organic compounds. Warburg's reaction was characteristically concise: 'Yesterday we could not have bought it for all the gold in the world; today we can have it for 2 Marks a pound.'

Krebs makes the point that Warburg's work on the purification and isolation of enzymes, besides adding tremendously to science, has led directly to the large-scale commercial preparation of great numbers of pure enzymes. These line the walls of the refrigerators of every biological and medical laboratory today. However that may be, the Warburg influence is felt in all such laboratories in multifarious ways.

Only in the field of photosynthesis was Warburg's record somewhat tarnished. Krebs deals with this part of the work lightly, but it is interesting to this reviewer for special reasons, and it was here that his strong physical background led him astray. At first Warburg performed a vast service to the field by introducing, in 1919, the use of the one-celled alga *Chlorella*. Comparison of the work with this little plant, easily grown in a standard milieu, pipetted in known amounts and highly adapted to the manometric technique, with the old apparatus using unstandardized leaves exposed to light for many hours, is like that between night and day. Many fundamental discoveries were made with this method, in widely dispersed laboratories. Warburg's own discovery of the photosynthetic reduction of nitrate, an important step towards understanding the cycle of nitrogen in the world, was marked by his characteristic clarity and simplicity. But his stubborn adherence to the physical notion that only three quanta of light must be needed for each molecule of carbon dioxide reduced was at variance with all the biologists' measurements, even when made with Warburg's own techniques.

Emerson, who had worked in Warburg's laboratory in the 1920s, and personally introduced Warburg manometers to the USA, was one of the protagonists. When Emerson reported on the long controversy at a Harvard biological colloquium, George Wald said repeatedly, 'But Warburg can't be wrong!' Eventually Emerson obtained funds to bring Warburg to his own laboratory at the University of Illinois to try to settle the matter. Warburg replied that he could not come without Jakob Heiss, his friend and manservant, so Emerson managed to get additional funds for this purpose. Warburg arrived, complaining that he had not been able to bring his horses! Within a short time he had elbowed Emerson out of most of his laboratory space, and, so far from coming to an agreement, relations grew so strained that Warburg soon left to spend the remaining time with Dean Burk in Washington. There I found him one day, working alone in a small laboratory with a set of the inevitable manometers. Later there was a large public discussion at Woods Hole, with James Franck, the physicist, participating, but it was to no avail. The biological system was just more complex, and less efficient, than physical theory had predicted, and Warburg was in fact wrong, (though he never admitted it!)

The two authors, with a sculpture of Otto Warburg between them, met on the occasion of the publication of their book at the Otto–Warburg House in Berlin–Dahlem on 14 March 1979.

In his feelings Warburg was evidently very pro-British. He went to London to have his suits tailored in Savile Row, he considered England 'the bastion of Old Europe,' he wrote to Lady Krebs about his great pleasure at the visit to Berlin of the Queen and the Duke of Edinburgh, and one of his most treasured honors was the honorary doctorate at Oxford. He was even for much of his lifetime a regular subscriber to *The Times*! Perhaps this was why he received this reviewer so graciously in 1935. It was just after the historic isolation of the coenzyme, and we discussed the recent developments and how nicotinamide might be acting, for two or three hours.

Krebs has the best possible background for providing a picture of Warburg's personal life, for he began his scientific career as an assistant at the Institute. When he told Warburg that he wished to get a PhD, Warburg replied that then he would have to leave at once. Indeed, Warburg was curiously uninterested in teaching or in raising the younger generation of scientists. When someone approached him about this he answered, 'Meyerhof, Theorell and Krebs were my students. Haven't I done enough for the next generation?' Though he sometimes described recent ideas and developments to the assistants he really made no attempt to let them develop their own concepts, and Krebs admits 'it was not permitted for them to discuss with him freely in a critical and independent way.'

Warburg was evidently something of a 'health nut.' In the 1920s he was up at 5.30 a.m., went horseriding, and was in the laboratory by 8 o'clock. The assistants were all expected to be in their places before then. By the 1950s, when Warburg was approaching 70, the hours had become 9 to 6, with a half day on Saturdays. He distrusted all grocery foods, and installed a large garden where Herr Heiss raised fruit and vegetables and kept hens, ducks and geese. They even baked their own bread to avoid the additives in commercial flour. Long before British and American medical communities announced it, Warburg had become convinced that smoking caused cancer. All these precautions stood him in good stead, for when he died at 87, Warburg had been active in the laboratory daily since his doctorate at age 28, except for four years' service in World War 1. His terminal illness lasted only a week. As the numerous portraits show, he retained his youthful figure to the end.

A phenomenal story, and a wonderfully enjoyable book, highly recommended to every aspiring young natural scientist. It includes, *inter alia*, a complete bibliography of all the 503 papers(!) from 1905 to 1970. A number of these do not bear Warburg's name but are from the Institute and written with his characteristic clarity and simplicity, and typed by his own hand. Anecdotes abound; some of them, like those about Hill and Barron above, are probably made public here for the first time. Thus, the book is not only a highly readable biography, an important contribution to the history of science, but it is also a valuable work of reference. The combination is unbeatable. An English translation is to be published by the Oxford University Press.

Professor Kenneth V. Thimann is the famous physiologist who isolated auxin, the important plant growth hormone, and developed its practical application. Born in Ashford, Kent, he went to Caltec in 1930, to Harvard in 1935, and has worked in California since 1965.

PROFESSOR ARTHUR J. BIRCH CMG, FAA, FRS

Reviews

The Basel Marriage

Paul Erni: The Basel Marriage. History of the Ciba–Geigy Merger. Neue Zürcher Zeitung, Zurich, 1979. 419 pp. SFr. 23.80.

The Swiss Chemical Industry is located in Basel, allegedly because here the Rhine quits Switzerland. A classical epithalamium, based on an episode in the history of this sophisticated but messy industry, the marriage of Ciba and Geigy, is the sort of whimsy which might well occur to a German–Swiss. The cover shows, against a background of 15th century Basel, a maiden in classical garb with arms wreathed in chains of peptides instead of flowers (Geigy?) and a handsome young man clad only in jeans (Ciba?). Such is the sexism to which this approach tends.

We are not deceived by the soft carpets and works of art in Swiss chemical firms into believing that they are other than profitable instruments for promoting high technology. Despite the excellent drawings of Hans Erni and the occasionally frivolous headings, the marriage described is as practical and contrived as those of the condottieri of Renaissance Italy in uniting their fiefdoms. As such the account is a fascinating, instructive, and as far as can be seen from the outside, an honest one of the course of the courtship, the consummation of the marriage, and the legal and emotional accompaniments.

The English and French versions read like elegant originals, rather than translations. Ciba–Geigy is to be congratulated on promoting the publication of a warts-and-all portrait in an elegant frame.

Professor Arthur J. Birch is a Member of our Editorial Board and Professor of Organic Chemistry at the Australian National University. For many years he has enjoyed in Basel his connection with the chemical industry.

LEON RAPHAEL

Reviews

The Crime and Punishment of I.G. Farben

Joseph Borkin: The Crime and Punishment of I.G. Farben. André Deutsch, London, 1979. pp. 250. £6.50. ISBN 233 97126 2

Joseph Borkin began to investigate cartels in 1934, following I. G. Farben's agreement with Standard Oil of New Jersey. As a lawyer concerned with upholding the US Anti-Trust laws, he shows in the present book the dangers of allowing any large organization to wield power. I.G. sought to conquer the world chemical industry and because it allied itself to Hitler with this aim in view, outstanding scientists and men of business acumen became so super patriotic, that they sacrificed morality and principle.

Some of their Jewish colleagues, forced into exile by the Nazi racial laws, were found positions in the US affiliate companies, but these refugees were used as men of straw to protect Standard–I.G. Farben from confiscation during World War II. This was part of a clever camouflage in which Swiss, Dutch and US nationals played a part. But the real men of straw were the victims of the concentration camps, who paid with their lives at I.G. Auschwitz. They were treated as raw material from whom every ounce of energy was drained until they died of exhaustion and brutality. Then their bodies were processed to extract fat, hair and gold fillings from their teeth. Yet not one gram of Buna rubber or synthetic oil was produced at Auschwitz, despite this huge human investment.

The story of I.G. Farben began over 100 years ago with the exploitation of Perkin's discovery of synthetic dyes in England. By the turn of the century, Germany's Big Six chemical companies dominated their industry worldwide. Corruption was already evident and Carl Duisberg, managing director of Bayer, was a devotee of the *Führer Prinzip*, long before Hitler. I.G. became a loose umbrella organization, an *Interessen Gemeinschaft* or community of interests, which involved cartel agreements in dyestuffs and other chemicals.

I.G.'s main success was Fritz Haber's synthetic ammonia process, which Carl Bosch applied on an industrial scale just before World War I. Haber was Jewish but his ultra patriotism led him to introduce chlorine and phosgene as poison gases on the battlefields of World War I. For this he was branded a war criminal and the scientific world protested when he was awarded the Nobel prize for his high pressure work in chemistry. Carl Bosch is the only character in this whole story who deserves our sympathy. His devotion to I.G. surpassed his patriotism and he remained loyal to his Jewish colleagues, defying the Nazis. He tried unsuccessfully to persuade Hitler not to expel Jewish scientists, arguing that it would be Germany's loss. As time passed, Bosch became more disillusioned and emigrated in 1940, predicting Hitler's downfall; but he died without seeing his prophecy fulfilled. Too late, he realized that if you do not destroy evil, evil will destroy you. Other I.G. leaders who knowingly collaborated with the Nazi murderers received prison sentences, after World War II, but were soon reinstated on the boards of I.G.'s successor companies.

It is astounding to read of the naïvety of highly placed executives in the Standard Oil Company like Walter Teagle and Frank Howard, who were charged by President Truman with trying to camouflage I.G.'s activities in the United States and for holding back the development of the war effort. The last two chapters of this book deal with legal proceedings and I.G.'s success in recovering its US interests twice, after they were confiscated during two world wars; Foster Dulles, for example, among other statesmen, gave I.G. full support on both counts. These chapters are rather drawn out and tedious and the real crime of I.G., as 'jackal to the Nazi lion' is overshadowed by what became, in effect, an Anti-Trust lawsuit. One is left wondering 'what punishment'! The ghosts of their victims are:

> 'Doomed for a certain term to walk the night,
> And for the day confined to fast in fires
> Till the foul crimes done in my days of nature
> Are burnt and purged away.'
>
> (Hamlet I, v.)

Leon Raphael, MSc, FRIC, is a senior chemist in the marketing division of Associated Octel and a freelance science writer.

DR AUGUSTIN WERNET

Reviews

Berichte zur Wissenschaftsgeschichte

Berichte zur Wissenschaftsgeschichte. Organ der Gesellschaft für Wissenschaftsgeschichte e.V. Wiesbaden, Akademische Verlagsgesellschaft 'Athenaion', 1978, Band 1, Heft 1/2.

The Society for the study of History of Science *Gesellschaft für Wissenschaftsgeschichte*, founded in 1965, decided to widen its activities with the publication of a specialized and interdisciplinary journal entitled *Berichte zur Wissenschaftsgeschichte*, Reports on the History of Science, whose first number (Heft 1/2) appeared in 1978.

In accordance with introductory comments by the Editors, the main part of the Journal consists of articles on specific subjects, previously presented by members of the Society at annual symposia. Part of the journal is reserved to Book Reviews, Communications, and News and Comments on important publications concerning the history of science.

The main subject of the first number concerns the general theme 'The beginning of new sciences in the Modern Age' (Das Entstehen neuer Wissenschaften in der Neuzeit). Laetitia Boehm is the author of the first article 'Science – Sciences – University Reform' (Wissenschaft – Wissenschaften – Universitätsreform) which covers historical and theoretical aspects of the 'Scientification of Knowledge' (Verwissenschaftlichung vom Wissen) and the organization of the science institutionalized at the beginning of the Modern Age (pp. 7–36). Fritz Schalk presents an investigation on the role of the Academies in the formation of the new sciences at the time of the Illuminism (Die Akademien und die Entstehung neuer Wissenschaften im Zeitalter der Aufklärung). Subsequent articles deal with the rise of History, Classical Philology, Clinical Medicine, Eugenetics and Chemistry.

The content of the material presented shows clearly that the Society and the Journal, following an old cultural tradition, mean by 'Wissenschaft' a variety of activities which, in the course of History, acquired the names of episteme, scientia, philosophia naturalis, and so on. This point is stressed by the President of the Society and Editor of the Journal, Professor Dr Fritz Krafft, who shows in the editorial the difference between the meaning of the German word as compared with the English (Wissenschaft/Science).

The new journal has an interdisciplinary character, open to the history of all scientific activities, with characteristics that differ from other similar German journals like the *Sudhoffs Archiv* (since 1925), *Techniksgeschichte* (since 1933), *Medizinhistorisches Journal* (since 1960) and the bilingual publication

Archive for the History of exact sciences, all of which are concerned with a certain specific scientific subject.

Dr Augustin Wernet is Lecturer at the Chemistry Department of the University of São Paulo.

DR ANTHONY R. MICHAELIS

Reviews

Otto Frisch's Autobiography

Otto R. Frisch FRS: What Little I Remember. Cambridge University Press, Cambridge, 1979. xi+227pp. £9.50. ISBN 0 521 22297 4

The only aspect I can find wrong with this famous scientist's life story is the title he chose for it – it should have been 'How much I remembered'. Frisch must have been gifted with an extraordinary memory, as there are few paragraphs in this book which do not finish with an anecdotal phrase, either human or scientific. Thus, recalling Liverpool during World War II one such sentence reads: 'At other times I practised the piano; some of the lodgers said the noise was worse than the guns'.

Much has been written about Frisch and his aunt Lise Meitner, and how they as the first interpreted correctly the splitting of the uranium atom as 'Fission', a term coined by Frisch himself; re-telling how he joined his aunt for Christmas in a small Swedish town where the right solution came to them, he finished the paragraph with these words 'That was the most momentous visit of my whole life'.

Frisch's life began in Vienna in 1904 and ended so abruptly in September 1979. It covered almost the whole history of radioactivity and he was singularly well qualified to write about it, as not only did he himself as an experimental physicist make many important contributions, but also because during his work in Berlin, Hamburg, London, Copenhagen, Birmingham, Liverpool, Los Alamos and finally back to England's Cambridge, he met, worked with, talked to, and even made caricature sketches of all the other famous nuclear physicists during those historic decades.

Few men can say at the end of their life, as Frisch has done in the last paragraph of his book: 'All my life

Otto Frisch (1904–1979) as drawn by his wife, Ulla Frisch.

I have been interested in the design of scientific devices, even more than in the results which I or others might obtain with their help. The firm we started for manufacturing SWEEPNIK (a machine to measure bubble chamber tracks) is still growing and giving me scope to do what I like best. I can do much of my work at home and interrupt it to play the piano whenever I like. I am a lucky man.'

Frisch's autobiography is an outstanding contribution to the history of nuclear physics, as seen from the inside by one of its very human and very gifted practitioners. It is highly recommended from all possible points of view.

Dr Anthony R. Michaelis is the Editor of *Interdisciplinary Science Reviews.*

The America Question: World War III?

Europeans remember that World Wars I and II both began in the autumn after wheat and potatoes had been harvested, in both cases a wise precaution. That such thoughts are now again freely discussed in the Federal Republic of Germany, and only slightly less frequently in Britain and France, would have been thought inconceivable a year or so ago. Then détente was fully operative and a balance of military power seemed firmly established for decades ahead.

World-wide Unease

Should the unthinkable war erupt, no single person on this planet would remain unaffected; it would be idle to speculate further. It seems far more profitable to enquire into the reasons for the sudden unease throughout the world and to locate some of the causative factors. One of the answers appears to be America's retreat from her position of pre-eminence during the last decade, a development watched with deep sorrow by her friends and with *Schadenfreude* by her enemies. But England needed her Dunkirk and America her Pearl Harbor before these two democracies awoke from their earlier dreams of false security.

Decline of Innovation

As almost all political decisions are today based on scientific and technological considerations, we must find out if American science and technology has really declined. Two eminent American scientists, Dr Simon Ramo and Professor Edwin Mansfield, analyse in this issue the decline of American innovation in recent years and ascribe this to a number of causes. In a further review, Professor Mel Horwich gives the reasons why no American supersonic transport (SST) was developed and why the production of synthetic fuels is apparently taking such a very long time in the United States. These articles, if taken at their face value, will give false comfort to her enemies.

'Hysteria of Absolute Safety'

Other examples of apparent decline in American attitudes to the progress of science and technology are not difficult to find. At a recent meeting of the Royal Society, what was described by Dr Lee A. du Bridge, President Emeritus of the California Institute of Technology as the 'hysteria of absolute safety' has shown itself in vast American legislation to assure extreme standards of air and water purity, to obtain total absence of all radiation, to perfect labyrinthine testing of new drugs, to ban fluorocarbons as spray propellants to safeguard the stratosphere, to outlaw saccharine and to facilitate the now likely permissive use of marijuana. Similar arguments were adduced by the Nazis in the 1930s to prove the degeneracy of England.

New Military Technologies

In the final analysis, it will be military science and technology which will decide World War III. In World War II it was American mass production of ships, tanks and aircraft which overcame Germany's military might and the atomic bomb which forced Japan to her defeat. As those in charge of military forces learnt the value of applied science and technology to improve existing and devise new branches of warfare, spending world-wide for this purpose now at a daily rate of US $1000 million, prodigious progress was achieved. Warfare in space and electronic warfare are the latest, both technologies in which America has shown scientific innovation leading soon to technological supremacy.

America's Electronic Supremacy?

If tanks in World War I and aircraft in World War II were decisive in many battles, then mastery of electronics will decide the next victor, the mastery that enabled American astronauts to walk on the Moon. It is now known that Soviet electronics were inadequate for this endeavour. A line of caution must, however, be inserted here, as over-sophistication in American aircraft electronics has apparently now reached such a stage that some aircraft, like the swing-wing bomber F-111D, cannot serve its purpose for over 60% of its life. This defect has been recognized by Defense Secretary Dr Harold Brown who is redirecting technology away from sheer performance in favor of less complex and more practical weapons which can be mass-produced.

'Crying Wolf'

The false alarms of the United States Missile Early Warning System which in November 1979 and again in June 1980 sent shivers around the world, are probably due to bad software design. This multi-computer system is today the most complex in existence and in addition to its real function – to warn America of approaching enemy missiles – it must also provide for: (1) training signals to maintain vigilance amongst its hundreds of experienced monitoring technicians; (2) similar or less sophisticated signals to train new operators; (3) test programs of the actual computer hardware to detect faulty components; and (4) integration into the system of new or modified computer hardware, as ground-based and satellite surveillance components are improved. All these must be synchronized with the existing system. False alarms may thus arise because information is incorrectly handled within the system and because it is out-of-date. The real danger lies in 'crying wolf' once too often!

Real Strength

America's real strength must be great if it allows the world to know about these false alarms, when all other countries would never publish such serious defence shortcomings. It seems that American Defense officials have learnt from American space scientists who also let the world share their agonizing early failures, as for example the Ranger program, which for seven out of nine Moon probes malfunctioned before success came in 1964. Let us hope that the Early Warning System can also be perfected, although here our hope must be that it will never be called upon to fulfil its proper function.

Will Peace Last?

If World War III is fought and decided in hours and days, the outlook for the free world may indeed be grave. But if it should last longer, then with human freedom of scientific thought and research, the ability to pool ideas in interdisciplinary collaboration between scientists, the military and national leaders – which led to Operations Research in World War II – then there can be no doubt of its outcome. *Si vis pacem, para bellum*!

Anthony R. Michaelis.

America's Technological Society: The Next Decade

DR SIMON RAMO
TRW Inc., Redondo Beach, California, USA

Evidence is presented that the US position in science and technology is slipping behind that of her competitors in Europe and Asia. The consequences are unacceptable in the economic, social and political fields, not only for the USA, but also worldwide. The reasons for the slippage are given as inflation, high tax policies, a too low investment rate in R and D and particularly governmental over-regulation. The proposed answer is effective cooperation between the private sector and the US Government; the production of synthetic fuel from coal is cited as an example of how such teamwork should be organized.

As the world enters the 1980s, virtually every nation perceives its economic strength and security as critically dependent on its science and engineering prowess. This includes the United States, today a giant but slipping technological nation, where concern is growing that its innovative and entrepreneurial spark is ebbing as in area after area its previously held leads are closed by other nations. Indeed, if the trends of the 1970s are not reversed, in the 1980s the USA will become a so-so technological country, one at best in the middle of the ratings among all countries as judged by how well it employs science and technology to meet national objectives.

PRODUCTIVITY LAG

In America a technological lag is especially penalizing because its economy and social structure are rooted in a generous availability of the fruits of technological advance. Political pundits aver that no solution to US domestic problems is viable that contemplates the average citizen accepting a significantly reduced supply of goods. It is equally unrealistic to expect disadvantaged Americans to abandon aspirations for the higher living standard the majority enjoy. If these are political truths, then scientific and engineering tools must be sharpened and employed vigorously because they are indispensable for a feasible approach to US domestic problems in the decade ahead.

Advancing technology is essential to America in the 1980s for more than domestic tranquility. International economic competitiveness by the US and its contribution to world stability and progress both hinge on the status of its science and technology. If world economic health is fostered by each nation supplying what it best can to other nations and trading with them for their most suitable items in return, then continuing advances in science and technology by the USA are fundamental to that nation doing its part. High technology product exports, such as aircraft, computers, petroleum equipment and chemicals, are presently the important contributors to the positive portion of the US trade balance. The only other strong contribution comes from agriculture and that is mainly because US fertilizer, farm machinery, pesticides and other farming and food technologies lead the world. To maintain America's position in the decade ahead in these areas requires continuing technological advance to increase productivity steadily.

Scientific research can teach how to apply the laws of nature more effectively. Engineering effort can lead to economically advantageous products whose manufacture would create new jobs to fight unemployment and increase exports to restore the US trade balance. Further advances can cut costs as a counter to inflation, create substitutes for materials in short supply and make possible meeting requirements with less harm to the environment. But these are merely possibilities. Is the USA acting so as to realize them fully? To this important question, the answer is *no*. Of course, even if America performed perfectly in selecting and implementing technological programs in the 1980s, this would not guarantee a healthy economy and a happy society. By handling non-technological matters badly, unemployment, inflation and other economic ills can exist all at once. Science and technology are not sufficient. But they are necessary. Without a deep, sturdy technological foundation US needs cannot be satisfied.

NATIONAL GOALS NEEDED

Many impressive lists have been drawn up of possible technological achievements for the 1980s. Electronic

CCC-0308-0188/80/0005-0173 $04.50/1

microminiaturization and further advances in semiconductors, computers, communications, cable, fiber-optics and satellite technology will create man–electronics partnerships for superior handling of information to make us all smarter at our jobs and more efficient in running our businesses and homes, controlling our autos, and being educated and entertained. Ocean resources will be exploitable for minerals, energy and food. Agriculture and industry will be made more energy-efficient and less dependent on hazardous processes. Weather will be more predictable and airline navigation and traffic control more reliable with added safety. New mass urban transit systems will be created to save energy, money and time of the users and cut pollution. New energy approaches will add to supply by the end of the 1980s. At least, we could easily make a strong case that says all these advances and many more *could* happen if the only limitations were the fundamental science and technology underpinnings.

But which possibilities flower and which droop, die or are neglected will depend not so much on the scientific and technological potentials as on the social-economic-political environment of the 1980s. These non-technological factors will control and limit the selection and speed of technological developments. As to appropriate social advance, in the USA in the next ten years it is not a matter of 'keeping up the good work' of the past ten because the work has not been good. In the last decade or two a severe mismatch has developed between the high potential of technological advance and the low rate of social-political progress.

The reason for this mismatch is not science and technology *per se*. We simply lack social organization to use these tools to the fullest. Critical and controlling are the interfaces of technological and nontechnological factors as we set objectives, examine alternatives and make decisions. These three vital activities are now almost absent or are being carried on with a lack of understanding of the process by the American citizenry. In choosing where and how to apply science and technology it would be helpful, for example, if we possessed clear national goals. When we find it hard to articulate and decide on what kind of society we want, it is understandably difficult to pinpoint the effective use of science and technology to help build it.

We cannot make satisfactory decisions without an understanding of tradeoffs and options. We need to be in a position to compare the good or benefit that can come from specific technological advance against the bad or the cost. As to our employment of technology in America, we can be likened to a group of inept carpenters. Equipped with strong and ever sharper tools, they use them clumsily, often getting their fingers in the saws, hitting each other's heads as they bring their hammers back and their own thumbs as the hammers come down. They are not sure what they are trying to build. Confused, yet sensing an unsatisfactory situation, which they would like to change but do not know how, they meanwhile blame the tools.

TECHNOLOGICAL SLIP

How serious is the US technology slip? What real evidence exists that we are getting behind, aside from the Japanese and European cars on the streets and the foreign-made TV sets and tape recorders in our homes? What must we do in the 1980s to reverse the trend toward inferiority we began developing in the 1970s?

To begin with, let us look at the measure most commonly cited when concern is expressed, productivity rise. For over a century the USA steadily increased its productivity. We were the world leader in inventing improved methods and building ever more productive facilities – until the last 10 or 20 years. By 1980 the USA has managed to become last in the listing of the world's industrialized nations in annual rate of productivity growth. From the mid-1940s until the mid-1960s, US productivity increased over 3% per year. Then between the mid-1960s and the mid-1970s this rate dropped to just over 2%. For the last half of the 1970s it sank to 1% and a negative figure is on our report card as we start the 1980s.

All in all, from 1950 to the present time, productivity in Japan rose at four times the US rate. In the mid-1960s it took some 25 man-hours to produce a ton of steel in a Japanese mill, twice that needed in the USA. By the mid-1970s the US figure improved to 10 man-hours, but Japan's came down to 9.

During the first half of the 1970s compensation per hour in the USA increased about 8% per year while productivity went up only about $1\frac{1}{2}$% annually. Wages constitute the major component in the cost of all production. When labor cost rises faster than

DR SIMON RAMO – a Member of our Editorial Board – is a recipient of the National Medal of Science, the highest award for an American scientist. He was until recently Vice Chairman of the Board and Chairman of the Executive Committee of TRW Inc. of which he is a co-founder. The 'R' in TRW stands for Ramo. He has been Chief Scientist of the USA's Intercontinental Ballistic Missile Program and Chairman of the President's Committee on Science and Technology. Dr Ramo is a fellow of the National Academy of Sciences, a Founding Member of the National Academy of Engineering, a Visiting Professor in Management Sciences at the California Institute of Technology, and a Fellow of the Faculty of the Kennedy School of Government of Harvard University.
Address: TRW Inc., One Space Park, Redondo Beach, California 90278, USA.

productivity the difference must find its way into prices. The diminishing productivity rises thus certainly have not helped curb US inflation. The total loss in goods and services that could have been available to America had we maintained the productivity increases of the mid-1940s was $300 thousand million in 1978 alone, well above the entire defense budget for the year plus the cost of all automobiles purchased.

TRADE DEFICIT

Our deteriorating trade balance is as alarming as our productivity lag. In 1978 the US trade deficit reached a record $30 thousand million and this was not simply the result of buying so much higher-priced OPEC oil. In 1978 the USA imported 25% of total energy consumed while Japan imported 95% of its energy requirements. In the same year, Japan enjoyed a $63 thousand million surplus in manufactures while the USA experienced a deficit of $10 thousand million in that category. The problem for us is not alone oil imports, although that hurts, but also that we are no longer as effective at converting materials and energy into manufactured goods as other nations. They have taken much of both the world and our domestic market from us. If they were not superior in manufacturing, then their more adverse situations in energy imports would have caused them to experience serious trade deficits. Instead it is we who have them.

We are now importing manufactures at two and a half times the rate of oil. Our biggest loss of trade position in manufactures has been in the category that includes heavy machines, light machines, cars and trucks, and aircraft. Only in the last category do we remain strong, with exports of $8 thousand million and essentially no imports. In automobiles and trucks we are now importing more than we export by almost $12 thousand million; in entertainment electronics we are off by $4 thousand million; in heavy machinery by over $6 thousand million. In light machinery, consumer goods or apparatus used to manufacture them, like typewriters, clothing, furniture, cameras and similar goods, we are running a deficit of over $12 thousand million.

REASONS FOR SLIP

Our growing disadvantage in manufactured goods stems partly from our lessened investment in research and development relative to it. Applied R and D in US industry, the total of that government sponsored and privately backed, has not kept pace with inflation over the last decade. The actual expenditure in such R and D, considered as a fraction of GNP, has fallen from a figure of over 3% to one just over 2%. During this period the R and D budgets, again as a ratio of the nation's overall domestic product, rose in France and Germany. If we look carefully at the makeup of our R and D programs as compared with other nations we discover that the lowered R and D in the USA has resulted from cutbacks in defense, space and nuclear R and D, all sponsored by the government, while industrial R and D has held up well. Apparently, in view of the negative US results in productivity and trade, our industrial R and D is not concentrated on those aspects of technology which have most effect on improving these two performance indicators. We know, of course, that our R and D is more focused on defense and the meeting of governmental environment and safety regulations.

Even though it is believed that new technology has accounted for over a third of past productivity increases, the USA spent only something like 1.8% of its GNP in the 1970s on non-defense R and D while Japan spent 2.1% and West Germany 2% of theirs. Private US industry actually employed 5% fewer scientists and engineers at the end of the five-year period 1970–75 than at the beginning.

CONCERN FOR CREATIVITY

We are also low in capital investment, another major influence on technology advance. If from the figures for capital investment backing up each worker in the USA we subtract that part of the investment made to comply with pollution and safety regulations, the net investment, expressed as a ratio of value added during production operations, has actually declined in the manufacturing sector of the US economy since the 1960s. In the same period this indicator rose in West Germany and Japan and for each this figure now exceeds ours. Overall, the USA reinvests only 15% of its GNP, West Germany 22% and Japan 29%.

But there is more to worry us. The fraction of US patents granted each year to non-US citizens has doubled in the 1970s. In the past five years foreigners have been awarded more than one third of all US patents issued. Meanwhile, the share of patents by Americans abroad has declined substantially. In the 1950s over 80% of the most significant technological advances brought to market were developed in the USA. By the late 1960s the figure declined to around 50%.

Until the late 1960s, a large number of new US technological corporations were launched each year. Stock issues to found such entrepreneurial enterprises then dropped to a trickle. Despite some recent recovery, the activity remains well below earlier levels. Such a fall-off should be a source of concern about creativity and international competitiveness when we consider that the key ideas for the following fields came from individuals not employed in large organizations: atomic energy, advanced electrical batteries, computers, cellophane, color photography,

the cyclotron, DDT, FM radio, foam rubber, inertial guidance, insulin, lasers, the Polaroid camera, radar, rockets, streptomycin, the vacuum tube, xerography and the zipper. The bulk of industrial R and D is done by the top 100 largest technological corporations, but a loss of the contributions from innovative small entities is bound to be penalizing.

SCIENCE WINNERS

If conspicuous evidence exists suggesting America is losing leadership in technology, we are perhaps still ahead in the pure science olympics, even if we do not win every event. We have more Nobel laureates than the rest of the world combined. We are still ahead in solid state technology, where we were first with the transistor, and large-scale integrated circuits. We lead as well in digital computers, digital communications and virtually every other aspect of information technology. Microminiaturized electronic circuitry now makes feasible very low priced, fast, high capacity and reliable handling of information basic to business, industry, government and the professions. Here, continued US leadership bodes well for future US productivity improvements.

The Soviet Union holds the world record for most hours for a man in a space capsule, but we have accomplished much more in putting space to use. Our communications, Earth observation, weather and planetary research spacecraft surpass USSR performances in these areas. We sent men to the Moon and back while the Soviet Union failed and could only land robot vehicles there.

In agriculture, we are blessed with a combination of topography, soil, weather, water, mechanical equipment, fertilizers, pesticides, herbicides, food preservatives, and packaging and distribution techniques substantially superior to the agricultural resources of other nations. Moreover, we have research ideas to pursue that, from seed to mouth, could greatly increase our potential to be a world exporter of food for which the rest of the world will have an increasingly urgent need. Food and nutrition may transcend energy as the world's No. 1 shortage problem in another ten years and the US position as a supplier may be like that of the Near East in oil.

SCIENCE LOSERS

However, there is also some bad news on the pure research front. Pioneering, basic scientific research, largely sponsored by government and accomplished by universities, has suffered from shrinking budgets in the late 1960s and early 1970s and growing bureaucratic handicaps since then. Even though the trend to lower budgets was reversed by the Ford administration, a reversal fortunately continued by Carter's, nevertheless, after compensating for inflation and the enormously added administrative burdens that now accompany research grants, US basic research backing is seen not to have kept pace as a fraction of the nation's GNP. Such research, generally regarded by everyone as the planting of seeds for future technological advance and economic strength, is increasing in other countries.

When productivity is discussed quantitatively, the concept of efficiency, i.e. output divided by input, dominates. As to efficiency *per se*, no doubt exists that we are falling behind. However, efficiency, while it is important, is not the only criterion of success in applying science and technology. We clearly are willing to sacrifice efficiency if necessary to attain certain other objectives – for example, less pollution. We will accept producing less food per acre if it means fewer cancer deaths. We will make do with less manufactured units per man-hour if the drop brings added safety.

It costs something to find out how to be safer and cleaner. We have to buy new equipment to decrease hazards and we have to use more man-hours to achieve the same output. Those funds and efforts could have gone instead into making the operation more efficient. There is a tradeoff here: greater safety and less pollution versus higher productivity.

Now, it is one thing to tolerate inefficiency to reach a goal unattainable if we allow efficiency alone to control. It is another to be inefficient because we are backward, or because we inadvertently put barriers in the way of innovative effort. It may be a desired social gain to trade more shoes for more time to walk barefoot on the beach. But going barefoot without choice because we have no shoes is something else again, especially if the beach is filthy. If we become inefficient enough we will lose the option to reach the very social goals we may have chosen because we saw them as superior to blind, all-out technological advance and a worship of efficiency.

THE TRADEOFF

Unfortunately, if our productivity falls far enough, then US industry will lose out in both domestic and world markets. Unemployment will go up, living standards will diminish and the poor will get poorer. Health and happiness will be downgraded as much as if we polluted more or were less safe. So we need to arrange the right tradeoff balances. Do we really have ways to do this? That is doubtful. The agencies that set safety and environmental standards are not required to consider the related productivity or cost consequences, even if a slight lowering of standards might hold a given hazard still within a sensible range while at the same time almost eliminating the productivity drop which overly severe standards might cause. We need to learn how to make these tradeoff decisions or the US technology lag will persist and grow during the 1980s and America will indeed lose its international competitiveness.

When it comes to doing the right things as against the wrong ones, it is encouraging that no evidence has surfaced to suggest that Americans are now basically any less innovative than in earlier times. 'Yankee ingenuity' is as present in America now as before. If there is less manifestation of inventiveness and ambition to break new ground it must be because something about the pattern of US society today is creating too many obstacles, providing too little motivation. Here, I am afraid we have more bad news.

A serious limitation to our wise utilization of science and technology is the public confusion as to the right roles for free, private enterprise, on the one hand, and government control, sponsorship and regulation, on the other. We have two routes for initiatives to get things done in the USA: free enterprise and government. The approaches are not mutually exclusive. In fact, the USA has a hybrid economic structure and we have a need for balance as we allocate assignments to private and government action. The free enterprise system – private capital at risk seeking a return – has served well for over two centuries to connect the needs and desires of the citizens with the capabilities of technology. This wedding has blessed us with an immense stream of products and services. Why does not the free market automatically solve the system and organizational bottlenecks just referred to? Why is the free enterprise route not still the practical answer to the challenge of putting science and technology to work sensibly and completely?

DISINCENTIVES

In the USA the biggest concentration of technological innovation expertise resides in the private sector. Here is to be found know-how in applying science and

technology to meet the requirements of the society. The industry has systems for arranging financial backing and experienced management to carry out successful implementations. Unfortunately, a straightforward, severe limitation on technological innovation in American industry now exists: lack of adequate cash generation. We could state this in various other ways – the real return on investment is not good enough, or the reward-to-risk ratio has deteriorated.

We could point to the fact that out of every dollar of reported earnings some 60 cents goes to taxes, higher than the nominal rates because inflation causes reported earnings to be overstated. Then some 20 cents must go to providing for true facilities replacement costs, since Internal Revenue Service rules do not allow for the real costs of depreciation. Of the remaining 20 cents, perhaps 15 cents are put out as dividends to the shareholders; they consider that too little to invest more. This leaves 5 cents for risk-taking on innovative new products and methods. We could also sum it up by saying bluntly that the critical limitations on technological creativity in America are inflation, disincentives through wrong tax policies, and over-regulation. The US government is rather heavily involved in all three of these factors.

Unfortunately, it is also true that an increasing fraction of the most needed and potentially beneficial technological projects call for investments beyond the net worth of most corporations, and large corporations in the USA now are essentially not allowed to team up to share the risk. For such projects, the 'risk-to-return' ratio is too great, the start-up losses are too high, the time to 'turn-around' to an eventual profit phase is too long, and the dependence of success on unpredictable political decisions is forbiddingly close. The market for these large-scale projects is far from being a free one and is instead a cacophony of semiautonomous and conflicting interests. Finally, many of these programs, such as development of ocean resources and full use of space, require a high degree of international cooperation, difficult to arrange.

REGULATORS

Improving the environment and increasing safety are meritorious objectives and the government has a necessary function to perform here. However, government regulations in 1978 alone cost industry well over $100 thousand million. That means those funds were not available for technological innovation to create jobs and compete internationally. Some of these expenditures were for purposes totally consistent with the desires of the American citizenry. However, the sum has become this huge because it covers expenses to respond to many highly questionable regulatory requirements. It also funds a massive government regulatory bureaucracy and counterparts in industry to deal with it. If no changes in regulatory trends occur, the 1980s will see rapid growth in all these effects. We wanted to increase safety and cut pollution but we have been slow in developing practical balancing in the process and have produced a system costly well beyond the intent.

Without meaning to do so, our regulatory climate discourages innovation in new product areas. The more innovative a new program the greater the business risk. Regulations create delays and uncertainties, making more difficult the satisfying of conventional return on investment criteria, already bad when high inflation exists. The easy option for US technological industry is to play safe, put available chips on incremental improvements and avoid trying for bigger steps.

Under these circumstances, no wonder new technological corporations now are hard to finance. It is a far more risky business today to start a new company based on innovative ideas. The novel product may be welcomed by the market, but the cost of doing business will be higher and the return on investment lower than in former years. Investors are more inclined now to be leery of speculative technological developments.

GOVERNMENT OR PRIVATE SECTOR

Can America return to a technology leadership position in the 1980s? When a serious national problem surfaces in America, some propose the government should stay out and let the free market handle it. Government, to them, is an enormously wasteful and incompetent bureaucracy. To other Americans government action is the only answer. As they see it, the private sector will not act in the national interest and will only seek 'obscene' profits. In actuality a large fraction of all important areas of technology already involve the government heavily, with little chance of a truly free market approach. Business leadership will not make investments unless they see an environment for a satisfactory reward-to-risk ratio. This they do not expect when the area of endeavor has become highly politicized.

For example, by all-out use of science and technology, and the full cooperation of government and private industry, we could have a plentiful supply of energy for over a century by utilizing our large deposits of coal. Existing technology can be advanced to desulfurize coal, produce liquid and gaseous fuel from it, mine it more safely, obtain a good deal of energy from it without mining *in situ*, and burn it more cleanly and fully to limit environmental harm. The resulting fuel would be higher priced than today's petroleum, but the payments would stay in the United States. Those who have studied the problem tell us that the benefits of a strong economy based on adequate energy would outweigh the negatives of higher energy prices.

INVESTMENT FOR COAL

Success in broader use of coal requires both the know-how of private industry and enormous capital investment. Will the private sector go ahead in the 1980s without government support? A profit-seeking corporation choosing to make a major entry into synthetic gas or liquid fuel from coal by new technology would have to commit funds in the thousand millions of dollars and decades might pass before a return on that investment could be realized. Indeed, such return might never eventuate, so great are the risks associated with unpredictable later government entry.

The complete system needed for a synthetic-fuel industry based on coal involves a host of private and public organizations that are semiautonomous and not readily directed by any one body, not even the White House or Congress: land owners, mine operators, labor unions, railroads, power generating and water supply utilities, numerous engineering and manufacturing companies, county and state governments, and many, many agencies in the federal government that deal with prices, environment, labor, antitrust and transport, to name only a portion. After ten years and $10 thousand million put at risk, with a flow of synthetic fuel ready for market, the private venturer would have to reckon on the US government's stepping in to clamp on price controls or set more severe environmental standards or file an antitrust case, or all three. Anticipating these circumstances, a totally private sector approach to greatly enhanced use of coal is unlikely.

Some Americans want a broad government program to accelerate technological advance in the United States. If we are beginning to slip dangerously, then they argue the government should do more – beyond such projects as a rocket to the Moon, intercontinental ballistic missiles and breeder reactors. Why should not the government, they ask, embark on a giant effort to increase technological innovation and productivity? Why is not this the Apollo program of the 1980s?

EXPAND RESEARCH

Technological advance is infinite in its varieties and dimensions. The new products that could be developed and the inventable production alterations are endless in number. However, the nation possesses only enough resources to carry through a small percentage of all that can be conceived. The trick is in assigning well the limited, available strength. Nothing about government, in past practice or theory, suggests it would be effective in sensing the market's real requirements, the consumer's true wants. This does not mean the government should do nothing about technological innovation. It should expand its sponsorship of scientific research in the universities. It should create university programs in advanced production technology, an area where the USA lags behind Germany and Japan. It should improve regulatory functions.

In industrial technological innovation, however, the US government can help most not by launching more programs, which will mainly take needed capital out of the private sector, but rather by cutting inflation and improving the general economy so that the private sector will have the financial means and incentives to perform more R and D of its own choice. Licking inflation, lowering capital gains taxes and decreasing the taxation on earnings, presently double through dividends, would especially encourage investment. Germany's and Japan's investment ratios are much higher than ours and so are their productivity increases while their capital gains taxes are zero. In Japan and Europe, government and the private sector are cooperative in the main and think of themselves as being on the same team. In the USA the relationship is usually an adversary one.

Presidential candidates and voters should understand that ours is a hybrid society, part free enterprise and part government control. If we are serious about using science and technology more fully during the next decade on behalf of the nation, then the name of the game is to assign the proper roles and missions to the government and the private sector and get on with meeting the nation's goals. Unfortunately, a single,

simple formula for allocation of responsibilities between the two sectors will not suffice. The arrangement has to suit the area of endeavor. Let us illustrate this by taking one specific project in the field of energy. What are the best roles for government and the private sector in producing synthetic fuel from coal?

SYNTHETIC FUEL

The creation of such a fuel source for the United States is certainly a controversial topic. First of all there are lots of other alternatives, each with strong advocates. For instance, we could mount a crash program to modify our homes, factories, transportation systems and the rest so as to greatly conserve energy. Many are convinced that a much smaller investment in conservation than in synthetic fuels from coal would save millions of barrels of oil a day, a level hard to reach with the synthetic-fuel plants. Moreover, coal liquefaction is a dirty process and critics argue that high output equates to an unacceptable level of pollution. Still others believe the mere removal of all price controls and allocations by government would lead to oil discoveries on American land or off our shores or elsewhere in the world, and would trigger the application of advanced technology to increase greatly the available flow from existing domestic petroleum deposits.

Solar enthusiasts are confident that if solar energy R and D is accelerated in all its dimensions – solar panels in homes and buildings, biomass, the harnessing of the winds and tides, ocean-thermal energy – then we could obtain at least as much as we could get from a tolerable number of coal-based plants with less investment and lower pollution. Synthetic fuel from coal or shale is also attacked as putting an undue burden on our water supply problems and on our transportation systems. Finally, nuclear proponents believe that when the emotional reaction to the Three Mile Island incident has decayed, additional safeguards have been introduced, and some of nuclear energy's loudest but least sincere detractors have moved on to other things, then it will be seen that the nuclear route is the cleanest, safest and most economical way to supply our energy needs.

But no one of these alternatives is a sure thing. It is wrong to launch a super-crash effort on synthetic fuel from coal that violates reasonable environmental standards and neglects alternatives. It is equally unwise to rule out the approach entirely. Accordingly, let us assume it would be sensible to provide ourselves with the option of synthetic fuel from coal, to refine quickly the know-how so that we can move fast on a next step of high production if we so choose, or can stop without huge loss if other approaches turn out better. How can we best arrange for a project along these lines?

A GOVERNMENT CONTRACT

To answer these questions, let us first recognize that the US government has little competence with which to perform details of a difficult technological effort. The government can attract a few outstanding people for the top jobs whenever it embarks on a mammoth project, but as to overall expertise there will be no comparison in competence as to laying out, designing, developing, researching and bringing to success a synthetic fuel project between those the government can recruit as in-house engineers and managers and those American technological industry can assemble. Next, let us observe that if the government is to fund a synthetic fuel development, it must get the money out of the private sector somehow. When that capital is extracted, that means it is not available there to invest in synthetic fuel or anything else.

With these two thoughts in mind, consider this proposal: the US government announces that, for the government's own use, it wants to purchase from the private sector liquid synthetic oil to be extracted from coal. The government states the quality and nature of the fuel it wishes to buy and the delivery schedule. It offers a ten-year contract with a price adjustment factor for inflation during that period. The government uses a formula for deciding on the winners with credit given both for low price and a start on deliveries as early in the 1980s as possible. The government sets standards as to safety and pollution that it agrees not to tighten without upward price adjustment. The government plans to obtain these requirements – say of the order of a half million barrels a day equivalent – from at least two sources. The government provides immunity from antitrust suits if companies wish to create a joint venture to bid on the proposal. The government provides a proper cancellation fee if it wishes to cancel part way through the ten-year period.

SYNTHETIC FUEL PROGRAM

If the American government were to issue this request for proposals, a number of firms of high competence and substantial financial backing would bid. The submissions would be sensible from the standpoint of the bidders because they otherwise would not submit them. Doubtless the prices quoted in the proposals would be higher than existing petroleum prices but, from what we already know of the technology and economics, not so high as to vitiate the program. The requirements for access to suitable land and water resources for the coal would be included in the proposals and ultimate contracts. State or federal land would have to be made available at particular locations at stated price ranges as part of any deals made. Obviously, the legislation creating the program described could include provisions for the designated government contracting agency to have some of the same powers to accelerate regulatory approvals that President Carter has envisaged for his otherwise quite different Energy Mobilization Board.

If this program were created, it would fully cover the requirement to get started in a meaningful way on synthetic fuel in the 1980s. It would set up the option to broaden the program later or keep it as a lower-level program. The worst that could happen, from the standpoint of the government, is that if OPEC oil did not rise enough in price during the next decade – and we would welcome that unlikely occurrence – the government would overpay somewhat for the fuel it would have purchased for its own needs.

From the standpoint of appropriate roles for both the US private sector and the US government, we notice that in this proposal the government is not at all involved in the technology, an area where it has the least contribution to make. The government creates a guaranteed minimum market for the output of the private sector. The government sets safety and environmental standards. The free enterprise industry takes a calculated investment risk, choosing the technology it favors. If it wins the competition it will direct its program. The government will not.

If the USA were to handle better the development of energy alternatives for the period ahead, it would help solve not only our energy problems but those of the world, since we are such a large energy user. But we should generalize: if, instead of technological breakthroughs that we then are inept at organizing to exploit, we were to break bottlenecks to our full use of science and technology in the 1980s, it not only would help our international competitiveness but the economic condition of others. World economic stability depends in great part on a competitive free market motion of materials, goods and funds, because then each nation can benefit by trade with others. For this condition to exist, a strong US economy is necessary, and basic to that is US technological strength, effective decision-making and government–private sector cooperation to use that strength. We do not have to be first in everything. We never were. But we need superiority in enough areas so that we can trade, prosper, cooperate and stimulate, thus adding to world economic and social progress.

IN CONCLUSION

We have put forth some of the evidence that suggests the US position in science and engineering is deteriorating. A severe technology slip by the United States, we have observed, carries with it unacceptable economic, social and political consequences in the 1980s, not only for the USA but the world. We have concluded that the US situation is not the result of a mysterious, sudden decay of the basic innovation talents of Americans.

It is rather to be found in inflation, wrong tax policies, too low an investment rate and over-regulation, factors in which the US government is dominant. We have suggested, however, that it is not a real-world answer to the problem to suggest halting the US government's involvement with science and technology, leaving the arena totally to free enterprise. Nor is it to turn everything over to the government and foster the removal of capital from the private sector in order to fund a government-directed crash effort to accelerate technological innovation. The realistic approach is to recognize that, in the USA, using science and technology most fully in the society's interest requires the participation of both free enterprise and government. The way to the most effective results is to define and assign roles to the two sectors most suitable for each area of endeavor.

If we wish to predict what will happen on the technology front in America in the next ten years, we should concentrate not so much on the list of attractive possibilities for technological advance. Rather, we should focus on the nontechnological social-political-economic factors, the pattern of decision-making and government–private sector relations. These are the controlling parameters in America. This is where the priority effort toward progress most needs to be in the 1980s. If the USA were to make this a guiding policy for the next decade, that nation would see the implementation of a much wider range of the right technological developments.

This review leans on the author's Keynote Address to the Annual Meeting of the American Association for the Advancement of Science in San Francisco on 3 January 1980.

The manuscript was received 25 March 1980

Innovation in the United States: Its State of Health

PROFESSOR EDWIN MANSFIELD
Department of Economics, University of Pennsylvania, Philadelphia, Pennsylvania, USA

Great caution is necessary to interpret the rough approximations of the rate of innovation as there are no accurate measures of it. The slowdown in productivity can be recognized and comparisons with other countries can be made. Returns from investments can be stated and possible reforms indicated. It is concluded that American technology policy cannot be divorced from economic policies. Hence improvements in the general economic climate may have more impact on US technology than specific measures to stimulate innovation.

This is a very brief review of what economists believe they know about the state of health of innovation in the United States. Unfortunately we lack any simple, accurate measure of the rate of innovation. All we have are rough approximations that must be used and interpreted with the greatest caution. Perhaps the measure that is most frequently used is the rate of productivity increase. However, it is affected by a great many factors other than the rate of innovation – namely, the business cycle, changes in the quality of inputs, and the rate of diffusion of new technology, among other things. Thus, it is hardly a very accurate measure of the rate of innovation.

Nonetheless, in trying to evaluate the health of US innovation, one cannot help but be struck by the pronounced slowdown of the rate of productivity increase in recent years.[1] According to the Council of Economic Advisers, output per man-hour increased by 1.0% during 1973–77, as compared with 2.3% during 1965–73. And between 1977 and 1978 it increased by only 0.4%. According to unpublished data that John Kendrick has made available to me, total productivity in manufacturing fell sharply from 1973 to 1974 and was only slightly higher in 1975 than in 1974. By 1976 it was not much above its 1973 value.

SLOWDOWN FACTORS

It is important to recognize that much of this slackening of US productivity growth is due to nontechnological factors. According to Kutscher, Mark and Norsworthy of the Bureau of Labor Statistics, one of the major factors has been the increase in the proportion of youths and women in the labor force. Output per man-hour tends to be relatively low among women and among new entrants into the labor force. During the late 1960s, women and new entrants increased as a proportion of the labor force. Based on the Bureau's calculations, this change in labor force composition may have been responsible for 0.2–0.3 percentage points of the difference between the average rate of productivity increase in 1947–66 and that in 1966–73.

Another factor responsible for the productivity slowdown is the rate of growth of the capital–labor ratio. During 1948–73 relatively high rates of private investment resulted in a growth of the capital–labor ratio (net nonresidential capital stock divided by aggregate hours worked in the private nonfarm sector) of almost 3% per year. After 1973 relatively low rates of investment resulted in the growth of the capital–labor ratio by only about $1\frac{3}{4}$% per year. According to the Council of Economic Advisers, this reduction in the rate of growth of the capital–labor ratio may have reduced the rate of productivity increase by up to 0.5 percentage points per year.

Still another major factor is increased government regulation. A variety of new types of environmental, health and safety regulations have been adopted in recent years. Because reduced pollution, enhanced safety and better health are generally not included in measured output, the use of more of society's resources to meet these regulations is likely to result in a reduction in measured productivity growth. Further, the litigation and uncertainty associated with new regulations may discourage investment and efficiency, and the form of the regulations sometimes may inhibit socially desirable adaptations by firms. According to the Council of Economic Advisers, the direct costs of compliance with environmental health and safety regulations may have reduced the growth of productivity by about 0.4 percentage points per year since 1973.

Although the factors cited above can explain some of the recent fall in the rate of productivity increase, they are far from being able to explain all of it. Thus, Edward Denison concludes that 2.1 percentage

CCC–0308–0188/80/0005–0182 $02.00/1

points of the 3.3 point drop between 1948–69 and 1973–76 in the rate of growth of national income per person employed remains in a residual called 'advances in knowledge and not elsewhere classified'. To some extent, this residual may reflect the effects of the big oil price hikes that occurred in 1974 and 1975. Some observers attribute a substantial proportion of this residual to this factor. To some extent, it may also reflect a fall in the rate of innovation, but we have very little hard evidence concerning the extent to which this is the case.

In an attempt to shed light on innovation's role in this context, some economists have tried to determine the extent to which output (and thus productivity) might increase if R and D expenditures in the United States were to increase as a share of gross national product. For example, Roger Brinner has concluded that if R and D investment were to return to its peak, mid-1960s share of gross national product by 1982, and then maintain this share, the potential growth of national output would be 0.3 percentage points higher per year. Other economists, like John Kendrick, agree with Brinner that the productivity slowdown has been due in part to the leveling-off of R and D expenditures (in real terms) during 1966–77.

COMPARISONS WITH OTHER COUNTRIES

In trying to evaluate the health of US innovation, it clearly is necessary to compare our performance with that of other countries. As is well known, the available evidence indicates that we have long maintained a significant technological lead in many areas of manufacturing. In recent years American analysts and policy-makers have expressed increasing concern that the US technological lead is being reduced, and in some areas eliminated.[2] At least three types of evidence have been put forward. First, the percentage increase in American labor productivity has been smaller than in Japan, Germany, France or the United Kingdom. In Japan, output per man-hour grew by 290% during 1960–76, as compared with about 60% in the United States. Christenson, Cummings and Jorgensen have estimated that the rate of increase of total productivity during 1960–73 was lower in the United States than in Canada, France, Germany, Italy, Japan, Korea, the Netherlands or the United Kingdom. Although the productivity level in other countries still tends to be less than in the United States, results of this sort have been interpreted as evidence of a decline in the American technological lead.

Second, R and D expenditures as a percentage of gross national product have increased in Japan, Germany and the Soviet Union, while they have decreased in the United States. But this evidence should be treated with caution. Although the percentage of gross national product devoted to R and D has increased in these countries relative to the United States, the US percentage has been higher than any country other than the Soviet Union, where the figures are not comparable, and probably inflated relative to ours. Also, much of the industrial R and D in some major foreign countries is done by US-based multinational firms. One-half of the industrial R and D in Canada and one-seventh of the industrial R and D in Germany and the United Kingdom were carried out by US-based firms in the early 1970s. And more fundamentally, a nation's rate of economic growth depends on how effectively it uses both foreign and domestic technology, and this may not be measured at all well by its ratio of R and D spending to gross national product.

Third, the National Science Foundation has reported a study which indicates that the United States originated about 80% of the major innovations in 1953–58, about 67% of the major innovations in 1959–64, and about 57% of the major innovations in 1965–73. Without knowing more about how the sample of innovations was drawn, it is hard to tell whether the apparent reduction in the proportion of innovations stemming from the United States is due to the sampling procedures. Also, some attempt might have been made to weight these innovations. Nonetheless, taken at face value, the results indicate a reduction in the US technological lead over its international competitors.

RETURN FROM INVESTMENTS

Among economists, there has been considerable concern for at least a decade over the rate of US innovation. A number of economists, on the basis of their econometric studies, suspect that we are underinvesting in research and development. To obtain clues as to whether there is such an underinvestment

PROFESSOR EDWIN MANSFIELD holds a chair of Economics at the University of Pennsylvania. A graduate of Dartmouth College, he received his MA and PhD degrees from Duke University. Before joining the University of Pennsylvania faculty, he taught at Carnegie-Mellon, Yale, Harvard and the California Institute of Technology. He has been a consultant to many public agencies and industrial firms, including the US Department of Commerce, the National Science Foundation, the Congressional Office of Technology Assessment, the OECD and the RAND Corporation. Professor Mansfield was the first US economist invited by the People's Republic of China to lecture there. He is the author of 18 books and over 100 journal articles.
Address: Department of Economics, University of Pennsylvania, 3718 Locust Walk CR, Philadelphia, Pennsylvania 19104, USA.

and, if so, where it may be most severe, economists have attempted to estimate the social rate of return from investments in various kinds of new civilian technology. These estimates provide some idea of what society has received from such investments in the past. The earliest studies of this sort were carried out concerning agricultural R and D. Until recently no such estimates were made for industries other than agriculture. To help fill this gap, my coworkers and I estimated the social rate of return from the investments in seventeen industrial innovations, which occurred in a variety of industries.[3] Most of these innovations were of average or routine importance, not major breakthroughs. The results indicate that the median social rate of return from the investments in these innovations was 56%, a very high figure. The median private rate of return was 25%.

Using econometric techniques, a number of studies have estimated the marginal rate of return from various kinds of R and D. My own studies indicated that the marginal rate of return was about 40% or more in the petroleum industry and about 30% in the chemical industry if technical change was capital embodied, but much less if it was disembodied. Jora Minasian's study indicated about a 50% marginal rate of return on investment in R and D in the chemical industry. In agriculture, Zvi Griliches found that the marginal rate of return from R and D was 53%, Robert Evenson found it was 57%, and Theodore Schultz found it was 42%.

Put briefly, practically all of the studies carried out to date indicate that the average social rate of return from investments in new technology in both manufacturing and agriculture has tended to be very high. Moreover, the marginal social rate of returns also seems to have been high, generally at least 30%. To economists this is a sign that there may be some underinvestment in R and D. As I have stressed elsewhere,[4] there is a variety of important limitations and problems inherent in each of these studies, and it would be unwise to put too much weight on them. Nonetheless, the fact that so many studies based on quite different techniques and data come to such similar conclusions cannot be ignored.

POSSIBLE REFORMS

To help deal with whatever underinvestment may exist in civilian technology, at least three kinds of mechanisms are suggested frequently: (1) more government grants and R and D contracts to industry and universities, (2) increased use of government laboratories for such purposes, and (3) tax credits for private industry. Our results indicate some of the problems in each of these mechanisms.[5] With regard to grants and contracts, there is the problem that benefits and costs of various kinds of R and D are very hard to forecast. Even major corporations find it difficult to use various forms of cost-benefit analysis for R and D project selection, although they have a benefit concept that is much easier to estimate than most government agencies do. To illustrate how far off benefit estimates are, the chances were about 50/50 that a new product's, or new process's, estimated discounted profits would be more than double, or less than one-half, the actual discounted profits in one major firm we studied. In a situation of this sort, estimates might be biased for a variety of parochial, selfish or political reasons, the consequence being a distortion of social priorities.

With regard to the increased use of government laboratories to promote civilian technology, our results suggest the difficulties in having R and D conducted by organizations that are not in close touch with the marketing and production of the product. It is very important that there be unimpeded flows of information and good coordination of R and D, on the one hand, and marketing and production, on the other. Otherwise the R and D is likely to be misdirected, or even if it is not, it may be neglected or resisted by potential users. This is a difficult problem for various divisions of a firm, and it would seem to be exacerbated if the R and D is done in government laboratories.

With regard to tax credits, it is evident that they would reward firms for doing R and D that they would have done anyway, that they would not help firms with no profits, and that they would encourage firms to define R and D as widely as possible. Some of these problems might be solved by a tax credit for increases in R and D, but many problems remain. For example, firms would still have an incentive to redefine R and D, and for firms that can appropriate little of the social returns from new technologies, R and D would still be unprofitable even if the tax credit existed. Further, R and D is only part of the total cost of innovation.

PROCEED WITH CAUTION

Given the many gaps in existing knowledge, it is difficult, if not impossible, to specify what combination of these (and other) mechanisms would be most effective in compensating for whatever underinvestment there is in civilian technology. But it seems likely that any such civilian technology program should be neither large-scale nor organized on a crash basis, that it should not focus on helping sick industries – merely because they are sick – that it should not get the government involved in the latter stages of development work, that a proper coupling should be maintained between technology and the market, and that the advantages of pluralism and decentralized decision-making should be recognized. Given the current uncertainties, which, as I have tried to indicate, are great, it would seem sensible to proceed with caution, and to build into any program the capacity and necessity to resolve many of the key uncertainties before too big a commitment is made.

In conclusion, it should be recognized that our nation's technology policies cannot be divorced from its economic policies. Measures which encourage economic growth, saving and investment, and price stability are likely to enhance our rate of innovation. And measures which reduce unwarranted regulatory burdens are likely to do the same. Our current technological problems are due in considerable part to developments outside science and engineering, and it seems likely that they can and should be resolved in large part by measures influencing the general economic climate in the United States. Indeed, improvements in our general economic climate may have more impact on the state of US technology than many of the specific measures that have been proposed to stimulate innovation.

LITERATURE CITED

1. For a more detailed discussion of the productivity slowdown, see E. Mansfield, 'Technology and productivity in the United States: developments and changes in the postwar period', Conference on Postwar Changes in the American Economy, National Bureau of Economic Research (1980).
2. E. Mansfield, Economic growth or stagnation: the role of technology, in Jules Bachman (ed.), *Economic Growth or Stagnation*, ITT Key Issues Lecture Series, Bobbs Merrill, Indianapolis (1978).
3. E. Mansfield, J. Rapoport, A. Romeo, E. Villani, S. Wagner and F. Husic, *The Production and Application of New Industrial Technology*, W. W. Norton, New York (1977).
4. E. Mansfield, Contribution of R and D to economic growth in the United States, *Science* (4 February 1972).
5. For a more detailed discussion, see E. Mansfield, Federal support of R and D activities in the private sector, in *Priorities and Efficiency in Federal Research and Development*, Joint Economic Committee of Congress, (29 October 1976). The following paragraphs closely parallel part of my talk on 'The Economics of Innovation' at the American Chemical Society in 1979.

The manuscript is based on a lecture given at the Fourth Franklin Conference in Philadelphia on 29–30 November 1979

Drugs for Tropical Diseases in the Third World

PROFESSOR DR MED. W.-H. WAGNER

Former Head of Department of Chemotherapy, Hoechst AG, Frankfurt am Main, Federal Republic of Germany

The relationship between the World Health Organization's Special Programme for reducing tropical diseases and the pharmaceutical industry is the author's main theme. The problems of malaria, schistosomiasis, filariasis, trypanosomiasis – both types – leishmaniasis and leprosy are surveyed with references to their chemotherapy, the search for new drugs and possible immunity. It is concluded that existing drugs are far from perfect, but until better ones have been discovered, they represent the best treatment for most of the diseases reviewed.

Even in infectious diseases there is a north–south gradient: whereas the intensive use of effective chemotherapeutics and other measures, such as hygiene, adequate medical care and enough hospital beds, is largely successful in checking the majority of infections in northern countries with a high standard of living, the opposite situation is observed in many tropical countries. In the so-called developed countries most bacterial infections have been reduced to a considerable degree and diseases caused by protozoa and helminths have nearly vanished, with the sole exception of trichomad-caused infections. On the other hand, in many countries of the Third World, tuberculosis, common bacterial infections, such as salmonellosis, shigellosis, as well as wound and other infections, are prevalent; tropical infections add to this serious situation.

It is the aim of this review to deal with the specific problem of drugs for tropical infections. Therefore, firstly, what are tropical infections?

GEOGRAPHICAL SHIFT OF DISEASES

In the relevant textbooks[60,63,117] many infections caused by blood and tissue protozoa, spirochaetes, spirilla, bacteria, viruses, helminths or fungi are dealt with as tropical infections. The majority of them occur mostly, but not exclusively, in tropical countries. In the strict sense of the word, tropical diseases are those which flourish in tropical or subtropical climates only. Many of them require intermediate hosts to maintain the complex life-cycles of the parasites which give rise to infection. Here a good example is schistosomiasis, an infection caused by minute trematodes (helminths) which inhabit the circulatory system of man. Intermediate hosts are snails living in warm water or moist soil or on vegetation.

Various forms of filariasis also belong to this group of infections. Causative organisms are certain nematode worms (filariae) whose adults inhabit the lymphatics, the connective tissues or the mesentery. They produce live embryos (microfilariae) which can cause pathological manifestations. Vectors and transmitters are mosquitoes, or 'buffalo gnats' (*Simulium* species) in the case of onchocerciasis, the most dangerous form of filariasis (river blindness). African trypanosomiasis (sleeping sickness) and South American trypanosomiasis (Chagas's disease), leishmaniasis, paragonimiasis and clonorchiasis (both caused by helminths) also belong in this group, as well as other infections. These climatically confined diseases – the classical group of tropical infections – and several other important infections have by now nearly or completely vanished in the highly industrialized countries and are today encountered predominantly in the tropics. Other infections, however, must also now be considered as true tropical infections.

Malaria, formerly widespread in Europe, has virtually disappeared in countries with a temperate climate and exists almost exclusively in the tropics. The vectors – *Anopheles* mosquitoes – occur in many countries with a temperate climate but no longer harbour the infective organisms, the plasmodia.

Leprosy, a bacillary infection, also belongs to the category of diseases previously encountered in countries with a temperate and even a cold climate, for example Norway; they are now met practically only in developing countries with a tropical climate.

Some other diseases are undergoing a geographical shift and are now prevailing in tropical countries. Tuberculosis, once so menacing in Europe, has lost many of its dangers and fatal consequences in the developed countries; this is due to appropriate measures developed during more than a century of investigation and, above all, thanks to successful chemotherapy. In the developing countries, however,

CCC–0308–0188/80/0005–0186 $09.00/1

it is still a high-risk infection – never more so than now.

These very few examples demonstrate that the term tropical infections is not only restricted to a few infectious diseases which require parasites with complex life-cycles and occur only in regions with high temperatures – tropical infections *sensu strictiori*. 'Tropical infections' covers more diseases than that.

SOCIO-ECONOMIC FACTORS SHIFT DISEASES

There are several factors which increase the significance of tropical infections. Certain tropical infections are increasing in numbers in certain countries despite all the efforts made. Schistosomiasis should once again be cited as an example. Irrigation, intended to make land available for cultivation, can lead to the spread of water snails which are intermediate hosts of this infection. The same is true of onchocerciasis. The vectors of this systemic helminthiasis, the simulia, prefer to live near reservoirs, either completed or under construction. The development of such irrigation projects therefore leads to the unintentional dissemination of the disease, as was observed for example in the Volta basin in West Africa.

An important reason for the changing patterns of tropical diseases[96] is the rapid increase in urban populations. This is not only encountered in Africa, where urban populations are growing at least twice as fast as the total population, but also in regions of Asia and South America. This phenomenon has serious consequences for the spread of disease – squatters, poor housing and water supply, and adverse effects on mosquito-borne disease.

The causative microorganisms in some diseases have lost their susceptibility to chemotherapeutic drugs hitherto used successfully. This phenomenon is only too well known from many bacterial infections. *P. falciparum*, the causative organism of subtertian malaria, the most dangerous form of this disease, has in some regions, for example in South-East Asia, become resistant to the 4-aminoquinolines, for a long time the standard chemotherapeutics.

There are some important interrelationships between health and socio-economic development which make the effective control of these diseases particularly necessary.[58] Poverty, malnutrition, low income, high birth-rate, illiteracy and disease are associated. Disease is both the cause and the result of poverty, and experience shows that the improvement of any one of these conditions, for example health, can have a positive effect on the other conditions. The vicious circle has to be broken.

There is another important reason for controlling tropical infections effectively and systematically: many of the developing countries are predominantly agricultural and therefore economically dependent on agricultural exports. But tropical infections reduce productive capacity and therefore inhibit socio-economic development.

WHO'S SPECIAL PROGRAMME

The World Health Organization is particularly concerned with the health problems of the Third World and has recognized the threat posed by this situation. WHO has wide experience in campaigns against specific diseases. A very good example is the most successful smallpox eradication campaign, but experience has also shown that the malaria eradication campaign which started in 1957 did not result in a similar success. On the contrary, despite many efforts, malaria is now increasing again in some regions where eradication was thought to have been achieved. Therefore, a few years ago WHO drew the necessary conclusions and, together with UNDP, set up the 'Special Programme for Research and Training in Tropical Diseases', thus acting on a broader base. Working on a global scale, this Programme, now co-sponsored by UNDP, the World Bank and WHO, is conceived as a long-term endeavour lasting twenty years or more. [See *Interdisciplinary Science Reviews* **3**, 220 (1978).]

The main objectives of the Special Programme are the development of new drugs, vaccines and biological methods of dealing with intermediate hosts and carriers, as well as simple diagnostic tests. These remedies should and must be new, effective and cheap. In order to avoid a fragmentation of its resources and activities, WHO has set up a list of six infections to be given priority; they are listed here in order of epidemiological significance: malaria; schistosomiasis; filariasis (including onchocerciasis); trypanosomiasis (including African sleeping sickness and South American Chagas' disease); leprosy; and leishmaniasis. These diseases are, with the exception of leprosy, not simple disease-entities but groups of different specific diseases.

They were chosen for the Special Programme because they include most of the major human parasitic infections encountered in tropical regions. They are predominantly infections of long duration and lead in many cases to chronic, disabling or

PROFESSOR DR MED. WOLF-HELMUT WAGNER was Head of the Chemotherapeutical Department of Hoechst AG from 1964 to 1978; he is Professor at the University of Frankfurt am Main (1974) and received the Paul Ehrlich Prize (1952). He is the author of many scientific and medical publications as well as member of several scientific societies.
Private address: Im Paradies 25, D-8993 Nonnenhorn, Federal Republic of Germany.

disfiguring conditions as in leprosy, American leishmaniasis, river blindness and elephantiasis (both of which are caused by filariae), urinary schistosomiasis and so on. Most of them are naturally transmitted only in the tropics and the majority are caused by parasites with complex life-cycles:

- malaria, trypanosomiasis, leishmaniasis, caused by insect-borne protozoa;
- schistosomiasis, filariasis, caused by snail- or insect-borne helminths;
- leprosy, caused by intracellular bacteria.

More than one thousand million people are endangered by these in Africa, Asia, the Mediterranean region, as well as in Central and South America. Malaria, schistosomiasis and filariasis each menace more than 200 million people.[57] These diseases, interfering in many ways with the development of tropical countries, have not been brought under control by the means and methods currently available.

THE ROLE OF THE PHARMACEUTICAL INDUSTRY

Within the Special Programme elaborated by WHO, the development of new chemotherapeutic preparations is of considerable importance. This is the task of the pharmaceutical industry with the facilities and the know-how to do so. In recent decades the pharmaceutical industry has been developing new remedies in the area of infections covered by the Special Programme. However, it is impossible to list them all here. Pharmaceutical research did not neglect this area but as the licensing regulations and the registration authorities became stricter, new developments in the field of tropical infections also were drastically hampered. On the other hand, and for many reasons, available drugs in the developing countries are not used as they should be.

The main instrument for achieving the aims of the Special Programme are the Programme's Scientific Working Groups, the SWGs. They are involved in research and development of new agents to deal with the above-mentioned diseases. They seek the collaboration of the pharmaceutical industries wherever this is helpful in attaining the goals of the Special Programme.

It is of interest to know to what extent the activities of WHO are matched by the available funds. The estimated sum for Part II, Research and Development, of the Programme Budget[115] amounts to a total of US $ 100 516 000 (1978–1981). This comprises 'the costs of the development of improved preventive, diagnostic, therapeutic and vector control methods specially suited to prevent, treat and control the selected tropical diseases . . . '.[115] On closer analysis it becomes apparent that research in just one specific area, or the development of a single new medicine through all the necessary – the preclinical and clinical – stages, will consume a not inconsiderable proportion of this overall sum. It is certainly not adequate to cope with the magnitude of the task and it can only be considered as a beginning.

Government agencies, as for example the relevant Ministries in the Federal Republic of Germany, have also intervened to support individual cases of research into tropical medicine by the pharmaceuticals industry. However, despite the goodwill of all, one has to observe that, at present, we are only at the first stage of collaboration. It has as yet barely begun and it will be years before the right way is found.

It is impossible here to go into matters of such practical importance as the correct use of suitable drugs, matters intimately connected with the training of doctors and their assistants as well as administrative, organizational and, of course, financial problems. But apart from this there can be no doubt of the need to develop new chemotherapeutic substances as the experience of recent decades has clearly shown that successful control of tropical infections cannot be achieved by existing means and technology.

WHO is aware of the need to have a close association with the pharmaceutical industry in the development of new chemotherapeutical agents and therefore seeks co-operation with industry. In this context it has to be considered that there is not only an increase in development costs of all drugs but also the uncertainty of return on investment for drugs to control tropical diseases. This continues to limit the research investment by the pharmaceutical industry. To overcome these fundamental obstacles, at least partially, the Special Programme is ready to 'make available the scientific and technical information and the needed facilities directly related to industrial needs for production development and application. To obtain the knowledge necessary for technical innovation, the programme is investing in research.'[109]

Five types of involvement with the industrial sector are envisaged:

- participation by industry-affiliated scientists in Scientific Working Groups;
- screening of agents through Scientific Working Group projects by laboratories of industry and non-profit institutes;
- contracts for technical services;
- clinical evaluation of new drugs and vaccines; and
- training of scientists and technicians.[109]

THE ARMAMENTARIUM

The available tools to control the six diseases are: prophylactic and therapeutic drugs, vaccines, vector control methods, pesticides and insecticides and, particularly, better hygiene. Here, drugs will be dealt with as the main topic. The existing armamentarium should be completed by drugs for mass treatment, including children, and devoid of toxic or unpleasant side-effects. Chemoprophylactic drugs are urgently needed. Further, in cases where resistant strains, for example in tropical malaria, have emerged, new and effective preparations are to be developed.

De Maar[57] describes the properties of an ideal control drug. It should be effective, preferably orally, in a single dose and fail to generate resistance; be stable in tropical climates without special precautions; be so safe that it does not require patient selection or patient follow-up; require a minimum of specialized supervision or skills in mass campaigns; be adaptable for integration into the existing health services; have an acceptable benefit–risk ratio in the country; be marketable at a cost that the developing countries could bear and which is competitive with other control methods; and finally, have a viable commercial market. Obviously, it is very unlikely that all these requirements can be met by any single drug.

Until now, no vaccines for the major parasitic infections have become available. A main obstacle to the development of antiparasitic vaccines up to now has been the lack of methods for large-scale parasite cultivation,[57] though this is now changing. Malaria is a good example in this connection. Also, the risk of environmental pollution and the development of resistance to insecticides of the target organism have proved to be big obstacles to better vector control methods.

Finally, it has been a commonplace to say that better hygiene, a more effective education and better understanding of the complex causes of the various parasitic infections could eradicate, or at least reduce to a certain extent, the incidence of these diseases. However, experience has shown that, despite many efforts, such endeavours are not very effective.

Following this general review of the problems of developing remedies for tropical diseases, the present situation of the six diseases named in the Special Programme must be considered.

MALARIA

Malaria is an infection caused by protozoan parasites; it is transmitted to man by female mosquitoes of the genus *Anopheles*. The parasites have a complicated life-cycle. Their development takes place in mosquitoes and in man (Figs. 1 and 2). The different stages of the life-cycle of the parasites may show different degrees of susceptibility to chemotherapeutic agents. There are four species which cause human malaria: *P. falciparum*, *P. vivax*, *P. malariae* and *P. ovale*. The most dangerous form is falciparum malaria (malignant tertian malaria or subtertian malaria); it causes a high fatality rate in non-treated persons, mainly in non-immune or insufficiently immune patients. Infants, young children and pregnant mothers are a high risk population in endemic areas.

Though massive antimalaria campaigns were successful in some areas, there are still 343 million people in endemic areas not yet protected by any specific antimalarial measures; most of these live in Africa.[108a] Previously, malaria extended over a much wider area and some decades ago, malaria was endemic in regions where more than two-thirds of the world's population lived. In 1940 more than 300 million people were estimated to suffer each year from clinical attacks of this disease. After 1957 a global malaria eradication campaign was undertaken which resulted in the disappearance of the infection in many countries. However, since 1967 the progress of antimalarial activities was adversely affected in Asia, Central and South America and South-East Europe, and malaria epidemics re-occurred in regions which had been freed from it. Responsible for this re-occurrence are inflation, economic and energy crises, mosquitoes' resistance to current insecticides as well as parasites' resistance to chemotherapeutic drugs, and the over-zealous banning of DDT.

Table 1 shows the world malaria situation in 1976, and Table 2 depicts the evolution of the malaria situation in the years 1972–1976 in various regions.[115a] The figures given are based only on cases that have been confirmed in a laboratory and reported by the malaria services and therefore represent 'a considerable understatement of the actual situation'.[115a] They also show that the malaria situation has progressively deteriorated in several countries, though in other countries where malaria was endemic, hundreds of millions of people have been freed from the potential risk of this infection.

It has been estimated that in tropical Africa alone, malaria causes the death of one million children under the age of 14 years. Thus, 'malaria continues to be a major public health problem in the majority of tropical countries'.[108a] Furthermore, malaria is a major hindrance to economic progress in the afflicted

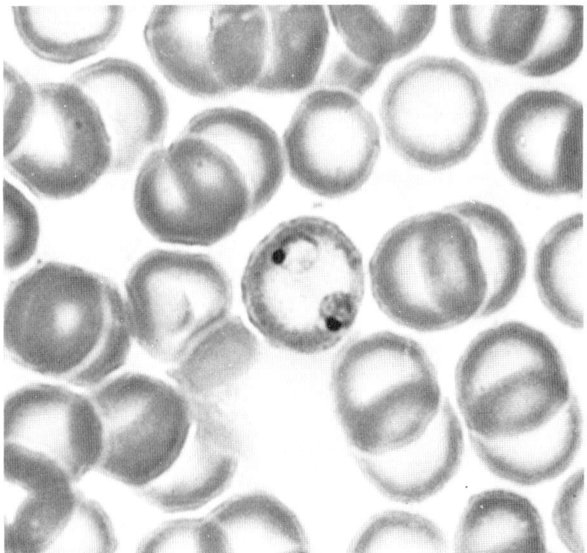

Figure 1. *Plasmodium cynomolgi*, the causative organism of rhesus monkey malaria (Giemsa stain; 1000×). Normal erythrocytes (red blood corpuscles) and parasitized ones (centre, two 'ring' forms).

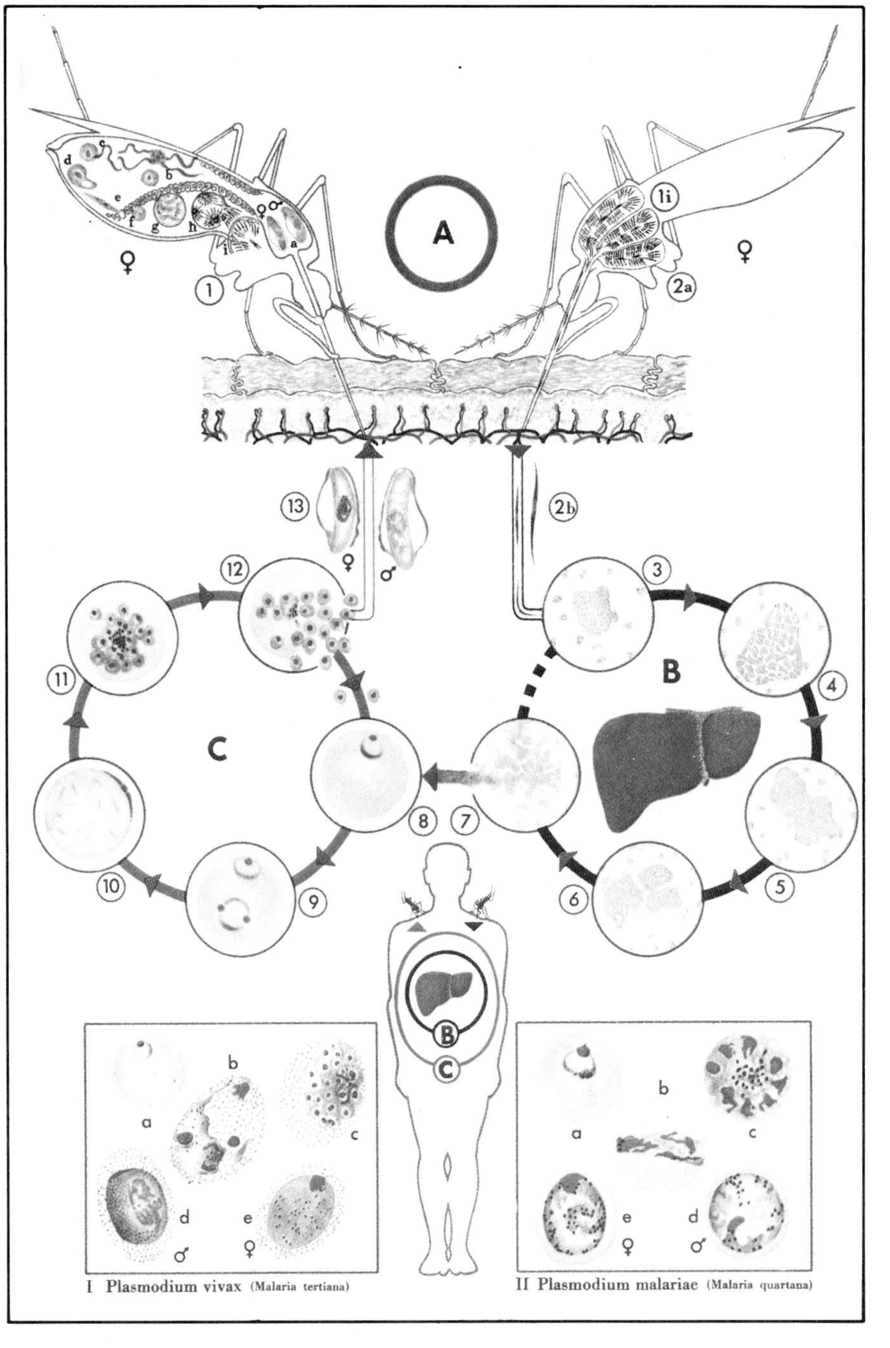

Figure 2. Life cycles of plasmodia, causative organisms of malaria. (A) Sexual development (gametogony) in the female *Anopheles* mosquito: (1a) gametocytes; (b) gametes; (c) fertilization; (d) zygote (so-called retort form); (e) ookinete; (f) oocyst; (g) sporocyst; (h) mature sporocyst after it has burst open; (i) sporozoites in the cells of the salivary glands. (2a) The biting *Anopheles* transmits sporozoites; (B, 3–7) Pre-erythrocytic development in the cells of the liver. (C, 8–13) Erythrocytic development (schizogony). Typical stages in the peripheral blood: (I) *Plasmodium vivax*; (II) *P. malariae*. (a) young ring-shaped schizont; (b) multinucleate schizont; (c) final stage of schizogony (the so-called morula (I) or 'daisy' (II)); (d) male gametocytes; (e) female gametocytes. (Courtesy Professor Piekarski, Ref. 71a).

countries not only due to morbidity and mortality directly caused by the disease, but also due to chronic impediment of health in rural populations.[110]

There are many causes for the inadequate efficiency of the malaria campaign: the rising costs of insecticides and larvicides for combating the mosquito vectors, inadequate development of health services and health manpower, poor awareness by the rural population of the basic principles of public health and other plural causes. But these constraints are not only specific for malaria but also relevant for other tropical infections.

MALARIA CHEMOTHERAPY

Successful chemotherapy depends on choosing the right drug for the right species of parasite at the right time. It is essential to make a correct diagnosis, particularly so if the patient is suffering from falciparum malaria. Chemotherapy against falciparum malaria is one of the most urgent problems and the possibility of resistant strains has to be considered when chemotherapy is started.

The aim of chemotherapy is eradication of the different stages of the parasite. In *P. falciparum*, where there is no persistent infection in the liver, the eradication of the asexual erythrocytic stages is the objective of treatment, namely a radical cure. With *P. vivax*, *malariae* and *ovale* infections, the situation is more complicated because there are persistent exoerythrocytic stages. For curing these forms of malaria, eradication of erythrocytic and exoerythrocytic plasmodia is necessary.

Table 3 shows a variety of antimalarial drugs currently in use; it also gives an impression of the

Table 1. The World Malaria Situation (January–December 1976)

Area	Total population (million)	Malaria situation (1976)
Africa north of the Sahara	83	235 cases (total number decreased from 1974 to 1975)
Africa south of the Sahara	335	284 million people live in areas where the risk of malaria is moderate to high. No notable change in the situation
America north of Mexico	239	No malaria
Central America	111	50 million people live in malarious areas (37% in areas of low endemicity) El Salvador and Nicaragua: high malaria incidence Honduras: 49 000 cases (1975: 30 000)
South America	220	61 million people live in areas with various levels of malaria risk
Europe, including Turkey and USSR	775	Turkey, outbreak with around 37 000 cases
Asia west of India	178	144 million people live in malarious areas, 95% of which are protected by antimalaria measures
Bangladesh, India, Maldives, Nepal, Sri Lanka	723	More than 6.2 million malaria cases
Eastern Asia and Oceania	530	281 million peoples live in malarious areas, of whom 71% are protected by antimalaria measures Burma and Phillipines: increasing numbers of cases

From Ref. 114a (abbreviated).

Table 2. Number of Autochthonous Malaria Cases (in thousands) Reported During 1972–1976, by Region (excluding Africa)

Region	1972	1973	1974	1975	1976
The Americas	284	280	269	356	379
South-East Asia	1920	2694	4210	5992	6539
Europe	21	13	8	12	39
Eastern Mediterranean	855	883	524	447	350
Western Pacific	171	203	170	197[a]	210[a]
Total	3251	4073	5181	7004	7517
1972 = 100	100	125	159	215	231

[a] Excluding China, Democratic Kampuchea and Vietnam.
From Ref. 115a.

Table 3. Antimalarial Drugs Currently in Use

Drugs acting on asexual parasites (schizontocides)	tissue forms of parasites	sexual forms of parasites
Quinine	Proguanil	8-Aminoquinolines
4-Aminoquinolines (e.g. chloroquine)	Pyrimethamine	(gametocytocidal activity against all species)
Acridines (mepacrine)	8-Aminoquinolines (e.g. primaquine)	Quinine, mepacrine and 4-aminoquinolines (gametocytocidal activity against sexual forms of *P. vivax, malariae, ovale*, but not *falciparum*)
Cycloguanil embonate (long-acting drug; embonate of a metabolite of proguanil)		
Pyrimethamine (closely related to the metabolite of proguanil)		
Sulphonamides		
Sulphones		

complexity of the chemotherapy of this infection. The most commonly used antimalarial is still chloroquine in cases of acute malaria of all kinds.

Immediate specific chemotherapy is especially necessary in falciparum malaria due to the severity of this infection. In uncomplicated and complicated cases chemotherapy aims at the eradication of asexual forms, and then the schizontocides (4-aminoquinolines, quinine, mepacrine) are used. If drug-resistant strains of *P. falciparum* are diagnosed, a combination of different drugs (pyrimethamine, sulphones, sulphonamides) is the chemotherapy of choice. Resistance of *P. falciparum* to 4-aminoquinolines, the most frequently used schizontocides, is already widespread in South-East Asia, South America and the Western Pacific regions; this phenomenon has, however, not yet been observed in Africa.

The many drugs available for chemotherapy of malaria may suggest that the problems in this field are solved. However, there are gaps to be filled, and new drugs are badly needed. Besides the lack of a well-tolerated, cheap drug with high activity against 4-aminoquinoline-resistant *P. falciparum* strains, an inexpensive long-acting schizontocidal antimalarial would facilitate treatment with a duration of the chemotherapeutic effect for a least one month after administration of a single oral dose.

SEARCH FOR NEW ANTIMALARIALS

In 1963 the Walter Reed Army Institute of Research, Washington, DC, started a major malaria research programme.[8,86] Particular emphasis was laid on the development of new drugs effective against infections with chloroquine-resistant strains of *P. falciparum* due to the severity of this infection and the problems involved in chemotherapy.

More than 250 000 compounds from many sources were investigated for their antimalarial activity during the following years, and it was demonstrated that certain 4-quinolinemethanol derivatives showed the most interesting properties. Besides other compounds of this series, compound WR-142 490, subsequently named mefloquine, displayed in animal experiments essentially equal activity against *P. falciparum* strains either resistant to chloroquine or pyrimethamine. Furthermore, it could be shown that this compound was more effective than chloroquine against chloroquine-susceptible strains. It was well tolerated in these experiments.[91] In principle, the findings in animal experiments could be paralleled by studies in human volunteers and field-type studies, and mefloquine is now in stepwise evaluation in human volunteers and in the field. If this compound proves to be not only active against many susceptible and resistant strains but also devoid of toxic side-effects in large-scale clinical and field trials, it could assume the role of a generally useful blood schizontocidal drug.

Three phenanthrene methanols, two pyridine methanols and a substituted *ortho*-cresol were also active or more active than chloroquine against infections with susceptible strains and were fully effective when tested against infections with chloroquine- and/or pyrimethamine-resistant strains.[90a] It cannot be decided as yet which of these compounds will be most useful to treat infections in human patients.

In this connection, it should be pointed out that success in chemotherapy is based not only on the search for new molecules with antiparasitic activity but also on biological progress, and malaria is a good example of this. For testing new compounds, animals are required (test-tube experiments are of no value) and birds, rodents, rhesus and other monkeys, infected with avian, rodent or simian plasmodia must be used. However, their plasmodia respond to chemotherapy in different ways when compared with human plasmodia. It was therefore real progress when a certain species of small monkeys (owl monkeys, *Aotus trivirgatus*) was found to be susceptible to infection with human plasmodia, in the first place *P. falciparum*.[90,119]

Another new compound from a series of acridon derivatives,[76,77] HOE 991 (Floxacrine), displayed better activity in rodent malaria than chloroquine against asexual parasites after subcutaneous application, and strains resistant against different antimalarials were also susceptible to it. In rodent malaria the compound showed a true prophylactic effect, and its injection prior to inoculation of mice with sporozoites protected the animals against infection. It is now undergoing thorough preclinical investigation (toxicology and pharmacology) to warrant stepwise evaluation in man.

MALARIA IMMUNITY

Besides the physical and chemical control of mosquito breeding, the spraying of houses to kill mosquitoes and antimalarial treatment of infected patients, there are still other ways to combat malaria. One is vaccination against this disease. In endemic areas most of the indigenous population may be regarded as semi-immune as after repeated infections they have acquired resistance to superinfection and thus achieved some immunity. In experimental animals, immunity can be acquired by the use of parasites attenuated in various ways.[100,111] For immunization of experimental animals, parasitized erythrocytes, parasite fractions,[1,5,17,18,87,89,101,118,120] merozoites,[7,14,15,26,29,62,95,97] and inactivated or irradiated sporozoites[64–66] have been used. Further, exoerythrocytic merozoites of avian malaria[44] have been tested for this purpose. In these different experimental systems it could be demonstrated that immunization of experimental mice, rats, owl monkeys and rhesus monkeys can be brought about by several species and stages of plasmodia.

A major characteristic of the immune response is the strict stage-specificity; sporozoite-immunized animals are fully protected against challenge with sporozoites, whereas they remain fully susceptible to infections with the erythrocytic stages. Inoculation with merozoites led to a very high degree of immunization. It could also be found that inoculation with irradiated erythrocytic parasites, non-infective but still metabolically active, leads to strong immunity to subsequent challenge. This method could be used for immunization against human malaria, *P. falciparum*; a prerequisite would be an appropriate source of *P. falciparum*-infected erythrocytes: 'it can be envisaged that a method of continuous propagation of *P. falciparum* in suspensions of human 0-type erythrocytes maintained *in vitro* could supply such an antigen'.[111]

Vaccination of humans has been performed with attenuated sporozoites of *P. falciparum* through the bite of irradiated mosquitoes; this procedure led to a species-specific resistance to challenge.[10,11] Besides other constraints, the practical implementation of human sporozoite vaccination meets major difficulties, for example, isolation of sporozoites from salivary glands of mosquitoes, and i.v. inoculation, the most important being the production of enough sporozoites for mass vaccination.[12]

The currently most promising form of vaccine against human malaria is that of blood-stage merozoites.[13] As early as 1961 Cohen[14] suggested that merozoite antigens play an important role in specific immunity against malaria. The isolation of extracellular merozoites required separation of parasitized from normal erythrocytes and this can be done by different methods.[112] It is now possible to cultivate continuously the erythrocytic forms of *P. falciparum* with high yields *in vitro.*[99] The vaccine can be applied i.m. in conjunction with Freund's complete adjuvant.

To summarize, the former pessimism about vaccines against malaria has yielded to a more positive attitude, based on demonstrations of immunization of various animal models.[51,112] The recent successes in continuously cultivating human malaria parasites and thus obtaining enough material for antigen production adds to this change of view. However, there can be no doubt at the moment that chemotherapy is far ahead of vaccination.

SCHISTOSOMIASIS

A group of chronic parasitic infections caused by blood flukes is termed schistosomiasis. These helminths (trematodes) of the genus *Schistosoma* produce disease by eggs laid by adult worms in the small blood vessels of the human host. Schistosomes have a complicated life-cycle: the adult worms, 10–20 mm long, live in the veins of the portal system and in the mesenteric and pelvic veins. They have separate sexes. The female worm deposits the eggs in the vessels of different organs causing tissue damage of various degrees. The eggs are finally excreted from the human body with urine and faeces; others remain in the vessels and the tissues. The oviposition may lead to urinary and dysenteric symptoms and eventually serious complications can occur in the chronic stage of the disease. For the completion of the life-cycle, freshwater snails serve as intermediate hosts. When mature, the larval stages (cercariae) leave the snail to attack the next suitable mammalian host. In man, infection as a rule is acquired by cercarial penetration through the skin.

There are four species of schistosomes which can cause the disease schistosomiasis. Table 4 gives an overview of the main symptoms of the different forms of schistosomiasis, their geographical distribution and the localization of the flukes in the human host. However, clinical features are much more multiform than indicated in this table, from malaise, skin eruptions, pain on urination, haematuria, to severe affections of the kidney, bladder, ureter, urethra, male and female genital organs, the alimentary tract and many other organs. Though it is not completely proven, there is much evidence for the assumption of an aetiological relationship between schistosomiasis of the bladder and cancer of this organ.

Schistosomiasis is a tropical infection in the strict sense of the word. The intermediate hosts of *S. mansoni*, *S. haematobium* and *S. intercalatum*, the snails, live only where the water temperature is between 25 and 35 °C. Thus, schistosomiasis is confined to tropical and subtropical countries.

Schistosomiasis is a true tropical infection in other respects also. As a disease prevailing in rural areas, it undermines productive capacity and thus inhibits socio-economic development. In many countries, not only in Egypt, primitive irrigation systems are used as a base for agriculture. If the water is contaminated by human excreta, schistosomal eggs reach the water system and the larval miracidia may hatch and eventually reach the snail. After maturation within the snail host, the cercariae are ready to infect man, having emerged from their intermediate host. Thus, water, an essential base for agriculture, may serve as a source for infection and schistosomiasis becomes an occupational hazard for peasant farmers. Children are also endangered to a very high degree by these epidemiological patterns of infection and, at the same time, act as important human vectors by urinating indiscriminately in fresh water. The close connection between irrigation systems, water snails, and

Table 4. Different Forms of Schistosomiasis

Species	Geographical distribution	Localization in the human host	Main symptoms and signs
S. mansoni	Endemic in Arabian peninsula; hyperendemic in many parts of Lower Egypt and the Sudan, many countries of the sub-Saharan zone of Africa, South America	Mesenteric veins of the portal system	Diarrhoea, schistosomal dysentery, spleno-megaly and other symptoms
S. haematobium	Old World only: endemic in Africa south of the Sahara, hyperendemic in the Nile Delta and other countries	Mainly in the terminal small veins of the bladder	Mainly urinary
S. japonicum	Far East: Philippines, foci in the Celebes, Thailand, Lower Mekong basin, People's Republic of China, some foci in Japan	Similar to *S. mansoni*	Similar to *S. mansoni*
S. intercalatum	Central and West Africa	Veins (venous plexuses of the rectum and pelvis)	Usually mild disease; 'bowel infection'

infection of the human host explains why schistosomiasis is spreading and increasing in many countries. As irrigation grows more extensive, the greater are the chances for the propagation of the intermediate hosts and for infection of man.

As to the prevalence of schistosomiasis, the global estimate extends to 180–200 million infections.[19] Mortality attributable directly to infection is low. In contrast, it is the multiplicity of chronic sequelae of this infection which characterizes schistosomiasis as a dangerous, chronic infection with many serious implications to public health and to the individual.

There are several current control measures.[19] Besides reduction in contamination of natural waters by human excreta and control of the snails serving as intermediate hosts by attack with chemical molluscicides, chemotherapy is a fundamental control measure.

SCHISTOSOMIASIS CHEMOTHERAPY

Control here means – and of course also in connection with other tropical infections – the reduction of the disease to a level where it is no longer a major health problem. Eradication specifies the total cure of all diseased patients and the prevention of transmission by chemotherapy. Therefore, chemotherapy in schistosomiasis has two main tasks: mass treatment, thus interrupting the transmission of the disease and preventing schistosomiasis in adults by their cure in childhood; and as a curative measure at individual level.

Chemotherapy in the form of mass drug treatment of infected populations would be the best way to eliminate adult schistosomes, thereby reducing or completely eliminating egg output from man. Thus, chemotherapy could play an important role in controlling the disease. However, schistosomicidal drugs available at present are too toxic for this purpose and the search for new and better tolerated schistosomicidal drugs is an imperative necessity. Such drugs could cover a far wider population in control campaigns than is currently possible. The search in this direction is, therefore, a most important aim in drug development.[33,114]

For individual treatment there are different drugs available,[55] as shown in Table 5. Chemotherapy with specific drugs against schistosomiasis dates back to the years before the First World War when emetine, not included in Table 5, was used first in Egypt. Its toxicity and low therapeutic value led to its abandonment. 2-Dehydroemetine,[117] developed a long time after the introduction of emetine, was less toxic than the former, but it too could not solve the chemotherapeutic problem. Antimonials have been in use since the early 1920s, and still are. It was a great advance when Miracil D[55] was found to be effective after oral administration against *S. mansoni*, as its molecule was devoid of antimony. Other schistosomicidals, also non-antimonials, were developed

Table 5. Chemotherapy of Schistosomiasis

Compounds	Class of compounds	Application	Side-effects and toxic manifestations	Efficacy
Tartar emetic	Antimonial	Intravenous	Nausea, vomiting, rheumatic and muscular pains	Therapy with antimonial drugs gives better results with *S. haematobium* than with *S. mansoni* and *S. japonicum*.
Stibophen (Fouadin, Neoantimosan)	Antimonial	Intramuscular	Electrocardiographic changes and others	In children there is better tolerance than in adults; however, relapses among them are more common
Anthimalin	Antimonial	Intramuscular, intravenous		
Triostam	Antimonial	Intravenous		
Astiban	Antimonial	Intramuscular		
Lucanthone (Miracil, Nilodin, Tixanthone)	Non-antimonial	Oral	Gastro-intestinal disturbances, giddiness, headache, insomnia	Therapeutic effect in most cases inferior to that of antimony compounds (*S. mansoni* particularly!); hardly any effect in *S. japonicum* schistosomiasis
Niridazole (Ambilhar)	Non-antimonial	Oral	Anorexia, vomiting, abdominal pains, side-effects related to the central nervous system (muscular tremors, psychotic episodes, but rarely epileptiform seizures)	Remarkable efficacy against *S. haematobium*, *S. mansoni* and *S. japonicum* as well
Hycanthone	Non-antimonial	Intramuscular	Hepatotoxicity	Similar to Lucanthone

after Miracil D, as for example Niridazole. All are still in use.

Besides these drugs there are now some most interesting new developments: Metrifonate, an organophosphorus anticholinesterase compound, is effective against *S. haematobium* but less active against *S. mansoni*. The drug is applied orally. Clinical trials confirmed the animal experiments; metrifonate was well tolerated.[20,27,32,47,48] Oxamniquine, a member of the novel series of 2-amino-methyltetrahydroquinoline derivatives,[82] displays good activity against *S. mansoni* in animal experiments but not against *S. haematobium*. Therapeutic effects are better after intramuscular than after oral application. Clinical investigations resulted in the cure of 95% of patients infected with *S. mansoni*. However, local pains after i.m. injection and adverse side-effects after oral application were observed.[28,49,59] Therefore, neither metrifonate nor oxamniquine meets all the requirements of an ideal schistosomicidal drug.[114]

4-Isothiocyanato-4′-nitrophenylamine (C 9333-Go/CGP 4540) displays activity against schistosomes in various animal hosts including primates after a single oral dose and proved equally effective against *S. haematobium*, *S. mansoni* and *S. japonicum*.[96a] Further tests have to determine whether this drug can be introduced into the therapy of human patients.

Besides other compounds, praziquantel should be mentioned in this context. This compound is highly effective against all schistosome species parasitic in man.[37,71,107] Inter-strain variations do not influence the efficacy of praziquantel contrary to other compounds like hycanthone[54] and oxamniquine;[28] the drug can be applied orally.

Clinical studies carried out in Zambia, Brazil and the Philippines indicate that praziquantel is highly effective after a single dose or a one-day oral treatment against *S. mansoni*, *S. haematobium* and *S. japonicum*. The compound appears to possess most of the desired criteria of the ideal schistosomicide. If the preliminary clinical observations are confirmed by further trials, population-based chemotherapy of *Schistosoma* infections may become a practical reality in the near future.[98] Further therapeutic trials against all the three above species are underway to determine the most suitable dosage against each.

The success of the fight against schistosomiasis is based not only on potent drugs, molluscicides and other means, but also on education to their proper use. The lack of understanding of the complicated life-cycle of the causative organisms and the ignorance of even the simplest epidemiological facts are severe handicaps for a correct and successful application of the available drugs. It is therefore essential to establish training centres where post-graduate training of doctors, social anthropologists, biologists, administrators and others can take place so that better knowledge in the use of different weapons can diffuse. This is one of the major tasks of WHO, not only in the case of schistosomiasis, but also for other diseases of the Special Programme.

Immunity against schistosomiasis is a common experience in patients of middle or old age when it usually takes a mild course, whereas in children it is almost always severe. One can therefore assume an acquired immunity in humans after infection. However, it is much too early to speak definitely of the possibility of actively immunizing a population. There are fundamental questions to be answered, for example the nature of acquired immunity in man, its degree of efficacy and specificity, and its duration.

Chemotherapy undoubtedly plays a decisive role in the fight against schistosomiasis, and the development of new and better drugs remains a most important goal, now and in the future.

FILARIASIS

The term filariasis embraces a group of human infections caused by nematodes, round filarial worms. Table 6 gives a survey of the most important forms of filariasis. In addition to the four main diseases there are other widespread minor filarial infections, caused by *Dipetalonema perstans*, *D. streptocerca* and *Mansonella ozzardi*. These filariae commonly infect man, yet disease is rare; they are therefore not dealt with in this context.

The incidence[23] of the main infections is as follows: bancroftian and brugian filariasis, 250 million cases; onchocerciasis, 30 million cases; and loiasis, 1 million cases. As for other parasitological infections, it should be emphasized that infection does not necessarily mean disease. Filariae are round worms with separate sexes, the adult females measuring up to 7–10 cm (*Loa loa*, *W. bancrofti and B. malayi*) and even 70 cm (*Onchocerca volvulus*), whereas the males are shorter. The adult worms, the macrofilariae, do not multiply in man; their maturation, after infection by the bite of the transmitting mosquito or simulium, may take 3–15 months and they usually have a long life, surviving in the human host for up to 15 years.

The female worms are viviparous in that they produce large numbers of larval stages, the microfilariae, which invade either the blood (*W. bancrofti*, *B. malayi*, *Loa loa*) or the eye (*O. volvulus*) and the subcutaneous tissue. The lifespan of the microfilariae is shorter than that of the adult forms, but they also can live in the human host for years. Their life-cycle is maintained by different vectors which infect themselves by biting infected man; after development within the vector the mosquitoes and simulia are ready for infection. Disease in the human host is caused either by the immature or adult worms or by the microfilariae. Table 7 shows the clinical features of the different forms of filariasis and of the disease-producing mechanisms of the causative organisms.

Table 6. Filariasis

Disease	Filaria species	Endemic areas	Vectors
Bancroftian filariasis: disease of the lymphatic system	*Wuchereria bancrofti*	Pacific regions, New Guinea, India, Sri Lanka, Thailand, Maldive Islands, Japan, China, Egypt, West, Central and East Africa, Madagascar, Eastern Mediterranean, South America	Mosquitoes of the general *Anopheles, Culex, Mansonia, Aedes*
Brugian filariasis: disease of the lymphatic system	*Brugia malayi*	Malaysia, East India, Southern India, Sri Lanka	
Onchocerciasis: disease of the skin and eye (river blindness, blinding filariasis)	*Onchocerca volvulus*	Tropical Africa south of the Sahara, West Africa (particularly the Volta basin), foci in East Africa, Mexico, Venezuela, Colombia	Black flies of the genus *Simulium*
Loiasis: disease of the subcutaneous tissue	*Loa loa*	West and Central Africa	Flies of the genus *Chrysops*

Table 7. Filariasis: Main Clinical Features and Disease-producing Mechanisms

Disease	Clinical features	Disease-producing mechanism
Bancroftian and brugian filariasis	Fever and inflammation of the lymphatic system, hydrocele, elephantiasis	Obstruction of lymphatic vessels by macrofilariae
Onchocerciasis	Skin lesions, nodules, lymphadenopathy, elephantiasis, blindness	Microfilariae cause skin lesions and river blindness
Loiasis	'Calabar' swellings affecting the face or the limbs	Macrofilariae

CHEMOTHERAPY OF FILARIASIS

The most prominent and dangerous clinical features of filariasis are elephantiasis and river blindness, the latter named after the habitat of the simulian vectors of onchocerciasis. More than 10% of the population in the worst affected African villages are blind as onchocerciasis causes the highest rates of blindness in the world.[23] As both macrofilariae and microfilariae can produce these afflictions, chemotherapy must attack both stages and kill the circulating microfilariae and destroy the macrofilariae, the immature and the adult worms. This may be dangerous in patients with a high incidence of worm populations as the filaricidal action can provoke allergic reactions.

The most extensively used drug is diethylcarbamazine (DEC)[41] which is active after oral administration. It is a microfilaricidal drug which is also effective against the macrofilariae of *W. bancrofti*, *B. malayi* and *Loa loa*, at least to some degree. DEC is well tolerated in the treatment of bancroftian and brugian filariasis and loiasis with only a few exceptions and has been used for mass treatment of *W. bancrofti* in the Western Pacific.[23] This does not, however, result in reliable results nor in the complete elimination of reservoirs of the parasite in human hosts. But in onchocerciasis patients treated with DEC dangerous hypersensitivity reactions occur, their severity being proportional to the number of microfilariae. DEC derivatives (HOE 28 637a and HOE 29 691a) have been synthesized and have displayed a higher tolerability in animal experiments.[113] Unfortunately, so far neither compound has been investigated in humans.

Suramin,[42,52] also known as Bayer 205, Germanin and Antrypol, was introduced in 1920 for the treatment of African trypanosomiasis and in 1948 it was by chance found also to be active against *O. volvulus*.[45,106] Suramin has to be administered parenterally (i.v., 1 g/week for about 5 weeks). The compound acts primarily on adult female filariae and then on adult males followed by slow clearance of microfilariae. It provokes some toxic reactions which are common in poorly nourished patients. Immediate reactions, nausea, vomiting and collapse, can be separated from later side-effects after 3–48 h: fever, photophobia, lachrymation and others.

Irritation of the kidney is the commonest and severest toxic reaction following administration of suramin. The very slow excretion of the drug is most likely due to a strong binding of plasma to proteins and to cells. In human patients, suramin exerts its action against the adult worms only after five weeks of therapy; however, it leaves many microfilariae alive. Inflammatory and allergic reactions caused by the death of the filariae add to the problems of chemotherapy with this drug. Though suramin is the only macrofilaricidal drug known up to now, its use is commonly assumed to be confined to individual treatment;[23] it has, however, been used on a wider scale in Venezuela,[85] Ghana[16] and Uganda[9] with only a few dangerously toxic side-effects. Obviously, suramin does not meet the requirements for a safe macrofilaricidal compound. However, more recently a dose regimen using a total dose not exceeding 3.8 g for an adult has been described[85a] which proved to be effective with only acceptable side-effects. Further clinical trials have to clarify whether this regimen can be recommended for community medicine. Nonetheless, at present there is no large-scale chemotherapy available for onchocerciasis.

Several compounds have been shown to possess efficacy against microfilariae, immature or adult worms in animal experiments and even in patients. Some of them are as follows: 33 258,[53] Thiacetarsamid,[67–69] Spirotrypan,[102–104] Mel W,[32] F 151,[30,31] and F 151/33 258.[21,22,24,34] The last five compounds are organic arsenicals or combinations with an organic arsenical. Due to local or systemic toxicity, none of these compounds could be introduced into treatment of patients or used on a larger scale.

The previously mentioned 4-isothiocyanato-4′-nitrodiphenylamine (C 9333–Go/CGP 4540) displays activity after oral application against macro- and microfilariae in animal tests. However, it is too early to arrive at an evaluation of the clinical usefulness of this compound.

There exists no safe filaricidal drug with reliable efficacy against adult worms and this is of special significance for onchocerciasis patients. Despite many efforts,[116] we have to face the same situation in the field of chemoprophylaxis. Therefore, 'there is a pressing need for a new and non-toxic macrofilaricide, particularly one effective against *O. volvulus*, and for a wide-range filarial chemoprophylactic'.[23] Though the development of a preventive vaccine is a promising goal, only basic studies have up to now been undertaken.

TRYPANOSOMIASIS

Trypanosomes are protozoa, the causative organisms of two different human diseases: African and South American trypanosomiasis.

African Human Trypanosomiasis

Sleeping sickness (Gambian or Rhodesian sleeping sickness) is caused by *T. gambiense* or *T. rhodesiense*. Transmission of the disease is brought about by the bite of infected tsetse flies (*Glossina* species). The disease is characterized by an early and a late stage. In the former, the blood-lymphatic stage, the patient suffers from skin eruptions, fever, headache, enlargement of lymph nodes and spleen, tachycardia, anaemia and so forth. After the invasion of the central nervous system by trypanosomes the late stage begins: mental dullness (hence sleeping sickness), tremor and other symptoms of impairment and irreversible damage of the central nervous system. In untreated cases, trypanosomiasis leads in almost all cases to death. The course of *T. gambiense* infections is more chronic than that of *T. rhodesiense* infections.

The incidence of the disease and its importance as a public health problem is shown by the fact that 10 000 new cases are known to occur yearly; 'sleeping sickness constitutes a permanent and serious risk to the health and wellbeing of at least 35 million people'[74,75] in Africa. Annual examinations of some 10 million inhabitants of endangered areas, control of tsetse flies and other means of control require large-scale efforts yet financial support is frequently insufficient. Control by medical surveillance, treatment of infected individuals and vector control during the past 50 years has resulted in a relatively low prevalence of the disease. Interruption of surveillance, however, due to political and military events, has led to new severe outbreaks as in Zaïre in the early 1960s, and now also in Rhodesia.

Sleeping sickness occurs in tsetse-infected areas, between the latitudes of 15°N and 29°S. There are some drugs available for human trypanosomiasis belonging either to the group of non-metallic or arsenical compounds. The majority of these drugs are ineffective in the treatment of infections of the central nervous system.

Suramin, the oldest drug for the treatment of African trypanosomiasis, may cause renal changes and other toxic side-effects. It does not penetrate into the central nervous system and therefore does not cure the late stages of the disease; it has a certain effect, however, in this situation and may act as a therapeutic predecessor of melarsoprol (see below). Pentamidine is also without effect on infections of the central nervous system; it may cause severe toxic effects and is of limited value against *T. rhodesiense*. It is a chemoprophylactic. Nitrofuran compounds, despite invasion of the central nervous system, display unpredictable therapeutic activity; toxic effects are common.

Melarsoprol (Mel B),[32] an arsenical compound, is not only effective against all stages of *T. gambiense* and the early stage of *T. rhodesiense* infection, but also displays curative effects in well-advanced cases of infections with *T. rhodesiense*. Prior to the introduction of Mel B, these cases were hopeless. The drug has, however, only a very narrow safety margin and may lead to severe toxic effects and even death.

During the past 20 years no new drugs have been introduced into the chemotherapy of African trypanosomiasis. Novel and better drugs are badly needed. They should be effective against all stages, early and late ones, of the disease, and also be effective against both *T. rhodesiense* and *T.*

gambiense type infection, they should be free from serious adverse side-effects, and their administration should be easy to handle.

Trypanosomiasis of cattle is widespread in many parts of Africa. Its causative organisms are *T. vivax*, *T. congolense* and trypanosomes of the *T. brucei* subgroup; they are non-infective to humans. One of the leading chemotherapeutic drugs effective against these infections is diminazene. It has been shown[3,46,87] to be very effective against early *T. gambiense* and also *T. rhodesiense* infections. However, due to toxic reactions (brain damage) observed in donkeys[4] and dogs,[50] diminazene is not recommended for the therapy of human sleeping sickness by its producers (Hoechst AG). However, in cattle, and apparently also in humans, toxic symptoms similar to those in the two animal species have not been seen up to now.

Chagas' Disease (South American Trypanosomiasis)

This is caused by infection with *Trypanosoma cruzi.* This organism, a protozoan flagellate, resembles in morphology *T. gambiense* and *T. rhodesiense*, but has a different life-cycle. Vectors are triatomine reduviid bugs of many species. Due to their habitat, in huts and walls, Chagas' disease is prevalent in rural areas. Poor housing and sanitary conditions favour transmission. Human infection is commonly acquired from the trypanosomes in the faeces of the bug while it bites man. Another important mode of transmission is by way of blood through infected syringes and transfusion.

There are three stages of the disease: the acute early stage, starting about two weeks after infection (skin lesion-chagoma, fever, enlargement of lymph glands, liver and spleen), followed by the chronic stage, 10–20 years later: severe heart damage and 'mega-syndromes' (irreversible dilatation of the oesophagus or the lower intestine) and other organ damage, and, finally the latent form. The death-rate is around 10%. Only in the acute phase can trypanosomes be found in the blood; thereafter, the parasite invades tissue cells where multiplication takes place until the cell bursts and the released trypanosomes invade further cells via the blood stream. At this stage, the parasites are less susceptible to chemotherapy.

Chagas' disease occurs in nearly all countries of Central and South America, particularly Brazil, Argentina, Venezuela and Chile. Though data on prevalence and incidence for most of these countries are scanty and incomplete, it is estimated that more than 10 million people are infected.[61]

Chemotherapy[39] must be considered as insufficient at present, despite many efforts. Drugs active against African trypanosomiasis, such as antimonials, pentamidine, suramin and antrycide, are ineffective in the treatment of Chagas' disease. The main problem is the therapy of the long-lasting chronic stage where the tissue forms dominate and cause additional problems. The drug must then not only be active against trypanosomes but must also first invade the cells which harbour the parasites.

Derivatives of several groups, bisquinaldines, arsenobenzenes, phenantridines and 8-aminoquinolines, which are active in infected animals and also in humans, had to be abandoned for several reasons. A thiosonicotinic acid amide, Hoechst 30 036,[78] was nearly as active as nifurtimox (see below) in animals experimentally infected with *T. cruzi*; it was, however, not developed further.

From the series of nitrofurans, nifurtimox proved effective in animal experiments[36,38,40,43] and in clinical trials.[108] The results were much better during the acute phase than in the chronic one.[39] But the compound is not devoid of side-effects. Another nitrofuran derivative, SQ 18 506, also proved active in animal experiments.[39] Ro 7–1051 benznidazole from the 2-nitroimidazole group displays very good activity against *T. cruzi* and the drug proved active against intracellular trypanosomes in tissue culture.[72,83] Patients with acute infections responded favourably to treatment with benznidazole[56] but it has yet to be seen whether the same holds true for patients with chronic infections. Unfortunately, both nifurtimox and benznidazole require administration in two-to-three-month courses for treatment, for acute and chronic cases.

Thus, the outlook for better chemotherapy in Chagas' disease is not too brilliant at present. WHO therefore recommended the following lines of research to be pursued: the development of new drugs; the improvement of existing drugs; and increased and more effective utilization of existing drugs.[61]

LEISHMANIASIS

Like malaria, filariasis and schistosomiasis, leishmaniasis is a complex of infections, caused by protozoa of the genus *Leishmania.* Three forms of leishmaniasis have been distinguished hitherto, visceral, cutaneous, and mucocutaneous forms, caused respectively by *L. donovani*, *L. tropica*, *L. braziliensis* and other leishmaniae. Table 8 demonstrates – in abbreviated form – the present classification.[73]

The disease vectors are female sandflies of the genera *Phlebotomus* and *Lutzomyia.* After sucking blood contaminated with leishmaniae and development of the protozoa within the vectors, the tiny insects, 1.5–3.5 mm in size, are infective. The protozoa are transmitted by the biting sandfly and are thus injected into the skin. Thereafter, the parasites invade the cells and multiply until the host cell is destroyed and other cells can be parasitized.

Kala-azar – visceral leishmaniasis – is a severe infection causing death in untreated cases usually

Table 8. Leishmaniasis: Different Forms

Organism	Geographical distribution	Type of country	Clinical characters	
L. tropica major	Africa, Asia and Europe	Rural	Wet cutaneous form	Old World
L. tropica minor		Urban	Dry cutaneous form	
L. donovani donovani	Indian subcontinent	Open country	Kala-azar, all ages	
L. donovani infantum	Middle East, North Africa, Europe	Open or lightly wooded	Infantile kala-azar	
L. mexicana mexicana	Northern part of Central America	Forest	Chiclero's ulcer, bay sore	New World
L. braziliensis braziliensis	Brazil	Forest	Espundia, non-diffuse, diffuse cutaneous	
L.b. panamensis	Central America, Northern South America	Forest		

From Ref. 73, abbreviated.

within two years; the Indian name *kala-azar* means increased pigmentation. The cutaneous forms of the Old World, the oriental sore, due to *L. tropica*, lead to the formation of chronic, granulomatous ulcers. They do not heal spontaneously and respond poorly to chemotherapy.

The cutaneous form caused by *L. mexicana* (bay sore and Chiclero's ulcer) usually affects the external ear, lasts untreated for many years and may cause partial or complete destruction of the ear. Mucocutaneous leishmaniasis (espundia) is characterized by nodular granulomatous tumours with destruction of the lips, soft palate, face, and the tissue of the nose.

'No estimates are available as to the number of people actually infected with the disease or at risk, but it is likely that the importance of leishmaniasis as a public health problem is underestimated.'[73] In China and India, visceral leishmaniasis was considered as one of the most important diseases. In 1937 its incidence in India was 1 per 1000 inhabitants; mortality was 6–15%.[2] In the Indian subcontinent in the 1950s about 77 000 newly infected cases were observed; after the use of DDT, however, in antimalaria campaigns, the sandfly population was markedly reduced and kala-azar disappeared from large parts of India. When the use of insecticides was stopped, the return of the disease was reported from several areas.[73]

For chemotherapy of visceral leishmaniasis, antimonials are the drugs of choice, for example pentostam. Pentamidine should be used if the former fails. Amphotericin B, an antimycotic, is also in use in rare cases; this drug may, however, cause severe toxic effects. There are also antimalarials, for example pyrimethamine, where alternative drugs are required.

From the foregoing, it is evident that the prevalence and the significance of leishmaniasis as a public health problem cannot easily be evaluated. However, the different forms of leishmaniasis are dangerous and crippling diseases, partially resistant to therapy. Thus, a search for better chemotherapeutics should be made.

LEPROSY

In contrast to the diseases mentioned above, leprosy is a bacillary infection; its causative organism is *Mycobacterium leprae.* In its morphology, this bacterium resembles the organism which causes tuberculosis but differs from it in that *M. leprae* until now has not been cultivated *in vitro*, that is, on artificial media like the majority of pathogens. Thus, the routine method for screening the inhibitory activity of substances to be tested in a simple *in vitro* method (evaluation of the minimum concentration to inhibit multiplication of the bacteria in the test-tube) cannot be performed with *M. leprae*, which is a serious disadvantage for the search for new and effective chemotherapeutic drugs.

Although, centuries ago, leprosy was endemic and even epidemic in many countries of Europe, it is now only endemic in the tropics and subtropics; in other regions, such as southern Europe, there are now only few cases, most of them having been imported from elsewhere. Endemic areas are found in Mexico, Central and South America, and in Asia. In some parts of Africa the incidence is high. There are no accurate worldwide data, and in 1972, 11–12 million cases were estimated to exist throughout the world.[88]

Leprosy is a chronic, communicable disease, naturally confined to man. It is a disease of the peripheral nervous system; skin and, less often, the mucosa of the mouth are involved and also eyes, bones and testes. Leprosy in untreated or unsuccessfully treated patients is very often a long-lasting, crippling disease. Further, the psychological consequences for the patient are often serious because of society's prejudice against the disease.

There are various clinical types of leprosy, the most important ones being the lepromatous (L) and the tuberculoid (T) forms. The L form is a malign type with high bacterial load and distribution of bacilli throughout the body. There is poor resistance to infection and in many cases uninterrupted progress of the disease. On the other hand, the T form is usually benign, the patient displays good resistance and is negative on bacteriological examination. In patients with a high degree of immunity the disease may even come to a standstill.

CHEMOTHERAPY OF LEPROSY

The oldest and still widely used drug is a sulphone, 4,4′-diaminodiphenylsulphone (DDS, dapsone); in most cases it is well tolerated. However, its main disadvantages are slow action and insufficient efficacy and, despite apparently adequate treatment, not all organisms are killed by DDS. It is of major importance that the drug is inexpensive as patients often require continuing treatment for 5–15 years, or even lifelong therapy for lepromatous type patients.

A useful alternative to DDS is thiambutosine, a member of the thiourea series of compounds. Thiambutosine is virtually non-toxic and is applied orally. Other thiourea derivatives have been described[70,105] as being active in mice infected with *M. leprae* but have not been investigated clinically.

Apart from other compounds of minor importance, rifampicin is to be mentioned as a highly active drug against *M. leprae*. This antibiotic is now one of the leading compounds in the chemotherapy of tuberculosis but its position in the chemotherapy of leprosy is not yet clear; experiences in Africa and Asia have shown a rapid development of resistance to this drug.[88] Clinical trials in Asia have been planned and will facilitate the ultimate decision on the value of this drug. The high cost of this antibiotic, however, must be emphasized, an important financial consideration in most of the countries where leprosy is prevalent.

Clofazimene (B 663)[6] is also a drug with efficacy against leprosy, having about the same effect as DDS. It belongs to a group of aposafranine dyes and, being a dyestuff, it provokes a red coloration of the skin and the urine. Other side-effects are rare.

Progress in the field of experimental leprosy, and above all the search for new chemotherapeutics, encountered many difficulties as it proved impossible for a long time to infect suitable experimental animals. New possibilities were opened up by Shepard's findings that *M. leprae* could be grown in the foot-pads of mice.[92,103] The antileprotic activity of DDS could be proven in this test.[79,93,94] Further, it could be demonstrated that, in patients with clinical resistance to DDS, this was paralleled by a resistance of the organisms to this drug in the mouse foot-pad test.[80,81] Another animal found to be susceptible to infection with *M. leprae* was the armadillo.

These important biological findings may facilitate the search for new, effective drugs for the chemotherapy of leprosy. However, the foot-pad test differs from other animal models for screening new compounds against pathogens in that it is very time-consuming because of its very slow generation time – the time in which one bacterium divides into two – for *M. leprae*, many days instead of minutes. Therefore. 'the further development of new compounds that appear promising in initial studies in mice will be very costly in terms of both time and money'.[88]

But new and better drugs are urgently needed; they must be effective against both susceptible and DDS-resistant *M. leprae*, be inexpensive, and should be well tolerated for long-term treatment; this must be envisaged even with better drugs than those currently available. The cost of treatment plays a decisive role in leprosy, yet research and development for new drugs, involving tests of efficacy, toxicology, metabolism, cancerogenicity, mutagenicity and so on, is extremely expensive, irrespective of the number of patients who benefit.

In view of this situation, a better use of existing drugs has to be investigated; an important problem in this context is the emergence of DDS-resistant strains of *M. leprae*, and this must be prevented at all costs. In the chemotherapy of tuberculosis the problem of resistance played a decisive role two decades ago; it was overcome by drug combinations. The prevention of leprosy by active immunization has to face the inability of culturing *M. leprae in vitro* so that the material (live or killed bacteria) is not available in the quantities necessary for producing a vaccine.

CONCLUDING REMARKS

Apart from the six diseases, or groups of diseases, of the Special Programme of WHO, there are other tropical diseases. To these belong certain helminthic diseases: clonorchiasis or Chinese liver fluke disease, caused by the trematode *Clonorchis sinensis*; fasciolopsiasis, caused by the trematode *Fasciolopsis buski*; and paragonimiasis, caused by *Paragonimus westermani*. Chemotherapy for these diseases is up to now insufficient, though a compound could be detected which is active in animals infected with *Clonorchis sinensis*;[25,84] results of clinical trials have not yet been reported. Other infections which should be mentioned are amoebiasis, caused by the protozoan *Entamoeba histolytica*, bacterial infections like tuberculosis, cholera and plague, as well as fungal and viral infections. Chemotherapy is either in a state of development, or in existence as for fungal diseases, amoebiasis and bacterial diseases, or does not at present promise success, as for viral infections. In this context the most important role of 'unspecific' diarrhoeas should not be overlooked.

It was the purpose of this review to describe some of the special problems of chemotherapy for tropical diseases. They were here restricted to the six most important groups of tropical infections, included in WHO's Special Programme for Research and Training in Tropical Diseases. Drugs were here defined as chemotherapeutic compounds which act directly against the parasite (Ehrlich's parasitotropism); drugs with other activities like corticosteroids and cardiovascular drugs which, though important, would only find a role in the management of complications and sequelae of an infection, are excluded.

Tropical infections have implications not only for the individual but in many cases for the population as a whole. The successful fight against tropical diseases may, therefore, change socio-economic development for the better and improve the general situation of the afflicted regions. At present, chemotherapy is still the most effective weapon for the treatment of these diseases. It is far ahead of active immunization, though in some diseases, most notably malaria, the outlook is no longer as gloomy as it was some years ago. This survey of the chemotherapeutical armamentarium available for the therapy of the infections of the Special Programme – though certainly incomplete – has shown that there is a need for improved treatment. The survey has also demonstrated that for the majority of diseases covered by the Special Programme, intense research has found new, active compounds. The most spectacular example of the size of this type of research has been the co-operation between the Walter Reed Army Institute of Research and the pharmaceutical industry and academic institutions; it resulted in the screening of more than 250 000 compounds. However, it also illustrated the immense difficulties of developing drugs superior to those existing at present.

The Special Programme has been launched to develop new and better chemotherapeutic drugs in co-operation with the pharmaceutical industry. It is to be hoped that the independent endeavours of the pharmaceutical industry will play a prominent role in the development of new drugs and that, in co-operation with WHO as the Executive Agency of the Special Programme, this important goal will be reached.

LITERATURE CITED

The references marked with an asterisk are here quoted by permission of the World Health Organization.

1. Anon., Immunopathology of nephritis in Africa. *Bull. WHO* **46**, 387–396 (1972).
2. N. Ansari, The leishmaniases as a world problem. *Sci. Rep. Ist. Super. Sanità* **2**, 63–67 (1962).
3. N. M. Bailey, *East African Trypanosomiasis Research Organization*: *Annual Report for 1966*, p. 83. Government Printer, Entebbe (1967).
4. W. P. Boyt, J. A. Lawrence, P. K. I. Mackenzie and R. L. French, The toxicity of diminazine in donkeys. *Rhod. Vet. J.* **1**, 75–78 (1971).
5. K. N. Brown, I. N. Brown and L. A. Hills, Immunity to malaria. Protection against *Plasmodium knowlesi* shown by monkeys sensitized with drug-suppressed infection or by dead parasites in Freund's adjuvant. *Exp. Parasitol.* **28**, 304–317 (1970).
6. S. G. Browne and L. M. Hogerzeil, B 663 in the treatment of leprosy. *Lepr. Rev.* **33**, 6–10 (1962).
7. G. A. Butcher and S. Cohen, Antigenic variation and protective immunity to *Plasmodium knowlesi* malaria. *Immunology* **23**, 503 (1972).
8. C. J. Canfield and R. S. Rozman, Clinical testing of new antimalarial compounds. *Bull. WHO* **50**, 203–212 (1974).
9. J. K. Cherry, The treatment of onchocerciasis. *East Afr. Med. J.* **37**, 550–558 (1960).
10. D. F. Clyde, V. C. McCarthy, R. M. Miller and W. E. Woodward, Immunization of man against falciparum and vivax malaria by use of attenuated sporozoites. *Am. J. Trop. Med. Hyg.* **24**, 397–401 (1975).
11. D. F. Clyde, H. Most, V. C. McCarthy and J. P. Vanderberg, Immunization of man against sporozoite-induced falciparum malaria. *Am. J. Med. Sci.* **266**, 169–177 (1973).
12. S. Cohen, Development of a malaria vaccine. *J. R. Soc. Trop. Med.* **71**, 476–478 (1978).
13. S. Cohen and G. H. Mitchell, Prospects for immunization against malaria. *Curr. Top. Microbiol. Immunol.* **80**, 97–138 (1978).
14. S. Cohen, I. A. McGregor and S. C. Carrington, Gammaglobulin and acquired immunity to human malaria. *Nature (London)* **192**, 733–737 (1961).
15. S. Cohen, G. A. Butcher and R. B. Crandall, Action of malarial antibody *in vitro*. *Nature (London)* **223**, 368–371 (1969).
16. O. F. Conran and B. B. Waddy, Effect of Antrypol treatment on blindness due to onchocerciasis. *J. Trop. Med. Hyg.* **59**, 52–57 (1956).
17. A. Corradetti, Particolari fenomeni immunitari nell'infezione de *Plasmodium berghei*. *Riv. Parassitol.* **11**, 201–209 (1950).
18. L. E. D'Antonio, I. D. Spira, R. C. Fu, D. M. Dagnillo and P. H. Silverman, Malaria resistance: artificial induction with a partially purified plasmodial fraction. *Science* **168**, 117–118 (1970).
19. A. Davis, Schistosomiasis. World Health Organization Special Programme for Research and Training in Tropical Diseases, WHO Document TDR/WP/76.8.
20. A. Davis and D. R. Bailey, Metrifonate in urinary schistosomiasis. *Bull. WHO* **41**, 209 (1969).
21. D. A. Denham and P. Mellor, The anthelminthics of a new compound 'E' (Friedheim) on *Onchocerca guttosa* in the cow: a possible tertiary screening system for drug action against *O. volvulus* in man. *J. Helminthol.* **50**, 49–52 (1976).
22. D. A. Denham, R. R. Suswillo, R. Rogers, P. B. McGreevy, and B. J. Andrews, Studies on *Brugia pahangi*. 13. The anthelminthic effect of compounds F 151 (Friedheim), HOE 33 258 (Hoechst) and their reaction product. *J. Helminthol.* **50**, 243–250 (1976).

*23. B. O. L. Duke, Filariasis. WHO Document TDR/WP/76.10.

24. D. Düwel, Studies in rodents and dogs with a new macro- and microfilaricidal compound. In *Proceedings of the 9th International Congress on Tropical Medicine and Malaria, Athens, Greece,* Vol. II, p. 117 (1973).
25. D. Düwel, W. Dürckheimer and M. Schorr, Über die Wirksamkeit von Bis-dithio-carbaminsäure gegen *Opisthorchis felineus* und *Clonorchis sinensis*. *Z. Parasitenkd.* **21**, 77–78 (1970).
26. J. A. Dvořak, L. H. Miller, W. C. Whitehouse and T. Shiroishi, Invasion of erythrocytes by malaria merozoites. *Science* **187**, 748–750 (1975)
27. D. M. Forsyth and C. Rashid, Treatment of urinary schistosomiasis with trichlorphone. *Lancet* **2**, 909 (1967).
28. R. Foster, The preclinical development of oxamniquine. *Rev. Inst. Med. Trop. São Paulo* **15**, Suppl. 1, 1 (1973)
29. J. Freund, K. O. Thomson, H. E. Sommer, A. W. Walter and T. M. Pisani, Immunization of monkeys against malaria by means of killed parasites with adjuvants. *Am. J. Trop. Med.* **28**, 1–22 (1948).
30. E. A. H. Friedheim, Therapeutic field trials in canine filariasis. In *Proceedings of the 9th International Congress on Tropical Medicine and Malaria, Athens, Greece,* Vol. II, p. 118 (1973).
31. E. A. H. Friedheim, Therapeutic field trials of a macro- and microfilaricidal agent in canine filariasis. *Bull. WHO* **50**, 572–574 (1974).
32. E. A. H. Friedheim, Trypanocidal and spirochetocidal arsenicals derived from *s*-triazine. *J. Am. Chem. Soc.* **66**, 1775–1778 (1974).

33. E. A. H. Friedheim, Chemotherapy of schistosomiasis. In *International Encyclopedia of Pharmacology and Therapeutics*, Sect. 64, Part 1, p. 29 (1973).
34. E. A. H. Friedheim and R. Cavier, Un nouveau composé trypanocide et filaricide: le F 151. *Bull. Soc. Pathol. Exot.* **66**, 531–533 (1973).
35. M. Gelfand, E. T. Reid and B. E. Simpson, *Trans. R. Soc. Trop. Med. Hyg.* **56**, 77 (1962), quoted in Woodruff (Ref. 117).
36. R. Gönnert, Nifurtimox: causal treatment of Chagas' disease. *Drug Res.* **22**, 1563–1642 (1972).
37. R. Gönnert and P. Andrews, Praziquantel, a new broad-spectrum antischistosomal agent. *Z. Parasitenkd.* **52**, 129 (1977).
38. R. Gönnert and M. Bock, The effect of Nifurtimox on *Trypanosoma cruzi* in tissue cultures. *Drug Res.* **22**, 1582–1589 (1972).
39. W. E. Gutteridge, Chemotherapy of Chagas' disease: the present situation. *Trop. Dis. Bull.* **73**, 699–705 (1976).
40. A. Haberkorn and R. Gönnert, Animal experimentation investigation into the activity of Nifurtimox against *Trypanosoma cruzi. Drug Res.* **22**, 1570–1582 (1972).
41. F. Hawking, Diethylcarbamzine. A review of the literature with special reference to its pharmacodynamics, toxicity, and use in the therapy of onchocerciasis and other filarial infections. WHO Document WHO/ONCHO/78.142.
*42. F. Hawking, Suramin. A review of the literature with special reference to its pharmacodynamics, toxicity, and use in the therapy of onchocerciasis. WHO Document WHO/ONCHO/78.143.
43. K. Hoffmann, Toxicological investigations of the tolerability of Nifurtimox. *Drug Res.* **22**, 1590–1603 (1972).
44. T. W. Hollbrook, N. C. Palezuk and L. A. Stauber, Immunity to exoerythrocytic malaria. III. Stage-specific immunization of turkeys against exoerythrocytic forms of *Plasmodium fallax. J. Parasitol.* **60**, 348–354 (1974).
45. L. van Hoo, C. Henrard, E. Peel and M. Wanson, Chemotherapy of onchocerciasis. Preliminary note. *Trop. Dis. Bull.* **45**, 456–457 (1948).
46. M. P. Hutchinson and H. J. C. Watson, Berenil in the treatment of *Trypanosoma gambiense* infection in man. *Trans. R. Soc. Trop. Med. Hyg.* **56**, 227 (1962).
47. C. James, G. Webbe and J. M. Preston, A comparison of the susceptibility to metrifonate of *Schistosoma haematobium, S. mattheei* and *S. mansoni* in hamster. *Ann. Trop. Med. Parasitol.* **66**, 467–474 (1972).
48. N. Katz, J. Pellegrino and J. P. Pereira, Experimental chemotherapy of schistosomiasis. III. Laboratory and clinical trials with trichlorphone, an organophosphorus compound. *Rev. Soc. Bras. Med. Trop.* **2**, 237 (1968).
49. N. Katz, Chemotherapy of *Schistosomiasis mansoni. Adv. Pharmacol. Chemother.* **14**, 1 (1977).
50. H. Kief, Department of Experimental Pathology, Hoechst AG, personal communication.
51. J. P. Kreier, The isolation and fractionation of malaria-infected cells. *Bull. WHO* **55**, 317–331 (1977).
52. G. Lämmler, D. Grüner and S. Chandra, *Bibliography on Suramin.* Institute for Parasitology and Parasitic Diseases of Animals, Justus Liebig University, Giessen, Germany (1976).
53. G. Lämmler and W. Raether, Chemotherapeutic experiments with a 2,6-bis-benzimidazole in cotton rats infected with *Litomosoides carinii.* In *Proceedings of the VIIth International Congress on Tropical Medicine and Malaria, Teheran* (1968).
54. H. G. Lee, A. W. Cheever and W. R. Fairweather, Influence of parasite strain on the chemotherapy of murine infections with *Schistosoma mansoni. Bull WHO* **45**, 147 (1971).
55. H. Loewe, Mittel gegen Trematoden. In *Arzneimittel. Entwicklung, Wirkung, Darstellung,* Band 5: *Chemotherapeutica,* Teil 2, edited by G. Ehrhart and H. Ruschig. Verlag Chemie, Weinheim/Bergstrasse (1972).
56. H. Lugones, B. Rabinovich, J. A. Cerisola, O. Ledesma and C. Barclay, Preliminary results of the anti-*T. cruzi* activity of Ro 7-1051 in man. In *Proceedings of the 3rd International Congress on Parasitology, München,* Vol. 3, pp. 1291–1298 (1974).
57. E. N. J. de Maar, The search for new tools against tropical diseases. *Ciba-Geigy J.* No. 3, 27–31 (1978).
*58. N. C. McEvers, Socio-economic considerations of the Special Programme. WHO Document TDR/WP/76.31.
59. J. E. McMahon, Oxamniquine (UK-4271) in *Schistosoma haematobium* infections. *Ann. Trop. Med. Parasitol.* **70**, 121 (1976).
60. H. Manson-Bahr, *Manson's Tropical Diseases: A Manual of the Diseases of Warm Climates,* 16th Edn. Baillière, Tindall and Cassell, London (1966).
*61. C. J. Marinkelle, Chagas' disease. WHO Document TDR/WP/76.12.
62. G. H. Mitchell, G. A. Butcher and S. Cohen, Isolation of bloodstage merozoites from *Plasmodium knowlesi* malaria. *Int. J. Parasitol.* **3**, 443–445 (1973).
63. W. Mohr, H.-H. Schumacher and F. Weyer (eds.), *Lehrbuch der Tropenkrankheiten,* 4th revised Edn. Georg Thieme, Stuttgart (1975).
64. H. W. Mulligan, P. P. Russell and B. N. Mohan, Active immunisation of fowls against *Plasmodium gallinaceum* by injections of killed homologous sporozoites. *J. Malar. Inst. India.* **4**, 25–34 (1941).
65. R. S. Nussenzweig, J. Vanderberg, H. Most and C. Orton, Specificity of protective immunity induced by X-irradiated *Plasmodium berghei* sporozoites. *Nature (London*), **222**, 488–489 (1969).
66. R. S. Nussenzweig, J. Vanderberg and H. Most, Protective immunity produced by the injection of X-irradiated sporozoites of *Plasmodium berghei.* IV. Dose response, specificity and humoral immunity. *Mil. Med.* **134**, 1176–1182 (1969).
67. G. F. Otto and T. H. Maren, Treatment of canine filariasis with trivalent arsenicals (*p*-arsenobenzamides). *J. Parasitol.* **31**, 17 (1945).
68. G. Otto and T. Maren, Filaricidal activity of substituted phenyl arsenoxides. *Science* **106**, 105–107 (1947).
69. G. F. Otto and T. H. Maren, Studies on the chemotherapy of filariasis. *Am. J. Hyg.* **50**, 92–141 (1949).
70. S. R. Pattyn and W. H. Wagner, Activity of compound TH 270 (*N*-*p*-isobutoxyphenyl-*N'*-*p'*-*a*-pyridyl-ethyl-phenylthiourea) on the experimental infection with *Mycobacterium lepraemurium* and *Mycobacterium leprae. Ann. Soc. Belg. Méd. Trop.* **52**, 55–61 (1972).
71. J. Pellegrino. F. F. Lima-Costa, M. A. Carlos and R. T. Mello, Experimental chemotherapy of *Schistosoma mansoni.* XIII. Activity of praziquantel, an isoquinoline-pyrazino derivative, on mice, hamsters and Cebus monkeys. *Z. Parasitenkd.* **52**, 151 (1977).
71a. G. Piekarski, *Medizinische Parasitologie in Tafeln,* 2nd Edn. Springer-Verlag, Berlin–Heidelberg–New York (1973).
72. A. Polak and R. Richle, Mode of action of the 2-nitroimidazole derivative benznidazole. *Ann. Trop. Med. Parasitol.* **72**, 45–54 (1978).
*73. P. de Raadt, Leishmaniasis. WHO Document TDR/WP/76.14.
*74. P. de Raadt, African trypanosomiasis. WHO Document TDR/WP/76.12.
75. P. de Raadt, K. van Hoeve, N. M. Bailey and E. N. Kenyanyñi, *East African Trypanosomiasis Research Organization: Annual Report for 1965,* p. 60. Government Printer, Entebbe (1966).
76. W. Raether and E. Fink, Suppressiver und kausalprophylaktischer Effekt von 7-Chlor-10-hydroxy-3-(4-trifluor-methylphenyl)-3, 4-dihydroacridin-1, 9-(2H, 10H)-dion (Hoe 991) bei der Nagetiermalaria. *Z. Tropenmed. Parasitol.* **28**, 268 (1977).
77. W. Raether and E. Fink, Antimalarial activity of Floxacrine (Hoe 991). Studies on blood schizontocidal action of Floxacrine against normal and drug-resistant lines of *Plasmodium berghei, Plasmodium vinckei* and *Plasmodium cynomolgi. Ann Trop. Med. Parasitol.* in press.
78. W. Raether, M. Schorr and H. Seidennath, Ein neues Thioisonicotinsäureamid mit Wirkung gegen *Trypanosoma cruzi* bei experimentell infizierten NMRI-Mäusen. *Z. Tropenmed. Parasitol.* **23**, 227–240 (1972).
79. R. J. W. Rees, Recent bacteriologic, immunologic and pathologic studies on experimental human leprosy in the mouse foot pad. *Int. J. Lepr.* **33**, 646–657 (1965).
80. R. J. W. Rees, *Br. J. Exp. Pathol.* **45**, 207 (1964).

81. R. J. W. Rees, Drug resistance of *Mycobacterium leprae* particularly to DDS. *Int. J. Lepr.* **35**, 625–638 (1967).
82. H. C. Richards and R. Foster, A new series of 2-aminomethyltetrahydroquinoline derivatives displaying schistosomicidal activity in rodents and primates. *Nature (London)*, **222**, 581 (1969).
83. R. Richle, Chemotherapy of experimental Chagas' disease in mice: beneficial effect of Ro 7-1051 on parasitemia and tissue parasitism. *Prog. Méd.* **101**, 282 (1973).
84. H. J. Rim, S.-S. Kim, S.-D. Song and S.-J. Lee, Anthelminthic activity of bisbenzadole against *Clonorchis sinensis* in rabbits and rats. *Korea Univ. Med. J.* **13**, 197–204 (1976).
85. A. Rivas, G. Gonzales, L. Zsogon, E. Rasi and J. Convit, Onchocerciasis in Venezuela. *Acta Med. Venez.*, Suppl. 1, 36 (1965).
*85a. A. Rougemont, J. Delmont, P. Ranque, M. Ducam and A. Prost, Traitement collectif de l'onchocercose par des faibles doses de suramine dans un village hyperendémique au Mali (premiers résultats parasitologiques). WHO Document WHO/ONCHO/79.147.
86. R. S. Rozman, Chemotherapy of malaria. *Annu. Rev. Pharmacol.* **13**, 121–152 (1973).
87. E. H. Sadun, B. T. Wellde and R. L. Hickman, Resistance produced in owl monkeys (*Aotus trivirgatus*) by inoculation with irradiated *Plasmodium falciparum*. *Mil. Med.* **134**, 1165–1175 (1969).
*88. H. Sansarricq and J. Walter, Leprosy. WHO Document TDR/WP/76.16.
89. R. H. Schenkel, G. L. Simpson and P. H. Silverman, Vaccination of rhesus monkeys against *Plasmodium knowlesi* by the use of non-viable antigen. *Bull. WHO* **48**, 597–604 (1973).
90. L. H. Schmidt, *Plasmodium falciparum* and *Plasmodium vivax* infection in the owl monkey (*Aotus trivirgatus*). I. The courses of untreated infections. II. Responses to chloroquine, quinine, and pyrimethamine. III. Methods employed in the search for new blood schizonticidal drugs. *J. Trop. Med. Hyg.* **27**, 671–737 (1978).
90a. L. H. Schmidt, Experimental infections with human plasmodia in owl monkeys—their contributions to development of new broadly active blood schizontocidal drugs. *Adv. Pharmacol. Ther.* **10**, 79–90 (1979).
91. L. H. Schmidt, R. Crosby, J. Rasco and D. Vaughan, Antimalarial activities of various 4-quinoline-methanols with special attention to WR-142.490 (Mefloquine). *Antimicrob. Agents Chemother.* **13**, 1011–1030 (1978).
92. C. C. Shepard, The experimental disease that follows the injection of human leprosy bacilli into foot-pads of mice. *J. Exp. Med.* **112**, 445–454 (1960).
93. C. C. Shepard and Y. T. Chang, Effect of several antileprosy drugs on multiplication of human leprosy bacilli in foot-pads of mice. *Proc. Soc. Exp. Biol. Med.* **109**, 636–638 (1962).
94. C. C. Shepard and Y. T. Chang, Activity of antituberculous drugs against *Mycobacterium leprae*. *Int. J. Lepr.* **32**, 260–271 (1964).
95. G. L. Simpson, R. H. Schenkel and P. H. Silverman, Vaccination of rhesus monkeys against malaria by the use of density gradient fractions of *Plasmodium knowlesi* antigens. *Nature (London)* **247**, 304–305 (1974).
96. G. Smith, Changing patterns of disease in the tropics. *Br Med. Bull.* **28**, 3–9 (1972).
96a. H. P. Striebel, 4-Isothiocyanato-4′-nitrodiphenylamine (C 9333-Go/CGP 4540), an antihelminthic with an unusual spectrum of activity against nematodes, filariae and schistosomes. *Experientia* **32**, 457–458 (1976).
97. G. A. T. Targett and J. D. Fulton, Immunization of rhesus monkeys against *Plasmodium knowlesi* malaria. *Exp. Parasitol.* **17**, 180–193 (1965).
98. H. Thomas, Progress in the chemotherapy of schistosomiasis. In *Proceedings of Symposium on Chemotherapy of Schistosomiasis, 7th International Congress of Pharmacology, Paris* (16–21 July 1978).
99. W. Trager and J. B. Jensen, Human malaria parasites in continuous culture. *Science* **193**, 673–675 (1976).
100. A. Voller and N. R. Rossan, Immunological studies on simian malaria. Immunity to challenge and antigenic variation in *Plasmodium knowlesi*. *Trans. R. Soc. Trop. Med. Hyg.* **63**, 507–523 (1969).
101. A. Voller and W. H. G. Richards, Immunity to *Plasmodium falciparum* in owl monkeys (*Aotus trivirgatus*). *Z. Tropenmed. Parasitol.* **21**, 159–166 (1970).
102. W.-H. Wagner, Die Chemotherapie der menschlichen Filariosen. *Antibiot. Chemother. Fortschr.* **3**, 343–397 (1956).
103. W.-H. Wagner, Neue Methoden der experimentellen Lepra-Chemotherapie. *Hautarzt*, **20**, 385–389 (1969).
104. W.-H. Wagner and W. Schulz, Chemotherapeutische Untersuchungen über Spirotrypan. *Z. Ges. Exp. Med.* **119**, 204–228 (1952).
105. W.-H. Wagner and E. Winkelmann, Tuberkulostatisch wirksame *N,N′*-diarylthioharnstoffe. 2. Mitt. *Drug Res.* **19**, 719–730 (1969).
106. M. Wanson, Contribution on the study of African onchocercosis (problems of prophylaxis in Leopoldville). *Ann. Soc. Belg. Méd. Trop.* **30**, 671 (1950).
107. C. Webbe and A. James, A comparison of the susceptibility to praziquantel of *Schistosoma haematobium*, *S. japonicum*, *S. intercalatum*, and *S. mattheei* in hamsters. *Z. Parasitenkd.* **52**, 169 (1977).
108. D. H. G. Wegner and R. W. Rohwedder, The effect of nifurtimox in acute Chagas' infection. *Drug Res.* **22**, 1624–1642 (1972).
108a. W. H. Wernsdorfer, Malaria. WHO Document TDR/WP/76.6.
*109. WHO Document TDR/AR(1)/77.14, Annual Report 1977: Association with the pharmaceutical industry in the development of new agents.
*110. WHO Document TDR/AR(1)/77.2, WHO Special Programme for Research and Training in Tropical Diseases. Annual Report 1977: Scientific Working Groups on Malaria.
*111. WHO Techn. Rep. Ser. 519, Developments in Malaria Immunology, Report of a WHO Scientific Group. Geneva (1975).
*112. WHO Document TDR/IM/76.3, Special Programme for Research and Training in Tropical Diseases: Report of the Scientific Working Group on the Immunology of Malaria, Geneva, 8–14 July 1976.
*113. WHO Techn. Rep. Ser. 542, WHO Expert Committee on Filariasis, Third Report. Geneva (1974).
*114. WHO Document TDR/MPD/SWG-Sch(1)/77.3, Report of the First Meeting of the Scientific Working Group on Schistosomiasis.
*114a. WHO Document E/iid. Rec. (1977), Information on the World Malaria Situation 1976.
*115. WHO Document TDR/AR(1)/77.15 Rev. 1, Special Programme for Research and Training in Tropical Diseases.
*115a. WHO Document A31/19.15.3 (1978), Malaria control strategy.
116. E. Wolf, Zur Chemoprophylaxe der Filariose: Experimentelle Untersuchungen an der *Litomosoides carinii*-Infektion der *Mastomys natalensis*. Inaugural Dissertation, Justus Liebig Universität, Giessen (1976).
117. A. W. Woodruff (ed.), *Medicine in the Tropics*. Churchill Livingstone, Edinburgh and London (1974).
118. M. Yoeli and B. J. Hargreaves, Brain capillary blockage produced by a virulent strain of rodent malaria. *Science* **184**, 572–573 (1974).
119. M. D. Young, J. A. Porter Jr. and C. M. Johnson, *Plasmodium vivax* transmitted from man to monkey to man. *Science* **153**, 1006–1007 (1966).
120. A. Zuckerman, J. Hamburger and D. Spira, Active immunization of rats with a cell free extract of the erythrocytic parasites of *Plasmodium berghei*. *Exp. Parasitol.* **21**, 84–97 (1967).

The manuscript was received 17 November 1978

Better Cities for the Future
Especially for the Third World

PROFESSOR RICHARD L. MEIER
College of Environmental Design, University of California, Berkeley, USA

In future competition between metropolitan regions basic human needs should lead to a novel urban form in the less developed countries as new resource-conserving technologies are becoming available. Water is best conserved as marketable commodities of different qualities, allowing recycling when economic. Organic contents of sewage can be returned to the city as fish, poultry, and pork; and 'grey' water could be channeled into intensive gardening. Thus, the bulk of perishable foods could be produced in the interstices of urban settlement. Transport in poor cities is most economic by bus and jitney, with bicycle and moped for personal vehicles and carts, vans and trucks for goods movement. Land use should remain mixed. Housing that is predominantly multistory walk-up would save time, energy, effort, space, materials and capital. New forms of participation in the community improvement process deserve imitation. A quality of life roughly equivalent to the best in the West should be achievable at one-fifth the consumption of fuel and materials.

Most of the world's cities are growing rapidly. And this occurs despite the fact that the supply of resources funneled to them is roughly constant – and scheduled to decline. The well established procedures for town planning must be expected to fail because price structure and social conditions of the future will be very different from the past. Until now the success of a community has been judged by its architectural display and by the energy-consuming industries – together they signify its political and economic power. In the future, a city will maintain its relative position among communities only if it manages to conserve resources without loss in the Quality of Life, the QOL. Since many qualities of human well-being require a social and cultural context, the conditions in East Asian metropolitan areas – especially South Korea, a society which is advancing as rapidly as any in the world and also likely to be able to afford investment in infrastructure – will be used to illustrate some of the above implications.[1]

FUTURE COSTS AND CRISES

An approach to such planning, drawing upon the principles of ecology to integrate the diverse requirements, has been put forward.[2] In it I adjusted the arguments to the worldwide shifts in energy price and availability, occurring in the 1973–1975 period. I demonstrated that technological solutions existed for the provision of all the basic human needs[3] required for the flood of urban immigrants expected over the next several generations, even if the global population of urban residents multiplied from approximately 1×10^9 persons to 7–10 times that number. Only a few new technological concepts are added to this contribution with the intent to lay out a path for the provision of essential services in water supply, sanitation, domestic space heating, transport and housing – in societies like Korea, Malaysia, the Philippines, Brazil, Mexico, Nigeria, Algeria, Morocco and other rapidly modernizing countries. In the instance of Korea, major commitments must be made very soon; the other countries are not far behind.

The changing situation is best put in the form of expectations regarding future costs of essential inputs. These societies will be subject to periodic regional and global shortages, superimposing periods of sharp price rises and failures in supply upon the long-term trends. A typical outlook for the future in those parts of the world as laid out by a global simulation model, for example, by Mesarovic and Pestel,[4] is summarized in Fig. 1. It omits the problems with respect to materials requirements, mainly non-ferrous metals, wood, cotton and rubber, too complex to be represented graphically. The expanding urban communities will be forced to draw resources of much lower grade requiring more energy to extract and thus resulting in substantial increases in cost; these are only somewhat diminished by the use of substitutes and better fitting product design.

Due to the diminishing costs of telecommunications, and continuing shifts to air transport, there will be increasing transactions between metropolitan areas. Enterprises will expand through these enlarged channels, causing high grade human resources to be redistributed. The flow of funds and investment capital has already overcome a number of barriers set up by the respective nations, so the money is very likely

CCC–0308–0188/80/0005–0204 $05.50/1

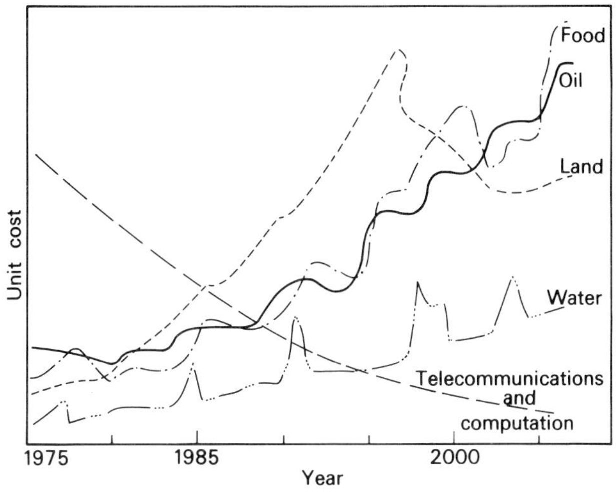

Figure 1. Expectations for Third World metropolitan regions: Crises. The crises in inputs, indicated by discontinuities in trends, paralyze many of the functions of a city. The impact of shortages, or escalations in costs, are more serious when they affect government and other regulatory and distributive institutions in a society. The best community design buffers the social system from such perturbations, most often with instruments employing telecommunications and computation.

to be invested in the best opportunities for either making more money or for achieving a better quality of life. If this declines in one metropolis in comparison to others with which it interacts, the supply of technological and cultural expertise diffuses away and the supply of mobile capital accompanies it. Employment by multinational organizations shrinks particularly rapidly. Beirut and Teheran are prominent examples of disastrous declines in quality of life; both metropolitan areas have become quite unattractive.

QOL—QUALITY OF LIFE

The techniques for the measurement of quality of life have developed very recently, with the best results obtained from responses to carefully constructed interviews. Campbell, Converse and Rogers[5] attempted to find domains of experience that would explain satisfaction with life and the level of happiness reported. A more tightly constructed study went on from there by posing the question, 'How do you feel about life as a whole?' The respondent was expected to place himself on a seven point scale that put 'Delighted' at the top, 'mixed' in the middle and 'Terrible' at the bottom. Then other conditions of life were questioned in the hope that these feelings could be explained. Andrews and Withey have extracted a remarkable amount of internal consistency and stability from these indicators.[6] Nevertheless, additional local surveys would be needed to discover how much weight should be assigned to income, education, diversity of friendships and so forth, all features that can be changed by reorganization of urban communities.

In Japan, the approach to such measurement of quality of life more nearly resembled social engineering. The domains were derived from the wisdom of Japanese civilization, which generated somewhat different questions, such as queries about 'something worthwhile to live for' and 'spiritual fulfillment', but also about conditions in the environment. Communities were graded from best to worst in a number of domains in order to produce a profile (Fig. 2). Representative communities, ranging from the fanciest new towns on the periphery of Tokyo to rural settlements hundreds of kilometers away, could be graded according to the profile provided. What differences would result from the installation of new facilities, such as, for example, modern telecommunications?[7] A surprising answer reported in conversations is that in a homogeneous society like Japan with as yet inadequate social security provisions, a 'free telephone' would greatly enrich the life of the bottom tenth of the population.

ECOLOGICAL STANDARDS

It is now accepted that new urban settlement must be ecologically sound. An urban ecosystem should be established which possesses internal stability, mobilizing resources to stave off threats of immediate collapse of its utilities, economy and polity, and is not likely to be stifled by accumulation of its own wastes. People expect cities to be robust, diverse, responsive and adaptive – properties introducing the subtlety of 'resource-conserving' qualities.

Prior unconcern for pollution is now in some instances being replaced by excessive zeal for cleanliness and protection from toxins. Uncritical borrow-

PROFESSOR RICHARD L. MEIER has been Professor of Environmental Design, teaching in the Departments of City and Regional Planning, Architecture and Landscape Architecture at the University of California, Berkeley, since 1967. The present interdisciplinary review is a sequel to both *Planning for an Urban World – Design of Resource-Conserving Cities* (MIT Press, 1974) and 'A stable urban ecosystem' [*Science* **192**, (1976)]; *Urban Futures Observed – In the Asian Third World* is in press. He is a natural scientist who has studied the impact of the new sciences on society and has become especially familiar with the development of the Thirld World. He teaches architects, planners and landscape architects at Berkeley.
Address: College of Enviromental Design, Department of Architecture, University of California, Berkeley, California 94720, USA.

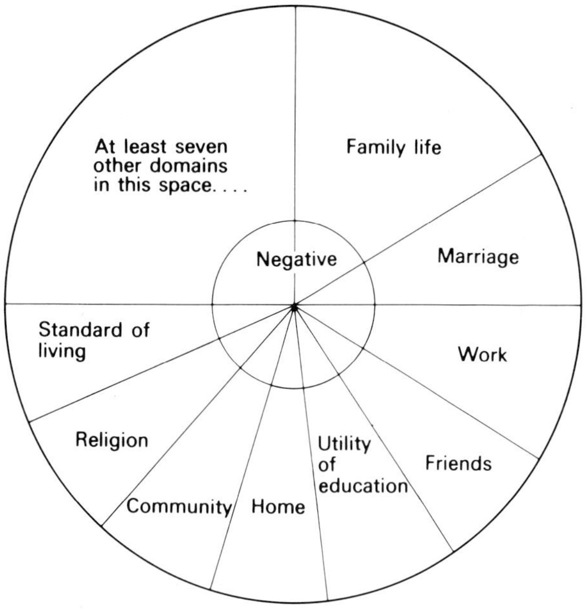

Figure 2. Quality of life: Domains of satisfaction. When people submit to a carefully structured interview intended to explore the sources of general satisfaction with life, the domains of experience which contribute can be weighed for the overall population. A community should support many niches which require quite different allocations of attention and resources on the part of individuals, but the means for urbanized communities are expected to resemble the distribution shown above. The distribution, from worst to best among all communities in a society, of an indicator can also be shown on the donut. This is a Japanese way of seing the 'whole'.

ing of American environmental standards constitutes a threat, since those quality standards were set at levels far stricter than the measured impairment of health and safety – they are intended to prevent violation of American esthetic norms and allay fears raised by the mass media.

Let me cite as an example Manila, where a strong group of environmentalist reformers insisted upon effluent standards no lower than those of the United States. A 1978 study of the investment required over the next planning period, suggested that it would exceed $\$2\times10^9$, as against $\$89\times10^6$ just to cope with growth at the present levels of service. The consulting engineers believed that basic health requirements could be met for the expenditure of $\$2\times10^8$. Rarely are the choices formulated in such a clear-cut fashion. Recent private communications suggest that the US Government is choosing to commit around $\$3\times10^8$, which means that a selected few streams and shorelines will be returned to their 'natural' state.

Recently, Rowe has provided a very balanced assessment of risks to health and safety, suggesting that for developing societies acceptable levels of danger from environmental and involuntary sources may be 10–100 times greater than those set in the United States (Fig. 3.).[8]

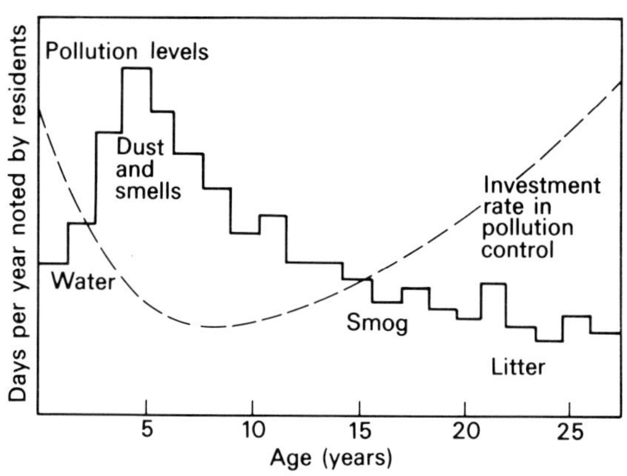

Figure 3. Control of pollutants in new communities. Some of the pollution control is incorporated in the homesites and vehicles at the time of construction and settlement. Later further installations are made, once the monitoring of the environment has established deviations from legislated standards. The levels of investment needed to mitigate air, water and land pollution in the maturing community may well be debated in the future, but in Europe and America at present they are rarely questioned, and a significant share of the available capital is being committed.

WATER CONSERVATION AND SANITATION

New ways of planning water and sanitation have not yet reached a consensus; many battles remain to be fought in the World Health Organization and professional bodies over the next decade or so as the implications are digested, but proposals for appropriate technology are already being entertained. [See also *Interdisciplinary Science Reviews* **3**, 177–232 (1978)].

The best strategy for design is to calculate minimum requirements – consumption just high enough to assure the survival of the community. A city can die from lack of water more quickly than it can from disease, starvation or failure of its transportation system. However, successful competition with other communities will require a careful matching of water supplies and treatment with the built-in demand from people, plants, and physical equipment that is markedly in excess of survival requirements. It is the competition that determines the actual quality standards to be employed and the dependability of supplies that must be sought. Visitors avoid cities with bad water and industries do not expand in locales where water supplies are insecure.

Cities can withstand brief shortages better than most engineers and planners imagine. Two multimillion metropolitan areas, Bombay and Hong Kong, were once reduced to providing water in the mains for two two-hour periods per week. Their strategy was to provide water for restricted periods during the day so as to reduce the waste through leakage; but later, as supplies became so short on some days they had to be skipped, until finally at the most critical period most days passed without a pressurizing of the grid. After the rains came, the cities bounced back to normal, except for a flurry of dysentery, within a month. An estimated 10 liters *per capita* per day were provided for direct consumption during the critical period; the real level of consumption can never be known because the statistical frameworks break down under those circumstances.

Neither Bombay nor Hong Kong were capital cities with large populations dependent upon continuation of central guidance. If Delhi, Bangkok, or Manila could not respond quickly to some regional catastrophe, the loss attributable to extreme water shortage could be multiplied many times. Therefore, for capital cities the planning should aim at delivering an absolute minimum of 20–30 liters *per capita* per day. This guarantee can be achieved through redundancy, such as several reservoirs and independent watersheds, and transport with tankers, barges, emergency pipelines, and tank trucks mobilized on short notice.

Desalination is not the answer to such crises, unless it is economic over the long run as well, due to the high capital cost. Dracones are certainly a cheap alternative to desalination if they are available. These huge floating rubber or plastic bags, resembling almost a sea serpent, were invented in Britain in 1956 and are now extensively used in the Greek Islands. Filled with fresh water from suitable supplies, they float and can be towed by tug to wherever they are needed, their contents being used as such or filtered first. Hong Kong people were heard to complain about their new computer operated desalination plant due to its high energy cost, but were grateful at the peak of the last critical drought that it could work to capacity.

For routine operations, which overcome the effects of seasonality of rainfall, moderate droughts and occasional floods, the institution of a 'quasi-market' appears to be the best policy. This means that prices rise automatically during times of shortage and will otherwise be kept at a level that will pay for all expenses, including external costs. Provision may even be made to allow for long-term contracts for future delivery. The nearest approach at the moment to the ideal of balancing supply and demand by means of price changes is to be found in Israel. There, as in the real world elsewhere, a number of compromises needed to be reached with vociferous special interest groups.

FOUR DIFFERENT WATERS FOR SALE

Imagine a quasi-market for water with three or four qualities as commodities. Each of the lower qualities can be upgraded through the expenditure of known quantities of capital and energy – labor and land being usually unimportant – so the ceiling price for the top two grades is known in advance and should be quite stable. The installations can be intermediate scale technology. Drinking water, the top quality, would be packaged and distributed at a fair price and for the sake of the poor it might be subsidized like rice or bread. The second commodity would be filtered, raw or fresh water; it would be distributed to points where bathing, laundry, food preparation and small scale manufacturing is carried out. The cleaner waste water can be filtered and recycled back for reuse sometimes for the same purpose, but usually for laundry and industry. Within urban districts the resulting 'gray water' can become a commodity transferred from one enterprise to another. It is the third commodity (Fig. 4).

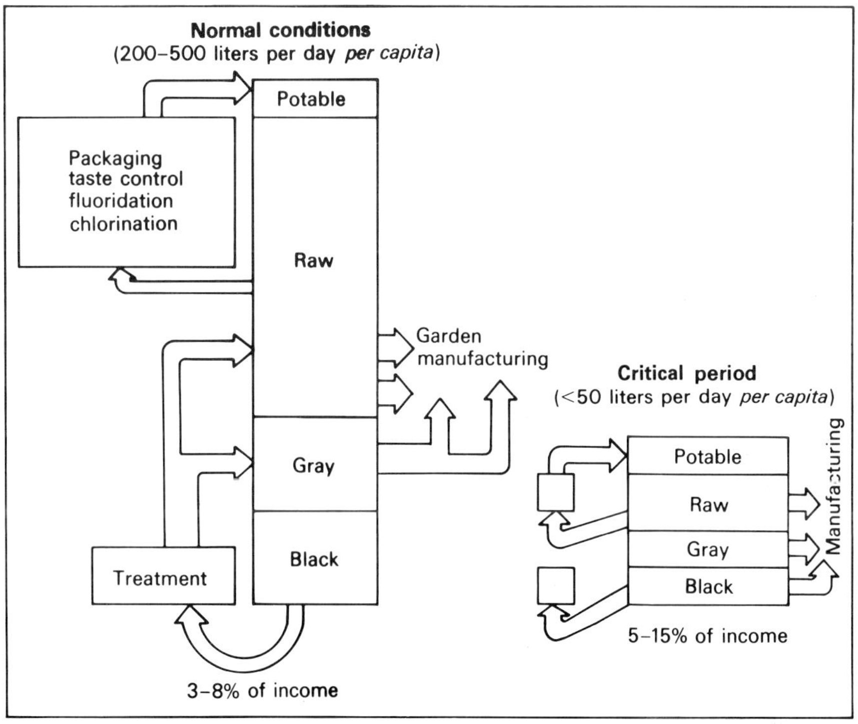

Figure 4. Water commodity production. The water and sanitation industry in resource-conserving urban settlement should be concerned with the production of four different water commodities. Note that when water becomes truly scarce the gardens dry up, and light manufacturing depends much more on gray water. Potable water consumption remains virtually unchanged. Gray water, sometimes even sewage, can be used for evaporative cooling purposes in an emergency, so outflow as sewage shrinks to a small fraction of raw water input.

Finally, the residual water can be used for flushing away waste. By then it becomes a fourth commodity but one with a negative value, because people must pay to get rid of it. Concentrated sewage can be converted in part to biogas and fish, while the water can be used for gardening, thereby significantly reducing the charge for removal.

When rainfall is sufficient, or much too plentiful, the city's storm drains must carry away the surplus. The sanitation treatment system should be kept separate, because dilution adds to costs.

WATER MARKET CONCEPT

The water market concept saves energy when it is applied, because only 5–10 % of the volume is chlorinated and fluoridated. Recycling is begun whenever the water price indicates that it is worthwhile; except in large industrial installations it is expected to require intermediate technology installed as needed. Recycling uses energy for pumping and materials to produce whatever grade of water is scarce. The water market can be extended to a whole watershed, or an even larger region, depending upon the possibilities of water export and import, but it is likely to be considered worthwhile only in periods of critical water shortage.

The market concept is readily understood by consumers, but as a practical matter a number of adjustments need to be made in city design to allow the system to function as intended. The changes include:

◇ Installation of water meters and spring loaded taps or faucets.

◇ Linking together the locations of activities such as bathhouses, laundries and toilets, food processing and local markets.

◇ Provision of space for kitchen gardens, greenhouses and hydroponic uses wherever land rent justifies this.

◇ High concentration sewage collection systems, varied according to the convenience that can be afforded. So, for example, luxury elevator apartments in Athens and Tokyo store human wastes in a tank with 2–3 days capacity for removal by tank-truck at night, while squatter settlements elsewhere may have public latrines delivering directly to a main sewer line.

◇ Provisions for readily modifiable piping, either flexible plastic or rigid polyvinyl, in preference to metal.

◇ Most fundamental is an educational program intended to make people change their water-using habits to fit the resource-conserving style of living appropriate for their locale.

The principal saving takes the form of security against loss of urban function and against illness. Packaged water, prepared under technical supervision and continuously analyzed, assures that waterborne disease will be minimized. The next saving is in the form of capital. Meeting the present water and sanitation standards is likely to account for 15–25% of the total cost for low income housing, a figure which can be cut substantially by fitting living patterns more closely to water supply conditions. An important saving may be shown up in the reduction of time required to convert raw land into habitable settlement. The energy economies become significant much later – after energy prices have risen further – and are primarily important for providing insurance against early obsolescence of the infrastructure.

The overall domestic consumption of water can then be planned at around 200 liters *per capita* per day. At that flow the reuse of water is quite limited. But if supplies should be restricted these planning and management provisions make it possible to keep the city functioning at a level of efficiency that to the outsider seems unimpaired. The industries and gardens cut back on consumption according to the changes in price. In the last drought in California charges for water used beyond a certain quota ran as high as ten times the normal rates, when the reservoirs were 10–20% full.

SEWAGE CYCLES

The treatment of sewage, leading to the production of industrial quality water and byproduct energy, is also opening new possibilities. Rapidly developing cities have exceedingly high land rents greatly adding to the cost of conventional treatment facilities. This is even more so for the high-rate oxidation-ponds, capable of harvesting the nitrogenous and phosphatic components in wastewater which could be recycled back into the urban ecosystem as fish, chickens or eggs (Fig. 5). The complete recycling loop was closed in Israel at the semiplant scale very recently.[9]

A novel approach to aerobic treatment involves mixing air with sewage and pumping it down a pipe to a depth of 100–300 m. Oxygen dissolves much more efficiently under those pressures, the oxidation requiring minutes instead of hours. When the liquid rises, bubbles form, producing an additional lift and greatly reducing the pumping cost. The process was developed by Imperial Chemical Industries of London as a byproduct of their single cell protein technology, and is being installed on a commercial basis in Canada.[10]

The new continuous flow anaerobic unit employs a bed of clay rings upon which a culture of filamentous bacteria build-up. The best conditions appear to be mesophilic, that is, at intermediate or warm temperatures and about 50–70% of the organic material can then be converted to standard biogas. Process development has been assigned top priority by the US Department of Energy and the work is being carried out at the Oak Ridge National Laboratories in Tennessee.[10]

Figure 5. A water and sanitation system to cope with scarcity. Attempts to conserve water, energy, biomass and capital for a city to be founded in the near future indicate the above sequence of technologies. Each one of them has already been linked into partial sequences before now, but the total system is unprecedented. it should evolve over time as the byproducts and outputs promise to be economic. It is important that the short term solutions should not prevent the further elaboration of this scheme, otherwise the city's ability to compete internationally over the long run will be constrained.

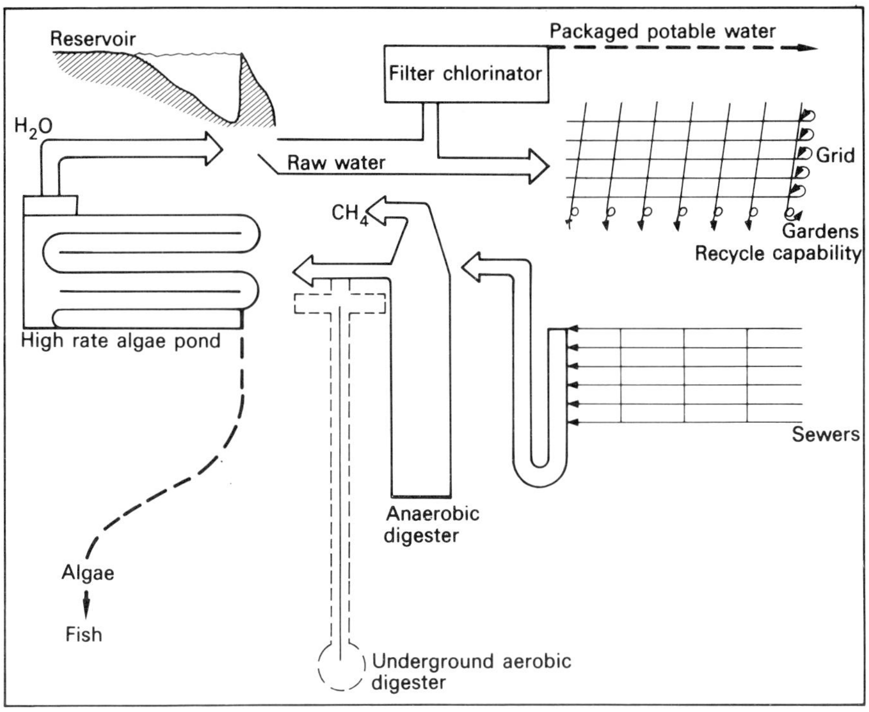

These technologies are complementary when water is scarce, and are fitted to temperate climates. One major advantage is that facilities at intermediate scale appear to be competitive, so the complete sewerage grid does not need to be planned in advance. The facilities could be installed in the basements of large buildings or in the small industry district of an urban neighborhood.

TRANSPORT, ACCESSIBILITY AND CIRCULATION

The movement of people and goods in cities of the future must be quite different from what we know at present. The total demand for liquid fuel by Third World cities is expected to rise beyond that of the already developed metropolitan areas in a time span that is less than a human generation.[11] This overtaking is expected to occur even though the *per capita* consumption in Third World big cities remains low, a half to a fifth that of the increasingly conservation-minded developed cities. It is the heavy pressure for urbanization noted everywhere in the less developed countries – the multiplication in numbers of the urban settlers – that pushes the Third World into the role of dominant consumer.

The prospects for hydrocarbon fuel scarcity are so imminent that one must design settlement patterns that will work smoothly using a third or less of the liquid fuel presently considered to be necessary for modern functions. If this is not done, the last half or three-quarters of the life cycle of buildings constructed in the 1980s will be lost and the overall ability of the metropolis to compete effectively on the world scene will be greatly diminished.

The cheapest and most flexible means of urban mass transport, expressed in energy as well as overall cost, remains that of the bus. Its all-weather characteristics are especially appreciated. The cost of subways, better suited to long trips, becomes comparable only when the value of human time saved from surface congestion is taken into account; the value of time becomes particularly significant for middle to higher income passengers moving longer distances. The latter may alternately be collected by radio-controlled minibus, or jitneys, for the trip to work or school, a service which is usually more efficient (Table 1).

A very strong demand for a personal vehicle is expected to continue regardless of the energy shortage. The bicycle is highly economical compared with the bus, because the energy source is food, which will remain still more valuable than liquid fuel but, except in the vicinity of China, its social status is low. Totally different bicycle designs and images are possible, and a few are beginning to be produced. A variety of stylish bicycles may be used for fair weather transport in the future and such examples are appearing in America and Europe.

A descendant of the bicycle, the moped or velo, is starting to become popular again in developed countries as a substitute for the second car in the household. If gasoline rationing should occur, its 30–50 km per liter consumption should make it very popular indeed in medium to low density settlement. Again, however, it is primarily a fair weather vehicle.

Electric neighborhood cars are feasible for those who can afford them at $3000–5000 apiece. They will be particularly appropriate to an era when nuclear reactors will produce most of the electric power for operating the city and its hinterland. Most dwellings and some shops can then have their fronts on 'slow-

Table 1. Relative Costs of Alternative Modes of Passenger Movement for 1980s (Korea)

Modes	Energy Cost (Calories per passenger per km)	Capital Cost (1980 $ per daily passenger)	Operating Cost (per passenger per km)	Time Cost (minutes per km)	Value of Passenger Time (per hour)	Total Cost ($ per km)
Walking	60 (food)	40	13	18.0	1	0.36
Bicycle	15 (food)	200	18	4.0	2	0.28
Moped (1.2 passengers)	60	400	17	3.0	3	0.40
Electric Car (1.2 passengers)	500	3 200	35	2.5	4	2.00
Bus	60	300	15	3.0	2	0.16
Rail Rapid Transit	60	10 000	15	2.0	3	5.60
Jitney or Minibus (5 passengers)	160	1 500	27	2.5	3	0.98
Taxi (1.5 passengers)	500	4 000	55	1.5	4	2.40

The assumptions in the crude calculations of expected comparative costs were as follows: (i) The value of liquid fuel will rise 50–100% over 1978 prices; (ii) the cost of electric power will rise another 30–50%; (iii) The interest paid on capital will decline, so 20% of existing value is expected to pay for both amortization and interest; (iv), Rail rapid transit has not been granted any credits for central city decongestion – the prime reason for installation in Third World metropolitan regions.

ways' which combine pedestrian, bicycle, moped, and battery-powered vehicle movement on a route 6–10 m wide. These dimensions fit the Asian neighborhoods very comfortably. At present in America the newly adopted standards call for an independent network of bikeways, but still more recent observations suggest that the traffic, ranging from bicycles to electric cars, will mix very safely with ordinary road traffic comprising automobiles, taxis, buses and trucks.[12] Combining intermediate transport vehicles with the full-scale road transport will save same land, perhaps about 2–3% of the total area, but it also simplifies planning and urban management and eliminates the need for many expensive overpasses and tunnels.

These intermediate technology vehicles are simple enough to manufacture, therefore most Third World countries can quickly incorporate the production within their own economies. Korea, the Philippine Republic, Taiwan, Indonesia, Thailand and India could accomplish it without hesitation, and others could follow within a few years. Therefore, the export market in intermediate vehicles and associated equipment is of short duration.

CITY COMMUNICATIONS

Telecommunications can substitute for urban trips to some degree, and should be encouraged whenever possible, because the energy and time savings are very considerable. Therefore, when laying out new land for urban settlement the telephone service should be planned as early as the water supply, the street grid, and the vehicular modes. Electronic data processes will link up the offices of the new urbanization with other urban centers around the world.

It is expected that communications satellites, microwave relays and fiber optics will combine to eliminate most of the complex underground network presently required for comprehensive telephone service. In the past the telephone has almost always followed electrification; by the end of the 1980s it should be possible to purchase a 'connection'. The quasi-market approach is advisable here, as in water and transport service, in hastily settled parts of a metropolis before a dependable electrical grid has been installed in the district. Portable telephones would also be feasible, although the military regimes in charge of many of the Eastern countries have prohibited the introduction of citizens' broadcast band sender-receivers, progenitors of the more flexible systems to come. These communications advances make possible strikingly novel forms of human organization in cities, capable of meeting human needs more promptly and precisely with far less effort. The new urban growth in the Third World has the opportunity of installing the advantageous new technology ahead of established metropolitan areas in the Old World.

The land use assignments that follow from this range of accessibilities available in the near future remain to be calculated from alternate quantitative models, but it is expected that households taking advantage of intermediate transport and communications technology can be more tightly packed and still leave as much latitude for self-determination of style, amount of living space, openness to kinfolk and neighbors, and an incorporation of special interests, such as gardens, pets, the fine arts, books and so on and so forth. The reason is that parking space can be largely eliminated even while personal convenience is retained or slightly enhanced as compared with almost universal private automobile ownership prevalent in the West (Fig. 6).

HOUSES AND NEIGHBORHOOD

Between 20% and 60% of the cost of a dwelling unit in a developing metropolis will depend upon land price or its share of infrastructure investments. The free market in land tends to anticipate future savings in time and convenience and incorporates them into

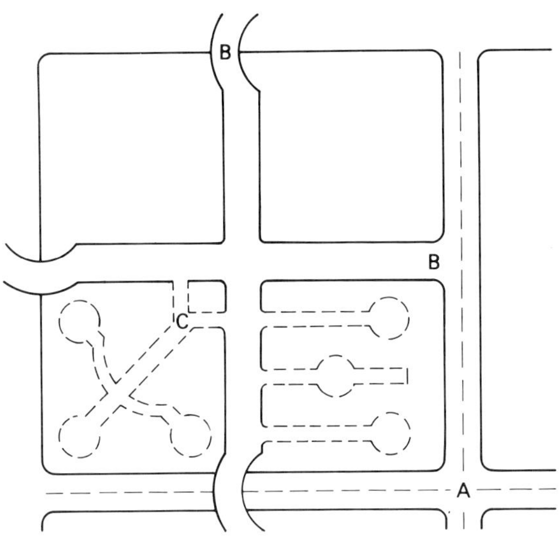

Figure 6. Potentials for intermediate transport. The possible combinations of intermediate transport technology with the existing network are quite numerous. In the above arrangement the A channels carry buses, trucks, taxis and other automotive transport. B might introduce jitneys and vans into the interior of the superblock, while C routes might be tree-lined 'slow ways' for pedestrians, bicycles, mopeds and handcarts. The merging of traffic is gradual, fitting what has recently been reported about the design of environments safe from accidents.

the selling price at the time of construction. Therefore, a transport system which increases the accessibility of land in various categories will also increase the supply, so the price should stay relatively economical. Extension of efficient transport networks into fringe areas of a metropolis should keep buildable land cost at the bottom end of this range. The principal factors leading to satisfaction with low cost housing are well integrated transport and communications networks. This is an argument understood by land economists, but it is rarely acknowledged by others.

The next factor in importance is interior space economy. Too often people are moved into a mass-produced unit with little opportunity to fit it to their own needs. Either some space remains unused, for example, the balcony that remains no more than an expensive decoration of the facade, or the sterile rooftop, while at the same time other needs, like the storage of clothes, toys and plants, or provisions for babies, remain unfulfilled due to lack of appropriate space. Houses put on the market are designed to meet the needs of average people; that leaves about half, who differ in significant ways from the mean, as misfits. These costs are only partly reflected in the market place, but they show up strongly in the measures of satisfaction that underlie quality of life.

Some people wish to socialize with their immediate neighbors, setting up friendly cooperative relationships. Others find their friends among colleagues and wish to minimize contacts with neighbors. These unneighborly types are members of nonplace communities who often congregate at some receptive central hangout. The metropolis should accommodate both extremes and many of the compromises existing in between. Organizing the occasions where people do things together, even if it is only talking and eating, is a great source of satisfaction for many people. In many developing countries family ties are so strong there is a need for clustering kinfolk together in a neighborhood. This need for propinquity is particularly strong for Chinese and South Asian families, once the extended family unit has broken up. The choice of appropriate locations for dwelling within a city depends upon the desires of prospective residents to form and maintain social bonds with other people.

Energy conservation, water saving, and the preservation of the qualities of the natural environment will have a strong effect upon future housing policy. Housing that can be relatively impervious to periodic crises in supply and keeps the risk of fire, flood, accident, street crime and theft at reasonably low levels, is what the great majority of urban dwellers hope to achieve. Often the prerequisites for risk reduction conflict with other values.

We can be explicit about solutions for Korea, where the climate approximates that of Western Europe. A traditional house for the 1980s might consist of two to three rooms with 40–60 m^2 floorspace for regular use by four to five persons. Walls would be hollow concrete block, the roof covered by glazed tile. Insulation needs to be added under the *ondol*, a stove that provides radiant heating from the floor level, as well as under the roof tiles, and on the north and east walls. Improvements on both the standard anthracite-clay briquettes and the stove used for heating the *ondol* have already been designed, and are coming onto the market. When the anthracite is fully depleted, locally produced charcoal may compete with imports.

A carbon monoxide detector and alarm should be installed to reduce the ever present fear of asphyxiation. During the humid summer fans would be used for cooling and ventilation. The amount of energy expected to be saved, as compared with the present, may be as high as one-third, although the level of comfort and security would be actually improved.

Solar energy should become economically feasible somewhat later. The *ondol* may be warmed by hot water from an insulated tank in the house. It is supplied from a central tank once or twice a day. The central tank can be heated by a solar collector, backed up with fuel, with the solar contribution expected to be about two-thirds or a little more. Such a system, after two years of learning, season by season, should save more than 50% of what is presently assumed to be adequate for that size house. Some community development organizations already established could easily take on the management and the program of public education required.

A third approach could provide the same levels of comfort and convenience, and would save 80–90% of the energy, but it is too radical to be accepted readily. It would use south-facing hillsides, including man-made hills, to install earth-sheltered dwelling units. One to three meters of soil would be placed on the roof, thus buffering the house against heat and cold. Cooking would have to go electrical, and the *ondol* would disappear. One of the most economical of ceilings is a concrete shell dome with extra sleeping and storage space on the second level. Windows would open to the south and west, doors would connect with a private yard and vehicle storage area, and the gate would lead onto a lane or 'slow-way'. The top would be a hilly mound with grass or shrubs, a vegetable garden, or a sun deck. The density of single family dwellings with associated circulation space, schools and shops could be as high as 100 units per hectare. The introduction of local markets, small-scale industries, public baths and other specialized public spaces, such as churches and parks would reduce this somewhat (Fig. 7).

This most efficient housing solution is likely to be accepted first by intellectuals, who would add many unique features, would build in small tracts and so the density of settlement might run lower, at 20–50 to the hectare. Very likely most houses would look like ancient Silla Dynasty tombs for noblemen and warriors when viewed from the top, but other shapes for the earth mantle will no doubt be attempted. Sand dunes covered with ice plant vines and north-facing ceiling level openings were chosen for middle class Mexico and Florida tourists, stopping at sites where desert and ocean meet. Other, more traditional, designs would produce comfortable conditions with low energy use during the winter, but earth-sheltered homes should be much more confortable in summer.

A NEW PLANNING PROCESS

Recall that the aim of the recommended changes is to enhance quality of life, even as the supply of resources declines. The solutions should be feasible now; the outlook for improvement should be continuously positive for anyone willing to put in the effort to help himself.

Some assumptions must be made regarding processes that produce satisfaction with life. They are assumptions because the confirmatory studies were undertaken primarily with North American cultures, a few related European and Japanese, but not in the urbanizing Third World. Small-scale inquiries suggest that findings based upon social science techniques are generally replicated in modernizing urban settlement, while indigenous tribal cultures respond less predictably upon arrival in cities. The partially supported assumptions may be stated as follows. People are satisfied

- if they are allowed to participate in face-to-face neighborhood, workplace and community decisions, even if their own arguments did not result in action, often even if they have not spoken at all.
- if the variety of available choices in some important features of life are increasing in number and diversity, the psychic income.

Figure 7. Earth-sheltered dwelling units. In continental climates earth-covered shells, with openings to catch the winter's sun and sealed against moisture, offer many opportunities for differentiation of form. Virtually all the heating and airconditioning energy normally required to operate a home or family business can be saved. Daylight can be beamed to the top and back by a light transmitter positioned at the top of the mound; it is operated with the aid of clockwork. The lane is designed primarily for pedestrials and intermediate transport modes, possibly 6 m wide. Landscaping opportunities are strikingly novel.

- if they can preserve, even if only symbolically, reminders of important collective achievements and experiences.
- if they are allowed, and enabled, to structure their immediate environment.
- if they feel that their social status, which usually depends upon property and education but may involve pilgrimages and other achievements, is improving, or can be improved through personal effort.

All the above propositions hold with a high frequency in Third World cities, but cannot be uniformly true. There are also other bases for satisfaction which apply to smaller proportions of the population. The most important of these other bases to promote acceptable quality of life should be discovered well before the settlement of a new community begins.

Assessment of the sources of satisfaction among the barely literate population can be carried out by means of consultation with the wisest of the social workers followed by a survey using brief, structured interviews. For the better educated portion of the new settlers the initial survey based upon interview should be followed by simple self-administered follow-up.

These social survey instruments might well be of the type devised by planners which identify comparative preferences for features of a home, qualities of neighborhood, services provided by community and aspects of employment, including the manner by which they can be fitted into a realistic budget constraint. In societies where gambling is important, it is also necessary to discover how people would spend their gains when they are lucky in acquiring unearned income. The choices indicated, when aware of the full spectrum of options, are most informative when planning for satisfaction over the long run.

SELF-ORGANIZATION

Catalogs based upon real possibilities for privately constructed dwellings may be provided to prospective urban settlers. The alternatives may range from contemporary developer-designed units to prefabricated alternatives through core assemblies to be completed by oneself with the aid of artisans; all the way to bare site-and-services, which allow one to build in one's own fashion, using available subassemblies on the market such as wall panels, windows, doors and closets. From their choices the market can be estimated and planned, by standard procedures, yet the quality-of-life approach must go much further.

The next overlapping stage is self-organization. In Korea, for example, churches are very important. In Song Nam New City they were built simultaneously with the houses, apparently stimulating much of the neighborhood cooperation. There are various Protestant denominations, Roman Catholic, a strong syncretic Buddhist-Christian synod and traditional forms of Buddhist and Confucian groups. For many of them singing together is rewarding, but this is symbolic of other modes of cooperation.

Also to be organized are businesses, health associations, sports teams, Boy Scouts, garden clubs and other voluntary groups. Self-organization is so important for satisfaction, that the process should be subsidized out of taxes. Thus the building of a church should be made easy, it usually is already, but also the various forms of cooperatives and shared neighborhood facilities should be expedited. Telephone service speeds up self-organization, a principal reason for its early and very high priority.

Another feature is the choice of neighbors. This is not quite as sensitive an affair as arranging a marriage, but compatibility with neighbors having closest contact, allowing arrangements about common walls, trees, vehicle parking, and cleanup responsibilities to be negotiated, can be very significant. Prior negotiations about neighboring have allowed greatly differing social classes, religious communities, and ethnic groups to live together peaceably even during public disturbances. Very likely it would be useful to introduce special subsidies or incentives to induce constructive face-to-face meetings and expand the use of the telephone.

What this policy implements is the rapid expansion of social networks, despite the extra effort of setting

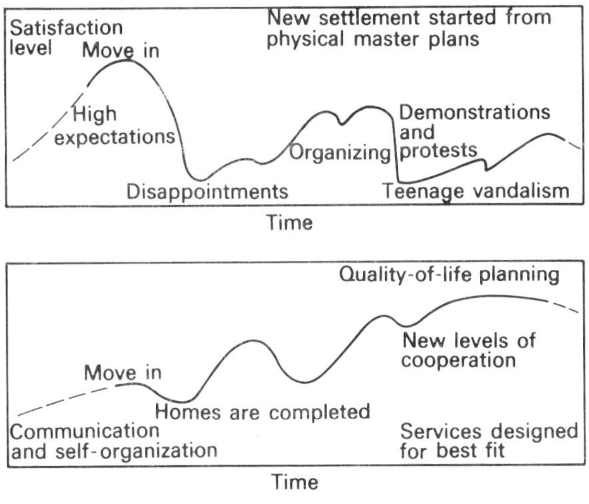

Figure 8. Expectations regarding alternative approaches to planning. Aiming to maintain levels of satisfaction adds to the competitive capacity of new communities. A reduced amount of central planning allows earlier start. Subsidies for cooperation and self-organization will uncover a cadre of local leaders and a capacity to solve mutual problems. The resulting social and cultural services should overcome the kinds of deficiencies that generate teenage vandalism, 'new town blues' and similar conditions arising from constrained social networks and resulting feelings of isolation.

up housekeeping in a new place. Many personal networks crystallize into busineses and nonmarket social organizations. The evolving social structure of the community can determine the future growth of services. The degree of subsidy for creating organization should depend upon the degree of consensus. If, say, two-thirds agree, and it is constitutionally feasible, then the startup cost for a new cooperative facility should be taken out of taxes; while if half consent, then the community should help promote the new activity through extension of credit or other devices. Promotion of self-organization and cooperation builds up a variety of leaders in the community who can get together to manage the next stages in its development and maturation. It also reduces the detail required in the land planning, since the planting of trees and shrubs, for example, could be undertaken by the residents with the help of horticultural and landscape consultants, assignment of space for the parking of intermediate transport could be worked out and many similar features treated as needed.

Most new communities are made up of two kinds of people: the members and clients of bureaucracies, firms and agencies who know they will be assigned to the new site and the opportunistic members of the urban poor who do most of the maintenance and service work. The latter are reinforced with surplus labor force from the countryside, many of whom will not stay in the settlements. Dissatisfactions arise soonest among the educated people and later their children. When complaints are voiced and formal protests produced, portions of the less educated population will find themselves in agreement. This normal process can interfere with the efficiency of the agencies (Fig. 8). An approach that enhances the quality of life should also contribute to stability, economic productivity and cultural advancements.

The overall process implies that the physical inputs of materials provide a setting for the development of human resources. The water and energy sustain the community and the transport and communications services expedite social transactions, allowing the urban system to produce satisfactions. Meanwhile the organizations – public, private and voluntary – enhance the quality and maintain the goodness of fit. The effective resource-conserving community must accomplish all of these tasks.

LITERATURE CITED

1. Korea '78: Catching up with Japan. *Far East Economic Review* **100** (21) 35–74 (26 May 1978).
2. R. L. Meier, *Planning for an Urban World – Design of Resource-Conserving Cities*, MIT Press, Cambridge, Massachusetts (1974).
3. J. McHale and M. C. McHale, *Basic Human Needs: A Framework for Action*, Transaction Books, New Brunswick, New Jersey (1977).
4. M. Mesarovic and E. Pestel, *Mankind at the Turning Point*, Dutton, New York (1974).
5. A. Campbell, D. E. Converse and W. L. Rogers, *The Quality of American Life: Perceptions, Evaluation, and Satisfactions*, Russel Sage Foundation, New York (1976).
6. F. M. Andrews and S. B. Withey, *Social Indicators of Well-Being: Americans Perceptions of Life Quality*, Plenum, New York (1978).
7. *Research Report Summaries '76*, Research Institute of Telecommunications and Economics, Tokyo (1976).
8. W. D. Rowe, *An Anatomy of Risk*, Interscience, Wiley, New York (1977).
9. G. Shelef, *Proceedings of the International Conference of Production and Use of Micro-Algae Biomass, Acre, Israel, September 1978*. National Council for Research and Development, Jerusalem, in press.
10. Earth-Covered Settlements Conference, sponsored by the US Department of Energy, Fort Worth, Texas, 17–20 May 1978. Proceedings in press.
11. R. L. Meier, S. Berman and D. Dowall, *Urbanism and Energy in Developing Regions*, Center for Planning and Development Research, University of California, Berkeley (March 1978). A report.
12. W. L. Garrison and J. F. Clarke, Jr, Report 78-1, *Prospects for Neighborhood Cars*; Report 78-4, *Studies of the Neighborhood Car Concept*. College of Engineering, University of California, Berkeley (October 1977).

The manuscript was received 1 May 1979

Chess with Computers

PROFESSOR DONALD MICHIE
Machine Intelligence Research Unit, University of Edinburgh, Scotland

The properties of chess as a domain for problem-solving by machine resemble those of other applications. Much conceptualization is needed from human practitioners but very little from new computers, allowing brute-force searches enormously larger than could be covered by the human brain. Because of its finite and modular structure and the availability of a large corpus of systematized conceptual knowledge bequeathed by the masters, chess constitutes ideal laboratory material for studying the relative effectiveness and computational costs of search-driven *versus* concept-driven approaches.

The history of computer chess begins with Charles Babbage and proceeds through Turing and Shannon to the present day, when tournament programs are beginning to challenge International Masters. In research studies, as opposed to tournament play, the focus has moved to the pattern-based rules mediating Master skill. Programming methods are required which will permit the incorporation of such rules on a large scale.

Understanding the nature, number and means of acquisition of chess patterns is seen as the key. Estimates have been made of the number of chess patterns acquired by a chess-master. Since these are in tens of thousands, accelerated machine uptake by use of inductive generalization is a necessity. According to initial observations, it is also a practical possibility.

One anticipated outcome of progress along this line is a methodology for using the machine to re-codify, in greatly improved form, human knowledge which at present enjoys only fragmentary, error-infested, inconsistent and unsystematic representations in textbooks. This kind of 'knowledge refining' belongs to the new discipline of Artificial Intelligence (AI).

THE CONCEPT OF ARTIFICIAL INTELLIGENCE

Artificial Intelligence workers sometimes complain that they are in a 'no win' situation. 'Would it be intelligent if a machine could read a newspaper and give you a summary of its contents?' enquires the AI scientist. 'Certainly!' concedes his critic. 'My student', replies the AI man, 'has just done that.' 'But how does his program work?' says the critic with an air of suspicion. After a spell with blackboard and terminal he decides that his suspicion was justified. 'So that's all! I don't call that intelligent.'

My AI colleagues find this understandably irksome. Is every implementation of this or that aspect of intelligence to be dismissed the moment the implementation is understood? It is of course agreed that there must be no trickery. A program which prints out just the headlines from a newspaper might achieve a surprisingly respectable summary. But even after tightening up the task definition it might still not require intelligence for a human to perform it. If so, it would be 'Type 0' and declared of no relevance to artificial intelligence research. Simple arithmetic is such a task; searching a dictionary is another.

Relevance begins when the task as defined cannot possibly be done by a human except on the basis of considerable intelligence. AI people have in the past lumped all these tasks together and have said 'If a machine ever does any such thing, then it is intelligent'. It is now becoming apparent that this is wrong, and that the category is compounded of two types. Type 1 demands intelligence from any human solver, but may be machine-soluble by some other route. Type 2 demands intelligence from any solving device whatsoever. In my use of the term 'intelligence' it is intimately related to the memory-saving trick underlying all cognition, namely use of conceptual models.

This is what the critic usually has at the back of his mind: the power of the human brain to conceptualize its task environment. Some tasks, he realizes, require a high degree of conceptualization, whether by brains or machines. An example would be watching baseball and preparing a newspaper report of the game. Even annotating a game of chess between strong club players would be regarded as a far more intelligent machine feat than a chess program's exploit in beating the players. The latter exploit would exemplify only Type 1. It certainly demands a high degree of conceptualization from the human player, as is evident from de Groot's[1] study. But in this era of fast processors and large memories the possibility exists of finding equally good or better solutions by other means. Indeed, chess computer programs already beat strong club players, who rightly regard their vanquishers as unintelligent.

CCC-0308-0188/80/0005-0215 $06.50/1

If we redefine the playing task as defeating Fischer or Karpov, then we almost certainly have Type 2. No amount of brute-force computation will then obviate the need for more or less elaborate conceptual representations of chess skill in machine memory. If we further tighten the performance criterion and demand that a correct move be always computed, that is, that the program be error-free in the game-theoretic sense, then we do not know whether the computation is feasible at all on a terrestrial scale, however 'intelligent' the machine. This depends upon the computational complexity of chess, at present unknown.

Consider a domain of great complexity – meteorology. Just as Type 1 tasks can be found, for example error-free play of elementary endings, within the larger Type 2 domain of chess, so Type 1 tasks of weather prediction can be found within the Type 2 domain of meteorology. An example is the preparation of one-week forecasts for the British Isles. It takes years to become a first-class forecaster, and not everybody has the intelligence to become one. But when the European Centre for Medium-range Weather Forecasting gained the benefit last year of an 80-million-instructions-per-second Cray 1 computer, no one doubted the Centre's ability to achieve one-week predictions with little to fear from unaided human competition. Eventually the aspirations of computer-aided forecasting may grow to the point that conceptual and heuristically structured models, of the kind beloved of AI people, will have to be integrated into their present supercalculational representations. By then the machines will have entered Type 2 territory and human forecasters, however intelligent, may be left behind.

Elsewhere[2] I discuss various issues of complexity which bear on what brains and machines can and cannot do. Ideal laboratory material for preparing the way for these topics can be obtained from computer chess, to which the remainder of this paper is devoted.

HISTORY OF COMPUTER CHESS

The first serious proposal to have a machine play chess was made in the 19th century by Charles Babbage, the British pioneer of digital computing. It was never executed. In the early years of this century the Spanish engineer Torres y Quevedo demonstrated an electro-mechanical device for mating with king and rook against king. The first experiments with the complete game were conducted in the later 1940s by the British logician A. M. Turing.

According to the testimony of I. J. Good[3], interest in the idea of mechanizing the play of intellectually challenging games had taken root in Turing's mind as early as 1941. After the war he collaborated with D. G. Champernowne and others to construct various paper machines embodying mechanized strategies for chess. Play was poor.

Of Turing's own chess prowess, International Master Harry Golombek[4] wrote:

> '. . . His marvellously inventive mind constantly delighted me but, curiously enough, though passionately keen on chess, he was a weak player; so weak that, upon his resigning to me, I was able to turn the board round and win against him with the colour with which he had resigned.'

Turing[5] refers to experimental tests. These were under way already in 1947, when he and Champernowne engaged with S. Wylie and myself[6] in an uncompleted paper machine experiment. He made subsequent attempts to complete the experiment by writing a program for the Ferranti Mark 1 machine at Manchester, and at the time of his death in 1954 he had active plans to continue the work.

In 1950 Claude Shannon published a paper[7] which went beyond Turing's proposals in important respects, though few of these extensions have yet been implemented. The Turing–Shannon 'lookahead–evaluate–minimax' paradigm still forms the basis of today's tournament programs.

During the same period de Groot's[1] study of the thought-processes of chess-masters revealed that their special ability does not reside in computer-like qualities of memory or of accurate, fast and sustained calculation, but from powers of conceptualization. This result had been foreshadowed by Alfred Binet's[8] investigation in 1894 of the ability of chess-masters to play many games of blindfold chess simultaneously. Binet concluded that in addition to *la mémoire* this accomplishment rested on *l'érudition*, the use of accumulated chess knowledge to form meaningful descriptions of board positions, and *l'imagination*, the ability to reconstruct mentally a position from a description. By *la mémoire* Binet did not intend the kind of photographic representation postulated in so-called 'eidetic' imagery. The point receives

PROFESSOR DONALD MICHIE has held a Personal Chair in the University of Edinburgh since 1967. Born in 1923, he was educated at Rugby School and Balliol College, Oxford. During the 1939–1945 war he was a founder member of Max Newman's team of code-breakers using the Colossus series of high-speed electronic computers. His working contacts at that time, particularly with Alan Turing and I. J. Good, gave him a lasting interest in computing. After a post-war career in genetics, he found his way back to machine intelligence and at Edinburgh in 1965 he founded the first-ever University course in the subject. He edits the *Machine Intelligence* series of volumes, of which nine have appeared to date.
Address: Machine Intelligence Research Unit, University of Edinburgh, Hope Park Square, Meadow Lane, Edinburgh EH8 9NW, Scotland.

emphasis from Reuben Fine's[9] introspective memoire on blindfold play: 'It seems to be impossible to think of the board without relationship to the pieces, and the pieces without relationship to the board' and '... this ability is related to the more general process of concept formation ...'. It seems that the chief significance of *la mémoire* lies in its use to give effect to the other two capabilities. Although machine play is now impressive, little progress has been made in mechanizing *l'érudition* and *l'imagination.*

The earliest chess programs were developed in the 1950s and have been reviewed by Samuel.[10] The modern era dates from the 1967 entry into human tournaments of the Greenblatt–Eastlake–Crocker[11] program. Play was at the level of a weak-to-middling club player, Class 'C', US Chess Federation (USCF) rating about 1500. Expert level begins at about USCF 2000, National Master at 2200, International Master at 2400, Grandmaster at 2500.

The first World Computer Chess Championship,[12] held in 1974 was won by the program KAISSA developed by V. L. Arlazarov, G. M. Adelson-Velskiy, A. R. Bitman and M. V. Donskoy. Standards of play corresponded approximately to USCF 1600. In 1977 the second World Championship was won by the program CHESS 4.6 of L. Atkin and D. Slate, with KAISSA second. Play rated about USCF 2000. Program improvements contributed to progress, but the major factor was order-of-magnitude increases in speeds of computing hardware to 15 million per second and 3 million instructions respectively.

Further large speed-ups are now available through the development of special-purpose chess circuitry, as in the CHEOPS device completed by Greenblatt and Missouris[13] at MIT, and the appearance of new main frames such as the Cray computer mentioned earlier. Tenders placed by American institutions for machines of highly parallel architecture with over 10^9 instructions per second capacity (NASA Ames Center), and on-line memories of 10^{16} bits (Lawrence Radiation Laboratory), add further indication that the next few years' hardware development may alone suffice to force the door to Master-level play. However, for scientific study of cognitive science and cognitive engineering, more ambitious goals of performance, or of computational economy, must be set.

SCIENTIFIC ASPECTS

Computer chess has been described as the *Drosophila melanogaster* of machine intelligence. Just as Thomas Hunt Morgan and his colleagues were able to exploit the special limitations and conveniences of the *Drosophila* fruit-fly to develop a methodology of genetic mapping, so the game of chess holds special interest for studying the representation and measurement of knowledge in machines. Its chief advantages are: chess constitutes a well-defined and formalized domain; it challenges the highest levels of intellectual capacity over a wide range of cognitive functions – logical calculation, rote-learning, concept-formation, analogical thinking, deductive and inductive reasoning, and so forth; a detailed corpus of chess knowledge has accumulated over centuries in chess instructional works and commentaries; a generally accepted numerical scale for performance is available in the USCF rating system; and finally, the game can readily be decomposed into sub-games which can be subjected to intensive separate analysis.

To clear the ground for further discussion of the knowledge approach, we must first review the attainment level and mechanisms of state-of-the-art chess programs and the reader should learn about a recent performance of CHESS 4.6 (see Appendix).

STATE-OF-THE-ART MECHANISMS

The fundamental Turing–Shannon machine procedure involves looking ahead from the current position along a branching tree of possibilities. Human players also look ahead, through concrete analysis, but on a severely restricted scale. According to de Groot, 30 positions mark the limit to what a Master normally holds in his lookahead memory, and there is general agreement that the upper bound may safely be placed at about one hundred positions. By contrast, chess programs commonly grow lookahead trees in computer memory comprising hundreds of thousands of positions.

If no rules for pruning unpromising branches were applied, branching of the lookahead tree would be around 30 branches per position in the mid-game. Without such pruning the search would accordingly grow almost by powers of 30 (the 'almost' is for the occurrence of confluences in the lookahead tree which could be detected and eliminated at some computational cost), and would soon be defeated by what has been termed the 'combinatorial explosion'. Pruning rules are of two types:

The first type, to which the term 'alpha-beta pruning' has been applied, preserves completeness. That is, it is guaranteed not to affect the move-choice finally extracted from the search. In favorable cases such pruning doubles the number of ply over which search can be made for a given number of positions examined in lookahead. (A ply is a half-move. A move usually denotes a complete 'White moves, Black moves' cycle.) A simple illustration of the rule's operation is shown in Fig. 1. The animating principle is that once evidence is to hand that a variation is not the best, no further computational work need be expended on determining by just how far it falls short.

The second type of pruning, sometimes called 'forward' or 'plausibility' pruning, suppresses branches when features detected in the position from which branches would be grown indicate that they are

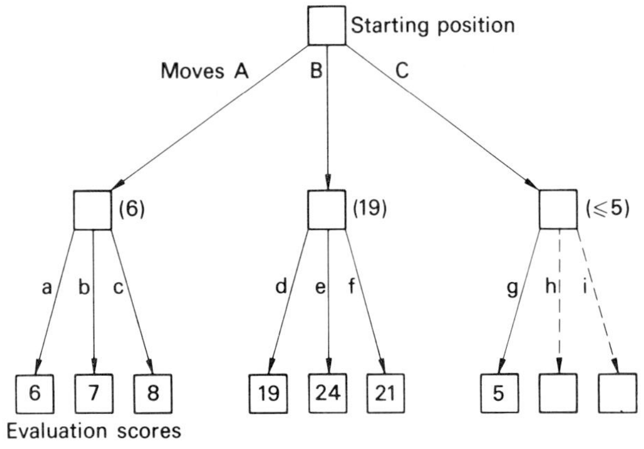

Figure 1. 'Toy-sized' search tree for illustration of the basic idea of alpha-beta pruning. The search here goes to level 2 and is conducted 'depth first'. Scores which have been backed up by the minimax rule are shown in parentheses and broken lines denote a part of the tree which the alpha-beta rule eliminates as unnecessary to explore. Such exploration could only alter downwards the score provisionally asssociated with the right-hand level-1 position, so that move B would be selected in any case. Move g is called a *refutation move* to move C.

Because the tree drawn here is only two-ply deep and the first refutation move was encountered late in the search, the economy appears small in this diagram. In the general case, when the rule is applied recursively level by level, savings are very significant. In the ideal case the number of positions requiring to be examined is reduced to the square root of the full number. Fig. 2 shows a slightly larger case, illustrating the minimax procedure on which alpha-beta operates.

likely to lead to bad continuations – for example, involving uncompensated loss of material. Forward pruning heuristics are many and various, and carry the risk of eliminating a line which continued search would have picked as the best. The strongest contemporary program, CHESS 4.6, makes no use of forward pruning.

All programs apply termination rules to halt growth of the tree beyond certain limits. The main factor in termination is the occurrence of quiescent positions, Turing's 'dead' positions, in which no captures or forced moves are in immediate prospect.

In the Turing–Shannon scheme the program applies an evaluation function to the terminal positions of the lookahead tree, labelling them with computed estimates of their strategic strength or weakness. The leading term of the function is derived from a count of own and enemy pieces assigning weights such as the conventional valuations: 9 for queen, 5 for rook, 3 for bishop, 3 for knight, 1 for pawn. Bonuses may be added for such qualifying factors as the two-bishop advantage, or the advancement of pawns. Mobility normally comes next. Its importance was established empirically by the British clinical psychologist Eliot

Table 1. Results of an Examination by E.T.O. Slater of 78 Arbitrarily Selected Master Games which Ended with a Decisive Result on or Before the 40th Move

After move	Winner's mobility (average)	Loser's mobility (average)	Difference
0	20.0	20.0	0
5	34.2	33.9	0.3
10	37.5	36.0	1.5
15	39.7	35.2	4.5
20	38.9	36.4	2.5
25	39.6	31.9	7.7
30	35.6	27.7	7.9
35	31.7	23.2	8.5

Slater[14] in 1950. His results are shown in Table 1. Other features contributing terms to the evaluation function are king safety, control of central squares or squares in the neighbourhood of the enemy king, pawn structure, rook control of files, and so forth. A tournament program's evaluation function typically consists of a weighted combination of thirty or more such terms.

After the terminal positions of the lookahead tree have been allotted values, these are backed up the tree by the minimax rule illustrated in Fig. 2. Starting from the bottom, each position for which it is White's turn to play is credited with the *maximum* of the values of its immediate successors, on the presumption that White wants to get the best for himself; the convention here is that large positive values are good for White, bad for Black. Each Black-to-play position is allotted the *minimum* of the values of its immediate successors. The wave thus spreads through the tree from the terminal positions towards the root, until all the immediate successors of the position currently on the board have received values. Playing White, the program selects the move which will give the highest-valued of these successor positions. Playing Black, it will select the lowest.

Various devices for eliminating redundant calculation and redundant storage can be used to improve this basic mechanism. The most far-reaching is KAISSA's use of a theory of influence to avoid repeated computation of sub-trees identical in all but inessentials. Allowing for overheads incurred in applying the method, the saving is approximately twofold.

The depth to which modern computer chess programs look ahead ranges over 6–15 ply, somewhat in excess of the typical depth of analysis by Masters and Grandmasters as measured by de Groot. Correspondingly, in sharp tactical situations machine play is now comparable with Grandmaster play, while meriting no more than 'C' rating for positional play. Machine tactical analysis is now so strong that decisions may sometimes be obscure to human onlookers even though soundly based. A case in point arose in the 1977 World Championship in KAISSA's losing game against Duke University's program DUCHESS, as shown in Fig. 3.

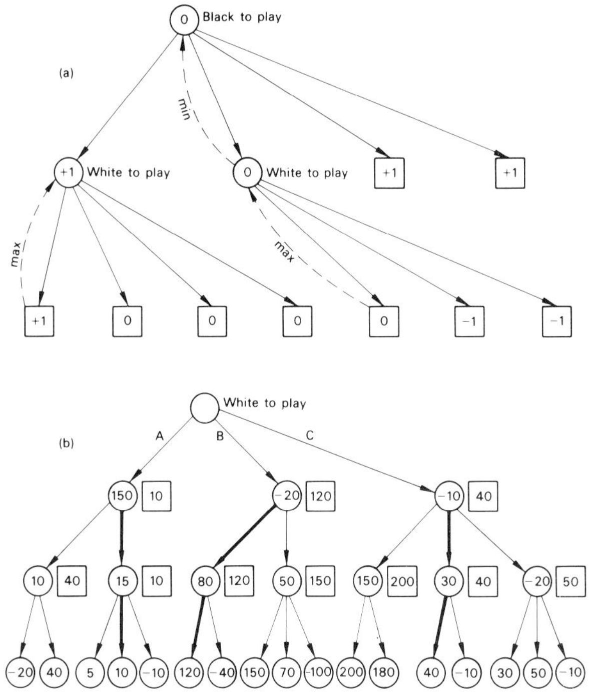

Figure 2. (a) The root of this two-level lookahead tree acquires a value by alternate application of the 'max' and 'min' functions. If alternation is extended backwards from all terminal positions of the game tree, the initial position of the entire game will ultimately be assigned a value. Terminal positions are shown as boxes.

(b) Lookahead tree in which the positions are marked with 'face values' assigned by an evaluation function (minus signs). Boxed values are values backed up by minimax from the lookahead horizon. Arrows with broken lines correspond to moves which would not need to be explored under the alpha-beta pruning regime: the move finally selected by White is B, as having the largest of the three backed-up values. Why not A, as having the largest of the three face values? Curiously enough, although supported both by intuition and by experiment, the rationale for the backup basis of move selection remains to be formally explicated.

LIMITATIONS OF THE TURING–SHANNON PARADIGM

The standard of overall play has yet to attain solid Master level. Some of the reasons are as follows:

Horizon Effect

Berliner[15] has pointed out that reliance on unaided Turing–Shannon procedures renders a program oblivious to all events beyond its lookahead horizon. Even though a trans-horizon loss, or a trans-horizon gain, may appear inevitable or obvious to the human spectator, the program plans from hand to mouth, short-sightedly sacrificing material to delay a loss which cannot indefinitely be averted. Alternatively it

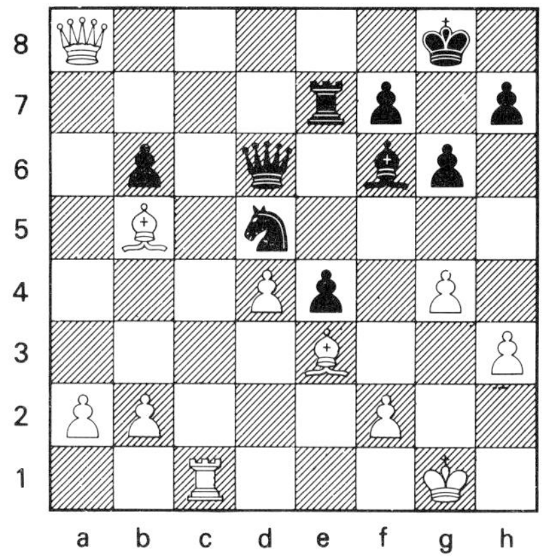

Figure 3. In this position KAISSA, playing Black, sacrificed a rook with 34. ... Re7–e8. Although this clever move greatly prolongs Black's life, since the alternative of moving the king opens the way to a 5-move mating combination by White, 500 spectators, including International Masters and a former World Champion, assumed that it was a blunder.

may forfeit an eventual large expectation by grabbing at an immediate small gain.

Here is a case as reported in *Computer Weekly* of 20 April 1978 from play against CHESS 4.6 at the Aaronson European Chess Congress, 1977. Mr L. Perry had the White pieces and play went as follows (algebraic notation):

English opening, transposing to Queen's gambit accepted.

	L. Perry White	CHESS 4.6 Black
1.	c4	Nf6
2.	Nc3	d5
3.	d4	d x c4
4.	Nf3	a6
5.	e4	b5
6.	a4	b4
7.	Na2	N x e4
8.	N x b4	e5
9.	Nc2	e x d4
10.	Q x d4	Q x d4
11.	Nf x d4	c3
12.	Bc4	Nd7
13.	b x c3	

Black: Chess 4.6

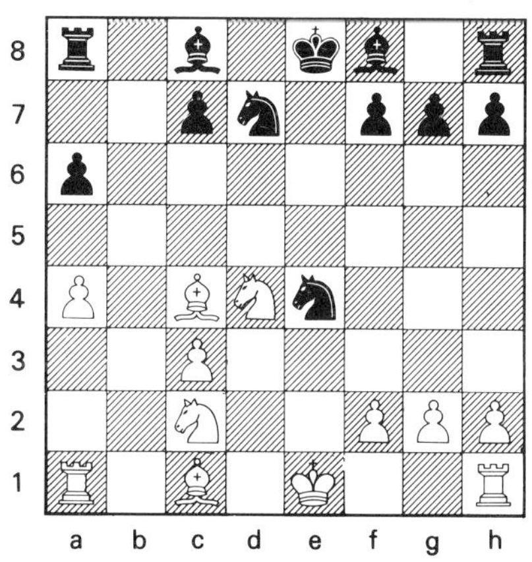

White: L. Perry
Position after 13. b × c3

At this point CHESS 4.6's seven-ply lookahead found a seven-ply combination forcing the gain of a pawn. But the program could not detect within this horizon that the spot in which Black's knight ends up is a death-trap. The next few moves went:

13. . . . Ne5
14. Bd5 Nd3+
15. Kf1 Ne x f2
16. B x a8 N x h1,
stranding the knight.

It might be asked whether the fateful Ne5 was necessarily based on lookahead of just seven-ply. Does not CHESS 4.6 sometimes analyse more deeply? Yes, but only along forcing sequences of captures and checks. The horizon position created by 16. . . N x h1 is quiet. Another 15 moves elapsed in the game before the knight was finally taken. But its role had been reduced to that of a passive onlooker. Black resigned on the 33rd move, almost immediately following its capture.

The human player sees that the knight will be trapped. He does this not by lengthy and detailed lookahead analysis but by perceiving familiar patterns remembered from past experience. Pattern-knowledge of this kind is fundamental to chess.

It is not to be supposed that the human's more far-sighted awareness is based on a longer lookahead in the ordinary sense of detailed analysis. We have already seen that the balance of tactical analysing power lies these days in the other direction. As is evident from a careful scrutiny of the examples in Fig. 4, the longer view is derived from processes of a different kind, involving reasoning about the broad properties of the position concerned. This type of 'conceptual lookahead' discovers, for example, that the bishop in Fig. 4(a) is lost without need to follow all the forward variations. Powers of abstract and causal reasoning have been implemented by Berliner in experimental laboratory studies, and a number of workers are studying the incorporation of chess pattern-knowledge into computer programs. Tournament programs, however, so far remain devoid of these capabilities.

Lack of Long-range Strategic Ideas

A Master plans at the conceptual level, linking the main milestones with detailed steps in a separate operation. Contemporary programs have no corresponding capability. In the end-game in particular, where long-range reasoning is at a premium, computer programs can flounder aimlessly, promoting small disconnected goals with no unifying strategic thread.

Lack of Locally Applicable Heuristics

By themselves advances in computing power are not expected radically to improve the situation described above. Remedies must take their departure from an appreciation of the chess master's ability to utilize very large bodies of conceptualized and indexed knowledge. In programs of the Turing–Shannon type, the sole significant repository of chess knowledge is the evaluation function, which embodies only those heuristic principles which have uniform application over the myriad sub-games which constitute the complete game of chess. Principles which apply in some kinds of situation, but not in others, cannot easily or compactly be accompanied by such a structure. The point can be illustrated by the results of submitting to CHESS 4.5, an almost identical earlier version of CHESS 4.6, three test positions from the ending king and knight against king and rook (KNKR).[16]

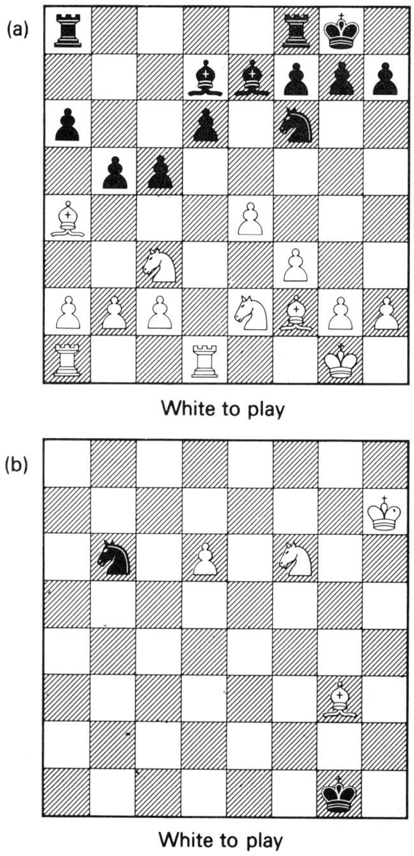

Figure 4. (a) With a five-ply lookahead White appears to escape loss of the bishop by advancing the pawn on the file. By the time Black has replied P x P in this branch of the lookahead the bishop loss lies on the far side of the horizon.

(b) Greed in going immediately for pawn promotion is rewarded by the tedium of playing out a bishop and knight mating sequence. Since the enemy knight is tied down, the bishop can move around to attack it via Be5–Bd4 and thereby enable smooth promotion.

PERFORMANCE OF CHESS 4.5 TOURNAMENT PROGRAM WITH KNKR

CHESS 4.5 was required to defend the weaker side of KNKR against a human opponent rated just over 2000 on the US Chess Federation scale, that is, an 'expert'. The program ran on a CDC 6400 machine, on which it was able to win the 1976 ACM Computer Chess Championship; more recently it has had highly successful trials on the much faster CYBER 176. CHESS 4.5's general evaluation function was used without allowing any adjustment or special tuning to the KNKR problem. Search depth was set to seven ply. Since forced variations are searched beyond this preset horizon, moves eight ply deep were occasionally searched in the present case. Under these conditions, CHESS 4.5 typically looked at a few tens of thousands of nodes per move and spent up to 120 sec per move, typically between 30 and 60 sec. The three trials are described below.

(1) A classical difficult defence described by Fine[17] and Keres[18] with the weaker side's king in the corner. CHESS 4.5 found correctly the move considered most difficult in the books, but then stumbled on the fourth move of the main 'book' variation, obtaining a lost position. Starting from the position of Fig. 5(a) with CHESS 4.5 as Black, the play went:

1. Rb2+ Ka1
2. Rb8 Ne2! (The move considered to be most difficult)
3. Kb3 Kb1
4. Re8 Nd4+? (Keres gives 4 .... Nc1+)
5. Kc3 Nb5+ (If 5 .... Nf3, 6. Re3 separating K and N definitely)

Play actually stopped here, but Black is lost as the further analysis shows: 6.Kb4 Nd4, 7.Re4 Nc2+, 8.Kc3 Na3 (If 8 .... Kc1, 9.Re2) 9.Re1+Ka2, 10.Re2+ Kb1, 11.Kb3, and White wins.

(2) Figure 5(b) shows another difficult position, with the weaker side's king on the edge. CHESS 4.5 playing Black found the only correct defence against the main line given by Keres (exclamation marks by Keres):

1. Rb7 Nh6
2. Rh7 Ng8!

In a game Steinitz–Neumann (1890, with colours reversed), Black wrongly continued 2..... Ng4 and lost after 3.Rh4 Ne3, 4.Re4! Nd1, 5.Rf4+ Kg7, 6.Rf3 Kg6, 7.Ke5 Kg5, 8.Kd4 Kg4, 9.Rf1 Nb2, 10.Rb1 Na4, 11.Rb4 winning the knight.

3. Rf7! Ke8
4. Rg7 Kf8!
5. Rh7 Ke8
6. Rf7 Nh6
7. Rf1 Ng8

This preserved the draw. Or, as also tried, instead of 7 Rf1:

7. Rg7 Kf8

(3) A further position (Fig 5c) has the weaker side's king in the centre, the easiest defence. CHESS 4.5 playing Black allowed its king to be driven to the edge resulting in a harder defence. This enabled the opponent to create mating threats, and after additional weaker moves by the program the king and knight became separated, leading to a lost position. In the game-record below, numerals in parentheses give the number of moves to knight-capture given best play. '?' denotes a mistake, that is, a White move which turns a win into a draw or a Black move which turns a draw into a lost position. (?) denotes an inaccuracy by White, that is, a move which turns a position winnable in a few moves into a position requiring many moves to win.

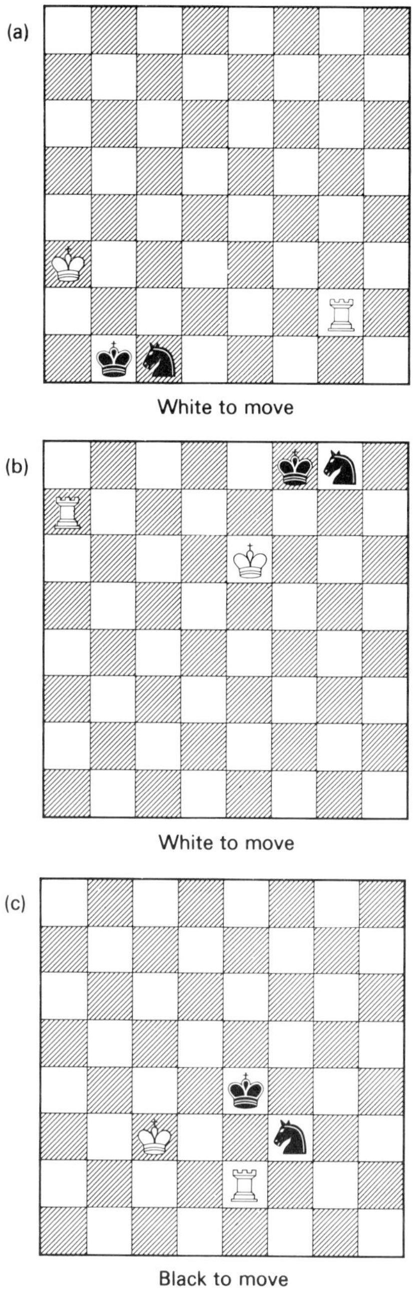

Figure 5. Three starting positions used for trials of the CHESS 4.6 computer program, as described in the text.

1. ... Kf5
2. Kc4 Nh4
3. Kd4 Kf4
4. Rf2+ Nf3+
5. Kd3 Kg3

6. Ke3	Ne5	
7. Rf8	Kg4	
8. Ke4	Nc4?	(After this Black is already lost; better is 8. . . . Ng6)
9. Rg8+ (10)	Kh5	(Now king and knight are separated)
10. Kd4?		(10.Rd8! wins)
	Nd6	(Drawn)
11. Ke5	Nc4+?	(Does not care about joining king and knight; Nf7+draws)
12. Kf4 (10)	Nb2 (8)	
13. Rg5+ (8)	Kh6 (7)	
14. Kf5 (11)(?)		(Rb5! wins easily)
	Nc4 (11)	
15. Kf6 (22)(?)	Ne3 (22)	
16. Ra5?	Ng4+	(Drawn)
17. Kf7	Nh2	
18. Ra1	Ng4	

The position is drawn again and play was terminated. Although starting from an easy drawn position, the program, lacking fundamental knowledge about KNKR ending, was wandering from one lost position to another. This last example shows that a sneaky player can, probably starting from any KNKR position, always induce the program to disjoin king and knight and get into a lost position. The program's 'global' evaluation function sees no reason why king and knight should not be separated!

It is interesting to observe that the program's opponent, although an expert, after achieving theoretically won positions never grasped the opportunity actually to defeat the program. This has a bearing on the level of difficulty of this sub-domain.

CHESS PATTERNS AND THEIR ACQUISITION

The underlying cause of CHESS 4.5's failure on the above-described test is the unsuitability of a unitary 'global' evaluation function to represent the fine structure of chess knowledge as we find it in standard expository works such as Reuben Fine's *Basic Chess Endings.* Tan[19] has remarked that chess 'may be considered as a mosaic of thousands of sub-problems, each requiring a different way of handling'. Contemporary research aims beyond the Turing–Shannon paradigm at the means for constructing specialized knowledge-sources for each of the thousands of sub-problems of the mosaic.

Illustrative of this style is Tan's own formalization of concepts for characterizing pawn structures. His program describes pawns-only positions in terms of 'islands' and 'fronts', forming pawn-relation graphs from which 'attack–defence diagrams' may be constructed. The aim is to generate an automatic summary of the position's dynamic potentialities.

Tan obtained his primitive relations from established formulations of chess theorists – hostile relations such as counterpawn and sentry, and friendly relations like duo and protector. In more complex subdomains there is no assurance that the relevant primitive notions will all have been explicitly identified and made available in the books. Attention therefore has been given to extending studies like Binet's and de Groot's to elucidate the basic components of the chess patterns on which Master skills are known to depend.

In tests of the recall and reproduction of chess positions, Simon and Chase[20] found that the relations of defence, proximity, and being of the same denomination or colour were all used as mental building-blocks. A particularly important relation for composing meaningful clusters of pieces was joint attack on the enemy king's position.

How many descriptive patterns does a Master store in memory? Two different approaches, by Simon and Gilmartin[21] and by Nievergelt[22] respectively, arrived at estimates in the range 10 000–100 000. This is comparable with human pattern-capabilities in some other contexts. There are, for example, approximately 50 000 Chinese ideograms.

The labour of introducing so large a number of chess patterns into computer memory would clearly be prohibitive if each pattern had to be specified in detail, with all its quirks and exceptions, by conventional programming. The labour could be eased if the machine system were capable of inductively acquiring patterns from playing experience or from user-supplied tutorial examples. This has not yet been achieved for any program, and chess systems capable of handling patterns and associated heuristic rules acquired from a human tutor are in their infancy, as Bratko, Kopec and Michie[23] reported. Inductive learning of patterns has been shown feasible for the limited case of the king and pawn *versus* king ending by Michalski and Negri.[24,25] It is not yet clear to what extent the methods used are capable of generalization to larger and more complex subdomains.

A PATTERN-BASED PROGRAM

A computer program into which a chess-master's pattern-knowledge can be incorporated manifests a new phenomenon. The machine is able to give explanations of its decisions expressed in concepts familiar to human players. The records reproduced earlier from trials of CHESS 4.6 with the king–knight–king–rook end-game represented everything humanly digestible which the machine was capable of reporting. It would have been possible to arrange additionally for output of the internal analysis tree underlying each move-choice; but since such trees typically consist of many tens of thousands of nodes, such an 'introspective record' would be of little value.

Knowledge-based programs studied in work on machine intelligence differ conspicuously in this respect. To develop such programs in general requires special computer languages tuned to the expression of descriptive concepts rather than imperative commands. Using the AL1 language developed at Edinburgh, Ivan Bratko embodied the master knowledge required for the defence of king and knight against king and rook in an Advice Table of the language. The following record was obtained when the system was given a position, theoretically lost in 27 moves, and required to defend it against a chess-master of USCF rating exceeding 2300. As might be expected of a machine representation specialized in this one particular end-game, it conducted the defence a great deal more skilfully than did CHESS 4.6. More significant was its ability to give an account of its thought-processes in terms comprehensible to others. Herein lay the crucial contrast between the new knowledge-intensive programs and the compact, knowledge-poor programs of the computer science mainstream tradition.

Two informal trials were made of the master's ability to win against the program, starting from a theoretically won, but difficult, position. For each trial the chess-master was given the following starting position. It is the start of one of the two longest wins in the KRKN domain discovered by Thomas Strohlein[26] from an exhaustive computation of the complete problem space of approximately 3 million positions.

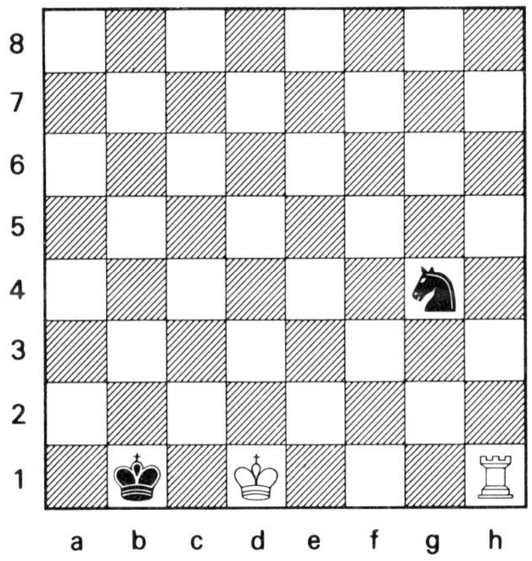

White to move

White can mate or win the knight in 27 moves against best defence

Game 1

1. Rh4!		(Optimal move in the minimax sense)
1. . . .	Nf6?!	(Suboptimal defence: the minimax path length is now 17 moves; optimal was 1. . . . Ne5)
2. Kd2	Kb2	(Both moves optimal)
3. Rd4?		(After this the position is drawn; correct was 3.Kd3)
3. . . .	Kb3	
4. Kd3	Ne8	
5. Rd7	Kb4	
6. Kd4	Kb5	
7. Re7	Nd6	
8. Kd5	Nc4?	= Draw agreed.

After a short post-mortem and analysis, the master tried again:

Game 2 (starting from same position as in Game 1)

1. Rh4!	Nf6?!	
2. Kd2	Kb2	
3. Rf4?		(This mistake brings the play from a 16-move win into a draw)
3. . . .	Nd7?	(Correct was Nd5. Position is now lost in 20 moves)
4. Kd3?		(Now position is drawn again. Winning line is 4.Rb4+ Ka3, 5.Kc3 with a win in 19 moves)
4. . . .	Kb3	
5. Kd4	Kb4	= Draw agreed

The comments made above are those of the Advice Table's author, Mr I. Bratko, a master-strength player. Below are the machine's own comments in the form of the 'motivations' of its moves.

Play was guided according to principles described elsewhere[16] by four self-administered 'pieces-of-advice' selected by the program from an 'advice-library' to meet each board situation as it arose. They were as follows:

Source of guidance for constructing strategy-tree: Advice 'HOLD 1'. *Phase of play*: Game 1, move 8.
Content of advice: Try to achieve a depth-4 position against all possible opponent's replies in which the opponent cannot win our knight on his next move. Our king is not to be left closer to an edge or corner than before. Our king is to be kept throughout at least as close to their king as our knight is, while avoiding a knight's-move relation between our king and our knight.
Source of guidance: Advice 'APPR-0'.
Phase of play: Game 1, moves 1, 2, 5, 6, 7; game 2, moves 1, 2, 4, 5.
Content of advice: Try out successor-positions in forward search according to the following criteria: prefer knight-moves towards our king and away from the enemy king, with a bias towards centralizing the knight. Lines may be investigated to a depth of 6, but at depth 4 consider only checking and attacking moves, and at depth 2 investigate only threats to our king or knight and moves that constrain the mobility of our knight. Find a strategy which either captures the enemy rook or, failing that, will avoid being mated, or losing or endangering or trapping our knight, while avoiding putting our knight into the corner or into the special pattern 'edge-loss', and not allowing it to drift away from our king or worsening in any other way the preference criteria given earlier.

Source of guidance: Advice 'APPR-1'.
Phase of play: Game 1, move 3.

Content of advice: Same as 'APPR-0' but dropping the embargo on allowing our knight to drift away from our king.

Source of guidance: Advice 'APPR-2'.
Phase of play: Game 1, move 4; game 2, move 3.
Content of advice: Same as 'APPR-1', but allowing investigation of lines which temporarily worsen the preference criteria, provided eventual restoration can be seen.

CONCLUDING REMARKS

Current endeavours centre round four foci:

- design of data-structures in forms not only suitable for representing conceptualized knowledge – descriptions, patterns, theories – but also convenient for the user to modify and increment interactively;
- incorporation into the program of powers of inductive inference, so as to be able to acquire new knowledge from tutorial examples, and also from internal computations;
- development of conceptual interfaces between program and human expert, facilitating both 'teaching' the computer program and testing what is taught;
- reversal of the direction of teaching and testing. The use of chess in computer-aided instruction experiments deserves study.

Relevant in this last connection is a finding reported by Bratko, Kopec and Michie.[23] After construction and validation of a suitable knowledge-representation for play of a given ending, this representation was translated back into English. The resulting instructional text was found to be more concise, more accurate, and easier to memorize and use than the master texts from which the knowledge had originally been obtained.

LITERATURE CITED

1. A. de Groot, *Thought and Choice in Chess*, edited by G. W. Baylor. Mouton, The Hague and Paris (1965). (Translation, with additions, of Dutch version of 1946.)
2. D. Michie, Practical limits to computation. *Research Memorandum MIP-R-116*. Machine Intelligence Research Unit, Edinburgh (1977).
3. Cited in B. Randell's 'Colossus', *IEEE Trans.*, forthcoming.
4. H. Golombek, *Observer Magazine*, 7 November 1976.
5. A. M. Turing, Digital computers applied to games. In *Faster Than Thought*, edited by B. V. Bowden, pp. 286–310. Pitman, London (1953).
6. J. Maynard Smith and D. Michie, Machines that play games. *New Sci.* **12**, 367–369 (1961).
7. C. E. Shannon, Programming a computer for playing chess. *Philos. Mag.* 7th Ser., **41**, 256–275 (1950).
8. A. Binet, *Psychologie des grands calculateurs et des joueurs d'échecs*. Hachette, Paris (1894).
9. R. Fine, The psychology of blindfold chess: an introspective account. *Acta Psychol.* **24**, 352–370 (1965).
10. A. L. Samuel, Programming computers to play games. In *Advances in Computers*, Vol. 1, pp. 165–192. Academic Press, London (1960).
11. R. D. Greenblatt, D. E. Eastlake and S. D. Crocker, The Greenblatt chess program. *Proc. FJCC* 801–810 (1967).
12. J. E. Hayes and D. N. L. Levy, *The World Computer Chess Championship*. Edinburgh University Press, Edinburgh (1976).
13. J. Moussouris, J. Holloway and R. D. Greenblatt, CHEOPS: a chess-oriented processing system. In *Machine Intelligence 9*, edited by J. E. Hayes, D. Michie and L. I. Mikulich, in press. Ellis Horwood, Chichester; John Wiley, New York (1979).
14. E. T. O. Slater, Statistics for the chess computer and the factor of mobility. In *Proceedings of the Symposium on Information Theory*, pp. 150–152. Ministry of Supply, London (1950).
15. H. J. Berliner, Chess as problem solving: the development of a tactics analyser. *Ph.D. Dissertation*, Carnegie–Mellon University, Pittsburgh (1974).
16. D. Michie and I. Bratko, Advice Table representations of Chess end-game knowledge. In *Proceedings of IIIrd AISB/GI Conference on Artificial Intelligence*, edited by D. Sleeman, pp. 194–200 (1978).
17. R. Fine, *Basic Chess Endings*. David McKay Co., New York (1941).
18. P. Keres, *Practical Chess Endings*. Batsford, London (1974).
19. S. T. Tan, Describing pawn structures. In *Advances in Computer Chess 1*, edited by M. R. B. Clarke, pp. 74–88. Edinburgh University Press, Edinburgh (1977).
20. H. A. Simon and W. G. Chase, Perception in chess. *Cogn. Psychol.* **4**, 55–81 (1973).
21. H. A. Simon and K. Gilmartin, A simulation of memory for chess positions. *Cogn. Psychol.* **5**, 29–46 (1973).
22. J. Nievergelt, Information content of chess positions: implications for chess-specific knowledge of chess players. *SIGART Newsl.* **62**, 13–15 (1977).
23. I. Bratko, D. Kopec and D. Michie, Pattern-based representation of chess end-game knowledge. *Comput. J.* **21**, 149–153 (1978).
24. R. S. Michalski and P. G. Negri, An experiment on inductive learning in chess end games. In *Machine Intelligence 8*, edited by E. W. Elcock and D. Michie, pp. 175–192. Ellis Horwood, Chichester; John Wiley, New York (1977).
25. P. C. Negri, Inductive learning in a hierarchical model for representing knowledge in chess end games. In *Machine Intelligence 8*, edited by E. W. Elcock and D. Michie, pp. 193–206. Ellis Horwood, Chichester; John Wiley, New York (1977).
26. T. Strohlein, Untersuchungen über kombinatorische Spiele. Dissertation for Dr. Rer.nat., Technische Hochschule, Munich (1970).

The manuscript was received 8 September 1978

APPENDIX

A Recent Performance by CHESS 4.6

The winner of the 1977 World Computer Chess Championship, CHESS 4.6, was given a board against the US Champion Grandmaster Walter Browne in a 44-board simultaneous display at Minneapolis the following year. Such a display imposes a severe handicap, so that Browne's unexpected loss to the machine occasioned surprise, but not incredulity, in the chess world. The machine's play showed interesting features which are discussed below by US National Master Danny Kopec in a commentary on the full game record.

The algebraic notation is used for describing the game, as elsewhere in this paper. For this, the ranks (horizontal rows) of the chess-board are numbered from 1 to 8 and the files (vertical columns) are labelled a, b, c, . . . , h, as has been done on the diagram of Fig. 3. The move denoted 'Re7–e8' in the figure's legend means 'the rook on square e7 moves to the square e8'. If the target square were already occupied by an enemy piece then the symbol 'x' would be used in place of '–', thus indicating that the move involved a capture. For pawn moves the piece-symbol is omitted, so that 'e2–e4' means 'the pawn on e2 moves to e4'. The symbol '+' denotes a checking move, that is, a move which threatens immediate capture of the enemy king. The symbols 'O–O' and 'O–O–O' mean 'castles on the king's side' and 'castles on the queen's side' respectively. In the common form of algebraic notation, designation of the source square is omitted, and also the '–' symbol. The three moves hypothesized above then become shortened to 'Re8', 'R x e8' and 'e4'. 'Rad8' means 'the rook on file a moves to d8'.

Mr Kopec's commentary follows:

In spring 1978, three times consecutively US Champion, Grandmaster Walter Browne, gave his second nationwide exhibition tour. Before stopping at The Twin Cities, Minnesota, Browne had accumulated an amazing record of only two losses and six draws in seventeen exhibitions. Due to its 5–0 sweep of The Twin Cities Open, the World Computer Chess Champion, Northwestern's CHESS 4.6 (written by Larry Atkin and David Slate), running on a Control Data CYBER 176, was allowed to be one of Browne's 44 opponents.

English Opening (by transposition)
White: Walter Browne (USCF rating 2560)
Black: CHESS 4.6/CYBER 176 (USCF rating 2070)

1.	d4	Nf6	
2.	c4	c5	
3.	Nf3	cd	
4.	Nxd4	e5	For this event CHESS 4.6 had been modified to 'think' during all the time it took Browne to get around to all 44 boards. This time, though certainly quite variable throughout the entire exhibition, was only a few minutes for the first six times around. Programs are now strong enough to be able to make real contributions to chess opening theory, and not just dubious, though interesting, moves. The move played, though not new, indicates a tendency for sharp play.
5.	Nb5	Bc5	The point of Black's previous move and it means that CHESS 4.6 would have met 6. Nd6+ with . . . Ke7! and not 6. . . . Bxd6 because the same position a tempo up after 5. . . . Nc6 could have been reached. After 6. Nd6+, Ke7, 7. Nf5+, Kf8 Black would stand well thanks to his fine development and the loose White knight on f5 which has already moved 5 times! In this opening, White can never play Bg5 because of . . . Bxf2+ and . . . Ne4+.
6.	N1c3		This move, played with hardly any hesitation, takes CHESS 4.6 out of its 'book'. It takes 2 min of processor time for the next move.
6.	. . .	O–O	
7.	e3	d6	
8.	Be2	a6	
9.	Na3		This knight will find it difficult to participate in the game.
9.	. . .	Nc6	Along these moves CHESS 4.6's evaluation function was telling it that Black stands the equivalent of a half pawn to the good and rising! This is based on its superior central influence and more active pieces. Note that the move 7. e3 has made White's QB very passive, though there was little choice.
10.	Nc2	Bf5	With an average of 4 min of processing time CHESS 4.6 is having little difficulty in predicting Browne's moves.
11.	O–O	Qd7	Now 12. Bd3 h4? 13. Nd5 Ng4?, 14. e4 was the predicted sequence, but it does not make much sense.
12.	b3		How else can White hope to find some life for his QB?
12.	. . .	Kh8	The first of a series of 'computer moves'. Though not bad in itself, the move indicates some difficulty in formulating a plan. Centralization of the rooks with Rfd8 and Rac8 would be the way a human master would proceed to build up this position. Of course, this is the aspect of chess where computers show their greatest deficiency. Such moves also explain how the initiative can change hands so

		suddenly and frequently in computer games.
13. Bd2	Rg8	
14. Na4	Ba7	
15. Bc3		Understandably Browne is trying to find some activity and at the same time trade off his passive N on c2. This is because Browne, as a GM, always tries to find the 'correct' move in a position. Had he better understood his opponent in this game, he would have made a waiting move (as David Levy frequently does against CHESS 4.6) when CHESS 4.6 might have weakened itself with . . . g5.
15. . . .	h6	A useful move, providing a haven for the QB and taking care of back-rank mates.
16. Rc1	Rad8	Now this pawn needed protection.
17. Nb4	Nxb4	
18. Bxb4		Now due to some technical problems CHESS 4.6 did not receive its full computation time and made a mild error:
18. . . .	Qc7	18. . . . Qe6 with the idea . . . d5 would have been much better.
19. Qe1		CHESS 4.6 still thinks itself a quarter pawn up, but with 19. Nc3 the game would have been roughly equal.
19. . . .	Bc5	
20. Bf3?!	Bd3	Now Browne starts to spend a long time at the board.
21. Bxc5		The only move. Now if 21. . . . Bxf1?, 22. Bb6 wins a piece, but of course CHESS 4.6 does not make such oversights!
21. . . .	dxc5	
22. Be2	Bf5	
23. f3	e4	
24. f4		In the last few moves the game has changed somewhat in character. Nonetheless Black's control of the d-file gives him an edge.
24. . . .	Bd7	
25. Nc3	Qa5	A typical 'computer excursion' on the Q-side, though it is not bad here and Browne must think twice before initiating his K-side attack. Black, however, finds good defensive resources there and thus 26. Nd5!? deserves attention.
26. Qh4		Played with an exaggerated flourish, yet Browne should have known that you cannot intimidate a computer program.

26. . . .	Bc6	
27. Rc2	b5	
28. g4	b4	
29. Nd1	Rd6!	Demonstrating the soundness of Black's last three moves. Now CHESS 4.6 expected: 30. g5 Nh7, 31. Nf2 Rgd8, 32. Bg4 Qb6, 33. Bf5 which is not best for either side.
30. Nf2	Rgd8	
31. Rd1	Rxd1+	
32. Bxd1	Rd6!	At this point CHESS 4.6 had used 2 h 44 min of computation time to Browne's 22 min.
33. Qg3		Perhaps best was 33. g5 Nh7, 34. gh gh (if 34. . . . Rxh6, 35. Qe7 looks good for White), 35. Qe7 Qd8 when a tough end-game would follow.
33. . . .	Qd8	Now CHESS 4.6 correctly predicts Browne's next 11 moves!
34. Rc1	Rd2	
35. g5	hg	
36. fg	Nh7	
37. g6		More solid was 37. h4, but then Black could safely snatch the QRP.
37. . . .	fg	
38. Qxg6	Qh4!	And now White is in great difficulty. Again Browne spent a long time deciding on his move. The end-game after 39. Qg3 Qxg3, 40. hxg3 would be hopeless for White.
39. Qf5	Bd7	This move leads to a won ending for Black, though even stronger was 39. . . . Ng5! with threats of Rxf2 and Nf3+. White's checks would quickly run out after 40. Qf8+ Kh7, 41. Qf5+ g6. Then if 42. Qf4 Qxf2+!.
40. Qf4	Qxf4	
41. exf4	e3	This wins a piece for two pawns. 41. . . . Nf6 was a more cautious way to retain good winning chances.
42. Ne4	e2	
43. Bxe2	Rxe2	
44. Nxc5	Bc8	Computer programs usually lack the special-purpose knowledge required to win end-games such as this one. They do not know what transpositions to allow and do not have the refined positional judgement which is required. This move gets Black into a little trouble. Much more lethal-minded was the move 44. . . . Bh3!? with ideas of mate. A continuation might be 45. Nxa6 Nf6, 46. Nxb4 Ng4, 47. Nd3

			Rg2+, 48. Kh1 Rxh2+, 49. Kg1 Rg2+, 50. Kh1 Rg3 etc.
45.	Rd1	Re8	
46.	a3!?		Browne knows that he must trade as many pawns as possible.
46.	...	ba	
47.	Ra1	g5?!	A dubious move because it offers to trade more pawns. Better was 47. ... Rf8, 48. Nd3 Rd8, 49. Ne5 Rd2! Then if 50. Ng6+ Kg1, 51. Ne7+ Kf8, 52. Nxc8 a2 and Black wins. Also good was simply 47. ... Re2.
48.	fg	Re5	
49.	b4?!		49. Nxa6 Bxa6, 50. Rxa3 would lead to the ending R+B+N *vs.* R which is a theoretical win, but here the White Q-side pawns might cause problems and it is questionable whether CHESS 4.6 would have the technique required in any case.
49.	...	a5!?	Again White could enter the above-mentioned ending, but under much less favourable circumstances.
50.	Nd3	Rxg5+	
51.	Kf2	ab	
52.	Nxb4	Ra5	
53.	Ke3		Here Browne misses a good drawing chance with 53. Ra2 (Not 53. Nc2 a2, 54. Nb4 Ra4!, 55. Nxa2 Be6 with Black winning). Then if 53. ... Bf5, 54. Ke3 Bb1, 55. Ra1 a2, 56. Kd4 White should still draw.
53.	...	Be6	
54.	Kd4	Ng5!	This piece ends up playing a vital role.
55.	Nc2		Browne offered a draw here which the Northwestern camp turned down in the 'interests of science'.
55.	...	a2	
56.	Nb4	Ra4!	
57.	Kc5		Browne declines his last chance for R *vs.* R+B+N. He originally intended 57. Kc3 which would have met the same response.
57.	...	Ne4+!	
58.	Kb5	Bd7+!	
59.	Nc6	Nc3+!	A pretty series of tactical moves.
60.	Kc5	Bxc6	
61.	Kxc6	Rxc4+	
62.	Kd6	Rd4+	
63.	Ke5	Rd1	and Browne resigned. An impressive finish by CHESS 4.6.

Key to special symbols used

! – excellent move;

!? – enterprising but possibly unsound;

?! – interesting, but probably unsound;

? – a mistake.

Mathematics, Music and Chess

HARRY GOLOMBEK, OBE
International Chess Master and Arbiter, Chalfont St Giles, Buckinghamshire, UK

Mathematicians, musicians and chess players were singled out by the British Foreign Office during the Second World War for the exacting task of breaking the German Enigma code at Bletchley Park. One of these cryptographers, the present author, has remained intrigued with the common factors linking these three noble activities of the human intellect: he suggests symbols of power and a logical line of thought. But an expert in one activity need not excel in the others, he concludes.

In 1941 I was transferred from the Royal Artillery to the British Intelligence Forces and told to report to an establishment known as Station X in the Forces and as the Government Code and Cypher School in the Civil Service. Once there, I was taken aside and informed in hushed tones that the particular section to which I was attached was devoted to the task of breaking the German Naval form of the Enigma Code. Further, I was told that I would have a probationary period of three weeks to show if I had any aptitude for the work or not.

I have occasionally wondered what would, in fact, have happened to me had I not revealed any skill at the work; for they could hardly have put me back into the Army for fear of my being captured and revealing what was, remarkably, the best kept secret of the war. Fortunately, this eventuality did not arise, and in three weeks I was delighted and suitably impressed to receive a letter saying that Mr Secretary was appointing me an administrative officer of the Foreign Office.

CODE-BREAKING AT BLETCHLEY PARK

Bletchley Park was practically a home from home for me. For one thing, I was not the only chess master there. When the war broke out in 1939 I was engaged in playing in a chess Olympiad in Buenos Aires, and of the English team of five, no less than three were at Bletchley Park: – C. H. O'D. Alexander, Stuart (later Sir Stuart) Milner-Barry and myself.

Nor were we the only chess players there, and it might be asked what was the connection between code-breaking and chess. The answer is that the objectives are much the same. In chess you want to discover what your opponent intends and how he is going to set about it; in cryptography you must try and discover what your unseen opponent is trying to say and how he is saying it. Long logical chains of thought are necessary in both cases and just as you will not get very far in chess without a plan, so you will have little chance of breaking a code without having devised a sound plan of campaign.

So I was happy to find my fellow chess players at Bletchley Park, but still happier to find that my second love, Music, was also strongly represented there. Two people I remember in particular were James Robertson, the opera conductor and Rowland Jones, a fine bass singer. Little wonder that we produced operas such as *The Marriage of Figaro* and *The Bartered Bride* at Bletchley with an almost professional expertise.

But the strongest group, in numbers and in importance, was that of the mathematicians. Many of these were later to become distinguished professors of mathematics and kindred subjects, such as Jack Good, Bernard Scott, Donald Michie,[1] Hilton, Joe Gillis[2] in Israel and in America Walter Jacobs.[3] Undoubtedly the greatest was Alan Turing with whom I had the pleasure of working during my early period at Bletchley Park; we were both breaking German naval codes.

Obviously, those who selected the personnel for Bletchley Park had decided from the start that their main sources were to be mathematicians, musicians and chess players; that they were right was proved by the fact that their method produced results.

MATHEMATICS AND CHESS

The link between mathematics and chess is fairly obvious. They have at least one element in common in that, in order to excel in either, one must possess powers of calculation. But in the use of this resemblance there lies a difference. Calculation might be termed the bread-and-butter of one vast area of mathematics. In chess it is only one of many attributes to be a master. Calculation is chiefly employed in the art of combination and therefore is a weapon of execution, but not necessarily basic to the art of chess itself.

But in some ways the processes of thought in chess and mathematics prove inhibiting for the ordinary individual. Understandibly enough, he cannot pack two lives into one or may be reluctant to make the

CCC–0308–0188/80/0005–0228 $01.50/1

immense effort involved in such arduous pursuits. I remember that, when we were undergraduates, Hugh Alexander and I were discussing this question and he said 'You are lucky in your subject'. He was taking mathematics and I was studying English and I knew he meant that mathematics, which was his subject, was so similar to chess as to make it difficult to duplicate his interests. However, being of extraordinary intelligence and industry, all who knew him realized he did manage to excel at both.

Perhaps the best example of a great chess player who was also a distinguished mathematician was Emanuel Lasker, world chess champion for 27 years (1894–1921) and yet a professor of mathematics of high quality.

MUSIC AND CHESS

The links between music and chess are less obvious; but they exist for all that. One striking mutual feature is the almost profuse existence of boy prodigies in both. The infant Mozart is parallelled in chess by the boy Reshevsky, or, even more strikingly, by the 15 year old grandmaster, Bobby Fischer. At this moment the Russians have a 15 year old player, Kasparov, whose game is like a grand master and in England we have a 13 year old player, Nigel Short, who is of near master strength.

In addition, music possesses in its history the practically unique case of a person who was the leading French operatic composer of his time and quite certainly the world's best chess player. This was François André Danican Philidor, one of the greatest chess theorists of all time. Then there were great musicians who were also fine chess players, like Prokofiev of minor master strength and David Oistrakh, a strong player.

SYMBOLS OF POWER

What then is the common factor among all three, mathematics, music and chess? By common factor I mean the similarity, almost the identity of the processes which work in the mathematical, musical and chess *minds*. To me this factor is a planned symbolism; this becomes abundantly apparent when one considers the best products of each of these three minds. All three use symbols for power and, to put it at its lowest level – though by lowest I also mean fundamental, perhaps a significant pun – the mathematician uses figures as units of power, the musician uses notes and the chess player uses the moves of the pieces.

How essential it is for the mathematical mind to work in this way was amply demonstrated to me during the course of my war work at Bletchley Park. We were constantly concerned with letter transpositions and it was essential to change the letters into figures so that one could calculate in numbers. For example, *e* became 5, the fifth letter in the alphabet, *M* adopted a new significance in my mind as being, at 13, half-way along in the alphabet and I even worked on a book of Capablanca's selected best chess games by giving each an index figure of excellence, in order to make an accurate selection from all his games.

But this basic resemblance in symbolism does not stop with the initial units. It is just as important to develop a line of thought in each field. In planning his logarithm or theorem, or even his complete theory, the mathematician will build it up as though it were a living organism with a beginning, a middle and an end. The beginning would correspond to a birth and the early development of a man's faculties, the middle would be most of his adult life and the ending would give the final attainment of his full powers, finishing off with his passing away.

The planning and execution of a game of chess – well – I have just described it. The three basic phases of a game are precisely the opening, the middle game and the ending.

In music too we find exactly the same process. Consider a sonata by, say, Mozart. It is divided up into precisely these three parts, both chronologically and logically. The first movement is an opening allegro, the second or middle movement is an adagio or an andante and the third and final movement is an allegro or allegretto.

I remember discussing a few years ago with John Lill, the concert pianist, the so-called music of Stockhausen and of modern composers like Messiaen. Much though I loved classical music I could not begin to like the formlessness of Stockhausen's pieces. Yes, said Lill, a piece of music must have architecture. So too must a game of chess and a piece of mathematics. It might be argued that this applies to every work of art or scientific theory. But the closeness of similarity in mathematics, music and chess is such that I cannot believe it is a universal and therefore haphazard likeness.

A WARNING

A few final words of warning regarding the significance of my ideas as outlined here. Just because a person is a great mathematician, it would be quite wrong to assume he or she was a great musician or chess master, or even to assume he was good in either of the other two fields. Turing, a mathematical genius if ever there was one, was also a keen chess player but such a weak player that I could give him the odds of a Queen and still win. Bobby Fischer, who bestrode the chess world like a Colossus, had little music and less mathematics (I had some difficulty in making him comprehend the workings of the calendar during his World Championship match in Iceland in 1972); and no doubt there are many composers and great musicians to whom mathematics and chess are closed books.

But the resemblance is there all the same and, as we saw at Bletchley Park in the Second World War, it can be put to good use.

LITERATURE CITED

1. D. Michie, Chess with computers. *Interdiscip. Sci. Rev.* **4,** in press (1979).
2. J. Gillis, The mathematical theory of epidemics. *Interdiscip. Sci. Rev.* in press.
3. W. Jacobs, Cipher machines and mathematics. *Interdiscip. Sci. Rev.* **2,** 227–239 (1977).

The manuscript was received 20 February 1979

Uncontrolled and Unfocused Growth

The US Supersonic Transport, SST, and the Attempt to Synthesize Fuels from Coal

PROFESSOR MEL HORWITCH

Graduate School of Business Administration, Harvard University, Boston, Massachusetts, USA

This review examines the unsuccessful life cycles of two large-scale, public/private, technological enterprises: the United States Supersonic Transport (SST) program and the United States effort to develop synthetic fuels from coal. More specifically, the SST program exhibits uncontrolled growth and the following evolutionary pattern: containment, fragmentation and explosion. Synthetic fuels development exhibits another pattern, that of unfocused growth: limited activity, premature expansion and uncertain takeoff. Four determinants of the life cycles of such enterprises are identified: the state of the technology, the rate and nature of growth, the character of the external environment, and managerial ability. Successful enterprises, such as the Apollo program, exhibit another life cycle: controlled growth.

A great deal of investigation has taken place on what has been termed the 'product life cycle' for a manufactured product. There appears to be a general pattern of evolution for industrial products as they move from early growth to a mature stage. The characteristics of the demand structure, the production processes used, the industry structure and the type of innovation changes practiced, appear to change significantly as this evolution occurs.

More importantly, the identification of distinct stages and of appropriate managerial responses to them is critical to the success of a firm. For example, in the US semiconductor industry, Texas Instruments and Hewlett-Packard practiced very different, but equally effective, strategies for exploiting the product life cycle.

A NEW KIND OF TECHNOLOGICAL ENTERPRISE: LIFE CYCLES AND THE MANAGERIAL CHALLENGE

The question posed for discussion is whether the concept of a generalizable life cycle and its strategic implications are limited to manufacturing firms and their products. Here, instead of a manufactured product, an entirely different kind of entity will be examined: the large-scale, public/private, technological enterprise.

Such enterprises are usually very large as they involve millions and often thousands of millions of dollars. Thousands of individuals from many different organizations take part in such an enterprise and they possess both public and private components. They are also usually established to accomplish some mission such as landing on the Moon or as ambiguous as implementing an overall energy program. Moreover, significant component parts of such enterprises have quite different cultures, assumptions, priorities and goals. Finally, only a fraction of each of the various participating organizations in such an enterprise is actually involved and the component organizations continue to maintain their separate identities.

The rise of these enterprises is a worldwide phenomenon. Moreover, their impact is growing. A representative list might include the following:

- The Alaska gas and oil pipelines, US and Canada;
- The Channel Tunnel project, UK and France;
- Fast breeder reactor programs, US, UK, France and elsewhere;
- Coal Liquefaction or gasification programs, US, UK and Republic of South Africa;
- Connecticut Resource (Solid Waste) Recovery Authority, US;
- The program to make the island of Hawaii energy self-sufficient, US;
- Various non-defense aerospace programs (Apollo, Space Shuttle, Concorde, US SST). US and elsewhere;
- Various national energy plans, US, Spain, Iran and Brazil;
- SITE: a rural TV education program using satellites, US and India;

CCC-0308-0188/80/0005-0231 $07.00/1

- The program to make liquid fuel – alcohol – from sugar cane, Brazil;
- The Jubail industrial complex, Saudi Arabia;
- Development of an iron and steel complex in Nigeria's 1975–1980 Five-Year Plan, Nigeria;
- Various European uranium enrichment projects.

One of the most striking features of such a list is the diversity of the endeavors, in the energy field, aerospace, transportation, economic development, and communication. They exist in the industrialized world and in the developing countries. They include a few successes, considerably more failures, and even more projects where the outcome is still unknown, ambiguous or undetermined.

The global rise of the large-scale, public/private, technological enterprises will almost certainly continue and will most probably accelerate. Society confronts an increasing array of large-scale, often technologically innovative needs which are beyond the technological, financial, political and managerial capabilities of any single organization. Even very large, single groups have generally limited resources or low risk profiles.

Moreover, during the latter half of the 20th century the trend to conceptualize problems and issues and to formulate solutions in ways that cross organizational, industrial, geographic and public/private boundaries has emerged as a dominant mindset for policy-making and decision-making. In part, this development arose because of a growing ability and willingness to view societal issues in an interdisciplinary fashion, to model or simulate complex problems and to use synoptic analytical and managerial techniques, like systems analysis. Enterprises are then structured to follow this approach. But severe challenges of implementation and management frequently surface. And, as we shall see from the poor success rate of two such enterprises, the way in which these managerial challenges have been met is not particularly encouraging. Still, the need for these enterprises is real. Therefore, examining the management problems is critical if many of society's massive needs are really to be fulfilled.

The evolution of two such enterprises will be discussed in detail and compared: the American SST program and the American attempt to develop synthetic fuels from coal. The reason for looking at the SST program is that it is clearly a failure in the managerial, but not necessarily the societal, sense and that it is terminated. With the SST program a full view of at least one type of evolutionary process, termed here uncontrolled growth, can be discerned and the relationship between its failure and its pattern of evolution can be fully explored.

A discussion of the American effort to develop synthetic fuels from coal offers another kind of opportunity: to discuss the evolutionary pattern of a large-scale enterprise in being. This life cycle, which is here termed unfocused growth, has thus far not been outstandingly successful. The ultimate outcome of this effort is still not known.

These analyses may thus help to highlight ways and means how synthetic fuels can play a more useful role in solving the United States' energy problems. This review also aims to offer suggestions for the general determinants of the life cycles of this increasingly important set of technological undertakings.

SST PROGRAM
PHASE 1: CONTAINMENT

The United States SST program[1] followed three phases: containment, fragmentation and explosion. Each of these phases exhibited very different characteristics, issues and sets of actors. In many ways part of the failure of the SST program was due to an inability of its managers to make the transition from one state to the next and to control the rapid expansion of the program.

The first, the containment phase, lasted from about 1960 to about mid-1963. It was characterized by a contained and predictable set of participants in industry, in the aerospace-oriented government agencies and in congressional committees.

Although the American SST effort was perceived from the beginning as a potentially significant aerospace endeavor, it began in a conventional enough fashion. Its origins go back at least to the late 1950s when various American aircraft firms such as Douglas, Lockheed and Boeing publicly began to urge the development of an American SST and sponsored limited feasibility studies. The Air Force and NASA also undertook limited SST-related research activities.[2] Even in these early years, the Federal Aviation Agency later called Administration, FAA,

MEL HORWITCH is Assistant Professor of Business Administration at Harvard University Business School. There he has taught courses on technological innovation, managing public organizations, production operations and energy economics. In addition to his teaching, Professor Horwitch served as a Peace Corps volunteer in Thailand and worked in both government and private industry. He has also been a consultant to numerous private firms and public institutions. He is author of the forthcoming book on the American SST Program, entitled *Clipped Wings* and of the coal section in *Energy Future*, Random House, New York (1979).
Address: Graduate School of Business Administration, Harvard University, Soldiers Field, Boston, Massachusetts 02163, USA.

clearly aimed to establish itself as the lead agency for any American SST program.[3]

Interestingly, all involved parties – the Air Force, the FAA, NASA and industry representatives – exhibited a remarkably optimistic consensus at congressional hearings on the SST in 1960. There was general agreement that the SST was technically feasible, that commercial introduction could occur during the 1965–1970 period, that the SST was important for national prestige, that the US should aim for a state-of-the-art steel–titanium Mach 2.7 SST – rather than a Mach 2 aluminium-based vehicle which was being stressed in Britain and France – and that government financial support was required and should be urged. While certain technical problems were mentioned, such as engine noise and the sonic boom, the hopeful testimony clearly indicated that most of the traditional aerospace interests in government and in industry considered the SST to be a next logical step in commercial aviation.[4]

Promotion of an American SST intensified during the Kennedy Administration and matched the mood of the country as a whole. The SST also fitted in well with the Kennedy campaign theme of getting the country 'moving again' and his emphasis on increasing aeronautical research.[5]

The new FAA Administrator, Najeeb Halaby, a pilot and lawyer with extensive experience in the aerospace industry, was an ardent promoter of an American SST under FAA direction. From the start he actively campaigned for the SST. Between July 1961 and June 1963 the FAA, with the assistance of NASA and the Air Force, contracted for a series of technical and economic SST feasibility studies. Congress appropriated $31 million for this effort. Halaby headed a Task Force on National Aviation Goals, and its report strongly supported SST development. He formed airline and industry SST advisory groups which strongly supported SST development.[6]

Meanwhile, in November 1962, the British and French announced joint development of the Concorde, a Mach 2.2 aluminum-based commercial transport. This development fueled Halaby's own SST demands. Halaby darkly warned Kennedy in mid-November 1962 that a successful Concorde could 'conceivably persuade the President of the United States to fly in a foreign aircraft!' A month later Halaby reminded Kennedy that the Concorde project meant that the SST would be a major factor in world aviation 'regardless of whether it will be a strong economic success or a marginal producer'. He declared that the crucial American SST decision 'is not "whether to proceed" but "how to proceed"[7]'.

By early 1963 Halaby's campaign was clearly working. In January, Kennedy asked the FAA to 'take the lead in preparing as soon as practicable' a report and plan of action for the SST. Kennedy also established a cabinet-level committee under Vice-President Lyndon B. Johnson to review the FAA recommendations.[8] The FAA's reports, submitted in mid-May 1963, naturally presented a plan which called for immediately beginning SST development. Johnson made similar suggestions and also urged that a Deputy Administrator for SST be established in the FAA. But within Johnson's committee there was some disagreement over the justification and structure of an SST effort.[9]

Halaby urged Kennedy to announce the SST program at the Air Force Academy and even presented an outline for such a statement. But it was probably Pan American's announcement on 4 June 1963 that it had placed an order for six Concorde aircraft which directly triggered Kennedy's move into SST development. In fact, at least one Pan American official later acknowledged that one motive for ordering the Concorde had been to pressure the US toward SST development.[10]

On 5 June, Kennedy announced at the Air Force Academy that the US would begin its own SST program. In letters to the House and Senate Kennedy laid down the ground rules for SST competition. Development costs would amount to approximately $ 1000 million. The manufacturers would contribute 25% of the total costs; and in no case would the government's contribution exceed $750 million.[11]

The FAA, of course, was more than ready. On 19 June 1963 the FAA released a comprehensive report which declared flatly that the FAA would run the SST program and presented an overall plan. Halaby announced that the actual competition would begin in August 1963. This competition period would run for approximately five months when, Halaby hoped, the government would select one airframe and engine manufacturer to build an SST prototype. However, Halaby continued, if a 'clearly "winning combination"' did not emerge, the government would select two airframe and engine manufacturers for an additional year-long detailed design competition.

Clearly, the FAA was anxious to begin prototype construction by the spring of 1964 if at all possible. Halaby glowingly and optimistically identified the forthcoming SST program with national prestige, national interest and the American way of life. He established a permanent SST organization within the FAA and in late July 1963 he appointed Gordon M. Bain, a former airline executive, to the new position of FAA Deputy Administrator for Supersonic Transport Development.[12]

SST PHASE 2: FRAGMENTATION

Then the next phase emerged and lasted from about mid-1963 to 1969. During this period the SST program was increasingly battered by bureaucratic jealousies, economic doubt, lack of managerial confidence, changing loci of policy-making power, technical uncertainty, differential and ambiguous Presidential support, a growing number of actors, the beginnings of criticism in the media, and the emergence of organized public opposition. The fabric which held the SST progam eventually settled in a variety of organizations and interest groups.

The earlier containment phase quickly began to collapse in the late summer of 1963. The manufacturers started to complain bitterly to high-level Administration officials that they could not meet Kennedy's 75%/25% government/industry cost-sharing formula. Their protest grew to such a pitch that Kennedy asked former World Bank president, Eugene Black, and Stanley de J. Osborne, chairman of Olin Matheson, to review SST program financing.

Economic Differences

The Black–Osborne report, which was submitted in December 1963, proved to be a significant milestone in the evolution of the SST program. Its authors went beyond the original financial question with which they were charged. The report basically agreed with industry that the 75/25 formula was overly burdensome and recommended a 90/10 arrangement and an easier overrun provision. But then the report, unlike the FAA, also expressed no immediate and compelling need to build an American SST in order to compete with the Concorde. It warned that a crash program would be 'dangerous, technically and economically'. Of even more significance was its recommendation to remove the SST program from the FAA and its rather strong critique of the FAA's SST management capability.[13]

The Black–Osborne report led to an examination by the new Johnson Administration of the whole SST decision-making structure. Johnson decided to establish in April 1964 a Presidential Advisory Committee (PAC) on Supersonic Transport. Defense Secretary Robert S. McNamara was appointed chairman. The other members were Treasury Secretary C. Douglas Dillon, Commerce Secretary Luther H. Hodges, the Administrators of NASA and of the FAA, James E. Webb and Halaby, CIA Director John A. McCone, Black and Osborne. Halaby unsuccessfully protested against the establishment of the PAC.[14]

Halaby's fears of the PAC were soon borne out. The PAC met four times in April and May 1964 and directed the FAA to let out contracts to two airframe and two engine companies for an additional six months. The committee also recommended that various sonic boom studies, already under the guidance of the National Academy of Sciences (NAS), be expanded. The FAA was already conducting sonic boom tests over Oklahoma City, but the data had still to be analyzed. Finally. over the strong opposition of Halaby, the PAC recommended that the Department of Commerce undertake a series of SST economic feasibility studies.[15] Control over such key issues as development scheduling, economic analysis and determining sonic boom effects had effectively passed from the FAA and had begun to spread to other parts of the government.

These events established the pattern for FAA–PAC relations during the rest of 1964 and early 1965. Halaby and Bain desired to award a single airframe–engine combination quickly and to get on with building an actual SST prototype. But McNamara kept insisting that more economic, technical and sonic boom data were needed before a single airframe–engine contract could be let.

In the spring of 1965 Halaby decided to resign and Bain resigned later in September 1965. In the summer of 1965 Johnson publicly announced that Boeing, Lockheed, General Electric and Pratt and Whitney would enter another design competition for eighteen months.

A new management team took over operational direction of the SST. Johnson appointed Air Force General William F. McKee as Halaby's successor. McKee had been commander of the Air Force Logistics Command, Vice-Chief of Staff of the Air Force and was currently NASA Assistant Administrator for Management Development. McKee eventually installed Air Force General Jewell C. Maxwell, who had been chief staff officer at the Air Force Systems Command Headquarters and who was currently commander of the Western Test Range at Vandenberg Air Force Base, as 'Director, Supersonic Transport'.

McKee and Maxwell were determined to regain the control of the SST program. The area of economic evaluation was particularly troublesome. The Commerce Department, which the PAC had designated as responsible for the SST economic evaluation, submitted its initial set of SST economic studies in the spring of 1965. As expected, the Commerce Department studies did not agree with the FAA's own economic findings. On the other hand, both the FAA and Commerce supported building an SST.

Meanwhile in March 1965, to the dismay of the FAA, McNamara organized a new SST economic evaluation group in the Defence Department. This small group produced a series of very pessimistic internal working papers and in August 1965 an equally negative comprehensive report. These analyses raised serious doubts over the number of people willing to pay higher SST airfares, especially given the advent of more economic subsonic jumbo jets, and concluded that the SST was economically feasible only if no sonic boom route restrictions existed. With route restrictions, according to this group, the government would not recover its investment and a 'reasonable' 15% return.

McKee and Maxwell retaliated by forming their own SST economics evaluation group and then contracted close to $1 million for a series of economic analyses and for several consultants. But even some of these new FAA-contracted studies, such as the Institute for Defense Analyses report on balance of payments, proved troublesome. The FAA eventually convened a panel of airline experts, hired additional consultants, and released its own SST economic feasibility report in 1967 to counteract various pessimistic economic studies.[16]

The Sonic Boom

McKee and Maxwell also confronted another growing issue over which they had limited control: the sonic boom. A host of organizations were already intimately involved with this matter. The FAA ran sonic boom overflight tests at Oklahoma City in 1964 and 23% of the individuals interviewed said that they 'could never learn to live with the boom'. The National Academy of Sciences, at the request of the PAC and of President Johnson, began to monitor and to analyze sonic boom data.

Other actors became involved when two new series of sonic boom tests were planned in late 1965. McNamara assigned overall operational responsibility to the Air Force. The White House Office of Science and Technology (OST) established another Coordinating Committee for Sonic Boom Studies to oversee the Air Force's sonic boom operations.

Maxwell and McKee were particularly concerned in 1966 over a proposed series of community overflight sonic boom tests along a corridor between Indianapolis and Louisville. They wanted at least to delay these tests. But McNamara and OST apparently favored having these tests. The FAA was worried that the potentially negative public reaction would antagonize Capitol Hill, especially representatives from the overflown districts. They also claimed that the community overflight tests would not have sufficient time to take full advantage of other ongoing tests and, in any case, the ultimate results, Maxwell argued, would not be available to the PAC until the end of 1966. Therefore, they were not useful for the final prototype source selection, which was to be completed by the end of the year. Johnson, meanwhile, consistently avoided the whole sonic boom issue. But ultimately he succumbed to McKee's appeal for postponement. The proposed community overflight tests never took place.[17]

The PAC was clearly the dominant body during the late 1965–1966 period and effectively the major SST policy-making body. It met regularly. At times its sessions were long and it dealt with a host of issues including economic studies, congressional opinion, the Concorde, the USSR TU-144 program, the sonic boom, cost sharing and source selection. It was almost unique in maintaining such high-level, active involvement. However, the fact that this key body was headed by an SST skeptic, McNamara, and that it was interagency in composition and perspective, further worked to fragment and to delay the program.

Technical Difficulties

McKee and Maxwell paid the closest attention to the Congress. Although the SST program received funding for fiscal 1966 and 1967, the first serious congressional opposition to the SST surfaced in 1966. In August, Senator William Proxmire, Democrat of Wisconsin, moved to reduce the SST funding request from \$280 million to \$80 million. He lost on the first SST roll call vote, 31–55. Congress appropriated the full fiscal 1967 \$280 million SST request in early September 1966.

The majority of everyone's time in the FAA SST Office really involved dealing with the manufacturers and directing the final source selection evaluation process. The final examination began in the late summer of 1966. In September the four competitors, Boeing, Lockheed, General Electric and Pratt and Whitney, submitted a total of approximately 22 000 pages of proposal information. Evaluating this data was the task of about 260 government experts from NASA, the Air Force, the Civil Aeronautics Board, and the FAA. The airlines also were independently evaluating the proposals with multi-company specialist teams.

The Boeing swing-wing design was selected at the end of 1966 over the Lockheed fixed-wing double-delta design; and the General Electric afterburning turbojet was selected over the Pratt and Whitney duct-burning turbofan. The FAA and the manufacturers agreed to a 90/10 formula for normal costs and a 75/25 cost-sharing formula for overruns. But Johnson did not go ahead and announce SST prototype construction until 1 May 1967, when all domestic airlines which continued to hold SST delivery positions agreed to contribute \$1 million in risk capital for each SST delivery position they had reserved.[18]

Instead of solidifying the SST program, the source selection decision ushered in a period of further fragmentation. The program was severely set back by a disastrous technical problem. The Boeing swing-wing design's crucial weight/payload ratio deteriorated. After a year's delay, Maxwell made what he called the 'biggest decision I ever had to make'. He told Boeing to start all over, and he gave Boeing until January 1969 to submit an acceptable SST design.

Boeing organized a broad attack on the design question. It analyzed various concepts and eventually settled on a fixed-wing, delta configuration with a horizontal tail. It appeared easier and cheaper to manufacture than the swing-wing. Maxwell reportedly exclaimed, 'Why it's pure vanilla.' The new design (Fig.1) was accepted in 1969 by all evaluation groups.[19]

Meanwhile, the SST program received another blow. The Department of Transportation was established in 1966 and the FAA became part of the new agency in early 1967. A new layer of bureaucracy was thus inserted between the FAA Administrator and the President. McKee could no longer bypass the PAC and go directly to the President in important SST matters. McKee left office in July 1968.

Anti-SST Lobby

Another critical threat to the SST surfaced in 1966: meaningful public opposition to the SST. Serious

Figure 1. The Boeing Supersonic Transport in its final design, (Courtesy: Boeing Company.)

criticism first appeared in the press. Then, in 1967, a significant organized anti-SST group was established by Dr William Shurcliff, a physicist at the Harvard University Linear Accelerator. He formed the Citizens League Against the Sonic Boom in March 1967. He began by placing full-size advertisements in such newspapers as the *New York Times*, the *Christian Science Monitor*, the *Wall Street Journal* and the *Washington Post*, and he initiated an intensive letter-writing campaign to government officials, newspapers, columnists and congressmen.

In many ways, the Citizens League was the antithesis of the large Washington bureaucracies which supported the SST. Its headquarters was a small room in Shurcliff's Cambridge, Massachusetts, home. Its staff consisted of Shurcliff, his sister and, at times, his son, for graphics. At its peak, the Citizens League had about 5000 members and annual expenses of about \$20 000. Shurcliff's records were up-to-date and immaculate. Word of the Citizens League spread through the newspaper accounts and advertisements and by word of mouth. Financing did not appear to be difficult. Whenever he needed money, Shurcliff merely mailed a few postcards which usually produced a quick and generous response. Shurcliff sent out newsletters and wrote a book entitled *Sonic Boom Handbook*.

Shurcliff kept prodding and writing. He was agressive, meticulous and effective. For example, throughout 1968 Shurcliff battled with the National Academy of Sciences Committee on SST-Sonic Boom over a statement in a subcommittee report which was issued in February 1968. He demanded a clarification. He threatened to take his case to the Academy's membership as a whole and perhaps to the press as well. In the end, the NAS backed away from its earlier wholehearted support of the controversial statement. In its February 1969 newsletter it issued a clarification which acknowledged that the original wording of the statement was by no means clear. In spite of the rather quiet ending to this affair, the NAS controversy was a real victory for the SST opposition. An entire group of scientific elite had been made aware of the SST's alleged hazards, especially those associated with the sonic boom.

In 1968 Shurcliff also gained support elsewhere, including the Departments of the Interior and of Health, Education and Welfare. These agencies issued reports which called attention to the alleged SST hazards for people on the ground and in the aircraft. By 1968 the growing criticism of the SST from parts of the Administration and from the Citizens League increasingly occupied Maxwell's time.[20]

In early 1969 the new Nixon Administration's SST policy was unknown. At this point Nixon could easily have killed the program upon taking office because of technical problems and the spreading opposition. Also, the FAA's control was particularly weak and vulnerable. Instead, Nixon established still another interagency body, called the Ad Hoc Review Committee, under James M. Beggs, Undersecretary of Transportation, to examine the SST program. Although the committee's summary report by Beggs was optimistic and favorable toward the SST, its four panel reports were more or less negative. Eventually the panel reports, the Beggs summary and various protest letters were all delivered to the White House in the Spring of 1969.

Maxwell resigned as SST program director in the summer of 1969. He felt that he had shepherded the program 'over the hump' because the prototype phase and design change had been achieved. But the outcome of the Ad Hoc Review Committee controversy and the increasingly organized and effective opposition, if anything, left the program even more fragmented and open to criticism than it had been in the mid 1960s. The attempt by McKee and Maxwell to gain control of all the key levers for running the SST program had failed.

SST PHASE 3: EXPLOSION

With Nixon's announcement in September 1969 that he would request $96 million for fiscal 1970 for the construction of two SST prototypes, the final phase – explosion – began.

The SST go-ahead decision resulted in an outpouring of protest on an unprecedented scale. Letters against the Nixon decision were sent to the White House and to the SST Office, now in the Department of Transportation. The Ad Hoc Review Committee material was leaked and inserted into the *Congressional Record* by congressional SST opponents in late October 1969.[21]

The Citizens League was no longer alone. Previously, many environmental groups had shied away from condemning the SST because the strong SST proponent, Senator Henry Jackson of Washington, was Chairman of the Senate Interior Committee. These groups believed that they needed Jackson's support on other conservation issues. However, Friends of the Earth, a spin-off of the Sierra Club, came out against the SST in late 1969.

Meanwhile, the environmental movement reached a kind of crescendo with the first Earth Day in April 1970. One SST activist call Earth Day a 'psychological watershed'. Other environmental groups also came out against the SST. Finally, in April 1970, bankrolled by an independently wealthy Maryland English teacher, a host of environmental organizations created the Coalition Against the SST. The anti-SST forces now had a broad organized base of support with a nerve center in the nation's capital.[22] The important center of action shifted away from the FAA and the federal bureaucracy to Capitol Hill and to national and local opposition groups.

During the latter half of 1969 and the first third of 1970 the SST program was doubly hamstrung because it lacked a permanent director. It was only in April 1970 that Nixon appointed a new SST director, William M. Magruder. Magruder was an engineer and former test pilot. He had worked for Douglas and Lockheed and had been second in command of Lockheed's ill-starred SST effort. Magruder's immediate task was to capture the $290 million which Nixon requested for fiscal 1971. His more general problem was somehow to defuse the snowballing SST protest across the country.

Spreading Opposition

However, opposition to the SST in the spring of 1970 spread rapidly. Economic and environmental criticism of the project was given by respected experts in congressional hearings and the press. Although the House voted 176–163 in late May to approve the $290 million SST appropriation, the anti-SST groups were encouraged by the closeness of the vote and predicted a Senate rejection.[23]

The summer and autumn of 1970 was a time of feverish activity for the SST opponents. New studies appeared, critical of the SST. Similarly, fifteen leading and diverse economists publicly opposed the SST. The Coalition had access to several anti-SST experts in economics, engineering, physics and political organizing. A number of anti-SST teach-ins and discussion sessions were organized.[24]

The SST proponents followed the lead of Magruder and Senator Warren Magnuson, Democrat of Washington. Magnuson delayed the SST Senate vote until after the November election which immediately allowed the SST to become a hotly debated campaign issue in several states, including Rhode Island, Iowa and Hawaii. In fact, many state and local anti-SST groups sprang up spontaneously and had little direct connection with the Coalition in Washington.

The anti-SST forces were ultimately victorious. In early December 1970 Senator William Proxmire, Democrat of Wisconsin and leader of the Senate SST opposition, offered an amendment to delete all SST funds from the Department of Transportation Appropriations Bill, which included the $290 million request for the SST. The amendment passed 52 to 41. House–Senate conferees grudgingly agreed that the SST appropriations vote would be divorced from the Transportation Department appropriations vote. In March 1971 both the House and the Senate voted to delete all SST funds.[25] The SST program was terminated and the final explosion phase had ended.

SYNTHETIC FUELS FROM COAL PHASE 1: LIMITED ACTIVITY

Like the SST program, the American effort to develop synthetic fuels from coal grew in scope and funding. However, unlike the SST, the synthetic fuels effort was much less focused. Moreover, the overall effect and outcome of this effort is still undecided. The life cycle for synthetic fuels is therefore termed unfocused growth. Thus far it can be easily divided into three periods: limited activity, premature expansion and uncertain takeoff.

The period of limited activity for synthetic fuels began in the early 1960s and lasted until the early 1970s. This phase was characterized by very limited funding and by the active interest and participation of a small government agency and a few, but growing, number of firms and other outside organizations.

Although coal gasification and liquefaction were not new technologies – after all, at the turn of the century many areas in the US used so-called town gas from coal and during World War II Germany had become very efficient in developing and using synthetic fuels – these uses for coal had never played a major role in technological innovation in the US coal industry until the 1960s. Up to that time, most of the technological advances in the coal industry were in mining.

From the beginning there were several alternative technologies available for the development of synthetic fuels. Throughout the US effort to develop synthetic fuels, work was done on both the so-called first and second generation technologies. For example, first generation gasification technologies, developed before the 1950s in Germany and elsewhere, include the Lurgi, Koppers–Totzek and Winkler processes. Second generation gasification processes, developed in the post-World War II period in the US and elsewhere, include the CO_2 Acceptor, Bigas, Hygas, Synthane, Cogas and the Slagging Lurgi processes.

Like the SST, although Consolidation Coal Company, the second largest coal producer in the US, had been working on liquefaction in the 1950s under its 'Project Gasoline' program, the real impetus and support for a nationwide R and D synthetic fuels effort came in the early 1960s from the Office of Coal Research (OCR), a small unit in the US Department of the Interior.

In spite of the relatively small scale of this effort, it ignited considerable enthusiasm and optimism. OCR signed a contract with Consolidation for making gasoline at competitive prices as part of Consolidation's ongoing 'Project Gasoline'. Consolidation's President, G. Albert Shoemaker, happily told a Wall Street audience in November 1964: 'Thus far we have demonstrated that our method is technologically feasible and in an area of such economic attractiveness as to encourage moving into the pilot plant stage in the near future.'

In 1965 Consolidation bullishly reported that Dakota lignite could be converted to pipeline quality gas and sold for nearly 2 cents per million BTUs less than natural gas. The next year Consolidation said it would build a coal gasification pilot plant to use Dakota lignite. In the mid-1960s, meanwhile, Consolidation went on to build its liquefaction 'Project Gasoline' pilot plant in West Virginia, with OCR funding. Shoemaker optimistically commented at the time: 'While our projections must be confirmed in the ... pilot plant, we believe this development offers the potential of creating a major new market for coal.' In early 1967 an OCR official likewise termed the costs of the liquid output of the pilot plant favorable. Later that year Consolidation reported achieving gasoline costs from coal at 10 to 11 cents per gallon.

Multiplication of Effort

Other firms were also working on coal conversion efforts during the first half of the 1960s, again with projects funded by OCR. By 1963 the Institute of Gas Technology (IGT) in Chicago was developing a gasification process called 'hydrogasification' under a program sponsored first by the American Gas Association and later OCR. In November 1965 the IGT director confidently expected pipeline gas from coal to be the basis of a viable new industry. He called for one or more commercial plants by 1975 and a series of third generation plants by 1985. In the spring of 1966 IGT enthusiastically reported to OCR that its hydrogasification process using $4 per ton of coal could produce pipeline quality gas at 58 cents per million BTUs, with the potential of cutting 15 cents off this price. Natural gas was then selling for 45 to 50 cents per million BTUs.

OCR had funded FMC Corporation's COED Project liquefaction and gasification process since the early 1960s. In 1963 FMC reported very encouraging progress to OCR. In 1967 FMC stated publicly that a commercial COED plant could become a reality by 1972. OCR also funded work by Gulf Oil, General Electric and the Atlantic Refining Company (later ARCO) during the early and middle 1960s.

The optimism continued. By 1967 OCR was still reported to be enthusiastic about its synthetic fuels research effort. OCR officials predicted a new synthetic fuels industry by 1975 and a flourishing one by 1980. One OCR official in January 1967 optimistically referred to OCR contractors' projections of the cost of gasoline from coal at 10 to 12 cents per gallon and pipeline quality gas at or below 50 cents per million BTUs.

Still, funds were quite limited throughout this first phase. In 1963, out of a total federal government energy R and D budget of about $330 million, only $11 million was spent on coal R and D. In contrast, over $210 million was spent on nuclear fission research alone. OCR's fiscal 1969 proposed research budget totalled only $13.9 million, with the bulk of the funds going to Consolidation Coal (by then a subsidiary of Continental Oil Company), FMC, IGT and Westinghouse. In spite of the early optimism, when the 1970s began, synthetic fuels from coal was still a long way from being a viable industry.

The flurry of enthusiasm over liquefaction and gasification in the 1960s, however, did have one important effect: it served as a powerful magnet for pulling oil and, to a lesser extent, gas firms into the coal industry. These firms found it hard to resist the temptation to see coal as a potential feedstock for their refineries, pipelines and petrochemical plants. A Texaco official stated in 1976 that his firm acquired coal reserves in the west during the 1960s only because of the optimistic prospects for synthetic fuels. Interestingly, an officer of the National Coal Association in 1966 called for a solid front of coal and oil interests to combat government-funded nuclear energy research. In late 1967 Exxon was acquiring western coal reserves and it began building up a coal research staff and formed a Synthetic Fuels Research Department. In 1968 Sun Oil Company was reported to be marshalling resources for a large-scale coal synthetic fuel plant in Wyoming.[26]

Natural gas companies also grew more involved. In late 1970 Columbia Gas established a subsidiary, Columbia Coal Gasification Corporation, as its focal point to develop competitive, pipeline quality gas. Texas Eastern Transmission and El Paso Natural Gas proposed separate coal gasification projects in New Mexico. American Natural Resources and Peoples Gas planned commercial gasification plants in North Dakota.[27]

PHASE 2: PREMATURE EXPANSION

The second phase of the US synthetic fuels effort, termed premature expansion, began in the early 1970s and lasted until about 1976. It was characterized by the emergence of continued and growing involvement by oil and gas firms in synthetic fuels R and D. Government funding for this activity also increased substantially. Finally, serious attempts were made to design and implement relatively large gasification and liquefaction projects. But by the end of this second phase practically all such large-scale projects had been significantly delayed, stalled, aborted or abandoned.

The continued interest of oil and gas firms, in spite of their disappointing experience in the first phase, was particularly striking. Of the thirty-six proposed US synthetic fuels projects from coal, which the US Bureau of Mines identified in mid-1976, gas companies managed fifteen and oil firms seven. Furthermore, of the five proposed liquefaction projects in this group, three were directed by oil firms. Moreover, a recent survey of past and forecasted research and development expenditures in the coal industry by oil firms found that petroleum firm coal-related R and D spending rose from about $5.5 million in 1971 to an estimated $42.4 million in 1976. In fact, oil and gas R and D dropped from about 91% of 1971 R and D expenditures by oil firms to 74%, while coal-related R and D rose from about 2% of all 1971 oil firm R and D expenditures to 10.5% in 1976.[28]

The implications of these developments are very significant. Petroleum firms accounted for a larger share of total private R and D expenditures in the US coal industry than their share of either production or reserve holdings. More importantly, they were generally performing a kind of R and D very different from that which traditionally characterized the industry. In essence, this research was less disjointed than the fragmented technological innovation process found in the mining equipment area. Much of the research involving the oil firms dealt with developing whole processes and with intergrating and coordinating different groups and organizations.

Still, the federal government remained the dominant influence and overwhelmingly largest financial supporter of synthetic fuels development. Although only $11 million was spent on coal R and D by the federal government in 1963, coal R and D expenditures in 1974 rose to $106 million out of $755 million for total government energy R and D expenditures. Nuclear fission R and D accounted for $470 million in that year. Nuclear fission's share of total government energy R and D dropped from 64% in 1963 to 62% in 1974 while coal's share rose from 4% in 1963 to 14% in 1974.[29]

The rise and fall of proposed large-scale synthetic fuel projects was another characteristic of this phase. One kind of project which gained favor involved the integration of mining and synthetic fuel processing activities. It also required the active participation of several private firms and federal, state and local governments.

Examples of such activities included the Wesco and ConPaso projects in New Mexico which began in the early 1970s. Essentially, both efforts proposed to mine massive amounts of coal off the Navajo Indian Reservation in the Four Corners region and then gasify the coal into pipeline quality gas to feed into nearby gas pipelines. Both projects involved the participation of a gas company and a coal company, and a utility in the case of Wesco. Both planned to use the proven Lurgi gasification process. Wesco was a joint venture of the Texas Eastern Transmission Corporation and Pacific Lighting Corporation. Utah International agreed to mine and sell its coal to Wesco. ConPaso was a joint venture of El Paso Natural Gas Company and Consolidation Coal. Both projects, however, were delayed indefinitely in the mid-1970s due to regulatory decisions, inflation, local Navajo opposition, lack of water access and other problems.[30]

Meanwhile, in North Dakota two other major natural gas companies, American Natural Resources Company of Detroit and Peoples Gas Company of Chicago, were also separately pursuing coal gasification activities at this time.[31]

Another ambitious but ill-fated endeavor was the Coalcon clean boiler fuel liquefaction project. Coalcon was the first US coal synthetic fuel project to be funded for eventual demonstration plant construction. Coalcon began in 1975 and possessed an extremely elaborate organizational structure. It was a joint venture of Union Carbide and Chemico, a subsidiary of Aerojet, which in turn was a subsidiary of General Tire Corporation. Coalcon was ostensibly supported by a consortium of private firms from the oil, chemical, metals and construction industries as well as state and local governments to fund the non-federal government share of Coalcon's development cost. The plan to build the demonstration plant fell through because of doubts over the technology developed, over ERDA and Coalcon management and the lack of firm support from the consortium. Union Carbide went on to assume total control of Coalcon and to continue research on a very limited basis. But the complex organizational package and the prime role played by an aerospace construction firm presented one model of future synthetic fuel projects.[32]

PHASE 3: UNCERTAIN TAKEOFF

In spite of the substantial increase in funding and scale for synthetic fuels R and D during the first half of the 1970s, by 1976 practically all of the large-scale undertakings had collapsed for one reason or another. Still, the trend toward larger government funding for R and D in this area continued as part of ERDA's long-term plans and later President Carter's National Energy Plan. Therefore, the critical question for this next phase. termed uncertain takeoff, is whether this continued growth is merely a continuation of the premature expansion phase, and will eventually suffer similar disappointing results, or whether this next phase marks the beginning of the solid emergence of a synthetic fuels industry.

Substantial funding will almost certainly be available. The fiscal 1979 Department of Energy's research and technology budget development program budget allocated $618 million or 23% of the total for coal and $905 million or 34% for nuclear fission. Moreover, $294 million was slated directly for coal gasification or liquefaction.

If anything, the projects during this third phase appear to be growing in size, number of participants and general complexity. For example, the two original separate gasification projects by Peoples Gas and American Natural Resources in North Dakota were merged into a single project in 1977. This project will use coal from a North American Coal Corporation subsidiary. In mid-1978 three other natural gas firms, Transco, Tenneco and Columbia Gas System, joined in sponsoring the same project.

Figure 2. Aerial view of the US Department of Energy's Solvent Refined Coal (SRC) pilot plant at Fort Lewis, Washington. The SRC process will grind and dissolve coal, and filter out the ash and sulfur, reconstituting the fuel either as a hot liquid or brittle solid. The pilot plant is one of four basic liquefaction processes necessary to gain data for a larger plant. The Fort Lewis pilot plant was designed to handle 50 tons of coal daily and produce 30 tons a day of solvent-refined coal. It is operated for DOE by Pittsburgh and Midway Coal Mining Company, a Gulf Oil Corporation subsidiary. (Courtesy: Department of Energy, Washington, DC).

Significantly, the Department of Energy devised a plan, the so-called 'all events tariff', which would allow pipeline companies to increase their rates to pay back the debt which they would assume for this project. However, this plan still has to be approved by the Federal Energy Regulatory Commission.[33]

Oil firm interest in synthetic fuels also continued. By 1977 Exxon was managing a $240 million ERDA–industry liquefaction project which involved the construction of a pilot plant in Baytown, Texas, by 1979. 50% of the $240 million will come from DOE and 50% from Exxon, the Electric Power Research Institute and the Phillips Petroleum Company. Similarly, ERDA also funded in 1977 $142 million of a $178 million pilot coal liquefaction plant in Kentucky. The plant was to be supervised by a subsidiary of Ashland Oil. Ashland, the Electric Power Research Institute, the Commonwealth of Kentucky, Standard Oil of Indiana, Mobil Oil, Continental Oil and other oil companies were supposed to contribute $36 million towards its construction. In June 1976 ERDA selected a group led by Continental Oil for negotiation of a contract to design a Lurgi high BTU coal gasification plant in Ohio. About a year later the Continental group received a $28 million ERDA contract to design such a facility, with expectations that ERDA and the Continental group would share equally the costs of the second and third phases, which would cost more than $300 million.[34]

Since the early 1960s there has been a renewed focus on developing synthetic fuels from coal. The life cycle of this enterprise first exhibited a limited amount of activity and then work on a more ambitious scale. At first glance there was apparently considerable progress. A variety of processes were developed, which involved both first and second generation technologies. In addition, the general evolution of this whole effort was toward greater complexity and size of projects and toward developing whole systems. The massive influx of government funding into coal R and D during the 1970s encouraged coal R and D projects which were increasingly complex and large, both in terms of the actual innovation and in terms of diversity and number of partners who participated in various consortia.

DETERMINANTS OF THE LIFE CYCLES OF LARGE-SCALE, PUBLIC/PRIVATE, TECHNOLOGICAL ENTERPRISES

Yet, in spite of all this activity and in spite of the heightened interest in synthetic fuels, because of the oil embargo and the alleged decline of natural gas, there are at present no demonstration or commercial synthetic fuels plants in the US. Is the American effort to develop synthetic fuels doomed to failure? Is it another SST? The answers to such questions are, of course, not yet known. But a comparison of the life cycles of these two enterprises may well offer some clues. Also, such a comparison may help in discovering the general reasons for the differences between successful and unsuccessful large-scale technological enterprises.

A comparison of the SST program and the American effort to develop synthetic fuels from coal leads to the conclusion that there are at least four determinants for the evolution of such large-scale enterprises: the status of the relevant technology, the rate and nature of the enterprise's growth, the nature of the relevant environment of the enterprise, and the ability of the enterprise's managers to apply successfully the technology, to control the enterprise's growth, and to protect the enterprise from harmful forces emanating from the external environment.

Clearly the SST program and the American synthetic fuels effort suffered from overoptimism regarding the relevant technology. For example, the crucial two-year delay from 1967 to 1969 in the SST program was primarily due to the failure of Boeing's swing-wing design to perform and to the consequent

Figure 3. The Hygas pilot plant for producing high-BTU gas (over 950 BTUs per 1000 cubic feet) from coal (foreground), the Steam-Iron pilot plant for producing hydrogen from coal (middle), and the Ash Agglomerator (U-Gas) pilot plant for producing medium-BTU gas (300 to 630 BTUs per 1000 cubic feet) from coal (right), near Chicago, Illinois. The Hygas and U-Gas pilot plants are jointly funded by government and industry and are operated by the Institute of Gas Technology. The Steam-Iron pilot plant is currently inoperative. (Courtesy: Institute of Gas Technology.)

need for Boeing to redesign its SST model. This hiatus gave the emerging SST opposition forces sufficient time to grow and establish themselves.

The state-of-the-art for synthetic fuels was even more uncertain, varied, undeveloped and confused. There was initially a natural lack of focus and an inability to concentrate on first or second generation technologies. Essentially work was done on both stages of technological development throughout the three phases of this whole effort. This mixed technological state led to a certain controversy over the type of innovation or projects which should be promoted: basic R and D, pilot plants or demonstration plants. This conflict led to other arguments in such areas as the choice of contractors and the level and nature of government financing. Again, delays resulted and continue to exist due to the variety of alternative technologies and lack of technological focus in this field.

The second life cycle determinant is the rate and nature of growth of such enterprises. Both the American SST program and the American effort to develop synthetic fuels suffered from the effects of harmful expansion. The SST was terminally exposed to an uncontrolled, cancer-like growth which ultimately overwhelmed and smothered the enterprise. The FAA failed to stop the successive dismantling of the effort and to recapture the scattered components. Active participation in key aspects of the SST spread to other parts of the federal bureaucracy, later to other organizations, such as the National Academy of Sciences, and finally to the public at large. The FAA was unable to piece together the SST program. Instead, fragmentation continued and accelerated. By 1970 the metamorphosis of the SST program was complete. The American SST was a highly visible political issue and it became a target for widespread protests, the object of a mass movement. It ultimately became a major symbol of the rising concern over the environmental and social consequences of technology.

The synthetic fuels effort suffered from another kind of growth – unfocused growth, as the amount of funding grew enormously in the 1970s. The number and scope of projects also increased. But a positive result from these activities has yet to be discerned. It may happen. So far, however, basic and pilot plant research on second generation technologies is continuing and large-scale efforts using first generation technologies in the US are stalled or nonexistent.

The external environment seriously affected the evolution of both enterprises. The SST program occurred during a period of profound change. The SST mirrored and accelerated this change. Protests emerged on a wide variety of issues, including the environment. And the opposition was frequently effective. Public interest groups no longer feared entering debates over technology. Evaluation of technologies increasingly considered social and environmental consequences as well as purely technological aspects. One key reason for the ultimate collapse of the SST was that it fell victim to this new feature – the rise of large-scale and effective public protests and involvement – of the external environment.

The American effort to develop synthetic fuels was similarly affected by the external environment. The extreme lack of general interest in the synthetic fuels field, until the oil embargo of 1973, partially accounts for the early limited funding and activity and the continued lack of focus in this effort. Moreover, after the oil embargo, the heightened general interest in energy in the external environment and alleged drive for partial energy self-sufficiency resulted in substantially increased funding and rapid growth of synthetic fuels activity. However, there was never significant interest in specifically synthetic fuels. Therefore, the lack of focus, albeit on a much grander scale, continued into the mid-1970s.

This brings us to the managerial weaknesses which were exhibited in both cases, when the top managers failed to develop or focus technological development in a timely fashion. In both cases the top managers failed either to protect the enterpise from harmful external forces or to realize that the environment was not particularly conducive for the successful promotion of the project.

With regard to the SST, the managers either did not recognize or could not prevent the outside opposition from terminally affecting the project. Ultimately, no effective way was discovered to buffer or protect the project from outside forces. The extreme transformation of the SST program's life cycle – from containment to fragmentation to explosion – largely took top SST officials by surprise, although they were highly successful managers in the aerospace field. The very radicalness of the SST's life cycle helps to account for the inability of consecutive SST directors and FAA Administrators to stave off continuing disintegration and eventual termination.

There really were no top managers for America's effort to develop synthetic fuels. Instead, there were only groups of weak coalitions working on separate activities. It is no wonder then that such groups or individuals were unable to combat the apathy in the environment and the resulting lack of focus.

The life cycles of the SST and the American effort to develop synthetic fuels offer two roads toward failure: uncontrolled growth through overstimulation of outside involvement and unfocused growth through lack of real involvement. Clearly, successful enterprises, such as Apollo[35] or various weapons systems, emerge through another kind of life cycle, controlled growth. These successful enterprises also possess appropriate or general environmental support and protection from strong bureaucratic, congressional or public opposition perhaps through a strong Presidential commitment or a national defense cloak. Without these conditions for success, most such large-scale, technological enterprises are vulnerable to the uncontrolled growth life-cycle pattern which the SST exhibited or the unfocused growth life-cycle pattern which US synthetic fuels development illustrated. The very size, variety and scope of such enterprises allows them to become easily intertwined with the interests of diverse groups and to mirror large parts of the mood of the external environment. Delays caused by technological inadequacy, rapid uncontrolled or unfocused growth, outside opposition or extreme apathy, and lack of managerial skill or control are very real threats to the success of this increasingly important method of developing and implementing critical and needed technologies.

LITERATURE CITED

1. For a more detailed history of the American SST program see Mel Horwitch, *Clipped Wings*, forthcoming (1980).
2. *Aviation Week and Space Technology*, p. 21 (26 *May* 1958); T. F. Cartaino *et al.*, *Supersonic Transports*, RAND Corporation Report (November 1958); Glenn Garrison, Supersonic Transport may aim at Mach 3, *Aviation Week and Space Technology*, pp. 38–40 (2 February, 1959); Lockheed studies Mach 3 vehicle as next step in transport field, *Aviation Week and Space Technology* p. 49 (12 October, 1959); Robert A. Bailey, The case for the Supersonic Transport, *Interavia*, No. 10, pp. 1234–1235 (1959); William S. Reed, Study predicts Mach 2–3 transport costs, *Aviation Week and Space Technology*, pp. 36–38 (23 November 1959).
3. Memo, Dale D. Davis to Director, Audit Staff, AU-1, (20 November 1964), 'Audit of the SST Program' and Attachment No. 1, 'Chronology of Events from December 1959 through December 1963' (FAA Archives); note, 'things to discuss with General Preuss (like voice) and attachments, 'Progress Report, Federal Aviation Agency Supersonic Planning Team' (1 April 1960) and 'Report on the First Meeting of Agency-Wide Supersonic Planning Team' (FAA Archives).
4. US Congress, House, *Hearings Before the Special Investigating Committee of the Committee on Science and Astronautics*, 86th Congress, 2nd Session (17–20 and 24 May 1960). Government Printing Office, Washington, DC (1960).
5. *Report to the President-Elect of the Ad Hoc Committee on Space pp.* 7–8 (12 *January* 1961) (*JFK Library, Transition Files, Box* 1072).
6. *Supersonic Transport Advisory Group. Supplementary Report* p. 7 (14 May 1963) (LBJ Library, Vice-Presidential Papers); US Task Force on National Aviation Goals, *Project Horizon*, Report of the FAA. Government Printing Office, Washington, DC (September 1961); Supersonic Transport Advisory Group, *Report* (11 December 1962); letter, Cook to Halaby (14 December 1962); letter, Halaby to Cook (9 January 1963); Supersonic Transport Advisory Group, *Supplementary Report* p. 7 (15 May 1963); letter, Cook to Halaby (15 May 1963) (all documents in LBJ Library, Vice-Presidential Papers); FAA, *Brief History of the United States Supersonic Transport Program*, Washington, DC (November 1964) (FAA Archives).
7. Memo, Halaby to President (15 November 1962) (JFK Library. President's Office Files); memo, Halaby to President (26 December 1962) (JFK Library, President's Office Files).
8. Memo, President to Secretaries of Defense, Commerce, Administrators of NASA and FAA, Chairman of CAB and Director of OST, 'The Civil Supersonic Transport' (21 January 1963) (JFK Library, President's Office Files); President to Vice-President (19 February 1963) (JFK Library, President's Office Files).
9. FAA, *Supersonic Transport Program* (Report) (14 May 1963) (LBJ Library, Vice-Presidential Papers); memo, GER to Vice-President (17 May 1963) (LBJ Library, Vice-Presidential Papers); memo, E. C. Welsh to Vice-President, 'Supersonic Transports' (25 May 1963) (LBJ Library, Vice-Presidential Papers); Gordon/Heller to Vice-President, 'Comments on the FAA report and recommendations on a US supersonic development program' (23 May 1963) (JFK Library, Sorenson Papers); memo, GER to Vice-President (17 May 1963), *op. cit.*; letter, Johnson to President (30 May 1963) (LBJ Library, Vice-Presidential Papers).
10. Memo, Halaby to President, 'The Commercial Supersonic Transport – The Next Steps' (3 June 1963) (JFK Library, Sorenson Papers); personal interviews.
11. Office of the White House Press Secretary, *Remarks of the President at Graduation Ceremonies United States Air Force Academy*, Colorado Springs, Colorado (5 June 1963) (JFK Library, President's Office Files); letter, the President to the President of the Senate and the Speaker of the House (11 June 1963) (LBJ Library, Vice-Presidential Papers).
12. FAA, *Supersonic Transport Development Program* (19 June 1963) (LBJ Library, Vice-Presidential Papers); *Transcript Remarks by N. E. Halaby before the American Institute of Aeronautics and Astronautics* (19 June 1963) Los Angeles, California (LBJ Library, Vice-Presidential Papers).
13. Letter, John F. Kennedy to Eugene Black (13 August 1963) (inserted in Black–Osborne report); Eugene Black and Stanley de J. Osborne, *Report on the Supersonic Transport Program* (19 *December* 1963).
14. Letter, Halaby to President (13 January 1963) (LBJ Library, EX CA Files); handwritten note, Halaby to Myer Feldman (undated, *ca.* 25 March 1964) (LBJ Library, EX CA Files).
15. *Memorandum for the President, First Interim Report of the President's Advisory Committee on Supersonic Transport* (14 May 1964) (LBJ Library, FAA Historical Documents).
16. US Department of Commerce, *SST*: *An Economic Analysis*, Part I, Executive Summary, Preliminary, Supplement I and Supplement II (9 March 1965 and 30 April 1965) (NTIS Nos. AD 655603, AD 655604 and AD 655605); US DOD Economic Task Force, *The Economics of a US Supersonic Transport* (Report of Findings to Date), (20 August 1965); Norman J. Asher *et al.*, *Demand Analysis for Air Travel by Supersonic Transport*, Vol. 1, Text. IDA, Arlington, Virginia (December 1966) (NTIS No. AD 652309); FAA, Economic Feasibility Report, *United States Supersonic Transport*, FAA, Washington, DC (April 1967) (NTIS No. 652313).
17. Letter, Maxwell to Golovin (9 February 1966) (FAA Archives); memo, Golovin to Hornig, 'Summary, Proceedings of the OST Coordinating Committee Meeting of April 1, 1966' (6 April 1966) (FAA Archives); Golovin (draft), 'Consideration of Possible Postponement of Sonic Boom Studies Community Overflight Program' 3 May 1966) attachment to memo, Golovin to OST Coordinating Committee Members and Participants, 'Agenda and Related Items, Meeting of May 9, 1966' (4 May 1966) (FAA Archives); *Aviation Daily* **160**, No. 30, p. 1 (13 June 1966); memo, Golovin to Hornig, 'Sonic Booms and the SST Development Program' (16 August 1966) (FAA Archives); Maxwell, 'Comments on August 16 Letter – Golovin to Hornig' (undated) (FAA Archives); letter, Maxwell to Golovin (15 September 1966) (FAA Archives).
18. White House release, 'Statement by the President on the Supersonic Transport' (29 April 1967) (LBJ Library, Oversize); White House News Conference transcript (29 April 1967) (LBJ Library, National Security Files); letter, President to Boyd (1 May 1967) (FAA Archives).
19. Charles J. V. Murphy, Boeing's ordeal with the SST, *Fortune*, pp. 129–132 (October 1968); memo, Vierling *et al.* to Maxwell, 'SST Program Integration Board Report' (13 February 1969) (FAA Archives); memo, Maxwell to Thomas, 'Results of Evaluation of the Boeing's Proposed Redesign of the SST (B–2707–300)' (14 February 1969) (FAA Archives).

20. John E. Gibson, The case against the SST, *Harpers* (July 1966); *Wall Street Journal*, p. 14 (28 July 1966); memo, Pinnes to Assistant for Plans and Programs, SS-305, '"The Case Against the SST" by John Gibson' (25 August 1966) (FAA Archives); William A. Shurcliff, 'A Brief History of the Citizens League Against the Sonic Boom' (25 January 1973) (Citizens League Files); Subcommittee on Physical Effects of the Sonic Boom, *Report on Physical Effects of the Sonic Boom* (February 1968). NAS, Washington, DC (1968);several letters to and from Shurcliff to and from the NAS, NAS Committee members and the SST Office in the Citizens League Files, FAA Archives and NAS Archives; NRC/NAS/NAE, *Minutes of the Governing Board*, (6 October 1968), Summary (NAS Archives); NAS/NRC/NAE *News Report* **XIX**, No. 2, p. 11 (February 1966); John C. Calhoun, *Report to the Secretary of the Interior of the Special Study Group on Noise and the Sonic Boom in Relation to Man*, US Department of the Interior (November 1968); HEW, *Supersonic Transport (SST) – Potential Health Hazards to the Crew, Passengers and Population* (undated), Consumer Protection and Environmental Health Services, HEW; personal interviews.
21. Congressional Record, House (31 October 1969), H: 10432–10446; various documents in folder labeled 'Ad Hoc Committee Files' including (1) summaries of Commerce, DOD, OST, NASA, HEW, DOI, Justice, Labor, Treasury and CEA views; (2) memo, letter, Dubridge to President (2 April 1969); (3) OST Ad Hoc Review Committee Report (30 March 1969); (4) memo, President to Beggs (19 February 1969); (5) panel reports to Ad Hoc Committee; (6) various letters and memos to Beggs dated 20 March 1969 (FAA Archives).
22. H. Lenhart Jr, *National Journal* **3**, No. 2 (9 January 1971).
23. *US Congressional Hearings before the Sub-Committee on Economy in Government of the Joint Economic Committee, Part 4: Supersonic Transport Development*, 91st Congress, 2nd Session (7, 11 and 12 May 1970). Government Printing Office, Washington, DC (1970); *New York Times*, p.1 (25 May 1970).
24. Lenhart Jr, op. cit.; Study of Critical Environmental Problems (SCEP), *Man's Impact on the Global Environment: Assessments and Recommendations for Action*, MIT Press, Cambridge, Massachusetts (1970); *New York Times*, p. 19 (20 July 1970), p. 11 (28 August 1970); personal interviews.
25. Lenhart Jr, op. cit.; *New York Times*, p. 1 (4 December 1970).
26. The sources for the discussion on the limited activity phase of US synthetic fuel development include: Technology Newsletter, *Chemical Week*, p. 69 (13 April 1963); Coal on the comeback, *Chemical Week*, pp. 63–64 (3 August 1963); Gasoline: coal's hope, *Chemical Week*, p. 23 (18 November 1964); Warming coal's future, *Chemical Week*, pp. 36–37; *Coal Age*, p. 55 (April 1965), p. 58 (October 1965); Dr Henry R. Linden, pipeline gas from coal: status and future prospects, *Coal Age*, pp. 64–71 (January 1966); Progress in coal research, *Coal Age*, p. 46 (April 1966); Technology Newsletter, *Chemical Week*, p. 110 (28 May 1966), R. T. Eddinger and S. K. Reed, COED research aims at oil, gas and char from coal, *Coal Age*, pp. 90–92 (January 1967); Coal knocks at refinery door, *Chemical Week*, pp. 29–35 Research, *Coal Age* (March 1967); Technology Newsletter, *Chemical Week*, p. 40 (30 December 1967); *Coal Age*, pp. 40, 42 (March 1968); Technology Newsletter, *Chemical Week*, (2 March 1968); p. 67 (23 March 1968), p. 59 (11 May 1968); Stronger case for crude from coal?, *Chemical Week*, p. 41 (15 June 1968); Business Newsletter, *Chemical Week*, p. 13 (29 June 1968); Coal gasification moves to big hardware, *Coal Age*, pp. 26–27 (December 1968); *Coal Age*, p. 50 (May 1969); Work begins on pilot plant for IGT coal-to-gas process, *Coal Age*, p. 44 (July 1969); US Congress, Senate, *Energy Research and Development – Problems and Prospects*, Prepared at the Request of Henry M. Jackson, Chairman, Committee on Interior and Insular Affairs, Serial No. 93–21 (92–56). US Government Printing Office, Washington, DC (1973).
27. Technology Newsletter, *Chemical Week*, p. 23 (23 December 1970); Patrick Crow, Prospects seen brighter for US LNG and SNG, *Oil and Gas Journal*, pp. 23–28 (16 January 1978).
28. US Bureau of Mines, *Projects to Expand Fuel Sources in Western States*, IC No. 8719 (1976); Jesse W. Markham, Anthony P. Hourihan and Francis L. Sterling, *Horizontal Divestiture and the Petroleum Industry*, Ballinger Publishing Cambridge, Massachusetts (1977); Donald Norman, *The Performance of Oil Firm Affiliated in the Coal Industry*, American Petroleum Institute Research Study (March 1977); US Congress, Senate, *Petroleum Industry Involvement in Alternative Sources of Energy*, Prepared at the Request of Frank Church for the Subcommittee on Energy Research and Development of the Committee on Energy and Natural Resources, Publication No. 95–54 (September 1977).
29. US Congress, Senate, *Energy Research and Development – Problems and Prospects*, op. cit.
30. *Pacific Lighting: Wesco*, Intercollegiate Case Clearing House, ICH No. 9–377–889, Boston (1977); *The ConPaso Coal Reserve Project*, Intercollegiate Case Clearing House, ICH No. 4–678–152, Boston (1978); FEA, *Western Coal Development Monitoring System*, Summer Quarter, No. FEA/G–77/306 (1 August 1977).
31. How to finance gas from coal, *Business Week* pp. 33, 36 (19 June 1978).
32. GAO, *First Federal Attempt to Demonstrate a Synthetic Fossil Energy Technology – A Failure*, No. EMD-77-59 (17 August 1977); Coalcon future in doubt as ERDA plans scaledown, *Coal Age* (January 1977); Coalcon $ shifted, *Coal Age*, p. 47 (May 1977); Synfuels: data gap imperils 'Coalcon' demonstration, *Science* **195**, No. 4273, p. 38 (7 January 1977).
33. National Coal Association, *Special Analysis: Federal Funding for Activities Concerned with Coal – Fiscal Years 1978 and 1979* (27 March 1978).
34. Crow, op. cit; Coal-liquefaction plant due at Baytown, *Oil and Gas Journal*, p. 36 (23 January 1978); ERDA aids research in liquefaction, *Coal Age*, p. 32 (March 1977).
35. For a discussion of the success factors in the Apollo program see E. Ginzberg, J. W. Kuhn and J. Schnee, *Economic Impact of Large Public Programs: The NASA Experience*, Olympus Publishing Co., Salt Lake City (1976); L. R. Sayles and M. K. Chandler, *Managing Large Systems: Organizations for the Future*, Harper and Row, New York (1971); Robert C. Seamans Jr and Frederick I. Ordway, The Apollo tradition: an object lesson for the management of large-scale technological endeavors *Interdisciplinary Science Reviews* **2**, No. 4, pp. 270 (December 1977).

The manuscript was received 20 August 1978

Oil and Gas in the High Arctic

The Islands of Canada

DR CHARLES R. HETHERINGTON

President and Chief Executive Officer, Panarctic Oils Ltd, Calgary, Alberta, Canada

The world's northernmost energy reserves, the Arctic Islands of Canada, lying between 72° and 80° North and extending for 1500 km from west to east, have been actively explored during the last 12 years for gas and oil. This demanded a completely new technology for the harsh Arctic climate which was successfully developed by Panarctic, a consortium of the Canadian Government and private industry. Although no oil has yet been found, vast gas reserves, sufficient for Canada's long-term requirements, have been located and could by the late 1980s flow to the south, when either ice-breaking LNG tankers, pipelines, or both have been constructed. Many different engineering techniques brought success to this remarkable interdisciplinary adventure.

The Arctic Islands lie north of the Canadian mainland about 3000 km north of Winnipeg, Manitoba. The climate is harsh Arctic making a frozen desert with low annual precipitation, permafrost up to 1 km deep, and total darkness for three months in the winter.

In the early 1960s the Government of Canada licensed the area for oil and gas exploration. No major oil company took an interest and no single individual company operating in the area could mount a meaningful exploration program. Panarctic Oils Ltd, a consortium owned 45% by the Government of Canada and 55% by a number of largely Canadian industries, was formed to take over the lands held under Government permits by numerous companies for the purpose of mounting an effective exploration program.

Panarctic has spent, or caused to be spent, almost $500 million in exploration of its lands (all values given in Canadian dollars). Large quantities of natural gas have been found, adding about 25% to the known gas reserves of Canada. There are also indications of possible commercial oil.

In the process of conducting this exploration program, my Company and its contractors have developed procedures, techniques and equipment for routine operations almost year-round, in this harsh Arctic environment. We have a main base at Rea Point on Melville Island where drilling supplies and fuel are stockpiled for each year's operation. The base is resupplied once a year by sealift when ice conditions during late August and early September permit navigation by ice-reinforced ships escorted by Canadian ice-breakers. Drilling rigs, fuel and supplies are transported by rear-loading Lockheed Hercules aircraft operating on frozen dirt airstrips constructed at each drilling location. Crews, fresh groceries and miscellaneous supplies are transported from southern Canada by Boeing 727 passenger/freighter configuration jet aircraft. Inter-island movement of men and light freight is by DeHavilland Twin Otter STOL aircraft.

For short rig moves of less than 160 km where terrain permits, the moves are made at lower cost by truck over winter roads and ocean ice, or by large-tire transportation vehicles over the snow without roads. Communication is by radio from drilling sites to the Rea Point base where signals are relayed by Anik Satellite to the Calgary telephone system.

Initially, geological prospects were drilled from the land. Over 70% of the prospective area, however, lies offshore in water up to 457 m deep. The ocean is ice-bound for almost the entire year. Geophysical work done over the ocean ice has located a large number of offshore geological prospects. Panarctic pioneered the development of techniques and equipment to drill offshore wells using modified conventional land drilling rigs supported on artificially thickened ocean ice platforms.

Nine offshore wildcat wells have been drilled as expendable wildcats and were abandoned after testing the producing formations. In the spring of 1978, Panarctic completed an offshore development gas well and connected it to shore by a flowline bundle permitting production of the well from onshore facilities.

The demonstration of a viable system for the completion and production of a subsea well under ice was an important step in the exploration and development of petroleum reserves in the Canadian Arctic.

ARCTIC GEOGRAPHY

The Arctic Islands of Canada encompass 800 000 km^2 of area prospective for the accumulation of crude oil and natural gas.[1] This is an area

CCC-0308-0188/80/0005-0245 $04.50/1

larger than the Province of Alberta which produces the bulk of Canada's conventional oil and gas. Oil and gas exploration over the past 12 years has proved the Arctic Islands to be a vast source of energy to meet Canada's future needs.

The story of Arctic Islands exploration is a fascinating one, but this story will be overshadowed by what will happen as the energy resources of this area begin to serve the energy needs of Canada.

In particular, large volumes of natural gas have been discovered and the trend in the discovery rate indicates that a gas reserve equal to the remaining gas reserves in the rest of Canada can be developed over the next 10 to 15 years.[2,3] Through the use of this gas, coupled with increased production of synthetic oil from the tar sands and heavy oil deposits of western Canada, the fuel requirements of Canada can be met far into the future. This places Canada in a unique position among developed countries as the only such country capable of supplying its own energy needs domestically.

Projects for bringing Arctic Islands gas to southern markets are well advanced and the rising cost of world crude oil with its effect on the value of energy is making these projects economically viable. One needs only to continue exploration at a meaningful pace and diligently pursue these marketing projects to assure ample energy supplies some ten years hence.

The Arctic Islands prospective for oil and gas are located north of the Canadian mainland between 72° and 80° North latitude. These islands extend from Banks Island in the west to Ellesmere Island 1500 km to the east.

The harsh Arctic climate has a low annual precipitation of 6 to 12 cm, making it a frozen desert with permafrost up to 1 km deep. The area is in total darkness from mid-November to early February and for three months during the summer the sun does not set below the horizon. The Arctic Ocean between the islands is covered with ice normally 2 m thick with pans of multi-year ice up to 17 m thick. Typical temperatures are −40 °C in winter and +5 °C in summer.

In June the land begins to thaw. The snow melts and the top 15 to 30 cm of soil melt, creating a quagmire that makes movement by land nearly impossible in most areas. By late August the ocean ice melts and breaks up in whole or in part in some areas. During this time the central islands are accessible by ice-reinforced ships. By September the land and the ocean freeze again and the area is accessible from southern Canada only by air.[4–6]

EARLY EXPLORATION

The terrain in the Arctic Islands is almost devoid of vegetation. The surface is not obscured by glacial residue and for several months during the summer much of the land is free of snow so that bedrock formations are clearly visible. Under these conditions the study of surface geology is helpful in predicting subsurface geological structures that might contain oil and gas. Many geological surface structures can be accurately mapped from aerial photographs. The Geological Survey of Canada conducted reconnaissance surveys of the Arctic Islands following the Second World War and these studies indicated highly favourable geological prospects for oil and gas. A number of geological field studies were commissioned by private industry to identify attractive lands.

On the strength of these indications, the Government of Canada issued exploration permits in the Arctic Islands in the early 1960s. These exploration permits were requested by, and issued to, various mining companies and smaller organizations. No major oil company took any particular interest in the Arctic Islands. In those early days, oil and gas prices were low, transportation costs out of the Arctic were known to be high and it was uncertain whether effective exploration could be carried out in this remote and inhospitable area. Certainly no single company, or even group of companies, holding exploration permits in the Arctic Islands could mount an effective exploration program.

In late 1967, Panarctic Oils Ltd was initially financed with $20 million to undertake oil and gas exploration in this area. The plan was for the various oil and gas exploration permit holders to turn over their exploration permits to Panarctic which would do the exploration. The original permit holders would retain certain working interests in the lands which would be at least partially explored.

One might well ask why these industrial participants and the Canadian Government should enter into such a costly and long-lead-time venture. The industrial participants were visionaries who expected

DR CHARLES R. HETHERINGTON, President and Chief Executive Officer (1970) of Panarctic Oils Ltd, graduated from the University of Oklahoma in 1941 and from the Massachusetts Institute of Technology with a Doctorate in 1943. He was a Consulting Engineer involved in oil and gas work and in 1952 became First Vice-President and Chief Engineer of Westcoast Transmission Company Ltd which built the first major natural gas pipeline in Canada. In 1957 he was Managing Director of Pacific Petroleums Ltd, the principal company supplying this pipeline with gas and between 1959 and 1967 he returned to the consulting field in the oil and gas industry with his own company, based in Calgary, Alberta. In 1967 he rejoined Westcoast Transmission Company Ltd as Vice-President and Director as well as Vice-President and Director of Westcoast Production Co. Ltd.
Address: Panarctic Oils Ltd, 703 Sixth Avenue S. W., Calgary, Alberta, Canada T2P 2H6.

to find oil and they foresaw a future increased value of oil if it were discovered. They were relying on much the same reasoning as John D. Rockefeller who said that he did not care where he found oil since if he found enough there would be some way to get it to market. The possibility of finding natural gas was not an important factor in the decision to explore since the value of gas at that time was not recognized.

The Government of Canada was interested in evaluating the oil and gas potential of the Arctic Islands, and it was interested in doing so with substantial Canadian ownership and control. The Government also wanted to reinforce sovereignty over the Arctic which it had passively assumed belonged to Canada, an assumption not necessarily accepted by other nations. A Canadian presence on the ground was needed. Finally, the industries were not prepared to enter into exploration without some Government assistance or participation. The Government therefore chose to join in the Panarctic industry/Government exploration consortium as a 45% shareholder.[1] Petro-Canada, the Canadian national oil company, now holds the Government's share interest in Panarctic Oils.

Panarctic Oils commenced exploration in 1968 and made its first gas discovery in 1969 at Drake Point on the Sabine Peninsula of Melville Island. Other discoveries followed and other oil and gas companies joined in the exploration effort, exploration peaking in 1973 with the drilling of 23 wells.

THE CHANGING EXPLORATION PATTERN

In 1971 the Canadian Government withdrew the order which had previously established the regulations for leasing land. At the same time the Government indicated that it would be introducing new leasing and royalty regulations which would be more stringent. After a two-year period to 1973 to unwind the exploration momentum that had been established, exploration in the Arctic Islands steadily declined to the drilling of nine wells in 1979.

The Government has not yet enacted leasing and royalty regulations. This uncertainty is disturbing to the industry and may have triggered the decline in exploration activity, but it is not the sole reason for the decline. Oil in commercial quantities was not found. Only gas was discovered in substantial quantities and the bulk of this was found by only one explorer, Panarctic Oils. Because gas must be transported in very large quantities, large marketing projects are required and progress on these projects was delayed indicating that marketing of gas reserves from the Arctic Islands could be many years into the future. These factors caused the interest of the exploration companies to wane.

The exploration climate is changing again and exploration drilling is expected to increase in 1980 for the first time in seven years. There are several reasons for this change. World oil prices are increasing so rapidly that marketing projects which a year or so ago might have been considered marginal are now indicated to be highly viable. Monetary returns to the producers of gas in the field appear attractive, and demand projections indicate that the gas will be needed sometime before the end of the 1980s. In the spring of 1979 Panarctic made a substantial gas find at a location called Whitefish, offshore between Melville and Lougheed Islands. A fresh find of substantial proportion always whets the appetite of the explorer and this find did just that.

Initially exploration was widespread throughout the Islands, testing a number of different types of geological plays. Exploration for gas has now been centralized between Melville and Ellef Ringnes Islands in what is called the Central Sverdrup Basin. This basin constitutes the high grade area of the Arctic Islands as it is known today.

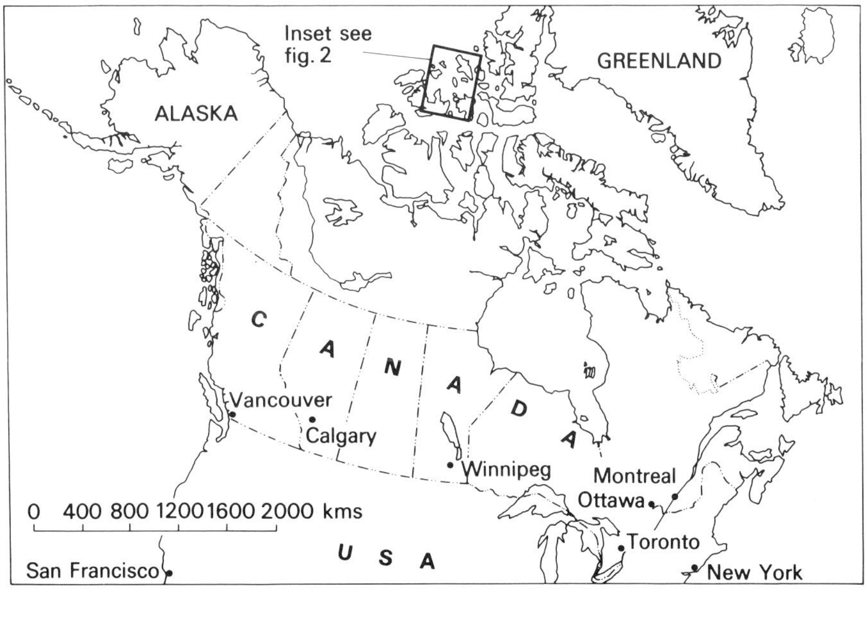

Figure 1. Location map of Canadian Arctic Islands.

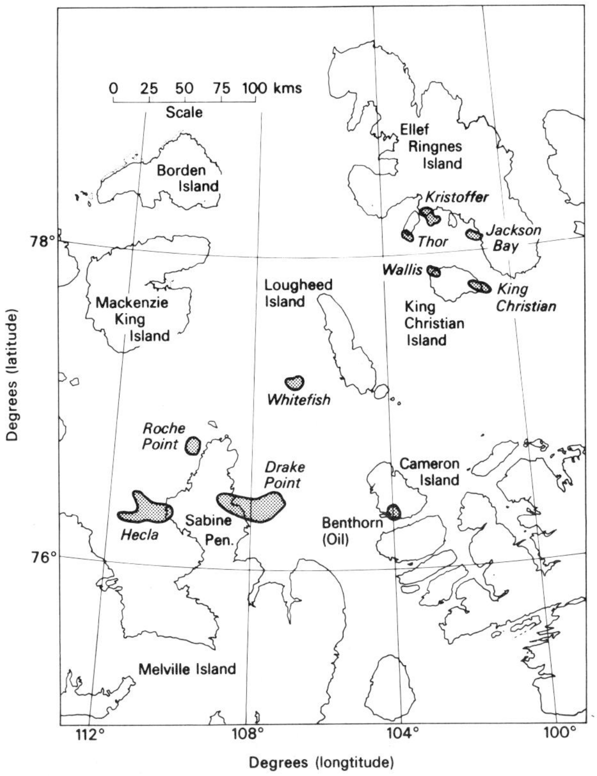

Figure 2. Map of Central Sverdrup Basin showing location of gas fields.

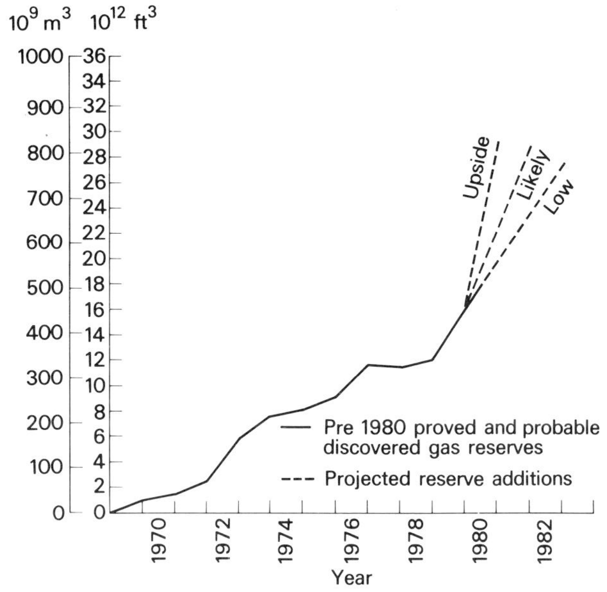

Figure 3. Arctic Islands marketable gas discovery rate.

EXPENDITURES AND DISCOVERIES

Panarctic Oils has expended or caused to be expended on its lands over \$500 million and exploration off Panarctic's lands has run to about \$150 million. Nine gas fields have been discovered, all within the Central Sverdrup Basin. Figure 1 shows the location of the Central Sverdrup Basin in relation to Canada, and Fig. 2 shows the location of the nine gas fields discovered to date. These fields contain an estimated 450×10^9 m^3 (16×10^{12} ft^3) of marketable gas.

An exploration program for the next two years is fairly well established for the testing of a number of large geological features that have been identified by seismograph work. These features appear similar geologically to the other features which have been drilled and proved productive. On the strength of expectations for drilling over the next couple of years, the discovery rate has been projected as shown in Fig. 3. A reserve of 850×10^9 m^3 (30×10^{12} ft^3) is indicated by 1981. There are also indications of possible commercial crude oil on Cameron Island, but additional delineation drilling is required before this can be proved.

HIGH ARCTIC OPERATIONS

Our experience over the last 12 years has identified the problems which distinguish operations in the high Arctic from more conventional operations elsewhere in Canada, and methods and equipment have been developed so that oil and gas exploration in the Arctic Islands can be carried on routinely almost every day of the year.

Operating problems in this harsh environment have been overcome by the use of proper equipment, accommodations, and the application of unique operating methods. Drilling rigs are closed in to provide workable conditions (Fig. 4). Daily flights with large aircraft change crews and deliver fresh groceries and freight. Fuel is transported to the rig sites, and entire drilling rigs are moved by large rear-loading aircraft (Fig. 5). Moving rigs overland and over the ocean ice is even more efficient where terrain permits. Camps are comfortable, if not spacious. During a short period in the summer the Arctic Ocean is navigable with ice-breaker support, and heavy supplies (Fig. 6) and fuel (Fig. 7) are shipped in once a year to Arctic bases for the supply of the next year's operations. Panarctic maintains a large base at Rea Point on Melville Island which receives these supplies by ship for transhipment by air to the various drilling locations within a radius of 1000 km.[7–9]

Since the land portion of the Arctic Islands constitutes only about a third of the area, many prospects lie offshore in ice-covered waters. Our Company has developed an effective, economic and safe method of drilling offshore wells with modified conventional drilling rigs supported on artificially thickened ocean ice. A camp is set up on the ocean ice as soon as the ice is about 1 m thick, usually in October. Holes are drilled through the ice and sea water is pumped on top of the ice. When the water is about 5 cm deep the pumps are turned off and the water is allowed to freeze. This flooding and freezing operation is

Figure 4. This 700 tonne drilling rig in 1978 drilled and completed the first offshore gas well ever completed below the ice of the Arctic Ocean.

Figure 5. The Lockheed Hercules rear-loading air freighter is employed to move drilling rigs from one location to another and to supply them with fuel and other drilling supplies during operations. Each drilling rig location has a 2000 m airstrip suitable for the Hercules and for Boeing 727 aircraft.

repeated until the ice is sufficiently thick to support a drilling rig. 6.5 m of ice thickness at the rig location, about 170 m in diameter, will support a 700 tonne drilling rig. A large amount of research and pilot experimentation was involved in evaluating the physical properties of ocean ice and in developing the mechanical design for the platform.

Since ice creeps under load, the free board on the ice platform decreases during the drilling period, but with a properly designed platform the rate of decrease also decreases.

Figure 6. Once a year in late August or early September the central Arctic is accessible by ice-strengthened ships which bring in heavy drilling supplies for the year's operations.

Conventional subsea blow-out prevention equipment and an offshore marine riser are employed. The marine riser is held in tension by hydraulic cylinders in the rig substructure to accommodate tidal movements, which in the Arctic are small – in the range of 1 m.

With this ice platform drilling technology we have the ability to drill offshore wells in very deep water, and wells can be drilled today at a fraction of the cost of drilling by any other offshore method.[10,14–19]

The bulk of gas reserves lie offshore and it was considered prudent to develop the technology and demonstrate that offshore wells can be completed on the ocean floor beneath the ocean ice and connected to production facilities on land for the taking of production. In the spring of 1978, Panarctic completed such a well offshore in the Drake Point field and connected the wellhead with flow and control lines to onshore production facilities (Fig. 4).

This offshore well completion project was somewhat akin to a major space project in that it was first necessary to assemble a group of specialists who would work out the technology and prepare a plan to do something that had never been done before. A group was assigned to the design of a new drilling rig

Figure 7. Ice-reinforced tanker off-loading fuel into 'bladder farm' fuel storage at Panarctic's Rea Point base on Melville Island.

that would be capable of handling, under Arctic conditions, the large subsea blow-out prevention and wellhead equipment. A group was assigned to design the ice platform for this large drilling rig and related equipment. A group was assigned to design the wellhead equipment and installation procedures, and a final group was assigned to develop the design of the flowline and control bundle and to develop means and procedures to install the flowline and control bundle on the ocean floor and connect them to the wellhead. The terms of reference called for the development of a system for installation in water up to 500 m deep with all operations conducted from the surface of the ice without the use of divers. This feat, involving the development of many new and unique techniques, firmly demonstrated that the offshore gas reserved can be produced commercially to the land.[11–13,20–25]

PRESENT EXPLORATION

Present exploration for gas is shifting almost exclusively to the offshore. A consortium of companies, of which Panarctic is the operator, was formed to take over some 12 million hectares of offshore acreage in the high grade area of the Central Sverdrup Basin. A large amount of seismograph work was done from the ocean ice (Fig. 8) which developed a large inventory of drilling prospects. My Company's system of ice platform drilling is being employed to test these drilling prospects.

Now, 12 years after initial exploration was commenced, one might again ask why would the industrial participants and the Canadian Government continue to invest what has now become very large annual sums of money ($60–70 million annually) in continued exploration of the Arctic Islands. While crude oil has not been discovered in commercial quantities, natural gas has become the explorationist's prize. The early visions of the industrial participants to profit from the discovery of oil have shifted to expectations for profit from the discovery and production of natural gas.

The Government of Canada is realizing its aims of developing oil and gas resources in the Arctic Islands with substantial ownership and control, along with reinforcement of sovereignty of the area. The exploration work done to date, with the discovery of very substantial gas reserves, indicates that the area has the potential for the discovery of even larger quantities of gas. If Canada is to achieve its aim of becoming self-sufficient by 1990, natural gas from the Arctic Islands is going to be required and the

Figure 8. This mobile drill is used to drill holes in the land or ocean ice to take a small explosive charge to obtain seismograph data on subsurface rock formations.

Government of Canada is interested in evaluating the gas potential of the Arctic Islands as a means to achieve this objective.

MARKETING PROJECTS

There are two projects in progress to market gas from the Arctic Islands. Petro-Canada, the Canadian national oil company, the Alberta Gas Trunk Line Company and Melville Shipping Ltd have applied for Government authorizations to construct and operate an Arctic Pilot Project to transport liquefied natural gas (LNG) out of the Arctic Islands in ice-breaking tankers to southern markets. The initial Project would deliver relatively small volumes of gas of about 7 million standard m^3 of gas per day (250 million ft^3 per day). This Project can be operational by 1985 and would provide production revenue to gas producers in the Arctic Islands at the earliest date.

The Arctic Pilot Project involves a pipeline from the Drake Point field to Bridport Inlet, a natural harbour on the southern coast of Melville Island. The barge-mounted liquefaction plant and storage facilities would be constructed in southern Canadian shipyards, and towed to Bridport Inlet. These floating facilities will be moored inside a protective dock structure which will also act as the LNG carriers' loading terminal. From this facility, the LNG will be transported in ice-breaking ships to southern markets on a year-round basis.

The most innovative and challenging portion of the Arctic Pilot Project is this shipping aspect. The year-round operations of two 335 m long ice-breaking carriers will be a world's first (Fig. 9). The ships will be built to Canadian Arctic Class 7, which means that they can make steady headway through more than 2 m of ocean ice.

The Project is being termed 'pilot' because it is one-tenth the size of any other project being planned for the Arctic. This reduced scale, the smallest economical size, will minimize construction delays, environmental impact and cost overrun problems such as those experienced by other large frontier projects. At the same time it will provide practical, proven operational experience for planning a future full-scale project.

The second project is the large diameter pipeline project of Polar Gas Ltd which has filed applications for Government authorizations to construct and operate a pipeline system out of both the Arctic Islands and the MacKenzie Delta area of Canada where gas has also been discovered (Fig. 10).

POLAR GAS PIPELINE

The Polar Gas Project, established in 1972, has expended some $70 million in developing the technological and environmental requirements to transport large quantities of natural gas by pipeline from the Arctic Islands to southern markets.

Participants in the Polar Gas Project are: TransCanada PipeLines, a Canadian corporation acting as the Project Manager; Ontario Energy Corporation, owned by the Government of Ontario; Petro-Canada, owned by the Government of Canada; Tenneco Oil of Canada Ltd, a subsidiary of a US company; and Panarctic Oils Ltd.

During the past six years, engineers and environmental and social scientists have undertaken intensive studies, gathered field information and evaluated existing data to build a comprehensive information base. Using these results, Polar Gas developed a plan for a natural gas pipeline system from the Arctic Islands to southern markets. A Polar Gas pipeline will cross under deep water Arctic channels, and while these crossings represent a formidable task, the technology is now available for their design and construction.

In order to select and adapt the best techniques for each crossing, scientists and engineers have conducted both office and field studies. Records of the duration and depth of ice cover are available and tide patterns, water depths, currents and bottom soil conditions have also been extensively studied. Current options for marine pipelining include utilization of lay barges, bottom pull from the ice, and tunnelling to protect the pipe in foreshore areas.

Land portions of the pipeline would be built with consideration for permafrost conditions in northern areas. In the permafrost zone the gas would be chilled to prevent frost heave. Aerial reconnaissance, soil sampling and other field work has been undertaken in support of the land pipeline activities.

Figure 9. An artist's conception of an ice-breaking liquefied natural gas carrier to transport liquefied natural gas year-round out of the Arctic Ocean.

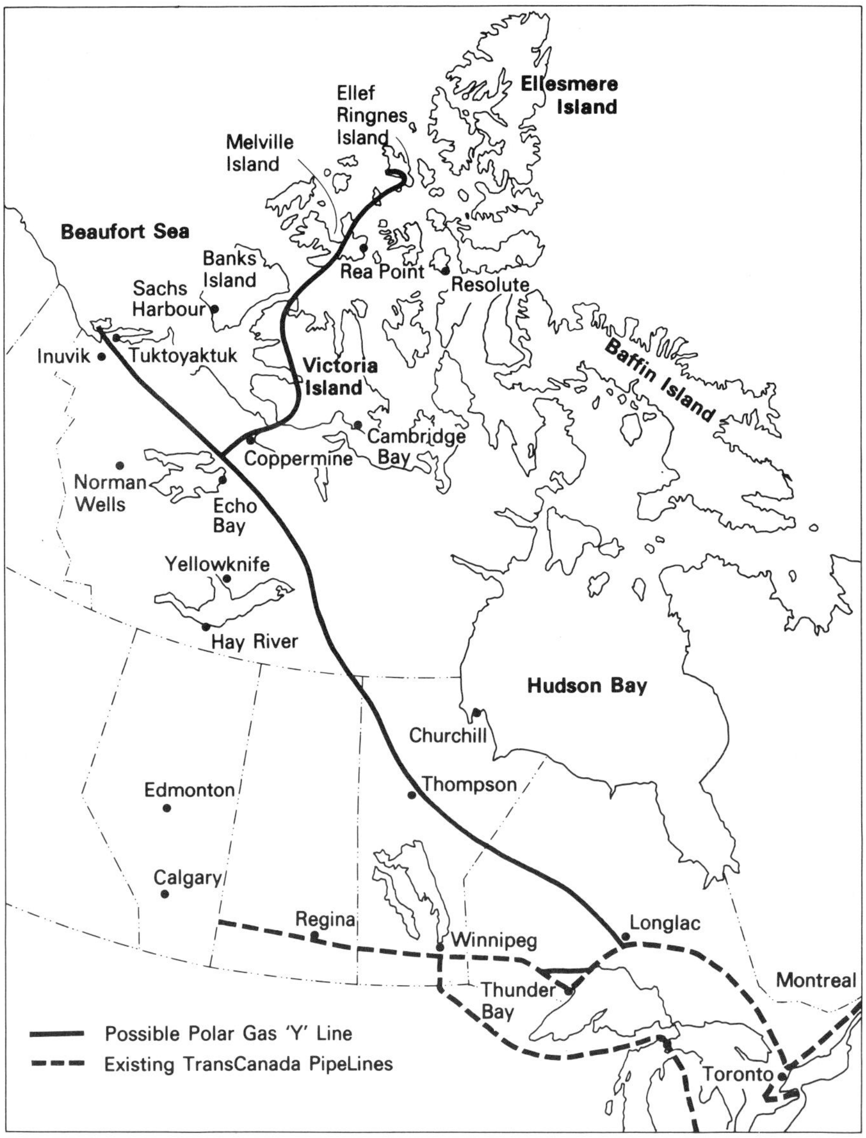

Figure 10. Map showing possible pipeline route for the Polar Gas Project to transport gas from the Arctic Islands to southern markets. (Note: The 'Y' Line displayed here shows one possible option for a line that would connect the Delta reserves in a common pipeline system. Comparisons with other 'Y' Line configurations, including a possible Mackenzie Valley 'Y' Line routing, are also being reviewed.)

In addition to its own study activities, Polar Gas will be able to draw on the experience of others and apply it to the Polar Gas Project. Pipelining techniques have already been successfully applied in an Arctic environment in 1978 by Panarctic Oils during the connection of a subsea well at Drake Point off Melville Island. In addition, advances have been made in welding techniques which would allow a pipe to be laid down to 700 m below sea level to be connected on the sea bed using one-atmosphere weld methods.

The current combined reserves of the Arctic Islands and the Mackenzie Delta/Beaufort Sea area have established current marketable reserves of approximately 600×10^9 m^3 (21×10^{12} ft^3). The volumes of gas needed to justify a large diameter pipeline system are now clearly in hand.

In mid-1979 Polar Gas advised Canada's National Energy Board that, following completion of 1979 studies, the Project plans to prepare and file a new application for a combined pipeline system to transport the natural gas reserves of both the Arctic Islands and the Mackenzie Delta to southern markets (Fig. 10). A final routing decision will be taken this winter by the Project. Currently, Polar Gas is studying comparisons of a number of 'Y' Line configurations. These include routings to the east of Great Bear and Great Slave Lake, cutting through Manitoba and Ontario to join the TransCanada PipeLines' system at Longlac. Also included in the studies are possible Mackenzie Valley 'Y' Line routings.

A Polar Gas pipeline will generate significant economic benefits for Canadians as well as connecting a large new secure energy resource which will make a major contribution to Canada's attaining its stated goal of energy self-sufficiency. While helping to achieve this goal, it will provide a major economic stimulus with long-range job creation potential. In addition to being an economically viable system for producers and transmission companies involved, Polar Gas will strengthen and assist in the development of Canada's north and it will provide thousands of millions of dollars in taxes, royalties and other revenues to governments. Based on completing and documenting current studies and applying to regulatory authorities in 1980, it should be possible to have the first gas flowing by the late 1980s.

CONCLUSION

Clearly, the Arctic Islands constitute the major supply source for natural gas to meet Canada's future requirements. In the course of continued exploration, oil will also probably be found.

In addition to these large gas reserves in the Arctic Islands, Canada is blessed with another very large energy resource – the tar sands and heavy oils in western Canada. Clearly, the use of Arctic Islands gas, coupled with increased production of synthetic oil from the tar sands and heavy oil deposits in western Canada, can provide the fuel requirements for Canada far into the future.

LITERATURE CITED

1. C. R. Hetherington, A story of Arctic exploration. *University of British Columbia Business Review* (1971).
2. D. C. Waylett, Natural gas in the Arctic Islands: discovered reserves and future potential. *J. Petrol. Geol.* **1**, No. 3, 21–34 (1979).
3. R. A. Meneley, Exploration prospects in the Canadian Arctic Islands. Paper presented to the Canadian Society of Exploration Geophysicists (May 1970).
4. C. R. Hetherington, Panarctic Oils Ltd. *Can. Petrol.* (October 1971).
5. C. R. Hetherington, Critical decisions now for prosperity in the 1980s. Panel Remarks presented to the Financial Post Conference on Oil and Gas, Calgary, Alberta (October 1978).
6. C. R. Hetherington, Address to the 1971 Annual Meeting and Convention of the Prospectors and Developers Association.
7. C. R. Hetherington and H. J. Strain. Innovations in petroleum exploration in the high Arctic. Paper presented to the 9th World Petroleum Congress, Tokyo (May 1975).
8. L. J. Franklin and K. G. Alexander, Exploration under the polar ice. Paper presented to the 5th International Ocean Development Conference, Tokyo (September 1978).
9. C. R. Hetherington, Oil and gas exploration and production in the Arctic Islands of Canada. Lecture to the Institution of Mechanical Engineers, London (October 1979).
10. C. R. Hetherington, Indian affairs and public lands. Presentation to the US House of Representatives Subcommittee, Washington, DC (April 1977).
11. C. R. Hetherington and L. J. Franklin, Offshore drilling and production in deep ice-covered waters. Paper presented to the 10th World Petroleum Congress, Bucharest (September 1979).
12. R. J. Brown, Connection of submarine pipelines in Arctic conditions. In *Proceedings of 4th International Conference on Port and Ocean Engineering under Arctic Conditions, St John's, Newfoundland,* Vol. 1, pp. 405–413 (September 1977).
13. D. M. Masterson, K. G. Anderson and A. G. Strandberg, Strain measurements in ice platforms using 3 m long resistance wire gauges. Paper presented at Workshop on the Bearing Capacity of Ice Covers, Snow and Ice Subcommittee, Associate Committee on Geotechnical Research, National Research Council of Canada, Winnipeg (October 1978).
14. D. J. Baudais, D. M. Masterson and J. S. Watts, A system for offshore drilling in the Arctic Islands. Paper presented to the 25th Annual Technical Meeting of the Petroleum Society of CIM, Calgary, Alberta (May 1974).
15. G. L. Hood, H. J. Strain and D. J. Baudais, Offshore drilling from ice platforms. Paper presented to the Society of Petroleum Engineers of AIME, Dallas, Texas (28 September–1 October 1975).
16. D. M. Masterson, D. J. Baudais and B. R. Wasilewski, Experience in ice platform construction for Arctic drilling. Paper presented to the CIM Annual Western Meeting, Edmonton, Alberta (October 1975).
17. G. L. Hood, H. J. Strain and D. J. Baudais, Offshore drilling from ice platforms. *J. Petrol. Technol.* (April 1976).
18. D. J. Baudais, D. M. Masterson and J. S. Watts, A system for offshore drilling in the Arctic Islands. Paper presented to the Offshore Technology Conference, Dallas, Texas (May 1976).
19. D. J. Baudais, D. M. Masterson and J. S. Watts, A system for offshore drilling in the Arctic Islands. *J. Petrol. Technol.* (March 1977).
20. D. M. Masterson, K. G. Anderson and A. G. Strandberg, Strain measurements in ice platforms using 3 m long resistance wire gauges. Paper presented to the National Research Council of Canada, Winnipeg, Manitoba (October 1978).
21. D. M. Masterson and A. G. Strandberg, Lond term loading analysis of floating ice platforms using finite element techniques. Paper presented to the National Research Council of Canada, Winnipeg, Manitoba (October 1978).
22. J. S. Watts and D. M. Masterson, Completion of an offshore well in the Canadian Arctic Islands. Paper presented to the Offshore Technology Conference, Houston, Texas (30 April–3 May 1979).
23. A. C. Palmer and D. J. Baudais, Design and installation of an offshore flowline for the Canadian Arctic Islands. Paper presented to the Offshore Technology Conference, Houston, Texas (30 April–3 May 1979).
24. G. L. Hood, D. M. Masterson and J. S. Watts, Installation of a subsea completion in the Canadian Arctic Islands. Paper presented to the 30th Annual Technical Meeting of the Petroleum Society of CIM, Banff, Alberta (May 1979).
25. A. C. Palmer, J. P. Kenny, M. R. Perera and A. R. Reece, Design and operation of an underwater pipeline trenching plough. *Géotechnique* **29**, No. 3 (1979).

The manuscript was received 27 November 1979

Book Review

PAUL STREETEN

Reviews

Disturbing the Universe

Freeman Dyson: *Disturbing the Universe* Harper & Row, New York, 1979. pp. x+283, $12.95. ISBN (USA) 0 06 011108 9 (Outside USA+Canada) 0 06 337004 6

Freeman Dyson, a physicist at the Institute for Advanced Study at Princeton, has written a fascinating book. It is the first in a series sponsored by the Sloan Foundation to encourage public understanding of science. Professor Dyson says:

'I am trying in this book to describe to people who are not scientists the way the human situation looks to somebody who is a scientist. Partly I shall be describing how science looks from the inside. Partly I shall be discussing the future of technology. Partly I shall be struggling with the ethical problems of war and peace, freedom and responsibility, hope and despair, as these are affected by science. These are all part of a picture which must be seen as a whole in order to be understood. It makes no sense to me to separate science from technology, technology from ethics, or ethics from religion. I am talking here to unscientific people who ultimately have the responsibility for guiding the growth of science and technology into creative rather than destructive directions' (p. 5).

A vivid illustration of this is the author's experience at Britain's Bomber Command during the War. He describes how technology had made evil anonymous. 'Through science and technology, evil is organized bureaucratically so that no individual is responsible for what happens . . .. Since the beginning of the war I had been retreating from one moral position to another, until at the end I had no moral position at all' (pp. 30–31).

Another sequence is that which led Robert Oppenheimer and Edward Teller from fear of Hitler or Stalin respectively, first to build the atom then the hydrogen bomb, to seek political, as well as technical power. And each in his own way came to grief over it (p. 91).

A common theme is the deadening hand of bureaucracy: Bomber Command killing the bomber crews and resisting simple improvements that would have saved them; the managers and accountants of the nuclear power industry killing the spirit of adventure that would have produced imaginative innovation and greater safety. 'Nobody builds reactors for fun any more' (p. 105); and under present circumstances it is impossible for any radically new type of reactor to receive a fair trial. The V2 rocket, a technically brilliant device, made no military or economic sense. The Gestapo had stopped the amateur efforts of the *Verein für Raumschifffahrt* and replaced the promising amateurs by professionals and bureaucrats. Operation Barrier in the Vietnam war was a clever engineering device (a system of electronic burglar alarms dropped by aeroplanes along the frontier) which completely disregarded the absence of an indigenous commitment to defend the country.

Dyson condemns bureaucracy and praises the small band of enthusiastic amateurs. But is not some form of management necessary to implement the ideas of the scientists? What form should it take? How can we avoid the blunders that Professor Dyson so vividly describes?

Another puzzle for a member of the public who wishes to understand science is the following. Dyson quotes with approval Einstein: 'One may say the eternal mystery of the world is its comprehensibility' (p. 50) (and he disagrees with Steven Weinberg's 'The more the universe seems comprehensible, the more it also seems pointless' (p. 246)). We begin to marvel that the electron on the oil drop should behave according to Dyson's complicated calculations. But then he quotes J. B. S. Haldane's famous remark also with approval, according to whom the universe is 'not only queerer than we suppose but queerer than we *can* suppose' (p. 191). How can both be right?

Both science and art have their origin in myth. Like Goethe's *Faust*, Dyson shares with the reader a search:

Dass ich erkenne, was die Welt
Im Innersten zusammenhält.

'The future is my third home, after England and America.' His vision of the future is humane, optimistic and imaginative.

Paul Streeten – a Member of our Editorial Board – is now Director of Studies at the Overseas Development Council in Washington; he has been Director of the Institute of Commonwealth Studies, Oxford, and Professor of Economics, the University of Sussex.

Nuclear Waste

The Battle for Gorleben

DR ANTHONY R. MICHAELIS, EDITOR

Spectrum House, Hillview Gardens, London NW4 2JQ, UK

Rioting and bloodshed to oppose the progress of technology are nothing new. Just as the destruction of textile machinery by the Luddites – handicraftsmen of Yorkshire – in 1812 has gone down in history as an example of utter futility, so the Battle for Gorleben, the proposed nuclear waste repository near Brunswick FRG, will become the perfect paradigm for the scientific and political ignorance of the protesting green ecologists. Neither nuclear power, nor the disposal of its waste in deep geological formations, are any longer a technological problem. Both have been fully proved. Of course all energy technology has its associated risks and Dr H. Inhaber on p. 305 of this Issue analyses such risks and their consequences. They are the real total costs of modern life in the West, which – if we want to contribute and enjoy it – we must accept.

The Space Option

Many different techniques for the disposal of nuclear waste from fission reactors exist in theory and are now seriously investigated; they must guarantee safety for thousands of years to render all end-products harmless for the biosphere. The Space Option, for example, is being studied by the Boeing Company under a $296 000 contract from NASA; it has been critically reviewed by D. H. Hinson (*Spaceflight* **22**, 182, 1980). He concluded that a soft-landing approach to sites on the Moon would be cheapest and safest. But for the time being, this technique is not considered safe enough. Disposal under the Antarctic Ice Shelf has also been seriously proposed, but the instability of the Shelf itself, and its possible sliding off the underlying rock, does not offer the required certainty of thousands of years.

Dumping in the Oceans

Disposal into the oceans has been practised since the beginning of the nuclear age – its 40th anniversary will occur in 12 months time. Dilute solutions of radioactive waste have been allowed to flow from Windscale, UK, and other plants into the sea, and free-fall containers have been deposited in both relatively shallow and deeper parts of the oceans. Unfortunately, leaks of these containers, as for example the recent ones near the British Channel Island Alderney, have already occurred and, if ever a recovery operation is to be carried out, one may well ask if these leaky containers can again be found or if their precise location is already lost for ever.

Buried in Ocean Sediments

There are two further possibilities of utilizing the oceans for the disposal of radioactive wastes, namely free-fall projectiles to penetrate the sediment and drilled holes through the sediment to deposit the waste into consolidated sediments and possibly the underlying rocks. These alternatives have been suggested in detail by P. D. Grimwood and G. A. M. Webb of the British Radiological Protection Board (*New Scientist*, 24 March 1977).

Stainless Steel Tank, the latest of a series to be installed at Windscale, Cumbria, UK. Here the cooling coils of a storage tank for highly radioactive fission waste products are shown before they are inserted into a newly completed tank. The tanks are double-walled and are located in concrete vaults which are themselves lined with stainless steel. Note the size of the workmen on top. (Courtesy British Nuclear Fuels Limited)

These proposals are now under active study, to assess the environmental impact of sea-bed disposal of high-level waste.

Leaks on Land

Disposal on the surface of the land and below it are the most obvious places of course. During the reprocessing of irradiated fuel elements of nuclear reactors any unused uranium and plutonium is recovered for further use, but other fission products cannot be economically separated and, as 'high-level waste', are at present stored in what was hoped to be suitable stainless steel containment vessels. Although satisfactory in most cases, here again leaks have occurred, perhaps the worst example being near the Lake Ontario Ordnance Works, USA. During World War II, about 20 000 tonnes of radioactive wastes were dumped there, partly in old water tanks and partly in ordinary drums. Even worse, because of the existing military security, false descriptions were affixed to these containers, so that now it is impossible to know what was stored where. Various United States Federal Agencies are hopeful of clearing up this dangerous mess by 1986.

Crisis of Confidence

It is perhaps not surprising therefore that as these radioactive leaks occurred and became public knowledge, a crisis of confidence arose between the nuclear establishment and those concerned about the environment. Assurances were no longer believed by the media and the public. Statistical investigations of leukaemia cases in children, and of cancer cases in employees in nuclear ship yards at Portsmouth, New Hampshire, as well as at the nuclear plants in Hanford, State of Washington, were given widest publicity in the USA. The adverse conclusions of these studies have been strongly disputed by many, including the International Atomic Energy Authority and the International Commission on Radiological Protection, an independent non-governmental expert body. But of course, once confidence was lost, it has so far not been possible to restore it. The quite separate question of accidents at nuclear power stations, their unpredictability and inevitability, has been discussed by me recently (*Interdiscip. Sci. Rev.* **5**, 79, 1980).

The Safety of Oklo

The last, and at present apparently the safest and most economical, disposal site for radioactive waste is below the surface of the Earth. But is there any evidence that radioactive compounds can remain undisturbed in geological formations for millions of years? It was the discovery of the Oklo natural nuclear reactors, operating for about 700 000 years and releasing 15 000 megawatt-years of energy, 1800 million years ago, which established that their fission products could lie immobile in the ground at the original reactor sites, to be unearthed by the scrupulous care of a present-day analyst (*Interdiscip. Sci. Rev.* **1**, 73, 1976). It is of course impossible to conceive of any man-made facility which will last long enough to allow for the thousands of years of decay of the longer-lived components of high-level waste. Only a natural geological formation, into which the waste can be sealed, offers safety and also the possibility to transmit accurate knowledge of its location to posterity.

Processing of Waste

Before burial in suitable geological strata, it is considered advisable to concentrate the liquid high-level waste, to evaporate it to dryness and finally to incinerate it to produce a fine grained mixture of radioactive oxides, the 'calcine' of the nuclear industry. Further processing is then necessary, for example by incorporating the calcine into a borosilicate glass matrix, now being developed in the Harvest vitrification process. As an alternative to this, Professor A. E. Ringwood of the Australian National University has produced a synthetic analogue of a natural igneous rock to provide specific crystal lattices into which the waste atoms can be incorporated; he has called this process SYNROC (*Geochemical J.* **13**, 141, 1979). Whereas glass as host for waste is being developed at Harwell, UK, at Harford, WA, USA, at Mol, Belgium and at Marcoule, France, SYNROC is also to be tried out in the USA.

US Salt Mines

Early in the 1950s the idea was born to dispose of radioactive wastes in deep and stable geological deposits, especially salt formations which by the mere fact of their survival for many millions of years have proved that they do not take part in the material exchange of the biosphere. Research and development programmes were started to prove this technology in the USA and the Federal Republic of Germany. In the USA the Lyons Mine, an old salt mine in Kansas, was chosen, and technical equipment and appropriate machinery were developed, manufactured and tested with highly satisfactory results. However, the Lyons Mine was closed in the early 1970s as the many boreholes around the site from previous activities were considered to reduce its safety. In 1980 another repository in the USA, the Waste Isolation Pilot Plant in south-eastern New Mexico was ready to be built, but because of adverse political clamour, its future is now uncertain. A panel of the National Research Council was recently set up to identify the social and economic issues to be considered in the selection of repository sites; its Chairman is Professor Angus Campbell, psychologist at the University of Michigan – Ann Arbor.

German Salt Mines

In the FRG, the abandoned salt mine Asse near Brunswick was purchased by the Federal Government and research work started in 1964; the burial of nuclear waste began in 1967 and more than 100 000 special drums were disposed of routinely without any difficulties until it too, in 1979, had to be closed down as political forces prevented an extension of its licence. Drs C. Salander and R. Proske together with Dipl.-Ing. E. Albrecht give a full account of this outstandingly successful 12 year operation in this issue of IRS (see p. 293). I myself went down the Asse mine whilst it was still in full operation. I could find to the best of my knowledge no scientific or other reason why it should not remain operational. It should also be noted that about 40 km north-east of the Asse, 2 km east of the German–German frontier, the German Democratic Republic is routinely disposing low and intermediate level radioactive waste into its salt mine at Bartensleben. Apparently no further details are known in the West about this repository.

Possible Failures

In a consideration of the safety of radioactive waste disposal in salt and other rock formations, M. Hill and P. Grimwood (*New Scientist*, 11 May 1978) have listed the possible reasons for a failure of the geological containment as natural rapid events, such as meteorite impact and volcanic activity; as natural geological events, such as erosion, climatic changes like glaciation and pluvial episodes; as geological events caused by the repository itself, such as thermal and radiation effects on the rocks; and as human actions, such as drilling for mineral resources, reservoir construction and nuclear war. The last category, human activities, supposes that future generations will be in complete ignorance of the existence, let alone the precise location, of geological repositories of radioactive waste. Although after about 1000 years radiation will be greatly reduced it will be reassuring to the present and future generations to keep the knowledge of its location alive as long as possible.

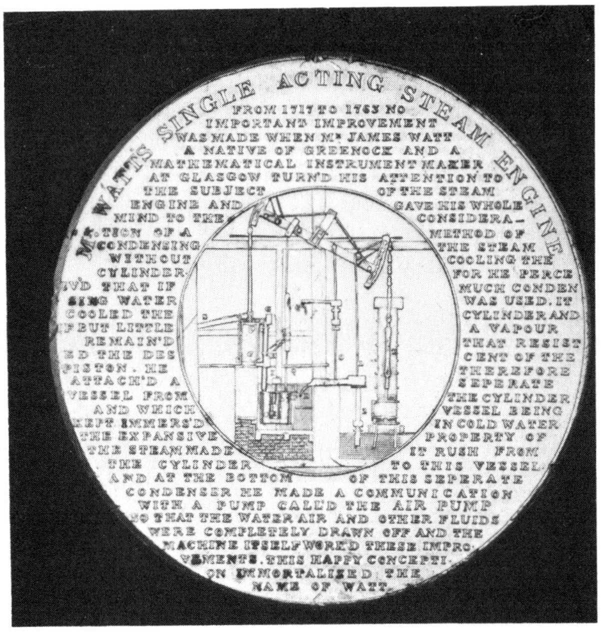

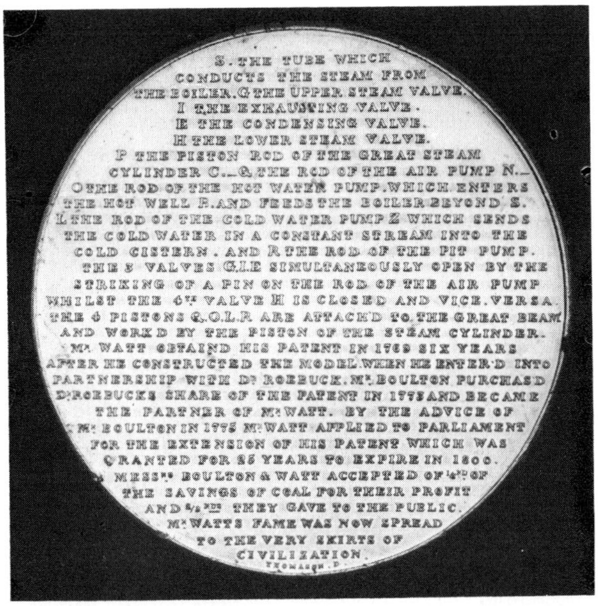

Informative Medals are nothing novel. This example, one of a series of 'Educational Medals', was struck in German silver towards the beginning of the 19th century, probably at Birmingham. Here reproduced full scale, 3 inches (75 mm), their purpose was to inform and to teach, as this full and highly detailed description of 'Mr Watt's Single Acting Steam Engine' illustrates.

Unless specific actions are taken now to transmit such information to future generations, the whole purpose of safe disposal could be rendered null and void by inadvertently returning radioactive waste into the biosphere. In view of what has already occurred at the Lake Ontario Ordnance Works after a mere 40 years it appears by no means too early to strike now suitable metallic tokens for transmission of information to the future, like medals. They are the only known means of safe communication with posterity over the centuries and even millennia (*Endeavour* XXIV, 72, 1975). It might well be done for Asse and each medal should give a precise latitude and longitude, depth of deposit, quantities of nuclear wastes deposited, their estimated half-lives and dates; they should be struck in as large a number as possible in the most permanent metals. They should be distributed world-wide to libraries and geological institutions.

Informing Posterity

By far the most imaginative solution to the whole problem of radioactive waste has come from the Deutsche Gesellschaft für Wiederaufarbeitung von Kernbrennstoffen, DWK of Hanover, Federal Republic of Germany, with its total system concept dealing with the back-end of the nuclear fuel cycle, the *Entsorgungs-Zentrum* (see p. 302). Sponsored by 12 public electricity utilities as an industrial activity but under strict Government control, its location is to be at Gorleben, within about 5 km of the German–German frontier, on top of a salt dome to receive the final nuclear waste. To prove the suitability of the site, a number of test drillings into the salt formations were undertaken during 1980 and these have led to most active, even ferocious, protests from often international groups. The site was occupied for weeks by protesters, but could be cleared by the police with relatively minor injuries to all concerned. The drillings are now continuing.

The Gorleben Site

This 'Battle for Gorleben' epitomizes the enormous emotion over nuclear waste which has been generated by environmentalists and politicians after the safe disposal in the Asse Mine has been proved over a 12 year period. One explanation may be that the risks involved were overestimated and not put in relation to other, routinely accepted, risks of our modern western society. For the 'nuclear have-not' countries, the question of waste disposal simply does not exist; as almost their total earnings of foreign currencies are now swallowed up by ever rising oil costs, nuclear energy is their only escape.

Enormous Emotion

Future generations are likely to be far less worried about the small amounts of radioactive waste which we shall leave behind us – provided they know their precise locations – than by the immense pollution caused by the burning of coal and oil and the increase of carbon dioxide in the atmosphere leading to possible changes of climate. They may accuse us also of having heedlessly burned up the valuable chemical resources of oil, instead of relying more on the alternatives, particularly nuclear power. If the public and politicians could better understand that the risks of nuclear waste disposal can be safely managed, then the correct decisions could decide our society's future from a basis of reason and interdisciplinary wisdom, instead of emotion-fanned hysteria.

Reason and Wisdom

Professor Dr Wolfgang Gentner, a Member of our Editorial Board since its inception, died in Heidelberg on 4 September 1980. He was not only one of Germany's foremost nuclear physicists, highly honoured in his own country and throughout the world, but also a very brave gentleman. His wise, interdisciplinary counsel will be missed by many.

Obituary

Freedom and the Responsibility of Science
Science and the Public

PROFESSOR DR REIMAR LÜST
President, The Max Planck Society, Munich, Federal Republic of Germany

Academic essays in Germany either start or end with a citation from Goethe. I cannot refrain from this tradition as I have in mind three lines which described the situation in his time and during the last century, but which no longer hold true in our changing world. This is not easy to accept, especially for a scientist. Until the beginning of the Second World War, the relationship between politicians and scientists could be characterized by the following statement from Goethe's *Xenien*:

> '*Forsche der Philosoph, der Weltmann handle! Doch weh uns: handelt der Forscher und gibt, der es vollzieht, das Gesetz!*'
>
> 'Let the philosopher research, the man of the world act. But woe to us: should the researcher act and the law-giver be he who executes it.'

Now it is obvious that such poetical phrases cannot be truly translated and I shall therefore try to circumscribe them.

SEPARATION OF DUTIES

There is first a clear separation of duties between the scholar and the man of public affairs. The scholar ought to do research, the man of public affairs should act, and Goethe adds the warning: 'Woe to us, if the law-giver also executes the law'. The scholar should abstain from interfering with politics or public affairs and it is stipulated that the man of action should deal with public affairs, both according to existing law. Goethe gives no indication as to who defines the law, but he makes the point that those who are active in handling the affairs of men ought to adhere to existing laws and not shape them according to their own ideas.

The strict separation – as expressed in Goethe's phrases – between research and practice, and between science and politics, originally characterized the relationship between scientists and those active in politics; at least this was the general understanding in Germany. Any possible communication between these two sectors was not mentioned – that seemed unnecessary. Indeed, the practical consequences of research results leading to tremendous technological developments and changing everybody's life were simply regarded as progress. Therefore the politician was in a very comfortable position because technical progress served the welfare of the community, although there was occasionally some resistance to progress that was too fast. There was no obvious necessity to communicate, neither for the politician nor for the scientist. For the scientist such a link was even more suspicious as a close relationship with politics endangered the freedom of his research.

Since those happy days we have found out that not every scientific discovery leads to a technical application which brings about a positive change for mankind. The discovery of nuclear energy has been a direct demonstration of forces that can be set free by scientific research but threaten human welfare and could lead to the total ruin of mankind.

We must accept today that communication between scientist and politician, and even more so between science and the public, is an urgent necessity. But this is not an easy task. I shall try to make a few comments on this situation in Germany and on its consequences. [See also *Interdiscip. Sci. Rev.* **4**, 6 (1979).]

Regarding the present relationship between the public and science, let me comment on three important aspects: the social relevance of research; democratization and bureaucracy; and the clamour against new technologies and for a clean environment.

RELEVANCE OF RESEARCH

At the end of the 1960s, there was a growing tendency to ask: 'How relevant is research work?' 'Relevant' had different connotations but the overall direction was clear: 'how beneficial is a particular research project for society, for our common benefit and for the social need of our times?'

These questions were first raised by some progressive politicians and then increasingly by the public. I

CCC–0308–0188/80/0005–0260$01.50/1

believe that we scientists were not quite innocent of this development. When asking for financial support for their research, scientists often argued that their special project would be valuable for technological, economic or social progress. Such arguments were dangerous if they lacked real substance or led to the belief that application and progress would be possible in a short time.

This was particularly true when money for big science projects was requested. In the USA it was common practice to point to a highly profitable 'spin-off' from research on almost every occasion. Finally, politicians and public realized that this spin-off did not come as fast as some applicants had led them to believe.

However, this short-sighted question 'How relevant is research?' did not remain a public issue for very long. Scientists could point out that the time scales for 'spin-off' from basic research, if any, might differ considerably from case to case and that the impact of certain discoveries could be substantial, even if there was no direct commercial profit. Thus, for instance, it was accepted that the work of Copernicus was socially relevant, although it did not have any technical or economic consequences.

Today, no responsible politician will use this question of relevance as a valid argument for or against basic research. But let us not make the mistake of thinking that these voices are completely silenced – at the moment they just do not have the support of the majority.

A somewhat different tune can be heard when there is a request for support of research which is 'useful' for human beings. This may sound good to the general public but it does not take into account the fact that in basic research it is never possible to foretell the outcome; furthermore, it leaves open the question as to who decides what is 'useful' for mankind.

DEMOCRATIZATION

The second problem which faces Germany's research policy is the impact of the process of democratization and, almost as its consequence, the impact of bureaucracy. This started at the end of the 1960s and, in German universities, the process of democratic participation in decision-taking has been introduced at all levels. I shall not particularize here as to why this came about and what the situation is like at present. As a result, at most German universities it has now become difficult to carry out quiet and undisturbed research. At the same time the influence of governmental administration has grown; this was due to the democratization at the universities against which parliaments have tried to regulate everything and to leave nothing to chance. This process is still going on and will not be easy to stop.

I should add that we are much better off in the Max-Planck Society. We are a private organization, although 90% of our financial resources comes from the government. We did not accept this undifferentiated principle of democratic participation. At our Institutes – at present we have 50 – the director is still fully in charge. He is obliged to consult all his scientific co-workers at his Institute when taking important decisions about common work. If he cannot reach a consensus, his decision is final. We have maintained the freedom to make our own priority decisions, for instance, when to start or to close down an Institute, and, more important still, the freedom of our own administration.

DISSENSION

Today, any public discussion of science and the clamour against it has also to include the influence of public dissension against new technologies, especially against nuclear power and the desire for preservation of an unchangeable environment. These dissenting movements have now become a factor of growing importance. Having started in Germany as *Bürgerinitiativen* – civic leagues – organized on a local level, these movements have now grouped themselves together to form a new political party, the *Grüne Partei*; it overcame the 5% hurdle for political representation in parliament in two recent state elections.

Again, it would be too much to analyse here the multifarious backgrounds of these groups. A number of motives that led to their formation, such as the fear and incomprehension of nuclear energy; however, this was only the tip of the iceberg. The Green Party is against any new technology and, therefore, against science; their rejection is of a very random and indiscriminate nature. Certainly, these movements do not represent the majority of citizens but they reflect certain feelings of the public, and naturally politicians have become sensitive to them.

Science and the freedom of science may well be under attack but I do not see a general anti-science movement. There is a general uneasiness about science; the prestige of science has been shaken and in certain intellectual circles it has become fashionable to speak about the dangers of science. Yet, as far as the general public is concerned, one can almost speak of schizophrenia concerning science. On the one hand, many people still believe that science can solve everything, while, on the other, science is regarded as the devil incarnate of our times.

Let me now discuss how, in the Federal Republic of Germany, we should accept certain responsibilities concerning the interaction between the general public and science. I shall try to make three points.

Scientists have a duty to provide information.
Scientists have to be especially careful in making any public statements.
Scientists bear a responsibility for their discoveries as consequent developments may either be positive or negative.

INFORMATION OF PUBLIC

Scientists have a duty to inform the public – including governments and parliaments – of new advances in research, of new insights and discoveries and must do this in an intelligible manner. As the natural sciences have had the strongest possible repercussions on the development of our society during the last hundred years, it has therefore now become so necessary that our society is made aware of the need, not only to understand all scientific development, but also be encouraged to take a greater interest in it.

The same problem also arises from the German educational system and in the scale of educational values: a lack of knowledge in the humanities and arts is considered to be more serious than absence of insight and understanding of basic science and technology. In this context let me refer to C. P. Snow and his idea of the two cultures. Without sufficient knowledge, how would one conduct a dialogue with scientists and engineers? And there is no point in questioning or examining the natural science by the methods of the humanities as social scientists or politicians are sometimes demanding.

DEMARCATION

Scientists must exercise particular care when making public statements. In my opinion, all too often it is not made clear that measurements always include a certain amount of uncertainty and that a statement is only valid within a given range of variables. This often leads to the public expecting more from a scientific answer than it contains, and the scientist may fail to define clearly the inherent limits.

Scientists have to make clear the demarcations of their scientific competence. Occasionally, scientists make public statements which may be well presented scientifically but lack specific conclusions. However, there are also politicians who often require scientific expert opinions, and there are scientists who provide such expert opinions, which have nothing to do with science and, if anything, are only common sense. Of course, nothing can be argued against common sense.

Scientists, as well as politicians, ought to realize by now that fighting over a problem through intensive discussions can lead to a decision. But, for each group, the way and the kind of discussion is quite different. Open discussions must be common both to politics and to science, but ending a scientific dispute by taking a vote runs contrary to the nature of science.

GENERAL WELFARE

Scientists bear a grave responsibility for their discoveries because scientific discoveries can either be positive or negative. Indeed, new scientific discoveries may be either useful or harmful to man. It is very rarely possible for anyone – even for a scientist – to foresee immediately the benefits or inherent dangers, especially in the case of discoveries in the basic sciences. There are many examples. Let me remind you of Einstein's discovery of the equivalence of mass and energy which, being made in 1905, led years later directly to the atom bomb. There is at present an intensive discussion on genetic engineering.

I believe that we have to make it clear, to one and all, that man must not put all his scientific insights and discoveries to practical use. Here the responsibilities of the scientist, the engineer and the politician meet. The core of the problems seems to be the need to arrive at an agreement: knowledge must be applied only to the extent to which the general welfare remains unharmed – this is the crucial issue. Prevention and restriction of free research provides no solution to this problem.

SCIENTIFIC OATH

I discussed once with Max Delbrück the possibility of asking scientists to take an oath similar to the oath of Hippocrates taken by physicians. Such an oath would commit the scientist to use all his abilities and all his thoughts to work on useful problems for society and to shield society against any harmful effects of science. Delbrück came to the conclusion that, for the scientist, such an oath would be impossible.

The original Hippocratic Oath required the physician to preserve the life of his patient under all circumstances. The highly developed clinical technology has now given rise to a number of problems, yet the purpose of the Hippocratic Oath is still well defined and the preservation of human life is clearly determinable. The benefit of science for society, on the other hand, cannot be determined equally clearly, and therefore it does not appear sensible to introduce such a Scientific Oath.

Let me conclude with the remark that all too often scientists have defended their science with the possible benefit it might bring to society. I do not contradict this argument. At a time when one is thinking of improving the quality of life, we should stress that research has much to contribute to this. For me, quality of life implies that man must be free to search for the origin, conditions and meaning of life, and that we have to try to answer these questions. These have always been with us and, to return to Goethe, I am convinced that it is the responsibility of each generation of scientists to contribute to the answers.

These views were expressed by the author – a Member of our Editorial Board – when he addressed the *Classe des Sciences* of the Royal Belgian Academy on 19 April 1980.

The manuscript was received 9 July 1980

Science and Civilisation in China

State of the Project

DR JOSEPH NEEDHAM, FRS, FBA
Director, East Asian History of Science Library, Cambridge, UK

To commemorate the 80th birthday on 9 December 1980 of Dr Joseph Needham – a Member of our Editorial Board – we are here publishing a somewhat condensed progress report of what has been rightly called 'the greatest single act of historical synthesis and intercultural communication ever attempted by one man'. After describing its contents, Dr Needham gives a preliminary outline of his methodology, how he has quarried Chinese literature and iconography, how he has immersed himself in China's living tradition to interpret correctly her past and how he has mastered the often century-old technical terminology of Chinese science – only thus could he build his oecumenical university. He concludes here with the hope that this vast study, so felicitously begun, will continue 'as limitless as is all history'.

The Editor

China signifies to Western man a vast territory, the cradle of a civilization old, rich and remarkable, of which he knows relatively little. It stands in his mind, with those of India, Islam and the Christian West, as one of the great surviving cultures, but of its achievement and significance he is almost entirely ignorant. He may well think of it only in terms of current events, at most those of the past half-century, and particularly with reference to China's resurgent influence in international affairs. He may believe (whether rightly or wrongly these volumes are showing) that among fundamental inventions, printing, paper and the magnetic compass originated in China, perhaps also gunpowder or spectacles or the potter's wheel; beyond that all is shadowy. He may have the impression that while the Chinese were a primarily practical people, capable of interesting technical innovations, nothing that we should call science developed among them. If he suspects that it did, he may doubt whether Europe was ever deeply affected by either. The purpose of *Science and Civilisation in China* was, and is, precisely to dissipate the shadows, break down the ignorance, clear up the misunderstandings, and bring together the different streams of human enterprise – to discover perhaps that they have been convergent rather than divergent. Such a task was formidable, yet at the same time necessary, valuable and urgent. All it needed was a group of people combining a knowledge of Chinese and Western languages with a grasp of science, technology and medicine, and an absorbing interest in the history of these things within the different cultures. Such a need is easy to state, but extremely difficult to meet in practice.

DESCRIPTION

As the pattern has worked out, the first volume of *Science and Civilisation in China* is essentially introductory. In order to prepare the reader for a study of the science, technology and medicine of a complementary culture, we begin by examining the structure of the Chinese language, reviewing the geography of the country and the long history of its people, finally by discussing the scientific and technical contacts which occurred through the centuries between East and West. The second volume proceeds to deal in detail with Chinese philosophy and the development of scientific thought, particular attention being given to the organic conception of Nature, and the place of the idea of natural law. The third volume begins the study of the specific sciences, starting with mathematics and going on to astronomy, meteorology and the sciences of the Earth.

In the fourth volume physics and all its related technologies are surveyed. Its first part covers the basic aspects of physics, treating also in more detail of physical acoustics, optics and magnetical science. Its second part describes the whole of mechanical engineering in traditional China, discussing the development of animal-power, water-power, water-raising machinery and wind-power, together with special sections on the prehistory of the steam-engine and aeronautics, and on the six centuries of hydromechanical clockwork which preceded the first clocks of mediaeval Europe. The third part is devoted to civil and hydraulic engineering, building technology, navigation, and the voyages which the Chinese undertook.

CCC–0308–0188/80/0005–0263$03.00/1

Joseph Needham FRS, FBA

The fifth volume concerns chemistry and the sciences and technologies related thereto. Its second part describes the Chinese origin of alchemy, discussing the specific idea of material immortality which gave rise to it in ancient times. The third part recounts the history of alchemy and early chemistry in China, ranging from the cinnabar elixirs of antiquity to the coming of modern chemistry. In the fourth part we describe the comparative development of chemical apparatus in East and West, the theoretical basis of Chinese alchemy, and the spread outwards from China of that quest for life elixirs which, passing to the Arabs, the Byzantines and the Latin West, gave rise to Paracelsian and all later chemical medicine.

The fifth part takes up the strange subject of 'physiological alchemy', peculiar to China, with the proto-biochemistry, the mediaeval preparation of hormones, which came out of it. In the first and sixth parts, not yet ready for press, many important branches of technology are described – metallurgy, ceramics, textile technology and the military arts, including the epic story of gunpowder and firearms. Then the biological sciences, including agriculture and medicine, occupy the whole of the sixth volume. Much of this is already written, though only one part, the third, which deals largely with acupuncture and moxibustion, has been published. At the end of all these chapters there follows a summarizing account of the special characteristics of science, technology and medicine in Chinese culture.

Throughout the work the Chinese contributions are considered in a comparative manner, that is to say, in relation to the achievements of the other great civilizations. To take a concrete example, Chinese scholars knew about Thang and Sung clockwork, but for lack of a comparative perspective they did not realize the significance of an escapement mechanism in the +8th century, when Europe had nothing of the kind until the +14th. Today it is quite apparent that in the history of science the Old World must be thought of as a single whole. The great paradox which then presents itself is: Why did modern science, the mathematization of hypotheses about Nature, linked to a consistent use of experimentation, with all its implications for contemporary technology, take its meteoric rise only in the West, in what is called the Scientific Revolution? This is the most obvious question, which many have asked in the past.

But there is another of quite equal importance. Why was it that between the 1st and the 15th centuries of our era, East Asian culture was much more efficient than the European West in gaining knowledge of Nature and applying it in purposes useful to mankind? Only an analysis of the social and economic structures of Eastern and Western societies, not forgetting the great role of systems of ideas, will in the end suggest an explanation of both these things. This is the subject of the seventh volume, which like the sixth will consist of several parts. Much of it has already been drafted, and there will be chapters on the social and economic structure of traditional Chinese culture, the world-outlook of the intellectuals, the place of the characteristic thought systems, including logic, and an analysis of the various factors, both sociological and intellectual, which at first stimulated, and then later inhibited, the development of science, technology and medicine.

METHODOLOGY

A few words may be appropriate here about the methods used in the work. Since this gigantic enterprise has been designed not only to raise, but also rather to settle, a thousand doubtful points in human history, it has been necessary to build up a massive system of information filing and retrieval. Since the project was begun long before modern computerization became available, the files and sub-files are real and palpable, arranged broadly in accordance with the spectrum of the sciences already mentioned. Files of photographic material accompany them. Since the successive volumes are the first of their kind in a Western language, massive bibliographies have had to be given in each one, and these remain for permanent reference conflated on cards in many card indexes. Besides them there are other special card indexes, for example one for Chinese technical terms, and another which constitutes a biographical glossary giving details of many thousands of Chinese scientific men, engineers and physicians during the past 3000 years. It should be understood that the files and indexes contain far more material than could be included in the volumes of *Science and Civilisation in China*, if only for lack of physical space. All these resources form part of the endowment of the East Asian History of Science Library.

Chinese Sources

Friends often wonder about the quarries from which the information is actually dug out. First come the original Chinese sources, overwhelmingly in number printed (since printing began there so much earlier than in Europe), but occasionally in manuscript, such as the Taoist and medical fragments found at Tun-huang, or the Han books recovered from tombs at Ma-wang-tui, or the late Fukien shipbuilding manual. Each one of the sciences has a long line of important books that were written through the ages, but besides these the lexicographic literature is not to be despised, since much important information which otherwise perished is contained in them. Moreover, the great series of dynastic histories, unparalleled in any other civilization, contains much on astronomy, calendar and acoustics, as well as thousands of biographies of men and women of achievement in science, technology and medicine.

What are called secondary sources are also valuable, that is to say, the books of those who have written the history of particular sciences already. Nearly all of these are in Chinese, and they cover only a few of the fields concerned. For example, there are good histories of mathematics, astronomy, entomology and medicine; but almost nothing at all for mechanical engineering, botany, zoology or pharmacology. Where they exist they are naturally of the greatest help in orienting oneself in the literature; where they do not, the work has to be done without such assistance. And then, since it is vital that all conclusions should be based on an adequate consciousness of what was going on in the other parts of the Old World, there is the non-Chinese literature to consider. Not only must one think of Japanese, Korean, Vietnamese and the other languages of East Asia; but also of Sanskrit and Urdu, Persian and Arabic, Greek, Latin and the successor tongues of Europe. Even as a whole our group cannot command all these, so we necessarily depend upon translations, aided however by a strong band of expert linguistic and cultural advisers.

Iconography

Textual evidence, however, is far from being the only resource for our work. There is the great field of iconography, of pictures, whether carved in stone in tombs and temples, or depicted on walls as fresco paintings, or inserted in books as woodcuts and other forms of reproduction. Long ago, Wang Ching-Ning and I built up an elaborate argument from iconographic evidence that the axial ('sternpost') rudder must have been an invention of San Kuo times (+3rd century) or earlier, and then after some years Lu Gwei-Djen and I found a Han tomb model ship at Canton which proved beyond doubt that it had been accomplished by the +1st century.

Archaeology thus came to the support of historical iconography, and indeed the role played by archaeological finds in our subject can hardly be overrated. Whether it be ratchet wheels or camphor sublimation apparatus in Han tombs, or labelled boxes of chemical substances in those of Thang date, or elaborate architectural models in the Ming, our enterprise demands the closest attention to the findings of Chinese archaeologists. Sometimes the results can be startling in the highest degree, as in the case of the incorrupt body of the Lady of Tai, who died in −168; a finding which showed that the adepts of the Han knew ways of preserving the human body indefinitely from decay – and not by mummification, nor tanning, nor freezing either.

Living Tradition

Lastly there is living tradition, for the study of which it is essential to have grown up, or to have lived for some time, in China; otherwise there is much in the books which would be difficult to understand. Just as one should have been trained to set up a vacuum distillation or carry out a titration oneself, so one should have sailed on Chinese ships in order to understand mat-and-batten sails, or known the small Chinese factories in which soya bean curd or soya sauce are made. Beyond that, in parallel to the non-Chinese literatures, stretches the almost infinite range of ethnography, including the accounts of travellers, throughout the Old World. If smallpox inoculation is still being practised among African peoples, if the characteristic Mongolian still is to be found in Ireland today, and if the Slavonic nations made use of finger pressure on acupuncture points, then we certainly need to know about it.

Travellers' accounts, moreover, are particularly valuable, whether it be the Turk Evliya Tchelebi describing in the +17th century Chinese metallurgical blowing engines in the Balkans, or Lazarist missionaries paying detailed attention to the procedures of deep borehole drilling for brine in early +19th century Szechuan. Of all this mass of material, the greater part is available to us in our headquarters, the East Asian History of Science Library in Cambridge. Beyond that we are free of the resources of the University, the Faculty, and the College Libraries in Cambridge; and beyond them again, the British Museum and many other libraries at home and abroad.

Terminology

Time and again it has been borne in upon us that when starting each new chapter one is faced with a situation of great confusion. In the past there have been so many misunderstandings and mistranslations, so many erroneous legends taken as facts, so many misdatings and misreadings, that one is tempted to speak in *New Yorker* terms of the Department of Utter Confusion. Then there is also the difficulty of the technical terms characteristic of the particular science or technology concerned, the necessary 'jargon' as it were, not only in the Western languages of modern science, but also in the Chinese of ancient and mediaeval science. It may take a certain time, for instance, to master for botany the phytographic language and the principles of nomenclature, but this is an essential preliminary to looking into Chinese plant science through the ages.

Here there are advantages and drawbacks. On one side there was a great stability in Chinese terminology over the ages, so that the 'yellow way' (*huang tao*) always means the ecliptic, and the 'fiery chemical' (*huo yao*) almost always means gunpowder. On the other side there was sometimes a failure to produce new terminology when the situation called for it, so that the 'fire arrow' (*huo chien*) which originally meant an incendiary arrow, came also to mean a fundamentally different thing, the rocket. Again, the first of all clock-drives for astronomical apparatus certainly needed a new name, and the hydromechanical clock too, but 'water-driven armillary sphere' (*shui yün hun i*) continued in use to describe it. In some cases the technical terminology is of extreme difficulty, as in Chinese medicine, where the translation problem is almost insuperable – almost, but we think not quite.

Skeletal Structure

Once the difficulties of the technical terminology have been overcome, then the facts start to fall into place, and a beginning can be made of seeing how they compare with what was happening in the rest of the world. It has often seemed to me that working out a chapter like this may well be likened to watching the development of a photographic negative in the red light of a darkroom; first everything is a grey chaos, then gradually it sharpens under one's eyes, and a thousand precise details emerge. Once the picture has become clear, and has been well set forth, one can have a good deal of amusement at the reactions of Western historians of science and technology, who tend to retire in good order when some Chinese priority has been demonstrated by maintaining that that was not exactly what they meant.

For example, if it can be shown that the magnetic compass and its use in navigation in China preceded the West by several centuries, then it is maintained that the only true, the only veritable, magnetic compass is that in which the needle is attached to the compass card – and that of course was European. Or again, if it is shown that there were 600 years of mechanical clock-making in China before the first clocks of mediaeval Europe, then it is maintained that the only real, the only authentic, escapement, is the verge-and-foliot – Europe's original response to China's hydromechanical clockwork. These reactions belong, we say, to the Department of Face-Saving Redefinition. But the basic edifice of assured fact remains. Modifications will come as knowledge increases, and we shall welcome them, but the skeletal structure is not likely to undergo much change.

The Unity of the Old World

These discussions about transmissions are the result, of course, of that dedication to the comparative which was mentioned before. The Old World, we can now see, was inescapably one unit, with resistances of various strengths to communication between the many emerging civilizations; and Africa was in the circuit too. Not everything travelled from east to west, though most things did; for example, the invention of eye-glasses, or spectacles for the assistance of myopia and presybyopia, unquestionably occurred in +13th century Italy, and spread quite quickly (like Amerindian tobacco later) all over Asia. We are not committed to any doctrinaire diffusionism, nor opposed to the possibility of independent invention, though we think that the further you go back the less likely that was.

We also think that when a discovery or an invention appears in one part of the Old World a long time after the same discovery or invention was made somewhere else, then the longer the time elapsing the more likely it was that diffusion occurred, and we feel that in such cases the onus of proof lies upon those who would maintain that independent invention took place. Only rarely can the intermediate stages in the transmission of a piece of scientific knowledge or technique be revealed. Sometimes an intermediate zone is indicated, diffusion taking place in both directions, east and west; such is the case for the water-mill, almost contemporary in origin in the −1st century; in Asia Minor (for cereal grinding) and in China (for metallurgical bellows), so that possibly the Persian culture area was the home of the original invention, which then spread in both directions. But the history of science and technology in that part of the world is still at a very rudimentary stage, and the problem remains open. So much for some of the interesting theoretical problems which we have encountered.

ORGANIZATION

From the beginning it was obvious that no single European and no single Chinese could cover the vast breadth of knowledge required for bringing this venture to a successful conclusion; collaboration was essential. No-one could tackle it singlehanded. So for the first nine years of writing I had the collaboration of Wang Ling (Wang Ching-Ning, now Professor of Chinese at Canberra, Australia), a friend whom I had met first at Lichuang in Szechuan, where he was a Junior Research Fellow in the History Institute of Academia Sinica, working on the history of gunpowder. Since 1957 my chief collaborator has been Lu Gwei-Djen, now Fellow of Robinson College, Cambridge, who came to join me after many years in the Secretariat of UNESCO in Paris.

Twenty years later it became gradually clear that even if we were all to live to the age of Methuselah or Phêng Tsu we could not accomplish the work yet to be done, so today our group consists of more than 20 collaborators scattered all over the world. Here we cannot but commemorate those who joined us earliest – Ho Ping-Yü (now Professor of Chinese at Griffith University, Brisbane) for alchemy and early chemistry, and Kenneth Robinson (late of the UNESCO Education Institute, Hamburg) for physical acoustics. A special debt of gratitude is owing to my wife Dorothy Moyle Needham, FRS, who was one of our war-time group in China, and though herself faithful to biochemistry, always gave full moral support to the project.

The Side-Dishes

One feature of the labour of publishing our material is that a series of monographs has been generated, side-dishes as it were from the groaning table of the main repast. Sometimes these have been overflows, with details too abundant for the principal series of volumes itself; sometimes they have been preliminary drafts needing finalization later. First came 'The Development of Iron and Steel Technology in China', done with Wang Ling in 1958, then two years later 'Heavenly Clockwork; the Great Astronomical Clocks of Mediaeval China', with the additional collaboration of Derek de Solla Price. In 1962 came 'The Pre-Natal History of the Steam-Engine', and in 1965 'Time and Eastern Man'. Five years later all of us combined to produce 'Clerks and Craftsmen in China and the West'. Just published is the work with Lu Gwei-Djen entitled 'Celestial Lancets; a History and Rationale of Acupuncture and Moxibustion'.

The main series of volumes is now being condensed by Colin Ronan into a set of six, with the title *The Shorter Science and Civilisation in China.* A parallel set of small paperbacks is also appearing in Chinese, but by a different group of abridgers.

A word may next be in place about the translations of the work itself. Two in Chinese and one in Japanese are in course of publication, and 10 or more volumes have already appeared. Subsidiary writings and excerpts have also gone into French, German, Italian, Spanish, Dutch, Danish and Singhalese. We are happy that our findings are thus reaching so many readers in Asia, Europe, Africa and Latin America.

The Financing

Here it would be inexcusable not to give a very brief account of the organizations which have financed our work during the past 30 years, though we could hardly mention all. After the expiry of Wang Ling's British Council grant, I supported him for some time out of my own Cambridge salary, but help then came from the Spalding Trust of Oxford, the Blue Funnel Line (Holt Family Fund) and the Leverhulme Foundation. Outstanding for its maintenance of

Lu Gwei-Djen over 15 years, as well as other aid, was the Wellcome Trust of London; while the Bollingen Foundation gave a useful grant permitting maximal illustration of the volumes. Some collaborators were able to come and work with us on sabbatical year salaries from their respective universities. In later times the project benefited much by support from the Lee Foundation of Singapore; as also from the Coca Cola Co. Inc. of Atlanta, Georgia, under its Senior Vice-President, the late Dr Clifford Shillinglaw, afterwards the first Chairman of our Trust in the USA. Our abiding debt to the Syndics of the Cambridge University Press remains, first for the backing they gave to a biochemist still untried as a sinologist, and then for their consistent financial support as publishers over the decades.

After the formation of the East Asian History of Science Trust (UK) in 1968, support on an altogether larger scale began to reach us. Coca Cola and the Wellcome continued their help, but we have also had generous support from the Ford Foundation, the Sloan Foundation, the Mellon Foundation and the National Science Foundation. It is now believed that sufficient funds have been collected to assure in due course the final completion of the project. At the time of the first establishment of the Library outside Caius College, we received timely help from the British Museum Library Ancillary Libraries Fund. But the financial means for the building of the permanent home for our library still remain to be found. We are confident that they will be.

With this, then, we may conclude our account of what *Science and Civilisation in China* is all about, its origins and background, its aims and objectives, some of those who have worked upon it, and the physical background or engine-room which we are desirous of perpetuating permanently for the benefit of all future workers on the subject. Over the years there will be much more original research to be done. The history of science, technology and medicine in China must take its rightful place alongside the history of institutions, and economic and social history, as a continuing study as limitless as is all history itself.

Editorial Note: One of Dr Needham's 'side-dishes' was 'The Evolution of Oecumenical Science: The Rôles of Europe and China', published in *Interdisciplinary Science Reviews* **1**, 203 (1976).

The United States Army Corps of Engineers:

Two Centuries of Interdisciplinary Engineering

LIEUTENANT GENERAL JOHN W. MORRIS
Chief of Engineers, US Army, Washington DC, USA

For over 200 years, since its foundation in 1775, the US Army Corps of Engineers has undertaken a vast variety of civil, in addition to its military, engineering tasks. Undoubtedly its most important has been, and continues to be, the construction of dams, the flood control, the bridging and the provision of locks for the American Continent's immense inland water resources, especially the Mississippi River. Other achievements, like the Apollo's Vehicle Assembly Building, the historical survey of the unexplored West, disaster relief and civil engineering works in Arabian countries, are relatively unknown, but here illustrated.

Richard Gridley, after a distinguished military career in the service of King George II of England, became the first Chief Engineer of the American Army in June 1775. His Royal Grants and Annuities were forthwith withdrawn and he was wounded at the Battle of Bunker Hill. As the 46th Chief of Engineers of the US Army Corps of Engineers, I am not writing to defend Gridley's decision but to describe the highlights of the sequel to that historic decision: 200 years of interdisciplinary engineering with major emphasis on the development of multi-purpose planning in water resources development.

Since the Revolutionary War the primary mission of the Corps of Engineers has always been to provide support to the Army of the United States and its allies.

American engineer troops have written a distinguished battle record in every war and campaign – from Bunker Hill to Vietnam. Our Engineer troops were first among American forces in action in World War I and suffered the first casualties. They were the first ashore in Normandy in World War II, clearing the beaches for the landing forces, and throughout that global conflict carried out as amphibious engineers three-quarters of all major landing operations. As military engineers, our assignment includes all of the traditional wartime support activities besides a major role in the construction of all Army and Air Force facilities at home and abroad. But, our role in assuring the defense and security of our country is not the subject of this article and I will not discuss the details of our military mission.

CCC–0308–0188/80/0005–0269$06.00/1

CIVIL WORKS

While contributing to the defense and security of the American people for over two centuries, the Corps of Engineers has also been given increasing responsibility for the planning and execution of important civil works in connection with the exploration and development of the United States. These civil works activities of the Corps of Engineers, past and present, make it unique among military engineers throughout the world.

Many people are surprised to learn of the variety and scope of the public works projects designed, built and operated by the Corps of Engineers since the Battle of Bunker Hill. To mention a few:

— development and maintenance of 35 000 km of inland waterways which carry about one-sixth of the Nation's domestic commerce;
— maintenance of the navigable depths of over 200 inland and coastal harbors;
— construction and operation of more than 250 locks on inland waterways;
— production of about one-quarter of the hydroelectric power produced in the United States;
— prevention of almost $100 000 million of flood damages by about 1000 projects including dams, floodwalls and levees which have returned about $8 for every $1 invested;
— construction and operation of over 350 man-made lakes which attract over 500 million visitors annually;
— over 70 of these man-made lakes are multi-purpose reservoirs supplying water for

Technical expertise was provided by army engineers during the Civil War, and examples like this enabled the Union forces to operate a national military railroad – a vital supply route.

hydropower, flood control, navigation, irrigation, recreation and conservation;

— construction of the Manned Spacecraft Center at Houston, Texas and the ground support and launch facilities in Florida for the American Space Program which placed men on the Moon;
— supervision of the construction and operation of the Panama Canal until the recent treaty was signed which will turn over its operation to the Panamanian Government;
— construction of the US Section of the St Lawrence Seaway and dredging and maintenance of the Great Lakes connecting channels to accommodate seaway traffic;
— numerous projects for beach erosion and hurrican protection, irrigation and water supply, salt pollution control studies and projects, acid mine drainage studies, major public works for Washington, DC, and small boat harbors for recreation and commercial fishing;
— exploration, surveying and mapping of the West;
— surveying of early transcontinental railroad routes and construction of the first railroad in the United States. At least 35 Army Engineer officers became railroad company presidents and 48 became railway company chief engineers.

There is a historical basis for the dual mission of the Corps of Engineers because they have been on the cutting edge of changing American goals, objectives, priorities and aspirations ever since the nation was formed. Since their establishment they have been involved not only in all the disciplines in the

LIEUTENANT GENERAL JOHN W. MORRIS, a 1943 graduate of the United States Military Academy at West Point, New York, USA, was Chief of Engineers of the US Army Corps of Engineers from July 1976 until he retired on 30 September 1980. As Chief of Engineers he directed the world-wide activities of the Corps of Engineers in military engineering and civil works. He commanded toops in World War II and in Korea and Vietnam. He also directed major regional water resources development programs before he became Director of the national and international civil works programs of the Corps of Engineers. He holds a Master's Degree in civil engineering and is a member of the National Academy of Engineering.
Address: Office of the Chief of Engineers, US Army, Washington, DC 20314, USA

engineering field, but also in the social and the so-called soft sciences, from archaeology to zoology.

Lieutenant General Sir Karl J. King, late Engineer in Chief, British War Office, said:

> It is of course beyond argument that in war there is an unlimited demand for major engineering works of all kinds. These examples comprise airfields, ports, railways, bridges, roads, water and oil supply and buildings of all kinds.... From this incontrovertible fact it follows that the military engineer must be trained in peace in all such major projects if he is to be an efficient military engineer.

ESTABLISHMENT OF CORPS OF ENGINEERS

The Continental Army Corps of Engineers was established on 16 June 1775 on the eve of the battle of Bunker Hill under the cannon bombardment from the British ships HMS *Lively* and HMS *Falcon* when the Continental Congress passed a resolution which provided for 'one chief engineer and two assistants at the Grand Army.'[1] George Washington appointed ex-British Colonel Richard Gridley, to be the first Chief Engineer.

After the British Government recognized the independence of the United States by the Treaty of Paris in 1783, the American Army was reduced to 80 men and the Continental Army Corps of Engineers was completely dissolved.

The French Revolution and the war between France and England presented grave international situations which President George Washington feared might involve the young nation in war. In 1794 the Congress authorized new fortifications along the Atlantic coast and established a new Corps (of Artillerists and Engineers). Since there were no native-born trained engineers in America at that time,[2] President Washington sought and obtained the assistance of French engineers to construct those fortifications, among them Etienne Rochefontaine and Pierre l'Enfant.

THE FIRST AMERICAN SCHOOL OF ENGINEERING

After the Revolutionary War George Washington, Thomas Jefferson and other leaders wanted the new nation to be able to train its own engineers. Some people, in fact, thought the Federal Government should set up a big national university where young people could learn all the branches of knowledge that would be needed in overcoming the wilderness and establishing a new civilization. However, the Constitutional Convention at Philadelphia in 1787 decided that education should be left to the States and localities.

When Thomas Jefferson became President, he showed Congress how it could get around the constitutional problem. It did not create a school; instead, it created a Corps of Engineers to consist of seven officers and ten cadets. Then it stated that this Corps 'constituted' a Military Academy.[3] Everyone seemed satisfied with this device, and the US Military Academy opened at West Point in 1802.

> The Military Academy was then instituted to train engineers, a step prompted by the almost complete absence of military and civil engineers in America. Engineering and related subjects were taught at West Point in a period when there were no other schools of technology in the Nation.[2]

For the next several decades the Army Engineers were the nation's principal practitioners of abstract and applied sciences as well as military subjects.[4] The Corps of Engineers had sole responsibility for developing national engineering skills for a generation. It was the only source of engineers in the United States until the Rensselaer Polytechnic Institute gave degrees in engineering after one year's course of instructions in 1835. By 1860, Engineer Officers had established science and engineering courses in at least 13 universities, including Yale, Dartmouth, Harvard, Michigan and the US Naval Academy which had been founded in 1845.

History has recorded the exploits of many graduates of the US Military Academy who distinguished themselves as outstanding military leaders or well-known explorers of the forbidding and hostile American West. Furthermore, the impact of graduates of the US Military Academy has influenced every segment of American society: military, economic, political, academic, religious, business and social. The civilian careers of the 2371 alumni of the US Military Academy who left military service for civilian life during the 19th century are very revealing because among them we find:

> one President of the United States; one President of the Confederacy; five Presidential or Vice Presidential candidates; 29 diplomats; 24 senators and congressmen; 223 high state officials, including 16 governors; 228 professors, including 87 college presidents; 236 corporation executives, including 87 presidents; 20 clergymen, including one bishop; 49 bankers or bank presidents; 179 authors; 352 merchants or planters.

CONTROVERSIAL EVOLUTION OF A NATIONAL ENGINEER DEPARTMENT

The Military Academy had hardly been established when it came under severe criticism and there was pressure to abolish it. Some charged that West Point violated the republican principles of the nation while others claimed that the technological training given to future officers, and its subsequent application to internal improvements, was a perversion of its original purpose.

In 1832, the Report of the Board of Visitors endorsed the curriculum at the Military Academy stating that it aided 'the sucessful prosecution of those great internal improvements, which in this enlightened age, the states are planning and executing.'

In the 1830s a Committee of the House of Representatives stated that the entire nation benefits from the training of Army Engineers to make them 'competent to superintend the construction of these... internal improvements which are to be the external bonds of our national union.'[5]

There were several splinter engineering organizations formed as the Congress groped to legitimatize a national engineering mission. By 1818 the United States Engineer Department, USED, consisted of the Corps of Engineers, the US Military Academy and the Corps of Topographical Engineers, all under the command of the Chief of Engineers. From 1838 to 1863 the Corps of Topographical Engineers and the Corps of Engineers were relatively autonomous under the Secretary of War until consolidation under the Corps of Engineers in 1863.

The changeable and fluctuating attempts to establish a national Federal engineering organization resulted from the political changes and national objectives which required an effective engineering capability in the Federal government.

For example, the Louisiana Purchase of 1803 doubled the territorial holding of the United States by adding an area about equal to the combined areas of Britain, France, Spain, Portugal and East and West Germany. The western frontier moved from the Mississippi River to the Rocky Mountains and extended from the Gulf of Mexico to British America, now Canada. The westward expansion was a driving force tempered only by the rampaging rivers, forbidding mountains and hostile Indians. But the dominant national imperative was the development of the resources of the nation and the exploration of the west.

The eastern seaboard on the other hand had to be secured. From 1806 to 1811 the military engineers designed and built 24 forts and 32 coastal batteries.

INTERNAL IMPROVEMENTS

Until 1824, engineer support to the nation's needs included the Military Academy, coastal surveys, scientific and pathfinding expeditions in new western territories, and surveys and construction of canals and roads. These were justified generally as necessary for national defense or the transportation of mail, both enumerated in the Constitution.

The General Survey Act of 1824 embodied the first decision and the first authority for a comprehensive internal improvements program. It authorized the President 'to cause the necessary surveys, plans and estimates, to be made of the routes of such Roads and Canals as he may deem of importance in a commercial or military point of view, or necessary for the transportation of the public mail....'[6] The Act authorized the President to establish a Board of Engineers for Internal Improvement and he named General Simon Bernard, special assistant to the Chief of Engineers, Colonel Totten who later became Chief of Engineers, and a prominent civilian engineer, J. L. Sullivan.

The Congressional Act of 24 May 1824 appropriated \$75 000 to improve navigation on the Ohio and Mississippi Rivers and 'committed river improvement to the Federal Government' through the Corps of Engineers.[7] Later appropriations became known as Annual River and Harbor bills.

President James Monroe, in his annual address of 7 December 1824, considered the Corps of Engineers as the national agency to accomplish internal improvements:

> With a view of these important objects, I submit to the consideration of Congress the propriety of enlarging both the Corps of Engineers, the military and the topographical... the more extensively these Corps are engaged in the improvement of their country, in the execution of the powers of Congress, and in aid of the states in such improvements as lie beyond that limit, when such aid is desired, the happier the effect will be.... By profiting of their science, the work will always be well executed; and, by giving to the officers such employment, our Union will derive all the advantage, in peace as well as in war, from their talents and services, which they can afford. In this mode, also, the military will be incorporated with the civil, and unfounded and injurious distinctions and prejudices, of every kind be done away.

By 1827, 35 surveys had been made by the Corps of Engineers, of 19 canals, 5 rivers in conjunction with canals, 10 roads and 1 railroad. There were canal surveys in progress throughout the nation. Clearly, the civil works mission of the Corps of Engineers was an idea whose time had come. It was established by need and by law.

NATIONAL SYSTEM OF WATERWAYS THWARTED

As a matter of fact, in 1824 the Corps of Engineers was the only Government agency with the engineering competence and personnel to supervise major public works undertakings.[9]

The 1824 Annual Report of the Department of War envisaged a national system of waterways. A letter to President James Monroe from Secretary of War John C. Calhoun, which emanated from the Corps and accompanied the 1824 Annual Report, contended that the unified development of our rivers 'would greatly facilitate commerce and intercourse among the States, while it would afford to the Government the means of transmitting information through the mail promptly to every part, and giving effectual protection to every portion of our widely extended country.[10]

Army Engineer officials believed that inland waterways should be considered as consolidated transportation systems, and that purposes other than navigation should be served. Congress was inclined to distribute river improvement funds widely on a political basis rather than as a part of a comprehensive plan of development. A national system for executing and financing major river projects was therefore not established at that time.

SURVEYING AND MAPPING THE WEST

While the young nation was fretting over the Federal role in water resources development and management, there was no reluctance to use Army Engineers, both from the Corps of Engineers and the Corps of Topographic Engineers, in the exploration and mapping of the West.

A selection of facts and quotations from their reports reveals the variety and extent of the surveys. Lt John C. Fremont and his legendary scout, Kit Carson, led three explorations to Oregon and California which earned for him the nickname 'The Pathfinder', and made him the idol of three generations of dime-novel-reading American boys and girls.

'The several expeditions from 1869 to 1879 inclusive, were enabled to extend the observations required, upon which to base a detailed topographic map, to embrace an area of 359 065 square miles. . . .'.[11]

'Twenty seven hundred geological specimens were collected during the season, also 4500 fossils, 1600 minerals, 67 mammals, 1450 birds, 1600 fishes, 280 reptiles, 3000 grasshoppers, 428 butterflies, 4500 wasps and bees, over 170 flies, 1500 bugs, 2000 ants, 500 shells, 600 worms and 19 lots of ethnological specimens'.[12]

US Geological Exploration of the Fortieth Parallel by Clarence King, lists 262 species of birds collected and listed.[13] The 19 000 botanical specimens collected and catalogued are listed in Volume V on Botany. Volume VII, titled *Odontornithes*, is a monograph of the extinct toothed birds of North America (1880) and included the remains of more than 100 different individuals of the Odontornithes.

The Pacific Railroad Surveys, besides locating four transcontinental railroad routes which were eventually used, produced a 12-volume report on tens of thousands of natural history specimens which helped to make the Smithsonian Institution one of the world's great museums. These included mammals, birds, reptiles, fishes, molluscs and crustaceans.

Lt George M. Wheeler who led one of the last Western surveys recognized that the secondary aim of his survey was to collect 'all the information necessary before the settlement of the country concerning branches of mineralogy and mining, geology, paleontology, ethnology and ruins.'[14] He shipped 61 659 natural history specimens to the Smithsonian Institution and developed 71 maps. By 1879 the Army had set forth 112 exploring expeditions.

EARLY FLOOD CONTROL

Mapping a river did not prevent it from going on the rampage. Early on, prevention of flood damages was regarded largely, but not exclusively, as a local responsibility and beyond the scope of the Federal authority granted in the Commerce clause or elsewhere in the Constitution. Flood control during the first half of the nineteenth century remained the responsibility of private enterprise and local and state authorities.

In 1850 Congress appropriated $50 000 'for a topographical and hydrological survey of the delta of the Mississippi River with such investigations as may lead to determine the most practical plan for securing it from inundation.'[15] Charles S. Ellet Jr, a renowned civil engineer, who had introduced the suspension bridge into the United States, was in charge of the surveys and his report dated 31 October 1851 was the first completed official study for the purpose of determining how to control destructive floods in the Mississippi Valley.[16]

He concluded that considerable obstacles existed in obtaining adequate flood relief by outlets or levees and recommended creating reservoirs by building dams in the upland areas of the streams tributary to the Mississippi River. Research approved by Congress continued for a decade.

In 1861 Captain Andrew Atkinson Humphreys and Lt Henry Larcom Abbott of the Corps of Engineers completed a detailed and scholarly *Report upon the Physics and Hydraulics of the Mississippi River*. Besides developing the science of river hydraulics, their report urged that levees should be relied upon for protecting all the alluvial bottom lands liable to inundation below Cape Girardeau, Missouri. The two Army Engineers recommended, as had previous Corps officials, that it was a Federal obligation to aid in the control of floods in the Mississippi Valley.

By 1858, the Mississippi River was seemingly confined within 2000 miles of levees with an average height of a little less than 3 m, which had been built by the states at a cost of about $40 000 000. But floods, that fall and in 1859, tore through the levees and a vast portion of natural deterioration, high waters and military destruction during the Civil War left the levees in shambles.

THE MISSISSIPPI RIVER COMMISSION ESTABLISHED

The opportunity to merge the work of the investigation of hydraulics in the Mississippi Basin, and the coalescence of those desiring flood control with the

Some major Corps-built projects. *Top left*: Dworshak Dam, Idaho, on the north fork of the Clearwater river. The concrete gravity dam is 717 ft high and has the capacity to generate 400 000 kW of hydroelectric power. *Top right*: Hydroelectric generators at Fort Randall, South Dakota. *Bottom left*: the 'Soo Locks' at Saulte Ste Marie, Michigan. *Bottom right*: the Mississippi, walled in by levees of concrete mats.

navigational interests to seek the best method of developing the river, resulted in Congress passing legislation on 28 June 1879, creating the Mississippi River Commission.[17]

To allay the fears of some congressmen that the proposed Commission would reflect only the opinions of the Corps of Engineers, the Commission included seven members: three from the Corps of Engineers, one from the Coast and Geodetic Survey, and three members from civilian life, two of whom had to be civil engineers. A Corps of Engineers officer was to serve as head of the Commission.

The composition of the seven-member Commission was to reflect the various theories and interests in river development. The organic law required it to develop a plan that would 'correct, permanently locate and deepen the channel, and protect the banks of the Mississippi River; improve and give safety and care to the navigation thereof; prevent destructive floods; promote and facilitate commerce, trade and the postal service.'[18]

The Mississippi River Commission held its first meeting on 19 August 1879. One of its more important initial decisions was to adopt a plan which would concentrate, rather than disperse, the waters of the river. It specifically rejected the outlet system proposed by some experts as 'quite impracticable'. The Commission concluded that 'the velocity of the current must be maintained' to prevent shoaling.

Significantly, the Commission declared that the levees were 'desirable, though not a necessary adjunct in the general system of improvement', although they were 'essential' as a protection against destructive floods. The initial recommendations included the closing in of the gaps in existing levees because the majority of the Commission believed that navigation of the river had seriously deteriorated because of the breaks in the earthen embankment. Their inference was that if the gaps in the levees were closed, the deposits formed under the influence of the crevasses would be swept away and the channel of the river correspondingly deepened. Its initial report therefore emphasized works for channel contraction and bank protection.

Two of the seven members of the Commission did not agree with the majority and held that 'when the increase of volume brings with it much coarse sediment, the slope may be increased instead of diminished, by increase of volume'.[18] On 3 March 1881, Congress appropriated $1 000 000 for use by the Mississippi River Commission, but included a strict prohibitive clause that no part of the appropriation could be allotted to levees.[19]

This constraint prevented the Commission from pursuing one of its charter obligations: flood protection. However after the devastating flood of 1882, the greatest on record to that date, Congress provided direct aid to flood victims, and authorized the Commission to 'repair or construct levees if in their judgement, it should be done as a part of their plan to afford ease and safety to the navigation and commerce of the river and to deepen the channel', although the Act stated that 'no portion of this appropriation shall be expended to repair or build levees for the purpose of reclaiming lands or preventing injury to lands by overflow.' The work of the commission was thus extended nearer to the organic act which created it.

It allocated more than one-quarter of its funds in 1882 for levee work, and in its report of 1 December 1882, it bluntly stated that 'the plan of improvement should include the control of floods in order to succeed in such general improvement as is described in the organic act creating the Commission, and in the River and Harbor Bill of 2 August 1882.'

From 1882 to 1890, the US Government funded through the Mississippi River Commission approximately $3 000 000 to assist local levee districts in the Mississippi Valley which spent $80 000 more than the Federal Government in the same period.

The flood of 1890 proved that these expenditures were insufficient, as the levees were too weak to stand an extraordinary high water stage. Congress, in an act of 19 September 1890, excluded the restrictive clause contained in previous acts prohibiting the use of Federal funds for levee construction.[21] A levee system became an integral part of the plan of improvement on the Mississippi River.

INTEGRATED, BENEFICIAL-USE PLANNING, 1890–1912

Near the end of the nineteenth century, the river development plans of the past were challenged by new approaches to the treatment of water as a resource. These were based on the slowly accumulating hydraulics data and theories of natural resources development and control which were integrated into popular thought. The new interests included control of flood waters for protection of lands against overflow, use of these waters to irrigate, generate hydroelectric power and aid navigation by maintenance of uniform stream flow. This new emphasis upon beneficial uses led to broader developmental plans for river basins that included protection of watersheds and headwater reservoirs, and coordination of the uses of water resources.

In 1888 Congress approved the establishment of an irrigation division within the United States Geological Survey. The Reclamation Act of 1902 led to the establishment of the Bureau of Reclamation in the Department of the Interior.

Unhampered by the legislative restrictions which Congress imposed on the Corps of Engineers, the scientists of the Geological Survey viewed water as a single resource with many potential beneficial uses. Broadened investigations of water conservation, flood reductions, erosion control, and watershed protection by the hydrographers of the Forest Service, established as part of the Department of Agriculture in 1906, promoted the values of headwater stream control. Integrated, multiple-purpose river basin planning for beneficial uses appeared to be a possibility in the early years of the twentieth century.

Growing popular demand that the Federal Government actively promote inland navigation encouraged those with multiple-purpose views to formulate developmental plans. Waterways associations and other special interest groups in the Mississippi Basin, confronted with increased railroad rates for inadequate service, demanded additional Federal aid for navigation of the inland waters.

BOARD OF ENGINEERS FOR RIVERS AND HARBORS

In an attempt to remove appropriations from politics, Representative Theodore Burton, Chairman of the House Rivers and Harbors Committee, proposed that Congress create a special Board of Engineers for Rivers and Harbors to approve or reject river development projects. Organized within the framework of the Corps of Engineers in 1902, the Board, consisting of seven officers detailed from the Corps, exercised a restraining influence over the inland waterway movement by carefully scrutinizing proposed projects before they were submitted to the Chief of Engineers. As a result, there were fewer, but more carefully selected projects. Today, the Board has an interdisciplinary staff of 55 persons, including 20 civil engineers, five regional economists, one transportation specialist, four environmental planners and one wildlife biologist. Since its establishment the Board of Engineers for Rivers and Harbors has reviewed about 9000 water-related reports and recommended no Federal action in 55% of the cases.

INLAND WATERWAYS COMMISSION

The next attempt towards the establishment of a national multiple-purpose water development policy was the establishment of the Inland Waterways Commission by President Theodore Roosevelt on 14 March 1907. The Chief of Engineers was one of the nine members of the Commission. The President's instructions to the Commission stated in part:

> The time has come for merging local projects and uses of the inland waters in a comprehensive plan designed for the benefit of the entire country. Such a plan should consider and include all the uses to which streams may be put, and should bring together and coordinate the points of view of all users of water.[22]

This was a succinct statement of the multiple-purpose and beneficial-use concepts of river basin developments for which officials of the new resources agencies of the Federal government had been striving. These concepts, if implemented, would revolutionize the resources policies and projects of the nation.

The preliminary report of the Inland Waterways Commission dated 26 February 1908, recommended the use of multiple purpose and multiple benefits in water resource planning and development. Unfortunately for the Roosevelt Administration forces, the entire attempt to recommend to the Congress proper and improved procedures for water resource development was poorly orchestrated and failed to obtain strong support in either house of the Congress. In fact, the members of the Inland Waterways Commission, denied any funds by Congress, themselves paid for the printing of their final report.

In spite of the multiple-purpose and multi-objective trends forcefully advocated early in the twentieth century, the nation would wait some decades to implement meaningfully comprehensive concepts in planning and managing the nation's water resources. Besides, due to emphasis on economy and efficiency, the administrations of Harding, Coolidge and Hoover, 1921–1933, marked time in conservation and water resource development efforts.

The River and Harbor Act of 3 March 1925, directed the Chief of Engineers to estimate the cost of making surveys of navigable streams. The list of recommended surveys was submitted to the Congress in 1927 and was printed as House Document No. 308, 69th Congress, 1st Session. On 21 January 1927 Congress directed the Corps to develop data on the '308' list of projects. In spite of major emphasis on navigation and hydropower, these surveys were the first comprehensive reports of the majority of the Nation's streams and 'formed the foundation for many of the multiple-purpose river basin developments in the United States'.[23]

THE NEW CONSERVATION, 1933–1965

The Great Depression of the 1930s during which one out of every three of the national workforce was unemployed, prompted renewed interest in large public works projects as a source of employment and as a stimulant to the economy. In spite of Presidential initiatives and support of multiple-purpose planning proposals, major water resources development projects during the Depression were aimed at getting people back to work. These included the Bonneville Dam on the Columbia river, Fort Peck Dam in Montana and Lake Texoma on the Texas–Oklahoma border, all Corps of Engineers projects, which were good, viable projects as planned and constructed.

Efforts to establish a national planning body for water resources development after the Inland Waterways Commission was abolished in 1912, were snarled in political controversy and ineffective. The last effort, the National Resources Planning Board, was dissolved in 1943. (Independent Offices Act, 26 June 1943.) Ironically, that same year, the Federal Government Agencies themselves recognized the inadequacies of non-coordinated efforts, and through the leadership of the Corps of Engineers set up the Federal Inter-Agency River Basin Committee to provide a means to 'cooperate more completely in the preparation of reports on multi-purpose projects and to correlate results to the greatest practicable extent' (Congressional Record 90:845). Many laws passed since World War II require full cooperation and coordination of the efforts of the various Federal Agencies involved in water and related land resources planning.

In 1965, the Water Resources Council became the means of strengthening all affected water and related land resources interests.[24] Reorganization and consolidation of Federal Agencies involved in water resources planning and management in 1939, 1945, 1949, 1955 and 1979 left the Corps of Engineers intact. The Water Resources Council remains today as the Executive Agency for the development of national criteria and policy in water resources development and management.

BUILDING TOMORROW TODAY

For over two centuries the Corps of Engineers has been building tomorrow today. A short tabulation of more recent Corps of Engineers missions illustrates the wide scope and variety of the activities of the Corps of Engineers beyond its historic water-related responsibilities for navigation, flood control, hydropower, water supply, recreation and conservation, beach erosion control, hurricane protection, irrigation, major drainage projects, small boat harbors, water quality, wildlife preservation and enhancement.

Under the 1899 River and Harbor Act the Corps of Engineers is responsible for maintaining the integrity of navigable channels, and no construction in these waters can be undertaken without a permit from the Corps of Engineers. The Corps also has limited

Activities of the Corps spanning over 100 years: the cast iron dome of the Capitol Building, Washington, DC, was built in 1857 and the training and launch bases for all Intercontinental Ballistic Missiles were built in cooperation with the Air Force in the 1950s and 1960s.

authority to remove hazards to navigation, including drift material and other debris.

Under Section 404 of the 1972 Federal Water Pollution Control Act, the Corps regulates all dredge and fill operations in 'the waters of the United States' including all marshlands, swamps and wetlands. The same act authorized the Corps to provide planning and engineering assistance to metropolitan areas in urban water management. Since 1970 the Corps of Engineers has provided flood plain management services to thousands of cities and towns.

In 1977 the Corps of Engineers became involved in a National Dam Safety Program. It had compiled an inventory of over 50 000 dams and began inspecting about 9000 of those whose failure would cause major damage and loss of life. The Corps of Engineers designed and built the bulk-mail handling centers throughout the United States for the US Postal Service. The Corps has been engaged in emergency recovery operations in all major natural disasters in the US for many years.

International operations of the Corps of Engineers include a many thousand million dollar nation-building program in Saudi Arabia, fully funded by that government. Some three dozen other countries have contacted the Corps of Engineers, seeking technical assistance in the development and management of their natural resources.

On the military side, the Corps of Engineers built the Intercontinental Ballistic Missile bases in the 1950s and 1960s and all other major missile defense systems since then. It is currently engaged in military construction support to NATO countries.

In support of the Israeli–Egyptian Peace Treaty, the Corps is constructing two large military airbase complexes in Israel's Negev Desert to replace two bases in the Sinai Desert which the Israelis must give up under the terms of the Camp David treaty.

ENVIRONMENTAL CONCERNS

The preceding decade of the 1960s saw a quickening of public and governmental concern for the deteriorating American environment. The Corps of Engineers was already among the lead agencies of Government working to protect the environment.

Shortly after the National Environmental Policy Act became law in 1970, the Chief of Engineers established an Environmental Advisory Board to provide him with a broad range of outside expertise

The contrasting roles of the Corps. *Top*: two members of the Corps making plans to replace a bridge destroyed by floods. *Bottom*: a Corps ranger examining a Canada goose nest in a wildlife development project.

and experience in environmental matters dealing with Corps projects. The Chief of Engineers also instructed all Division and District Engineers to give more attention to environmental concerns in planning water resource projects.

Thus before and after enactment of recent Federal environmental laws, the Corps has worked to redirect national attention to programs that balance the objectives of development and conservation, rather than merely emphasize development.

ORGANIZATION OF CORPS

The Office of the Chief of Engineers in Washington, DC, is assisted by a highly decentralized field office organization of 14 Divisions world-wide, 40 Districts and five research laboratories. Each District deals directly with the public and handles the design and construction of projects. The research laboratories support Facilities Engineers at Army installations as well as civil and combat engineers. The laboratories represent a broad base of various engineering disciplines with emphasis on research.

The Engineer Topographic Laboratories, located at Fort Belvoir, Virginia, develop mapping techniques and hardware which will provide the soldier with timely, precise and detailed information about his area of operations.

The Coastal Engineering Research Center, also located at Fort Belvoir, conducts research which results in new, improved methods for protecting the Nation's shorelines, inlets and harbors from erosion and storms, and for keeping navigation channels open.

The Panama Canal. The supervision of the construction of the Panama Canal and, after completion, its operation until recently, was undoubtedly the greatest single engineering achievement of the Corps during the last 200 years of civilian works. Reproduction of a lithograph by Joseph Pennell, showing the construction of the approaches to the Gatun Lock on the Canal (Library of Congress Archives).

The Construction Engineering Research Laboratory, located at Champaign, Illinois, solves problems encountered in the design, operation and maintenance of temporary and permanent facilities.

The Cold Regions Research and Engineering Laboratory, located at Hanover, New Hampshire, studies the phenomena of cold regions and devises methods for living and working under these conditions.

The Waterways Experiment Station, located at Vicksburg, Mississippi, has an extensive mission that includes hydraulic modeling, environmental and water-quality studies, weapons effects, soils and pavement testing and design, and developing improved concrete technology.

CONCLUSION

The Corps of Engineers today, with its 850 military officers and 43 000 civilians, is a far cry from Colonel Gridley's handful of men at Bunker Hill.

Its members today include scores of professions besides all disciplines of engineering: 384 general biologists, 50 fish and wildlife biologists, 108 biological technicians, 65 ecologists, 75 foresters and forestry technicians, besides agronomists, ornithologists, zoologists, archaeologists and probably more landscape architects than any other organization anywhere. Its history is indeed two centuries of interdisciplinary engineering.

LITERATURE CITED

1. W. S. Holt, *The Office of the Chief of Engineers of the Army*, p. 1. Johns Hopkins Press, Baltimore (1923).
2. M. Cunliffe, *The Nation Takes Shape*, University of Chicago, Chicago (1959).
3. Statute 132 (March 16).
4. W. T. Jackson, *Wagon Roads West*, p. 210. Yale University Press, New Haven, Connecticut (1965); also Ref. 1.
5. W. T. Jackson *Wagons Roads West*,Yale University Press, New Haven, Connecticut (1965).
6. Statute 22.
7. 4 Stat. 174.
8. Senate Document 1, 18th Congress, 2nd Session.
9. I. Y. Shermerhorn, *The rise and progress of river and harbor improvement in the US, Journal of the Franklin Institute*, CXXXIX (1895).
10. American State Paper, II, Commerce and Navigation 1934—700–701.
11. C. G. M. Wheeler, *US Geographical Surveys West of the Hundredth Meridian*, Vol. 1(1889).
12. Ref. 11, p. 74.
13. C. King, *US Geological Exploration of the Fortieth Parallel*, Vol. IV: *Paleontology*.
14. Newcomb, *Reminisences* p. 220–221.
15. Statute at Large IX, 523, 539.
16. *Report of the Secretary of War Communicating Reports in Reference to the Inundations of the Mississippi River*, 20 January 1852. Senate Ex. Doc No 20, 32d Cong., 1st Sess., 1852 Serial.
17. US Statutes at Large XXI, 37.
18. Minority Report, pp. 21–23 of Preliminary Report on the Mississippi River Commission Exec. Doc. No. 58, 46th Congress, 2nd Sess. 1880.
19. US Statutes at Large, XXI, 468, 474.
20. Report of the Mississippi River Commission, in *Annual Report of the Corps of Engineers*, pp. 2745–2783 (1882).
21. US Statutes at Large, XXI, 426, 450.
22. Preliminary Report of the Inland Waterways Commission, Sen. Doc. No. 325, 60th Congress, 1st Session, 15, Serial 5250 (1908).
23. 49 Schad, op. cit., p. 26.
24. P. L. 89–90, July 22 (1965).

The manuscript was received 12 August 1980

Science and Theology

Diverse Resolutions of an Interdisciplinary Gap by the New Priesthood of Science

EILEEN BARKER

London School of Economics and Political Science, London, UK

In an age where the trend towards secularization has been slowed down, new religions and ideological movements are competing with the more traditional positions to be found in society. Those who are anxious to justify their own particular position share only one criterion for comparison with those holding other positions: a belief in the epistemological efficacy of science. A new priesthood of scientists has emerged to offer justifications in the name of science for a variety of ideological positions and to explain to a bewildered public what the real relationship is between the various disciplines of science and those of theology or moral and political philosophy. This paper describes the variety of positions taken up by the new priesthood, typifying these as Fundamentalism, Orthodoxy, Liberalism, Modernism, Agnosticism and Atheism.

Ever since science and theology came to be regarded as separate disciplines there has been no general agreement as to what the relationship between the two might be. Are they competing sources of knowledge? If so, which wins? Or are they complementary, and, if so, how? This paper deals with part of the process whereby potential oppositions between the two disciplines are reconciled in the contemporary Christendom of the industrialized West.

THE APPROACH

Before proceeding further two points must be clarified to avoid any misunderstanding. Both points are related to the fact that the author is a sociologist.

First, at no point in the paper is there any claim to distinguish between truth and falsity in either science or religion, nor is there any attempt to define either discipline other than in the terms offered by, or acceptable to, those who do themselves form the data of the study. That the definitions of 'true science' and its relation to theology do in fact differ from one another is, indeed, exactly what the study is intended to highlight. The sociologist's is a descriptive not a prescriptive discipline. He or she can, at most, aspire to an expertise in documenting the expertise of others even, perhaps especially, when it is found that the experts disagree. Should any sensitive scientist, reading this paper, snort in disgust at the author's fallacious assumption as to what science is, it would mean that he had missed the point of the exercise. It is what is allowed to pass or is accepted as science by a consumer public and what is produced as science by those who are socially defined – through their degrees, teaching or research posts – as scientists, that defines the boundaries of what will be called science in this paper. The author has argued elsewhere[1] her own position in so far as scientific epistemology and other philosophical stances are concerned; but these are matters irrelevant to the present context.

Second, the aim of this paper is to indicate a broad spectrum of beliefs rather than to document details and qualifications for which there can be no space. Perhaps such an approach can be justified through the analogy of a picture. Viewed from a distance, fine distinctions merge into one another and are lost, but at the same time there emerge patterns and relationships which are missed when one peers too closely. Ideally, of course, one would wish to see the picture from as many perspectives as possible. On this occasion there is room only for a distant viewing. The picture to be presented will certainly be drawn from a study of detailed distinctions, but it is painted in bold and possibly crude strokes. It consists largely of a series of what sociologists call Ideal Types.[2] These are analytical tools used mainly for comparisons. Although they would obviously be useless if they did not bear a strong relationship to reality, Ideal Types are presented in over-clear terms in order to provide a template, as it were, whereby different phenomena can be recognized and compared with one another.[3]

Part of the problem of having human beings as the object of one's study is that, for various obvious reasons, individuality becomes a far greater methodological nuisance, or challenge, than it is for the physical or even the biological and medical sciences. This does not mean that we cannot see

CCC-0308-0188/80/0005-0281$05.50/1

patterns or make generalizations, but that these must, of necessity, be of a far more tentative and less clear-cut nature than the patterns and generalizations to be found in the natural sciences. In short, the smudgy boundaries of reality are acknowledged, and now ignored in this paper; a practice which has its uses and its limitations – and, it might be added, one that is not entirely unknown to other sciences.

The caricature with which I begin is of the general social context in so far as it provides a consumer public for a producer priesthood, anxious to explain the relationship between the disciplines of theology and science.

THE CONSUMER DEMAND

For perhaps half a century it looked as though the death of God was imminent with – and immanent in – the development of modern science. In many ways this could be seen as an act of parricide. Protestantism had been digging its own grave. It had been religious values, attitudes, and beliefs that had conceived and cradled the science of the 17th century[4] but by the end of the 19th century it began to look, at least to an influential proportion of the educated élite, as though science could cut the umbilical cord, having safely imbibed such sensible qualities as the belief in rational, objective inquiry and the existence of an ontological reality which obeyed regular, observable laws. Furthermore, science could dispose of the mythical paraphernalia which had surrounded such beliefs in the first place. Darwinism, through the disclosure of natural selection as the mechanism for man's evolution, provided the clue whereby even humanity could become the object of empirical scrutiny. To talk of the supernatural would be to admit to a science of the gaps. The discipline of theology described merely lacunae which were waiting to be filled by the discipline of science. Nothing was in principle unattainable. It might take time but ultimately the ultimate would be revealed by rational appraisal of the Book of Nature.

Such views do of course still exist, sometimes in highly sophisticated forms, but recent years have witnessed phenomena which indicate a curious reversal of what had seemed to some such an irreversible trend. While on the Eighth Day God may have said 'Let there be Secularization', it began to look as though He could have something else up His sleeve for the Ninth.

The science which was heralded with such messianic zeal began itself to contribute to its own demise – at least in its role of Provisor Absolute. Science, especially, but by no means exclusively, to an increasingly vocal youth, lacked something. It was incapable of reaching the transcendental heights of humanity. It reduced, it materialized, it objectified, it dehumanized. It understood quantity not quality; it did not understand love. It labelled, it determined, it atomized; it did not understand individuality, freedom or consciousness. The beauty of creation was lost in the destruction of Hiroshima. Auschwitz laughed at progress. Innocence became ignorance, relevance was lost in relativity and the ineffable became an obfuscating formula. Science had sought, and its discovery was loss. It was carrying Western civilization like a sorcerer's apprentice beyond material satisfaction to satiation, while the rest of the world, denuded of natural and spiritual resources, was being bled dry in the wake of a technological fervour, worshipping the false gods of modernism.

New religions emerged; old religions revived. The spiritual, the soul, the 'more than . . .' was affirmed. Jesus was rediscovered and proclaimed alive in corners where his name had been long forgotten. The privileged young first sought to transcend the authority of technological civilization in the freedom of antinomianism and then they sought freedom from anomie in dogmatic fundamentalism. Rationalists held burial services. Salvation was individual or social, public or private, in this world or the next, but salvation – to salvage that which science had denied from that which science had confirmed – became a burning necessity. The New Age might take the unalienated form of perfect socialism or the sinless form of the second coming or it might just be the celebration of true, spiritual selfhood through the realization of the Invisible Religion,[5] but the New Age was at hand. Scepticism of scepticism had led to faith in faith.

But now we find ourselves faced with an *embarras de richesses.*

The wide range and scope of the mushrooming beliefs and practices available in the 1970s pose the uncomfortable question of truth. What criterion of selection is available? What counts for knowledge in modern society?

EILEEN BARKER has enjoyed an interdisciplinary career. She took her Advanced Level General Certificate of Education in science then went into the theatre for five years. This was disrupted by five years of childbearing and motherhood. In 1970 she took a First Class Honours degree at the London School of Economics, was immediately offered a lectureship in sociology and has been on the staff there ever since. She has lectured in universities in Britain and North America and published on a wide range of philosophical, sociological and religious subjects. Her research interests are science and religion and the Reverend Moon's Unification Church.
Address: The London School of Economics and Political Science, Houghton Street, London WC2A 2AE, UK.

And of course there is only one universally acceptable answer: Empirical Science. There is only one élite which has universally acknowledged privileged access to truth: that of the Scientific Expert.

Those of us who have been brought up in an age of science, however disillusioned we may be with its uses and limitations, will be unlikely to admit publicly or privately that what we believe will go against its findings. Minimally, if we are to transcend the limits of rationality to enter the realm of the non-rational, we must not be seen by ourselves, or others, to be irrational; paradox is paradoxically allowed and even welcomed;[6] minimally if we go beyond the limits of science and empiricism we must not be seen to be unscientific. Science and logic must at least allow the truth that transcends their boundaries.[7]

Today one can find practically all theologies, and indeed secular ideologies, demanding from science certainly permission, frequently justification and sometimes proof.

THE NEW PRIESTHOOD

The consumer aspect is not discussed further in this paper. Instead what follows is a description of something of the variety of ways in which members of the contemporary scientific community meet the consumer demand through the supply of reassuring resolutions of any possible interdisciplinary tensions which could arise from competing epistemological and metaphysical claims of the two disciplines. This they do in the role of a new priesthood which can stamp a scientific seal of approval on the different theological and ideological packages that are on offer. Naturally the very variety of positions does create problems, and indeed the priests themselves are greatly concerned about the false images of science that abound. It is the duty of the priest to reveal a proper understanding of the potentialities of a true science which provides the necessary, and in some cases sufficient, pathway to the greater truth of ultimate knowledge. Like the priesthoods of yesteryear, the new priesthood can fulfill the function of providing cognitive support for our beliefs, and cognitive scorn for the beliefs of others, and, as of yore, it can prove that it can neither prove nor disprove any such beliefs.

The new priesthood provides the producers of 'speaking as a scientist, I can assure you . . .', 'science has clearly shown . . .', 'the belief, vindicated by modern science, that . . .' or merely 'there is absolutely nothing in science contrary to the suggestion that . . .'. These are phrases which have, I am continually assured, brought untold relief to the minds of many, torn by the schizophrenia of an age uncertain as to how to bridge the gap between the kinds of knowledge to be found in a discipline sanctioned for its ultimate epistemological authority and the kinds of knowledge to be found in a discipline sanctioned for its higher ultimate authority.

THE TYPOLOGY

After various Procrustean efforts to force the apparently idiosyncratic diversity of positions concerning the relationship between science and theology into neat taxonomic matrices, it began to appear both plausible and useful to classify the scientists according to how they viewed the truth or falsity of the Bible with regard to both general spirit and particular detail. From this it was possible to produce six Ideal Types along a slightly wobbly continuum with unreserved acceptance of the spirit and detail at one end and unreserved rejection of both at the other end.

The six types were called Fundamentalism, Orthodoxy, Liberalism, Modernism, Agnosticism and Atheism. These labels were chosen as they suggest fairly obviously the sort of position each type would take along the continuum. However, it must be borne in mind that the use of such familiar labels should not be taken to mean that all the attributes generally associated with these terms need in any way form an inherent part of the definition in the present usage. And again it must be emphasized that these are Ideal Types. Several members of the new priesthood straddle two or more positions and often those who fall into the same category according to this perspective make strange bedfellows from another perspective.

At the extreme ends of the continuum there fall, logically, two types which are irrelevant to this essay. These are those Fundamentalists and Atheists who see no need to reconcile the disciplines of science and theology. They accept as blind faith, or blind unfaith, the truth or the falsity of the Bible without resorting to any scientific justification. For them there is no need to find a bridge between the disciplines.

FUNDAMENTALISM

For the Fundamentalist scientist,[8] the Bible is true in both spirit and detail. It can be accepted as an historical and scientifically accurate textbook.[9] God does not contradict Himself so there must be compatibility between His truth as revealed in the Scriptures and in Nature. The Bible can be trusted because it contains absolute, eternal revelation, while scientific knowledge is relative to time and society, as has been shown frequently in the past when apparently well established facts have been proved false.[10] Evolutionary theory, for example, is the result of bad science in which facts are ignored and metaphysical speculation is accepted as empirical reality.[11] It is up to the Fundamentalist scientist to

Ideal Typical Characterization of the Scientists' Positions *vis à vis* Christian Religious Dogma

Ideal type	Beliefs about Bible: Spirit	Beliefs about Bible: Detail	Beliefs about Bible: Source	Attitudes to Evolution	Relevance of Science to religious knowledge	Sources of true knowledge	Work to be done	Key orientations	Organization examples in UK
Fundamentalism (Creationists)	True	True	God	Special Creation as in Genesis account. Evolution is false	Relevant: Bible verifiable (and verified) by science	Biblical Revelation (and science)	Show the Bible to be the correct textbook of science Show where false science, especially evolutionism, is wrong Prove lack of discrepancy at all levels	Acceptance of Scriptural Truth	Evolution Protest Movement; Newton Scientific Association
Orthodoxy (Evangelicals)	True	*(True)	God	Genesis account interpreted in the light of evolution. God is the originator and sustainer of Creation	Relevant: Bible vindicated through science and interpreted with the help of science	Biblical Revelation and Science	Reconcile science and Scripture by showing lack of discrepancy at deepest level Evangelize	Search for Scriptural truth	Research Scientists' Christian Fellowship; Victoria Institute
Liberalism	True	*(False)	*(God)	Theistic Evolution	Irrelevant: Bible and science concerned with different questions	Science or Biblical Revelation (depending on the question)	Show lack of discrepancy because of different spheres, so Bible not threatened by science Show importance of (non-scientific parts of) Bible to biological/social man	Religious Humanism	Science and Religion Forum
Modernism	*(True)	False	*(Man)	Evolution is God's method of reaching His ultimate goal	Irrelevant: in asking different questions Relevant: in importance of Natural Theology	Science (as the Book of Nature) and some mystical or personal revelations and Eastern philosophy	Discover direction of evolution, design, purpose, etc. through knowledge of (expanded) science and consciousness Assist in evolutionary process towards God and/or spiritual fulfilment	Teleological evolution, Purpose, Design	Scientific and Medical Network
Agnosticism	*(False)	False	Man	Evolution. God = Primary Cause by definition	Irrelevant: (as with Liberalism)	Science and Secular philosophies (mainly ethical and individualist with spiritual enlightenment)	Explain religious beliefs (origins, functions and persistence) through science Find secular, but spiritual, substitute for God	Secular Humanism	British Association for the Advancement of Science
Atheism	False	False	Man	Evolution occurred by chance	Relevant: Bible refutable (and refuted) by science	Science and Secular philosophies (mainly social or political and perhaps Utopian)	Disprove God, explain away religious beliefs and behaviours and show their danger Realize scientific ethic of the future.	Scientific Truth	British Society for Social Responsibility in Science

show that Genesis has nothing to fear from the facts of nature and that 'evolution is a faith, not a fact'.[12]

A spectacular growth in relative, if not absolute, terms of the Fundamentalist priesthood has been most visible in North America where the 'Science-Textbook Controversy'[13] has hit several states. This is a controversy in which Fundamentalists have argued for, and in some cases been granted, equal teaching time in schools for both special creation and evolution. In Britain most people, including scientists, are unaware of the growth of creation science. On hearing of it they tend to express incredulity or amused tolerance. However, in 1977, at the same time as a local authority in Dallas was insisting that the story of Adam and Eve must be taught to students as history,[14] an English local authority, and later an industrial tribunal, upheld the dismissal of a religious education teacher for doing that very thing.[15]

The Institute for Creation Research was established in the United States in 1970. It 'seeks to educate scientists, teachers, students and the general public on the scientific evidence supporting Biblical creationism'.[16] It is, it claims, the first known time in history that an educational and research centre has been founded strictly on creationist principles. Current projects include a search for Noah's Ark on Mount Ararat, research for fossil anomalies and field studies on inverted geological sequences; publications range from a monthly newsletter to attractively presented High School texts and 'technical monographs on specific evidences for recent creation'.[16]

In Britain the Newton Scientific Association[17] was founded in 1972 and as a result an increasing number of scientists, several of professorial status, are now publicly writing or speaking as scientists on the unscientific nature of evolutionary theory. Members of the Association are very careful not to resort to the Bible but to stick rigidly to secular references in their public statements. The other British society for creationist science, the Evolution Protest Movement,[18] was founded as long ago as 1932 with the stated aim of demonstrating that the theory of organic evolution is not in accordance with scientific fact. In the past this movement has tended to rely quite heavily on the Scriptures, but it now has younger officers and speakers engaged in scientific research and the movement has begun to present a more scientific image both in its literature and at its meetings.

The arguments put forward by the Fundamentalist priesthood range from the very simple to the very complicated. Many of the members are undoubtedly specialists in their own fields and can certainly succeed in blinding the layman with science. The areas covered range from taxonomy[19] to the Noachian flood,[20] from palaeontology[21] to cytochemistry[22] from the second law of thermodynamics[23] to probability theory.[24] Variation and natural selection within a species is never denied, but what is strongly contested is that natural selection can be shown to have operated between the species.[25] There is absolutely no evidence that one species – or, more accurately, one kind – could have evolved from another. Apparent evidence to the contrary, whether it be of the evolution of the horse[26] or of Piltdown Man[27] is exposed as either inconclusive or downright forgery.

Theologically conservative science, whether concerned with creation or the Biblical record in general, eliminates an area of ambiguity and uncertainty between the demands and criteria of the present and the inheritance of the past. Perhaps it is not crucial for the majority of lay Creationists whether or not the arguments put forward by the scientists are grasped. What does matter is that there are those who undoubtedly possess membership of the new priesthood who claim that the heritage of the ages is validated by the criteria of the present. The Creationist scientist may not be able to persuade those of his colleagues who are entrenched in the evolutionary paradigm that the Genesis story of the origin of the world is literally true, but he can certainly provide proof for those who need and desire such proof that all is not lost. Old faith can appeal to the new knowledge, ancient certitude to modern proof.

ORTHODOXY

The scientists of the Orthodox priesthood walk along a very delicately balanced tightrope, carefully avoiding falling into too literal or too liberal an interpretation of the Scriptures. The relationship between science and faith is for them a real, living problem. It is a tussle that must continuously be resolved at each point – not by accepting the literal truth of the Scriptures and then finding the scientific backing for this position as they believe Fundamentalists will do, nor by dividing faith and science into completely separate compartments as Liberals will do. Instead they must relentlessly search for particular truths, while holding the general truth that both science and the scriptures are correct, or rather that science is one way of understanding the nature of the God who wrote the Scriptures. The Scriptures must be interpreted at a level which does not fall into simple apologetics nor yet dogmatically stick to an over-naïve acceptance of a primitive interpretation.[28]

The Orthodox scientist is likely to consider that there is a lot of information for the scientist in Genesis but it is not necessarily scientific information. The Books of Nature and of the Scriptures each offer criteria by which the other can be interpreted. Neither must hold precedence over the other. Take, for example, Biblical references to Noah's Ark. The evidence would seem to be in favour of there having been a flood, but while the Orthodox priest will certainly delight in sharing this knowledge with the Fundamentalist, he does feel it his duty to point out some practical difficulties in accepting a literal interpretation of the story. How would Noah have got some 7000 odd species into a space roughly 450 × 75 × 45 feet? What about the range of temperatures and humidity, quite apart from the range of food that would be needed.[29] It is not that the discipline of science is correcting the Bible, but that God, through science, is correcting our interpretation of the Bible. The role of the priesthood of science is one of heavy responsibility towards a laity without direct access to the truths revealed to the new priests.

But it is not merely particular details that the Orthodox priesthood has to get straight. It is just as important to show the lay public that the existence and usefulness of science does not rest on philosophical assumptions that threaten a theistic belief. True science cannot prove that the world was created by chance,[30] nor that man is a clockwork machine.[31] A principle of causation does not imply that we do not have free will, nor that prayer is useless.[32] With a proper understanding of what the concepts of a scientific discipline really entail, what their status and their limitations are, one can see that they complement, rather than contradict, the concepts of theistic theology.[33] Whether one is concerned about the status of miracles,[34] of cosmological theories[35] or of Heisenberg's uncertainty principle,[36] so long as the priesthood can point out how the two disciplines employ different approaches or different ways of looking at the same thing, then we can relax and learn how much the scientific discipline offers the theist; how it adds to, rather than detracts from, our theological understanding.

The Victoria Institute is the oldest of all the organizations specifically concerned with science and religion.[37] It was founded in 1865 'to promote investigation of the relationship between science and Christian faith at a time when many people believed science to be an enemy of revealed religion and irreconcilable with it'. The other main organization in Britain which falls into the Orthodox category is explicitly and exclusively evangelical with a membership of men and women who have at least one degree in science and who are working in some kind of scientific research. This is the Research Scientists' Christian Fellowship which was founded in 1945 by a small group of research students, many of whom are now eminent professors in British universities.[38] In North America the organizational equivalent is the American Scientific Affiliation.[39]

LIBERALISM

Liberalism accepts that the spirit of the Bible is true and that its source is Divine, but it believes that many of the details are mythical or else need heavy reinterpretation. Such stories as that of the creation in Genesis are to be taken as parable rather than literal truth. It is the Bible which must be reinterpreted in the light of scientific discoveries rather than science in the light of scriptural teachings. It is accepted that there are Bible stories which contain statements about natural history which 'are factually wrong. But these . . . should be regarded as allegorical. . . . This need in no way lessen their spiritual truth, which is concerned with matters that come outside science'.[40] The problem of compatibility is removed by denying the problem in the first place, for while the Liberal Scientist will illustrate that certain things are not true, the truth and importance of the Bible is seen to lie not in claims to factual accuracy, but to spiritual guidance. Science cannot threaten theology as the two disciplines are concerned with fundamentally different areas, not just with different approaches as for the Orthodox, and to confuse these different areas is to miss the real point of both Divine revelation and science.

This does not mean that the Liberal priest will not seek some kind of synthesis between the two spheres. It is indeed his duty to suggest how this might be accomplished, once the two disciplines have been carefully delineated and analysed for their separate contributions to knowledge.[41]

The Science and Religion Forum was set up in 1974 because 'a group of scientists were concerned to relate their scientific knowledge and modes of study to their religious faith and practice'. It meets once a year for a residential conference and publishes papers and literature which will 'enable the general public, including the churches, to remove misconceptions about, and acquire an informed understanding of, the relationship between religion and science'.[42] It is also hoped that the association can build up 'a core of scientists who would be available for consultation by the churches when scientific issues are raised in relation to Christian beliefs and practices'.[42]

The real role of the Liberal scientist is to reassure the faithful that there can be no direct conflict between science and the Bible; the Bible, however, might tell us how to use science. The Forum has, on the whole, concentrated on practical problems such as Man's Responsibility for Nature,[43] or Religious Education[44] where the main problem is seen to be to communicate to the child that science has not disproved the Bible and that belief in the Bible can provide a positive, helpful foundation for living.

The Liberal scientist belongs to the priesthood of Trouble Shooters who have access to knowledge of both sides. Through judicial mediation he can keep the boat from being rocked by extremists who insist on an 'either-or' confrontation and who do not understand the fundamentally different functions of the two disciplines.

MODERNISM

While the Liberal priesthood accepts responsibility for showing that science does not exclude a spiritual reality, it will also insist that for science to assume 'an internal urge or a Life Force directing the course of evolution is inadmissible'.[45] The Modernist priesthood on the other hand will consider it perfectly legitimate – indeed its duty – to use science in the search for knowledge of the moving spirit or spiritual forces which, teleologically, provide creation with a transcendental direction.

With Modernism the two disciplines of science and theology are to be collapsed into one. Both share the same approach and subject matter toward the same end.

The Bible does not stand as any ultimate source of knowledge or authority. 'The orthodoxy of the past was based on the science of the past; and if the science alters and expands, so also must the orthodoxy'.[46] Biblical knowledge represents the struggle of man to grow nearer to God and the 'mistakes' it has made are the mistakes one could expect from men as they slowly evolve in knowledge of God and nature. Modernism enjoys a strong flavour of Spinoza's *Deus sive natura*, either being but aspects of the one wholly interrelated whole.

Natural theology is at its strongest in Modernism – it does, of course, also appear in a far weaker form in Fundamentalism. The priesthood must not only show how the Bible has been 'rightly wrong' in terms of the struggle of evolution, it must also discover the design and purpose of the life force through its knowledge of nature. The scientist can act as the midwife in the playing out of the Grand Entelechy. It is his task to show us the next step forward and help us in its realization.

As with any kind of naturalism, interpretations of the meaning of nature can vary, but whatever the interpretation, modern science is offered as a basic,[47] possibly the only, means of exploring the true, non-materialistic side of man whose life can be given meaning through understanding his place in a wider, spiritual Order and Purpose.

The relativism in this position is obvious. Religions other than those in the Judaeo–Christian tradition and especially those from the East – are seen as different but possibly equally legitimate means towards the final goal. Many are the keys to the kingdom of Heaven. It is even possible for the Modernist to be seen as straddling the theistic/non-theistic line, depending on one's definition of God or religion.[48]

It is within this category that one can find the priests absorbed in the exploration of the Aquarian Frontier,[49] and those esoteric groups which experiment with meditation, transcendence, ESP or LSD in attempts to discover the goals and direction of the universe. It is here too that one finds the research institutes which reintroduce to scientific appraisal phenomena that conventional science has spurned or rejected.[50]

The Religious Experience Research Unit was founded by the biologist, Alister Hardy in 1970 in order to explore man scientifically as a religious animal.[51] The Scientific and Medical Network came into being a few years later, declaring that its members 'share the conviction that the contemporary framework of thought in science and medicine should be widened to include certain ideas that go beyond what today is considered orthodox.'[52] Galileo, Pasteur and Einstein are cited as examples of the unorthodox. Members explore such questions as 'Do the order and design that pervade the universe imply purposeful influences that impinge upon the continuum of interchangeable matter and energy, or not?' The network organizes conferences and seminars with titles like 'Science, Mind and the Spirit of Man',[53] which attract a wide variety of persons from both established and fringe science as well as medicine.

For those who see the evolutionary process as meaningful and an unfolding of some divine revelation, Teilhard de Chardin has obviously provided an extremely powerful perspective with his conception of 'convergent integration' towards an 'Omega point'. The Teilhard de Chardin Centre for the Future of Man[54] is not intended to be primarily a scientist's organization, but several of the officers are well known scientists and the Centre is used as a platform for expounding and expanding the priesthood's scientific inquiries into emergent evolutionism.

Whether he be biologist, chemist or physicist, or any combination of these, whether he calls himself a monist, a dualist or a pluralist, whether he is guiding us through an exploration of the inner space of our consciousness or the outer space of the cosmos,[55] whether he is offering us the understanding of our destiny in this world or a glimpse of other worlds that exist beyond our own dimensions of space and time,[56] wherever he leads us, the Modernist priest is intent upon taking us beyond the reductionism of orthodox science. Thus, through his endeavours, the expanding discipline of science encompasses, in a holistic integration, the material and spiritual concerns of the contracting discipline of theology.

AGNOSTICISM

The Agnostic priesthood shares certain characteristics with Modernism, but without any celebration of supernatural teleology. Religion is seen as a natural phenomenon, but not necessarily, or even probably, a true one. There is no acceptance of the postulate that the Bible contains any revealed truth, nor does a search for knowledge of God have much point as there is probably no God for whom to search. God may be the First Cause but this is a matter of little interest as such a claim would probably be little more than definitional. The moral teachings of the Bible and the Christian Ethic are not likely to be rejected out of hand, indeed they are quite likely to be secularized and then elaborated.

So long as the Bible is only seen as an ethical phrase book it can perform a positive function. Like the Atheists they will explain religious beliefs – their origin, persistence and their functions – with reference to our biological heritage, but in so doing the Agnostic will acknowledge the importance of the spiritual side of life. While man arose from the apes, it does not follow he is nothing but an ape. This priesthood goes beyond the Modernist's midwife role for man in that it sees future evolution as being crucially dependent upon men.

It is thus not only necessary for man to find some spiritual sustenance from a demystified world, he must also accept responsibility for the demystification process in order to clear the way for a secular naturalism. As with Liberalism, there is a belief that ethics and science belong to different spheres, yet it is allowed that 'oughts' should not be completely unrelated to the world of nature. Nature provides not the content but the context within which morality is lived. It is this context which the scientist can describe and explain. The Agnostic is likely to find virtue in a pragmatic eclecticism and in an eclectic pragmatism.

Perhaps the clearest example of how Agnosticism can straddle the boundaries between liberalism, modernism and atheism is to be found in Julian Huxley who was a Humanist with a capital *H*, believing that while religion of some sort is probably necessary, it has not necessarily always been a good thing. However the 'emergent religion of the near future could be a good thing. It will believe in knowledge . . . it should be able, with our increased knowledge of mind to define men's sense of right and wrong more clearly, so as to provide a better moral support . . . and to focus the feeling of sacredness on fitter objects. Instead of worshipping supernatural rulers it will sanctify the higher manifestations of human nature, in art and love, in intellectual comprehension and aspiring adoration, and will further emphasize the fuller realization of life's possibilities as a sacred trust'.[57]

Sir Julian's new religion was an Evolutionary Humanism which was based, and could only be based, on our understanding of man and his relations with the rest of the environment, taking account of man as part of a comprehensive evolutionary process in the psychosocial stage of which he cannot avoid playing a decisive role.[58] The idea that a supernatural overriding mind would be responsible for any part of the evolutionary process is dismissed. Darwin showed that no supernatural designer was needed as natural selection can account for any form of life and the idea that God installed spirit and soul and mind into man is just against the facts – this is nothing but an arbitrary theological postulate.[59]

Not all members of the Agnostic priesthood are as positive as Sir Julian, who is very near the Atheist category at this point. It is indeed possible to be positive about being negative in relation to our knowledge while at the same time enjoying an almost mystical appreciation of its potentialities. J. Bronowski, for example, found comfort from the Principle of Uncertainty, or the Principle of Tolerance, as he called it. This denies the arrogant dogma that claims an absolute truth. Science can give us knowledge of our lack of knowledge and save us from the ignorance that dehumanizes man. This ignorance that dehumanizes man is the ignorance that results in the atrocities of Hitler's Germany and the ignorance that retreats from knowledge to 'Zen Buddhism, into falsely profound questions about, are we not really just animals at bottom; into extra-sensory perception and mystery'.[60] Bronowski saw man as nature's unique experiment to make the rational intelligence prove itself sounder than the reflex. Knowledge is our destiny. Self-knowledge, at last bringing together the experience of the arts and the explanation of science, waits ahead of us.[60]

Among the organizations in which one could locate members of the Agnostic priesthood are the British Association for the Advancement of Science[61] which is now deeply concerned with ethics, the Council for Science and Society[62] (a small self-appointed group of custodians formed in 1973) and the Science Policy Foundation[63] which aims to promote 'the scientific investigation of science and technology as social phenomena'. While these bodies are placed in the Agnostic category, it should not be assumed that their members do not hold any religious beliefs. The organizations themselves, however, function as secular ones with a secular concern for social and ethical matters. Here the priesthood is of socially concerned scientists who, as scientists, understand more than the rest of us how the world works and how we can best manage to live our lives in it.

Whatever the particular message of the Agnostic, he, like the religious scientists, denies the necessary denial of a spiritual part of man by modern science and indeed offers, through an enlightened science, a hope to light the path we make for ourselves in the future. While this is a secular path, it embodies many of the emotions and attitudes that few sociologists would have difficulty in recognizing as religious[64] in spite of the fact that Agnostics share much of the cognitive position of the Atheists.

ATHEISM

Atheism not only denies the existence of God, but also believes that science can prove God had no hand in the creation of the world, life or man. The Atheist priesthood shares with Fundamentalists the belief that there can be no incompatibility between the claims of the Bible and Science, for the Bible is making claims which are, in the Popperian sense, scientific.[65] But while Fundamentalism would choose the Bible if necessary, but actually manages to verify the Bible through science, Atheism chooses science and finds that through science the Scriptures are not only refutable, but refuted. Creation is nothing but a chance occurrence into which no purpose can be read. Not, that is, until the actual realization of this fact by man allows him to take his future into his own hands and choose his destiny.

Atheists emphasize the importance of knowledge, particularly scientific knowledge. Spiritual experience is more likely to be explained away than explored for its own sake. The negative rather than the positive aspects of religion are stressed and while the Agnostic will search for an 'honest functional equivalent' for religion, the Atheist will work for the elimination of such dysfunctional elements messing up the Brave New World of tomorrow. An almost ideal typical Atheist is Jacques Monod who, in his book *Chance and Necessity* explained how God was a myth, why man believes in this myth, why he should reject it and what he should put in its place.[66]

Monod's argument centres around the three concepts of chance, necessity and teleonomy. All life, modern science shows, evolved by chance. Nothing new occurs except by chance. There is no divine or transcendental or historical plan which is slowly being worked out. The origin of all species is the result of pure coincidence as in the random mutations which result in modifications of the genetic text. Once something living is produced, it reproduces itself of necessity, replicating the parent organism.

Monod does not deny that evolution has a direction. He believes it has, but this direction is due to what he calls a teleonomic filter which 'chooses' what can go on living. Everything that gets thrown up by chance does not necessarily go on to reproduce itself. In order to pass through the teleonomic filter it has to be internally viable and it has to be able to survive in the existing environment. Evolution, as illuminated by modern molecular biology, involves not planned revelation but chance creation, teleonomic selection and necessary reproduction. The religious factor itself first appeared in our ancestors' genes by chance. Because of the positive cohesive functions which it performs for the group, it passed through the teleonomic filter and has, of necessity been transmitted from one generation to the next ever since.

But if the discipline of theology is based on no more than a chance perturbance which has no meaning or ultimate truth attached to it 'where then shall we find the source of truth. . .? Only' Monod tells us, 'in the sources of science itself, in the ethic upon which knowledge is founded, and which by free choice makes knowledge the supreme value – the measure and guarantee for all other values'.[67]

Monod's is not the only explanation that is on offer from the Atheist priesthood. The discipline of science has several ways of explaining away the discipline of theology[68] and indeed of taking over not only its erstwhile epistemological, but also its erstwhile moral functions. Members of this priesthood frequently have a reverential acquaintanceship with the scepticism of Hume and not infrequently we can find them declaring that they accept the logical impossibility of passing from an 'is' to an 'ought'.[69] Being truly accredited priests, they then proceed to do the impossible with the aid of science.[70] For the consumer public which turns to the Atheist priesthood, such a bridge is, interestingly enough, rendered all the more effective through the denial of its potential existence.

On the whole, Atheist scientists do not bother to organize. The British Humanist Association[71] and other related organizations certainly employ the new priesthood in their attacks upon religious education, the anti-abortionists and so on, but individual members of the Atheist priesthood are generally quite happy to proclaim the error of man's religious ways without any institutional support. And so far at least the majority of priests have not agreed on a sufficiently common interpretation of the Book of Nature to unite under an organizational banner.

Unless, that is, they have a strong ideology to provide 'the sacred canopy'. There have been various groups of Socialist and Communist scientists, the most powerful and successful of which is the, now strongly Marxist, British Society for Social Responsibility in Science.[72] Like the Christians they do not reject science in itself but they want science to be truly 'for the people'.[73] It is Bourgeois science, they scientifically show, which is the type of science that produces all those negative aspects referred to at the beginning of this paper. The alternative science and technology which is offered tends not to be that of the USSR, which is defined as nothing but state capitalism, but rather the science and technology of Maoist China.[74]

CONCLUDING REMARKS

The world has always had its specialists, particularly in societies where alternatives present themselves to any marked degree. Modern, scientific society with its intensive division of labour and with a specialization such that few people can know more than a very little of what is going on in any one field creates a need for, and a reliance on, specialists to a quite unprecedented degree. Liberal individualism may allow us to have a hot line to God, but we need a specialist to tell us which God to plug into, and how to turn the switch.

The technological power and eerie mysticism of modern science, coupled with the fragmentation of modern society with its careful compartmentalization of life and dependence upon the expert to tell us what to accept or reject, has created a public ripe for new mysticisms clothed in scientific jargon. From scientology to encounter groups, from Calcutta to Wimbledon, there has mushroomed a plethora of –ologies and –isms covertly and overtly making capital out of authority borrowed from science. The Biblical literalist, the Evangelical revivalist, the political visionary and even the slightly perturbed old priesthood of the established theologies, turn to the new priesthood for reassurance that their beliefs have not been left behind in the wake of the revolutionary revelations of science. The new priesthood has not been found wanting. Sometimes with formulae, sometimes with rhetoric, but always with science, reassurance is dispensed.

We can all relax. Our beliefs are safe from the ravages of science after all. It is only They who are wrong. But perhaps, now that we know what is what, something ought to be done about Them?[75]

Acknowledgement

This paper is drawn from a study conducted with the help of a grant from the Nuffield Foundation to whom I wish to express my gratitude.

NOTES AND LITERATURE CITED

1. E. Barker, The challenge of social science to religious faith: Can sociological explanation threaten theology? In *Sociology and Theology: Alliance and Conflict*, edited by J. Orme-Mills, Chapter 1 (1978); Value systems generated by biologists, *Contact* **55/4**, 2–13 (1976); Apes and angels: reductionism, selection and emergence in the study of man. *Inquiry* **19**, 367–399 (1976); Facts and values and social science, in *Science and Absolute Values*, Proceedings of the Third International Conference on the Unity of the Sciences, pp. 757–776. I.C.F., New York (1976); Confessions of a methodological schizophrenic. *Institute for the Study of Worship and Religious Architecture Research Bulletin*, University of Birmingham (1978).
2. M. Weber, *The Methodology of the Social Sciences*, edited by E. A. Shils and H. A. Finch, pp. 41–43. Free Press, New York (1949).
3. It should perhaps be added that the essay is not restricted to a mere Ideal Typical exposition. Concrete references, mainly, though not exclusively from Britain, are given. These, however, are kept to a minimum and they have been selected to provide illustrative examples rather than a definitive categorization of all the available data.
4. The historical literature on the subject is vast, but for two books written by authors who are both natural scientists and theologians, see R. Hooykaas, *Religion and the Rise of Modern Science*, Scottish Academic Press, Edinburgh (1972) and S. I. Jaki, *The Road of Science and the Ways to God*, Scottish Academic Press, Edinburgh (1978).
5. T. Luckman, *The Invisible Religion*, Macmillan, London (1967).
6. E. Robinson, *Tolerating the Paradoxical: A Scientific Approach to Religious Experience*, Religious Experience Research Unit, Oxford (1978).
7. These arguments are elaborated in E. Barker, Biologists on religion: Contributions offered by contemporary biology to religious knowledge in the face of the changing image of modern science, in *Religion and Social Change*, 253–284, C.I.S.R., Lille (1975).
8. A more detailed account of the Fundamentalist scientists can be found in E. Barker, In the beginning: The battle of creationist science against evolutionism. In *On the Margins of Science*, edited by R. Wallis, Keele University Press (1979).
9. H. Morris, *Studies in the Bible Science*, p. 108. Baker Book House, Grand Rapids, Michigan (1966).
10. P. A. Zimmerman (Ed.), *Darwin, Evolution and Creation*, p. 10. Concordia, St Louis (1959).
11. E. Andrews, *Is Evolution Scientific*? Evangelical Press, Welwyn (1977).
12. R. T. Clark and J. D. Bales, *Why Scientists Accept Evolution*, Chapter 8. Baker Book House, Grand Rapids, Michigan (1966).
13. D. Nelkin, The science-textbook controversies. *Scientific American* **234**(4), 33–39 (April 1976).
14. *Sunday Telegraph* (6 February 1977).
15. *Daily Telegraph* (7 February 1977); *The Times* (21 May, 9 July, 27 July, 30 July 1977).
16. *Institute for Creation Research* leaflet (available from 2716 Madison Avenue, San Diego, California 92116, USA).
17. Information available from the Secretary, Newton Scientific Association, 2 Westoe Road, Edmonton, London N9 0SH, UK.
18. Information available from the Secretary, Evolution Protest Movement, 13 Argyle Avenue, Hounslow, Middlesex. TW3 2LE, UK.
19. H. D. Dean *et al.*, *The Living World*, p. 22. Creation-Science Research Center, San Diego (1971).
20. J. C. Whitcomb and H. M. Morris, *The Genesis Flood*, Presbyterian and Reformed Publishing House, Nutley, New Jersey (1961).
21. S. Baker, *Bone of Contention*, pp. 14, 24. Evangelical Press, Welwyn (1976).
22. H. Enoch, *Evolution or Creation*, p. 7. Evangelical Press, London (1967).
23. B. Davidheiser *et al.*, *The Beginning of the World*, p. 6. Creation-Science Research Center, San Diego (1971).
24. S. Ellacott, *Mathematical Problems in the Evolutionary Model*, pp. 1–7. Newton Scientific Association Report, London (1977).
25. E. Minter (Ed.), *Evolution and Christian Thought Today*, p. 129. Paternoster Press, London (1961).
26. H. Enoch, Ref. 22, Chapter 6.
27. F. Vere, *Lessons of Piltdown*, p. 51. Evolution Protest Movement, Stoke, Hampshire (1959).
28. D. M. MacKay, *Science, Chance and Providence*, Chapter 1. Oxford University Press London (1978).
29. G. S. Cansdale, A universal flood: Some practical difficulties. *Faith and Thought, Journal of the Victoria Institute* **98**(2 and 3), 61–68 (1970).
30. D. M. MacKay, Ref. 28, Chapter 2.
31. D. M. MacKay, *The Clockwork Image*, Inter-Varsity Press, London (1974).
32. D. M. MacKay, Ref. 28, Chapter 3.
33. O. R. Barclay, *Reasons for Faith*, p. 33. Inter-Varsity Press, London (1974).
34. J. N. Hawthorne, *Questions of Science and Faith*, Chapter 8. Inter-Varsity Fellowship (1972).
35. R. L. F. Boyd, The universe around us. *Where Science and Faith Meet*, Chapter 4. Inter-Varsity Fellowship, London (1953).
36. M. A. Jeeves, *The Scientific Enterprise and Christian Faith*, p. 139. Tyndale Press, London (1969).
37. Victoria Institute brochure available from the Assistant Secretary, 130 Wood Street, London EC2V 6DN, UK.
38. Information available from The Secretary, Research Scientists' Christian Fellowship, 38 De Montfort Street, Leicester LE1 7GP, UK.
39. Information from American Scientific Affiliation, 5 Douglas Avenue, Elgin, Illinois 60120, U.S.A.
40. D. Lack, *Evolutionary Theory and Christian Belief*, p. 113. Methuen, London (1973).
41. A. R. Peacocke, *Science and the Christian Experiment*, Oxford University Press, London (1971).
42. From a Science and Religion Forum press release. Information available from the Secretary, Dr G. Price, Department of Liberal Studies in Science, University of Manchester, UK.
43. Science and Religion Forum, 1976 annual meeting.
44. Science and Religion Forum, 1977 annual meeting.
45. Ref. 40, Chapter 5.
46. Anon – quoted by Surbury in Ref. 10, pp. 172–173.
47. For example, see W. H. Thorpe, *Science, Man and Morals*, pp. 135 and 143. Methuen, London (1965); A Hardy, *The Divine Flame*, p. 43. Collins, London (1966); E. W. Sinnot, *The Biology of the Spirit*, p. 11. Gollancz, London (1956).
48. It would be hard to decide, for example, whether the Spirit referred to by E. W. Sinnot, *The Biology of the Spirit*, p. 122, Gollancz, London (1956) was recognizably the God of the Judaeo-Christian tradition.

49. T. Roszak. *Unfinished Animal*: the aquarian frontier and the evolution of consciousness. Faber and Faber, London (1976).
50. In the United Kingdom for example there is the Society for Psychical Research (1 Adam and Eve Mews, London W8 6UG) and the College of Psychic Studies (16 Queensbury Place, London SW7 2EB), while in the United States there are many University departments undertaking research into the Paranormal. The Committee for the Scientific Investigation of Claims of the Paranormal can be contacted through Box 29, Kensington Station, Buffalo, New York 14215. It must be stressed of course that not all of the scientists involved in such investigations belong to the Modernist priesthood. Some are very sceptical, but there are many who are convinced that science can go beyond 'normal science' to enter realms normally reserved for a theological discipline.
51. Sir Alister's position is well described in *The Biology of God*, Jonathan Cape, London (1975). The Religious Experience Research Unit is now under the direction of Edward Robinson and it has published several books on its findings. Inquiries can be made to the Religious Experience Research Unit, Manchester College, Oxford, UK.
52. Information sheet about the network which is available from the Secretary, Lake House, Ockley, Nr Dorking, Surrey RH5 5NS, UK.
53. Title of the 1977 May Lectures held at the Royal Society of Medicine.
54. Information available from the Centre, 81 Cromwell Road, London SW7, UK.
55. C. Curling, Inner and outer. *The Teilhard Review* ***XI***(3), 106 (*October* 1976).
56. W. H. Thorpe, *Animal Nature and Human Nature*, Chapter 9. Methuen, London (1974).
57. J. Huxley, *Essays of a Humanist*, p. 92. Pelican, London (1966).
58. Ref. 57, pp. 26–27.
59. S. Tax and C. Callender (Eds.) *Evolution after Darwin* Volume III, *Issues in Evolution*, pp. 45–46. University of Chicago Press, Chicago (1960).
60. J. Bronowski, *The Ascent of Man*, p. 43, Chapters 11 and 13. British Broadcasting Corporation, London (1973).
61. Information available from the British Association, Fortress House, 23 Savile Row, London W1X 1AB, UK.
62. Information from Council for Science and Society, 3/4 St Andrews Hill, London EC4 5BY, UK.
63. Information from Science Policy Foundation Ltd, Benjamin Franklin House, 36 Craven Street, London WC2N 5NG, UK.
64. For example see S. Budd. The humanist societies: The consequences of a diffuse belief system. In *Patterns of Sectarianism*, edited by B. Wilson, Heinemann, London (1967).
65. K. Popper, *The Logic of Scientific Discovery*, Hutchinson, London (1968).
66. J. Monod, *Chance and Necessity*, Collins, London (1972).
67. Ref. 66, p. 166.
68. For example, D. Morris, *The Naked Ape*, Corgi, London, pp. 179–188 (1968). C. Waddington, *The Scientific Attitude*, Pelican, London (1948); J. Z. Young, *An Introduction to the Study of Man*, p. 253. Oxford University Press, (1971).
69. D. Hume, *A Treatise of Human Nature*, p. 521. Pelican Classics, London (1969).
70. For example, C. Waddington, *The Ethical Animal*, Atheneum, London (1961) and *The Nature of Life*, Unwin, London (1963); R. Sperry, Bridging science and values: a unifying view of mind and brain. In *The Centrality of Science and Absolute Values*, International Cultural Foundation, New York (1976); Ref. 66.
71. Information from British Humanist Association, 13 Prince of Wales Terrace, London W8, UK.
72. British Society for Social Responsibility in Science, 9 Poland Street, London W1, UK.
73. The British Society for Social Responsibility in Science quarterly is called *Science for People*, the American counterpart, *Science for the People*.
74. H. Rose and S. Rose (Eds.), *Ideology in/of the Natural Sciences*, (two volumes), Macmillan, London (1976).
75. E. Barker, How Can They? *Times Higher Educational Supplement* p. 5 (22 July 1977).

The manuscript was received 20 November 1978

The Asse Salt Mine
The World's Only Test Facility for the Disposal of Radioactive Waste

DR CARSTEN SALANDER and DR REINHARD PROSKE
Deutsche Gesellschaft für Wiederaufarbeitung von Kernbrennstoffen mbH, Hannover, FRG

DIPL.-ING. EGON ALBRECHT
Gesellschaft für Strahlen- und Umweltforschung mbH, Braunschweig, FRG

The Asse Salt Mine near Brunswick in the Federal Republic of Germany has since 1965 been used to develop and to test suitable techniques for the safe disposal of radioactive wastes. Test disposal of low-level wastes started in 1967. During this large-scale experiment about 141 000 200-litre drums of low-level wastes were successfully disposed of until November 1978. Starting in 1972 techniques for the disposal of intermediate-level wastes were successfully demonstrated over a five-year period by disposing of 1298 200-litre drums. As a result of these tests and the accompanying research and development programmes, especially for the disposal of high-level wastes, one can now conclude that reliable techniques exist for the disposal of radioactive wastes at a full-scale industrial repository. This was planned within a German *Entsorgungszentrum*.

All nations operating either nuclear power plants or using radionuclides for medical, industrial and scientific research are confronted with the necessity of safe disposal of the resulting radioactive wastes.

Since the early 1950s a number of concepts for the solution of this problem have been developed. Among these are the transmutation of long-lived radionuclides; disposal in deep space; disposal in and under the Antarctic ice shield; sea disposal; and geological disposal. The transmutation of long-lived radionuclides into stable or short-lived nuclides has been discussed extensively; it led to the conclusion that the separation and transmutation process is ineffective and that the risks created by the operation of the required additional facilities exceed the risk of a geological repository for radioactive wastes. Space disposal and radioactive waste disposal in and under the Antarctic ice shield turned out to be too expensive and combined with too many risks.

Another disposal option, still under investigation in the United States,[16] is the storage of high-level radioactive wastes in boreholes deep in the sea bottom. Since 1967 the disposal of low-level radioactive solid wastes by sea dumping has been organized and supervised by the Nuclear Energy Agency (NEA). In a region of the northern Atlantic Ocean these wastes are dumped into a depth of about 4500 m.

LAND DISPOSAL

Shallow land burial is used in the United States for the disposal of low-level wastes. In arid areas huge trenches, about 60 m long, 15 m wide and 10 m deep, are excavated and solidified low-level wastes in drums are dumped into them. These drums are then covered with soil.

The techniques of sea and land dumping can only deal with low-level wastes and high-level wastes must still be kept in cooled tanks. There are, however, suitable media which can receive all types of radioactive wastes, namely certain geological strata. Thus, the main emphasis of present research and development concentrates on geological formations like granite or salt.

As long ago as 1957, a report by the US National Academy of Science[1] recommended the disposal of radioactive wastes in deep and stable geological formations. Among those suitable for such disposal, salt formations were especially commended. Because of the high population density in the Federal Republic of Germany and its relatively high rainfall leading to water tables generally very close to the surface, such methods for the storage or removal of radioactive wastes from the biosphere as surface storage or shallow land burial could not be applied in

CCC–0308–0188/80/0005–0292$06.00/1

Germany. Thus, when a programme for the disposal of radioactive wastes was initiated in Germany at the beginning of the 1960s, the only possibility was disposal of radioactive wastes in deep geologic formations.

In a 1963 report[2] on different possibilities for safe disposal of radioactive wastes, experts of the then *Bundesanstalt für Bodenforschung*, the German Geological Survey of Hanover, recommended to the then responsible *Bundesministerium für wissenschaftliche Forschung*, the Federal Ministry for Scientific Research, to use salt formations. The basis for their recommendation was the favourable geologic situation in northern Germany, the available great experience of salt mining for over 100 years and the favourable properties of existing salt formations.

ADVANTAGES OF SALT MINES

Groundwater is the only significant transport medium in a geologic system. Therefore the most important specification for the disposal of radioactive wastes is to exclude them from such circulating water, which could transport radionuclides from the disposal site back to the biosphere. Salt formations fulfil this requirement, as their interior has no connection to the groundwater outside the salt structure. This is due to the plasticity of rock salt under pressure from the overburden of geologic strata, thus preventing cracks and fissures in the salt. Therefore no groundwater can intrude into the salt formation.

Although different kinds of salt are highly soluble in water, geologic salt formations are extremely stable and persistent; for example the age of the Permian salt deposits in northern Germany is more than 200 million years. Due to the geomechanical stability of rock salt, large cavities can be excavated and these remain firm and unchanged, even without any man-made support, over long periods of time. Furthermore rock salt has great thermal conductivity compared to other rock-forming minerals and this allows dissipation of heat generated from high-level wastes. In consequence, salt formations are also highly suitable for the disposal of this type of radioactive waste.

One year after publication of the above-mentioned recommendation by the German Geological Survey, the Asse Salt Mine was shut down for economic reasons. The Government of the Federal Republic of Germany purchased the Asse Salt Mine in 1965 and transferred it to the *Gesellschaft für Strahlen- und Umweltforschung mbH*, München, in order to develop and test methods for the disposal of radioactive wastes in salt formations; this task has been carried out in close cooperation with the Karlsruhe Nuclear Research Centre. In due course the research work on the development and testing of techniques for radioactive waste disposal showed excellent results.

Thus, based on this experience gained from experiments at the Asse Salt Mine, a concept evolved at the end of the 1960s for the management of the radioactive wastes from a nuclear fuel cycle. This concept relies on the successful operation of the test disposal facility at, and in, the Asse Salt Mine.

THE ASSE SALT MINE

Geologic Setting

About 250 to 220 million years ago the whole area of northern Germany, Denmark and the North Sea was covered by a shallow sea, part of a large ocean, and separated from the Atlantic Ocean by a land barrier. During subsequently occurring evaporation of the sea, salt crystallized out and sedimented. Several

DR CARSTEN SALANDER is a Member of the Board of Directors of the Deutsche Gesellschaft für Wiederaufarbeitung von Kernbrennstoffen mbH, the DWK. Born in Bremen in 1933, he studied Nuclear Physics at the Universities of Göttingen and Heidelberg, and from 1958 to 1970 worked in nuclear ship reactor development; he was responsible for the reactor of the German nuclear ship *Otto Hahn*. From 1970 to 1973 he served as First Counsellor for Scientific Affairs at the Embassy of the Federal Republic of Germany in London. In 1974 he joined Preussenelektra, Germany's second largest electricity utility, and led its department for nuclear power plants. In this position he took part in setting up the German Company for Reprocessing of Spent Nuclear Fuel, the DWK; in 1977 he became one of the four members of its board of directors.

DR REINHARD PROSKE is Coordinator and Planner of the DWK for Research and Development work at the interface between radioactive waste conditioning and its disposal. Born in 1949, he was educated at the Technical University of Clausthal. Before joining the DWK in 1978 he worked in the field of safety analysis of geologic repositories for radioactive wastes. Address: DWK, Hamburger Allee 4, D-3000 Hannover 1, FRG.

DIPL.-ING. EGON ALBRECHT is in Charge of Operations at the Asse Salt Mine. Born in 1920, he studied geology and later mining at the Universities of Göttingen and Aachen. From 1953 to 1969 he worked in the management of several potash and rock salt mines and then became managing director of the technical department of the Institute for Underground Disposal of the Gesellschaft für Strahlen- und Umweltforschung mbH, München.
Address: Institute for Underground Disposal, GSF, Heussallee 4, D-3300 Braunschweig, FRG.

flooding and evaporation cycles produced salt layers about 1000 m thick, containing rock salt, potash salt, anhydrite and clay in different strata. During later periods in the Earth's history these salt layers were covered by other sediments—sand, clay and limestone—and reached a thickness from 2000 to 4000 m.

Salt has a lower specific gravity than the strata of the overburden and also shows a plastic behaviour. As the overburden was not uniformly distributed above the salt, pressure gradients were generated and the salt flowed to weak points of the overburden and created so-called 'salt cushions'. Further uprise of the salt cushions led to a penetration of the overburden and 'salt domes' were built up. The main uprise of the salt took place about 120 million years ago: about 10 to 20 million years had to elapse to build a salt dome.

Since then the size of the more than 200 salt domes known in northern Germany (see Fig. 1) has not substantially changed; and this in spite of the great geological or meteorological events which have occurred since then.[3] In fact the very area where the salt domes are now situated was flooded by a shallow sea for several million years. Later on, the formation of mountain ranges began, the Alps were folded and the Harz Mountains rose more than 3000 m. Heavy tectonic activities also took place in northern Germany. At the end of the Cretaceous, a period about 65 million years ago, the land rose and northern Germany became dry land. But only for a geologically short time. During the next 60 million years

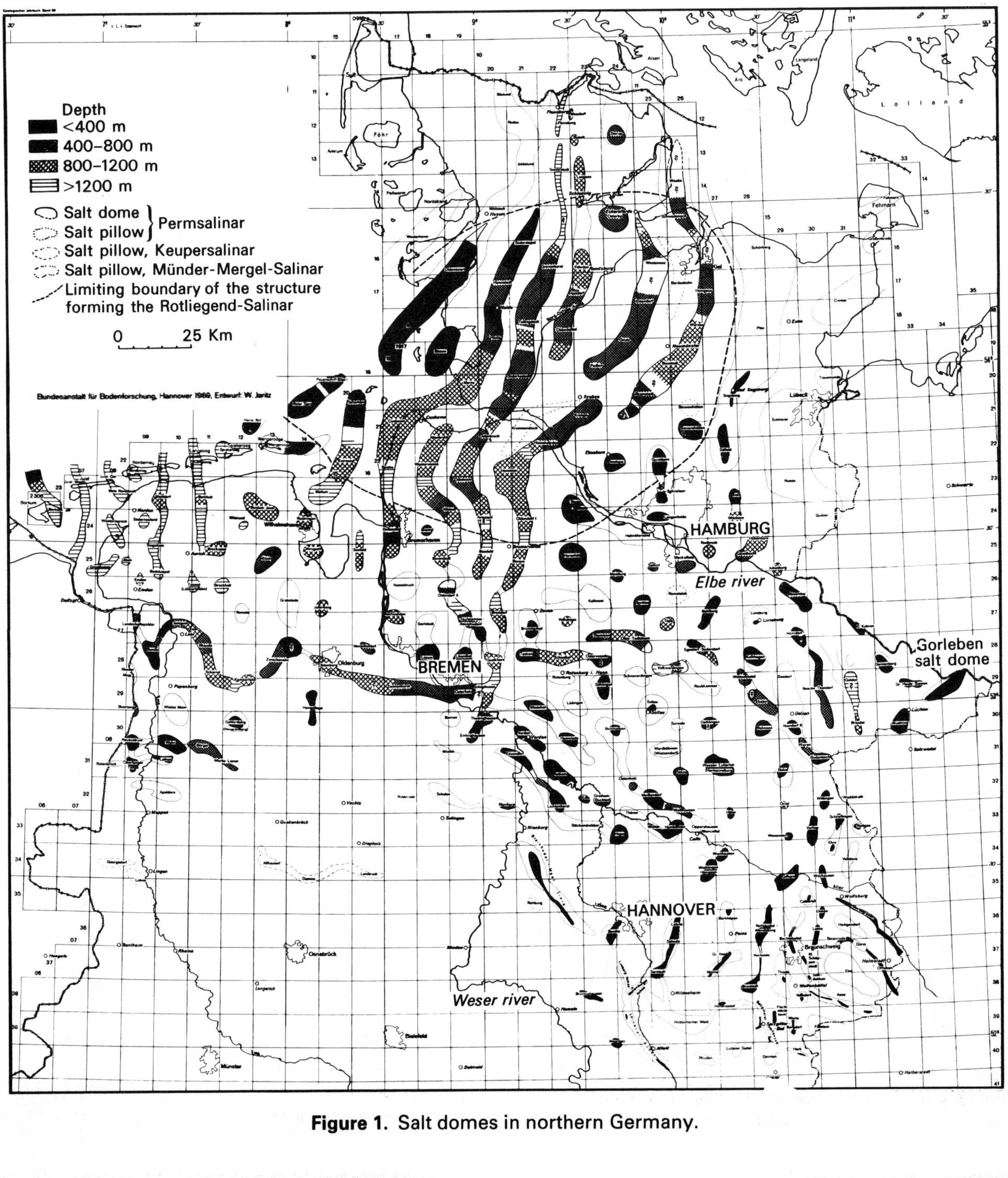

Figure 1. Salt domes in northern Germany.

the region was flooded again several times and for a very long period of time. During the Tertiary period heavy volcanism occurred in Germany, huge faults were created, and the Alps rose once again. In the Quaternary the salt domes were thrice covered by huge layers of ice during the ice ages. Most of the salt domes have endured these geological events without any significant change in their size.

Figure 2 shows a cross-section through the Asse Salt Dome, a geological structure which is relatively simple. The salt dome consists mainly of salt rocks of the Stassfurt Series (Older Halite, Na2, and potash salt, K2) and of the Leine Series (younger Halite, Na3). On both flanks of the salt dome lie strata of the youngest evaporation cycle of the Aller Series: clay, T4; anhydrite, A4; Youngest Halite, Na4. The salt dome itself is covered with geologic strata which seal it from groundwater, a fact which has guaranteed its existence in its present shape for over 100 million years. This proves that solution of salt has occurred only on an extremely small scale. The base of the salt dome is at a depth of at least 1500 m. Furthermore, one can deduce from the age and from the position of the salt, as well as from the overlying strata, that the process of dissolving any salt has been restricted to the uppermost parts of the salt dome, leaving the 'leaching residues' atop of the salt dome.

Mining Activities in the Asse Salt Dome

The first mining of the Asse Salt Dome started in 1899, when the shaft 'Asse 1' was sunk to a depth of 375 m and potash salt was mined on three levels, directly below the top of the salt dome. In 1906 water

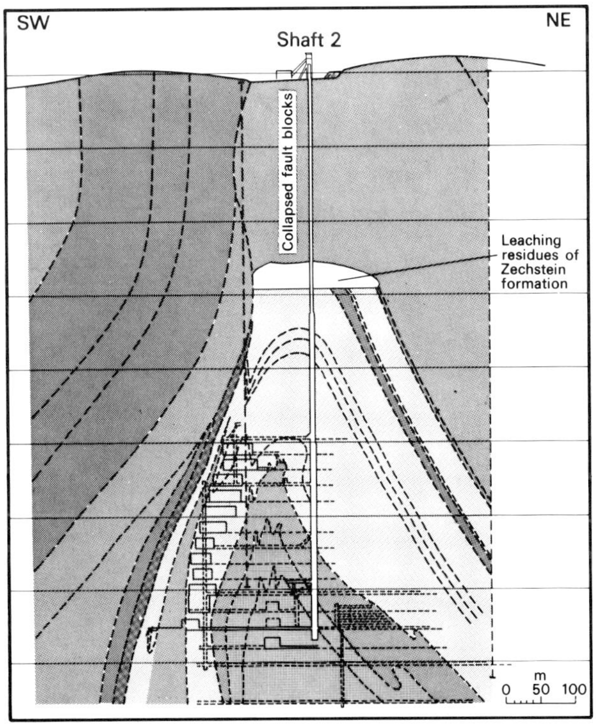

Figure 2. Cross-section through the Asse Salt Dome.

flowed into the workings, brought about by improper mining of the potash salts, thus destroying some of the covering watertight layers; flooding of the whole underground workings and the shaft occurred.

From 1906 to 1908 a second shaft, 'Asse 2', was sunk to a depth of 765 m at a distance from 'Asse 1' of about 1400 m. At the beginning only potash salt in the Stassfurt stratum (K2) (see Figs. 2 and 3) was mined; this was done deep in the core of the salt dome to exclude any danger of flooding. Later on, mining of rock salt started in the Older Halite (Na2) and after some years extended to the Younger Halite (Na3) on the southern flank of the salt dome which today forms the existing undergound workings described below.

In order to provide the Asse Salt Mine with two separate shafts at a distance of 3400 m from 'Asse 2', shaft 'Asse 3' was sunk to a depth of about 725 m, starting in 1911. However, this mine was never worked, as a crisis in the German potash industry occurred during the 1920s. Because of a lack of maintenance, surface water flooded the shaft and the water level stands today at about 40 m below the surface.

Underground Workings

During the first operation of the Asse Salt Mine, 26 rooms were excavated for potash salt along the northern flank of the salt dome (see Figs. 2 and 3). When this came to an end, all these rooms were backfilled and thus cannot be entered today.[4,5]

In 1916 the mining of rock salt began in the Younger Halite on the southern flank of the salt dome (see Fig. 3). During these operations 131 rooms, with a volume of about 3.6 million m^3, were excavated, on 13 levels starting from a depth of 750 m up to 490 m. A typical room has a length of 60 m, a width of 40 m and a height of 15 m. Between rooms on one level pillars are left standing with a width of 12.5 m and between the different horizontal levels strata with a height of at least 6 m are left untouched (see Fig. 4).

TEST DISPOSALS OF RADIOACTIVE WASTES

Low-level Wastes

On 4 April 1967 the first tests started by disposing of the first 200-l drum of solidified low-level waste on the 750 m level in rooms situated in the Younger Halite on the southern flank of the salt dome (see Fig. 3). Later on other rooms in the Older Halite on the 725 m level were also used to continue the test programme and up to November 1978, about 141 000 drums of low-level wastes have been deposited within the scope of the research programme. During these early operations the disposal technique was as follows: First, the drums containing the low-level wastes were piled up in horizontal layers from wall to

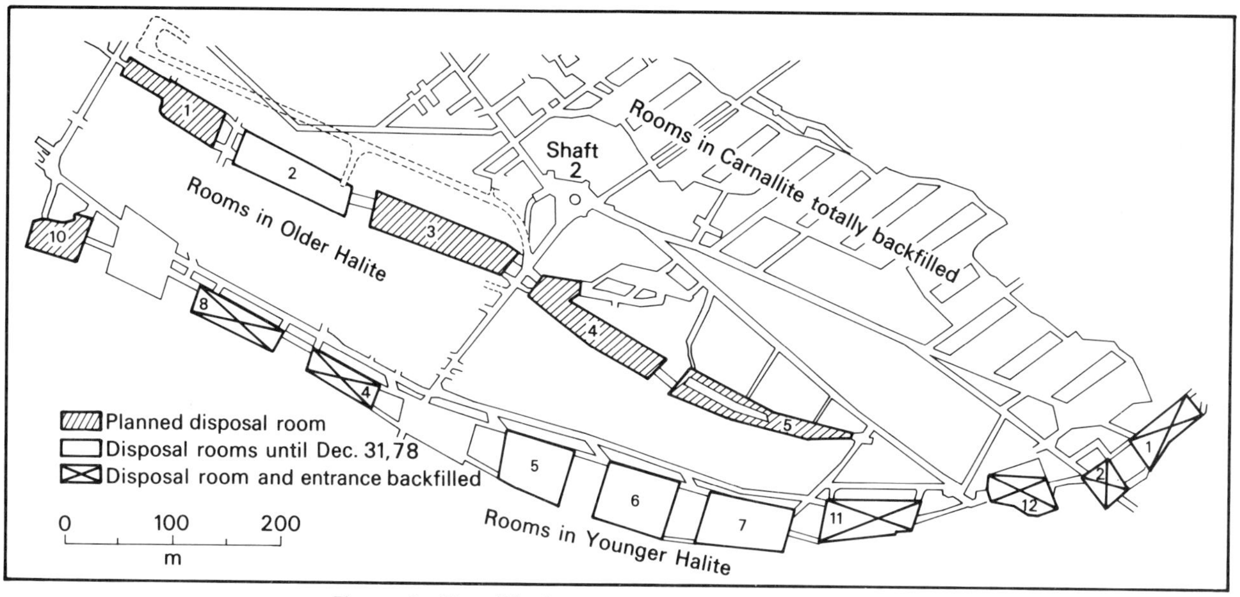

Figure 3. Simplified ground-plan of the 750 m level.

wall with ten drums on top of each other (see Fig. 5). Then the entrance to the rooms was closed and sealed with concrete and salt.

Today's disposal technique has this sequence:[4,6] A new entrance to the disposal room, directly below its ceiling, is excavated. From there a ramp of loose salt is banked up into the room. A scoop transports the drums from the shaft to the disposal room and dumps them on to the ramp of salt (see Fig. 6), the drums rolling down the ramp without being destroyed.

At short time intervals the drums are covered with loose salt and the disposal action continues. The empty space being left is then filled by piling up lost concrete shielding containers (see Table 1) going from the end of the room back to the entrance (see Fig. 6). After finishing the operations in one room, the entrance is sealed with concrete and salt.

This newly developed technique has several advantages compared to the piling-up of drums which was used before:

◇ The rooms are completely refilled and thus support the stability of the system of underground workings.
◇ In case of a flooding of the mine there will be enough salt to saturate the water so that it loses its dissolving capacity and will not be able to touch the radioactive wastes.
◇ The dose rate received by the disposal personnel is reduced, from 100 mrem/month in 1969 to 50 mrem/month in 1975, while the amount of wastes disposed of increases by a factor of three.
◇ The necessary dressing of the ceiling of the rooms can be carried out more easily.
◇ A higher efficiency of the disposal action can be achieved.

Disposal of low-level radioactive wastes was stopped at the end of 1978. In the near future a new licensing procedure will be initiated in order to continue these disposal activities, but as this

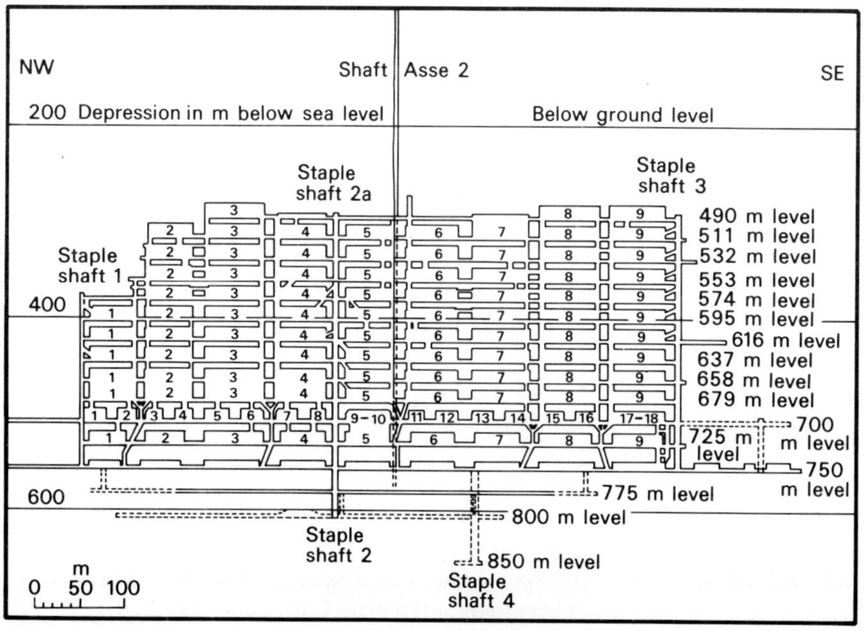

Figure 4. Longitudinal cross-section of the Asse Salt Mine.

Figure 5. Stacked 200-l drums with low-level wastes.

procedure will presumably take several years, an additional licence is now being sought for interim storage of radioactive wastes at the Asse Salt Mine.

Intermediate-level Wastes

The nature, treatment and packaging of intermediate-level wastes is very similar to that of low-level wastes. However, because of their higher radioactivity and consequently higher dose rates, these wastes can only be transported and handled within a shielding container. For the same reasons it is not possible to dispose of intermediate-level wastes in a room accessible to man. Thus, a disposal technique was developed and tested which guarantees a sufficient shielding of the wastes during transport and disposal operation. This technique was tested from 1972 until March 1977 when the licence for this activity was not renewed. A bulk shielding container holding seven 200-l drums of intermediate-level waste was used to transport the wastes to the Asse Salt Mine. In a shed adjoining the shaft, the drums were reloaded into a single shielding container and then brought down to the 490 m level.

Figure 7 shows this disposal operation. Above the sealed disposal room on the 511 m level the charging room was excavated on the 490 m level and there all technical equipment was installed. The two rooms are separated by a horizontal salt stratum of 6 m thickness, penetrated by several boreholes. The single shielding waste container is transported to the charging room and set down atop the charging borehole. Under full radiation protection the 200-l drum is then lowered from the shielding container into the disposal room.

To make sure that no radioactive dusts or aerosols can escape from the disposal room, a ventilation system is installed which generates a vacuum in the disposal room and filters all air before returning it to the atmosphere of the mine. From 1972 to March 1977, 1298 drums with intermediate-level wastes were disposed of in this way without any difficulty.

Asse Cavity Project

Although during the testing of the above-described techniques for the disposal of intermediate-level wastes no difficulties arose, this technique had one logistic disadvantage: the shielding container weighing about 9 t had to be transported through the shaft to the charging room and back again to the shaft shed. In order to avoid this disadvantage the following concept for the disposal of intermediate-level wastes in a separate cavity was developed.

A cavity with a volume of 10 000 m^3 was excavated at a depth of about 1000 m at the bottom of a specially drilled and completely separate isolated shaft. This shaft and the cavity at its bottom are a completely closed system, situated within the Asse Salt Mine. The concept foresees an unloading of the 200-l drums with the intermediate-level wastes from the shielding container in a hot cell situated in the shaft shed. They are then transferred to the hoisting cage, which has a direct connection with the hot cell. The cage lowers the drum without any shielding down to a separate discharge room at a depth of 926 m.

Figure 6. Optimized technique for the disposal of low-level wastes (above). Stacking of lost concrete shielding containers (below).

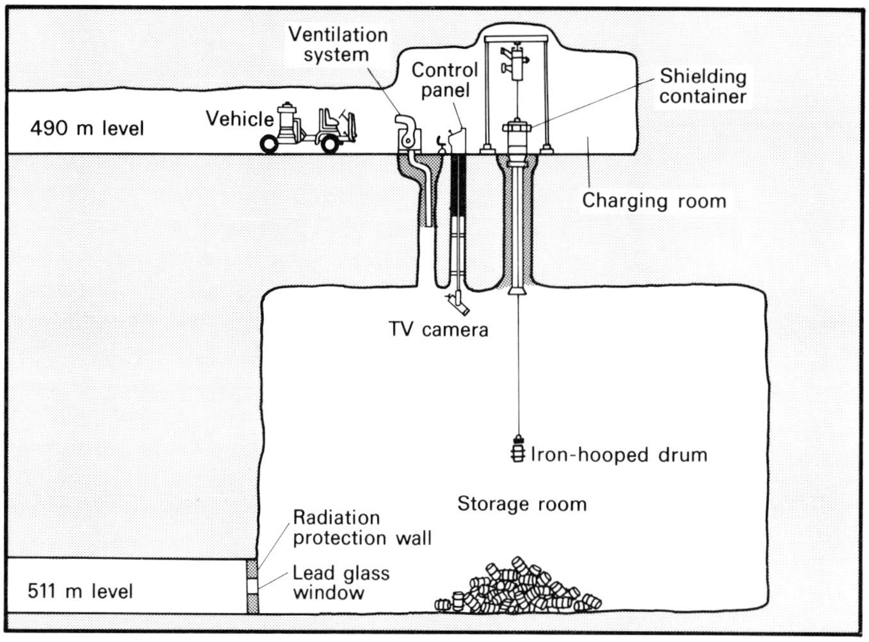

Figure 7. Scheme of the disposal technique for intermediate-level wastes.

From there the drum will be dumped automatically and will fall free to the bottom of the cavity. This facility was to start operation in 1979.

Another interesting cavity project, still in a preliminary stage of laboratory development, is the *in-situ* solidification of radioactive wastes. With this technique, liquid low-and intermediate-level wastes are mixed with a solidifying agent to form a slurry and then pumped into a cavity; there, this mixture solidifies. Investigations on material properties, rock mechanics, nuclear technology and safety considerations are under way.

AVR Fuel Elements

The AVR (*Arbeitsgemeinschaft Versuchsreaktor Jülich*) pebble bed reactor at the *Kernforschungsanlage* Jülich is a prototype High Temperature Gas Cooled Reactor; it was formerly run with carbide fuel elements whereas today oxide fuel elements are used. The old carbide fuel elements of spherical shape, diameter 6 cm, will not be reprocessed. After working out safety levels, it was decided to carry out a single experiment, disposing of the limited number of 100 000 of these elements in the Asse Salt Mine. One thousand of the fuel elements will be packed into one canister at Jülich. These canisters meet all safety requirements of impermeability to gas, corrosion and mechanical strength; the canisters are then put into shielding containers and transported to the Asse Salt Mine.

At the 750 m level of the Asse Salt Mine four boreholes were driven to a depth of 33 m, each borehole holding 25 cans with a total length of 25 m. This emplacement technique will be very similar to that for intermediate-level wastes.[7] After filling a borehole, it will be sealed with a plug of concrete, bitumen and silica-gel, thus assuring that no gaseous tritium or krypton-85 can escape. The activity of a can at the time of disposal is 22 000 Ci and heat production will be only 70 W; this is negligible.

Because of the 22 000 Ci activity per canister and the absence of heat production, these irradiated AVR fuel elements can be defined as intermediate-level wastes. In contrast to other intermediate-level wastes, these burnt AVR fuel elements still contain nuclear fuel and gaseous fission products; however, their amounts are relatively low. High-level wastes are also expected to be disposed of in boreholes. It is believed that this test is an important step towards the disposal of high-level waste. However, the absence of public acceptance has so far prevented the start of these operations.

High-level Wastes

In order to develop a technique for the disposal of high-level wastes, many experiments have already been carried out. One concept foresees the disposal of vitrified high-level waste in boreholes arranged in a hexagonal geometry; computer calculations have been carried out investigating the dissipation of heat generated by these high-level wastes and the effect of this on the mechanical behaviour of the salt rocks[8]. Laboratory and *in-situ* tests, simulated by electrical heaters, have concentrated on the effects of heat production on salt (see Fig. 8).

The results of these calculations, laboratory and *in situ* tests[8,9] have shown that no natural constraints exist which would limit the use of rock salt for the disposal of high-level wastes.

During 'Project Salt Vault', carried out in the United States,[10] today's techniques demonstrated the possibility of handling high activity wastes in underground workings. Up to now no disposal test with real high-level wastes could be carried out as at present there exists nowhere a representative amount of

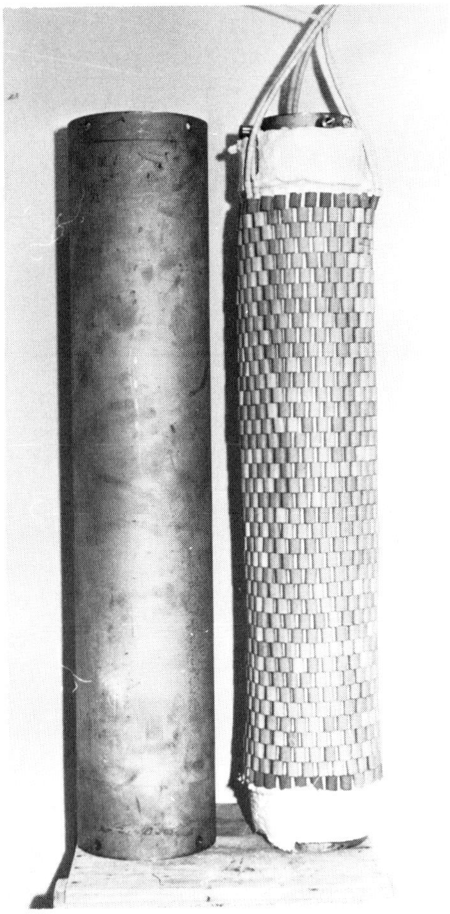

Figure 8. Electrical heater used in temperature test field III at Asse Salt Mine.

vitrified high-level waste for the performance of such an experiment.

Within a research and development programme of the European Community it is planned to perform a test for the disposal of real high-level wastes but in such a manner that the retrievability of the wastes is assured. This large-scale experiment is planned to start at the beginning of the 1980s.

DEFINITION OF ACCEPTANCE CRITERIA

Based on the first operations and the experience gained shortly after the recommissioning of the Asse Salt Mine, conditions for test disposal activities were worked out; they are continuously being revised.

It should be pointed out that these conditions only apply to the special circumstances of the Asse Salt Mine and the technical experiments performed there. They are in no respect applicable to a newly to-be-built full-scale repository.

Normally, low- and intermediate-level wastes are only accepted as solids. Either the wastes themselves are solid, or they have been solidified with concrete or bitumen. Low-level wastes can be delivered in different containers (see Table 1), whereas intermediate-level wastes must be delivered in a 200-l drum, inserted in a shielding container. The dose rate at the surface of the drums in the case of the low-level wastes must not exceed 200 mrem h^{-1} and at a 1 m distance, 10 mrem h^{-1}. For intermediate-level wastes these values apply to the surface of the shielding container and to a distance of 1 m, respectively. Normally, these values are far from being reached.

Another basic condition for low- and intermediate-level wastes is that the amount of nuclear fuel contained in the wastes must not exceed 15 g per drum. Table 1 shows the permitted activity per waste canister of low-level waste, whereas Table 2 gives the permitted activity per 200-l drum of intermediate-level waste.[11,12]

INDEPENDENT RESEARCH AND DEVELOPMENT

The Asse Salt Mine not only serves as a facility for the developing and testing of disposal techniques but also

Table 1. Permitted Activity for a Low-level Waste Container in Curies (Ci)

Container:			Sheet metal drum or drum with reinforced seams, lid with screwed clamping ring		Sheet metal drum or drum with reinforced seams with complete concrete lining, lid with screwed clamping ring Iron-hooped drum or drum with reinforced seams with flanged lid		200 l drum inserted into a 400 l drum interspace cast with concrete	Lost concrete shielding with inserted container
Volume:	200 l		400 l	200 l	400 l	200 l	max. 200 l	
Maximum		A	not permitted	not permitted	0.001	0.001	0.01	0.02
permitted	category	B	not permitted	not permitted	0.05	0.1	0.1	0.2
α-activity		C	0.2	0.4	2	4	4	10
Maximum		A	not permitted	not permitted	0.1	0.1	1	2
permitted	category	B	not permitted	not permitted	1	2	2	5
β/γ-activity		C	5	10	5	10	10	25

Waste Category A: Solid dry wastes without treatment in any form: no powders, granulated materials or soluble products originating from liquid effluent decontamination plants.

Waste Category B: Solid dry wastes in any form, packed in tight and stable containers, for example tin or plastic containers, with a maximum volume of 30 l. The space between inner and outer containers has to be filled completely with a suitable binding material (percentage of binding material at least 30 vol%, wall thickness of binding material to outer container at least 50 mm).

Waste Category C: Wastes immobilized in suitable binding materials like slurries, evaporator concentrates, ion exchanger, ashes, non-deformable wastes of activated or contaminated pieces.

Table 2. Permitted Activity per 200-l Drum of Intermediate-level Waste

Nuclide spectrum of the waste		Permitted β/γ-activity per 200 l drum (Ci)	
β/γ	α (Ci t^{-1})	Bitumen	Concrete
Mixture of fission products	10^{-2}	200	50
with <5% activated products	5	100	50
Mixture of fission products	10^{-2}	80	100
with 5–30% activated products	5	40	100
Mixture of fission products	10^{-2}	40	200
with 30–95% activated products	5	20	200
	10^{-2}	30	400
Activated products with <5% fission products	5	15	200
Activated structure material	10^{-2}	50	1000[a]
and radioactive sources	5	25	500[a]

[a] These values may be exceeded if additional shielding assures that the dose rates do not exceed 200 mrem h^{-1} at the surface of the shielding container and 10 mrem h^{-1} at a distance of 1 m.

as an excellent location to investigate all kinds of problems affecting the safety of radioactive wastes in a salt repository.

The range of the research and development programmes is wide indeed. In addition to the above-mentioned temperature tests for the disposal of high-level wastes, there are *in situ* measurements, laboratory tests and computer calculations of mechanical behaviour of salt rocks in order to predict any movement of underground workings. Geological investigations are also carried out in order to explore more fully the structure of the Asse Salt Dome. Hydrological measurements of all known aquifers in the surroundings of the salt dome try to determine water flow rate and flow direction.

Water inflow is defined as the maximum credible accident of a repository in a salt dome. Although this event is a very improbable one, much research has been done on the interrelationship between water, salt and radioactive wastes in case such an event occurs. The results of these investigations are the basis for theoretical safety considerations, verified in part by model experiments in the laboratory and in the Asse Salt Mine itself.

The practical results from the full-scale experiments on the disposal of radioactive wastes, as well as the results of other research and development programmes, will form the basis on which to plan and construct an industrial full-scale repository. The operation of the Asse Salt Mine and the results from its research and development programmes show that the knowledge exists for plans for construction of such an industrial repository using today's technology and with full consideration of all safety aspects.

THE WASTE REPOSITORY WITHIN THE *ENTSORGUNGSZENTRUM*[13]

In the late 1960s the Federal Ministry for Research and Technology initiated an integrated system for the handling of spent nuclear fuel and for the management of all radioactive wastes. This integrated concept led to the so-called *Entsorgungszentrum* which is planned to take over all spent nuclear fuel from German nuclear power stations.

The centre therefore includes (see Fig. 9) a facility for the storage of spent nuclear fuels, a reprocessing plant designed for a throughput of 1400 tU/year, facilities for the production of nuclear fuel from recovered uranium and plutonium, plants for the conditioning of radioactive wastes, and finally a repository for the safe disposal of these wastes.[14]

Having the waste repository on the same site as the other nuclear facilities of the Centre—it is the main advantage of the integrated waste mangement system that there is no need to transport radioactive wastes outside the site—requires that the site for the erection of such a Centre must be situated directly above a suitable salt dome. Thus, after a lengthy selection of several sites above salt domes in northern Germany, on 22 February 1977 the Prime Minister of the State of Lower Saxony named the salt dome near Gorleben as the site for the construction of the Centre (see Figs. 10 and 11).

According to the Atomic Law of the Federal Republic of Germany, the DWK, *Deutsche Gesellschaft für Wiederaufarbeitung von Kernbrennstoffen mbH* (German Company for Reprocessing of Spent Nuclear Fuel), is responsible for planning and construction of all facilities of the Centre except the waste repository; this comes under the jurisdiction of the *Physikalisch-Technische Bundesanstalt*, PTB (Federal Physical-Technical Bureau). Therefore, on 31 March 1977, DWK applied for a licence for all the facilities falling within its authority and presented an extensive safety report to the Lower Saxony Ministry of Social Affairs, the licensing authority for this area. In July 1977 the PTB applied for a plan verification procedure in order to license the repository.

After thorough investigation of the safety reports and the technological aspects of the whole concept, the German Reactor Safety Commission and the Radiation Protection Commission prepared a report[15] in late 1977 stating that the concept of the *Entsorgungszentrum* can be realized allowing full consideration of all safety aspects. This report also observed: 'The Gorleben salt dome has a gross volume of 80 km^3 down to a depth of 2000 m. This extensive space ensures that the disposal of low- and intermediate-level wastes is possible there.' The size of the salt dome also gives confidence that large rock salt sections for the disposal of high-level wastes can be found. This has still to be confirmed by deep exploratory boreholes, the first three of which have

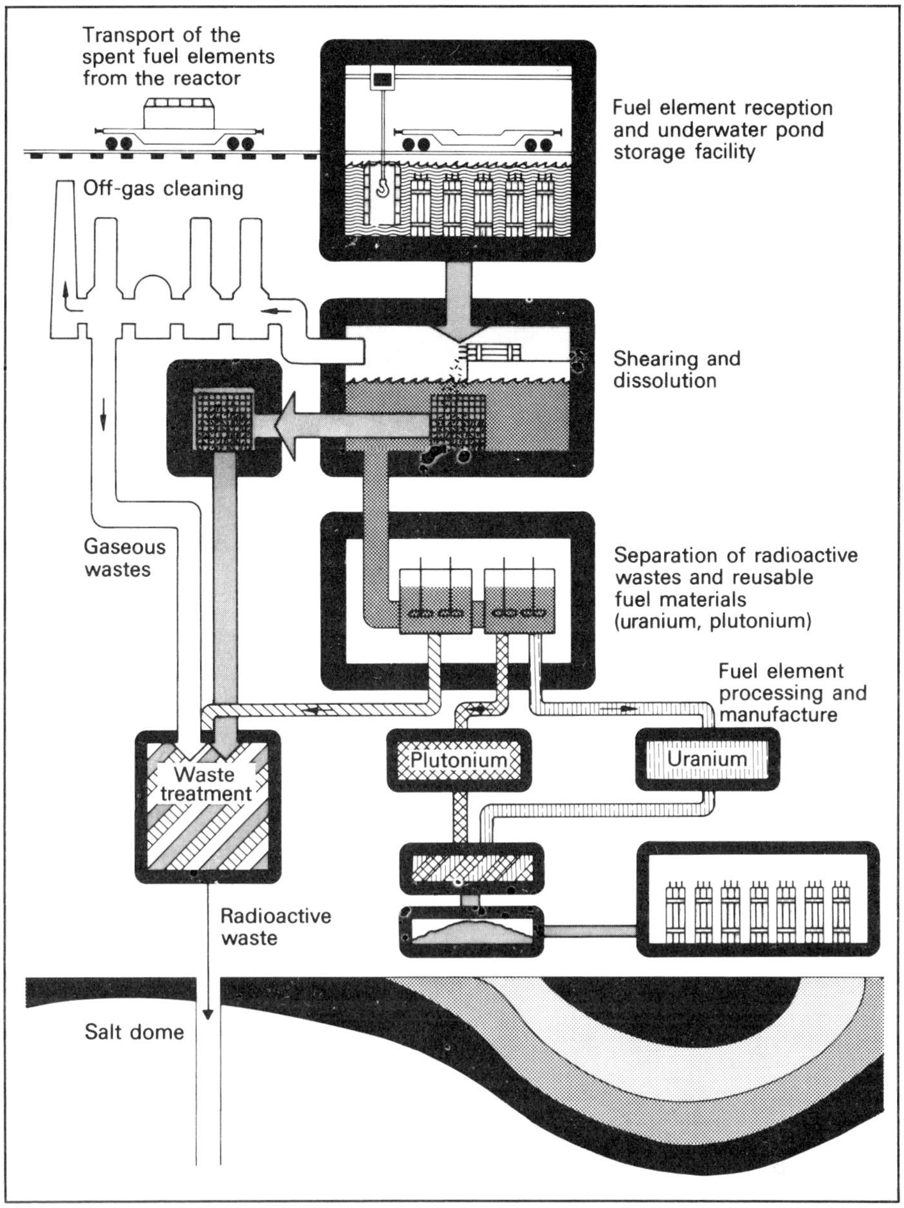

Figure 9. Process steps in the *Entsorgungszentrum*.

been licensed; drilling operations started on the first of these in November 1979.

The concept for a repository within the Centre relies on the experience gained from the research and development programmes as well as the test disposal operations carried out at the Asse Salt Mine. Thus, for the new repository only techniques embodying current state of technology are used.

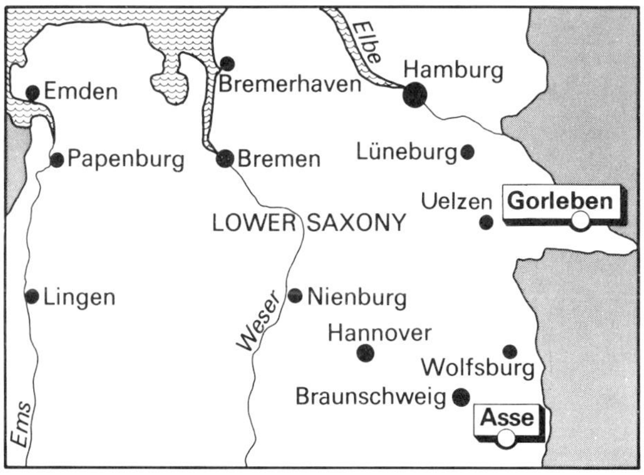

Figure 10. The site of the planned *Entsorgungszentrum* at Gorleben and the Asse Salt Mine.

As far as plans go at present, low-level wastes will be disposed of by using the dumping or piling-up technique. Intermediate-level wastes will be disposed

Figure 11. Artist's view of the planned *Entsorgungszentrum*.

of in inaccessible rooms or boreholes whilst under full radiation shielding during the whole operation. Vitrified high-level waste, amounting only to about 130 m^3/year, will be deposited in boreholes according to the principles developed at the Asse Salt Mine and confirmed by temperature tests and computer calculations.

It will be observed that this new repository is to be mined in a virgin salt dome after extensive geological and geophysical exploration and that it will create much smaller underground workings than exist at the Asse Salt Mine. In addition, all underground workings which are no longer needed for waste disposal will be refilled, so that only a minimum of open space will exist deep in the core of the salt dome. Before final decommissioning of the repository all spaces left open, such as galleries and shafts, will be properly refilled so that there will be no need for any further supervision.

PUBLIC RELATIONS AND POLITICAL DEVELOPMENTS

A somewhat strong opposition exists in the Federal Republic of Germany to the use of nuclear power. This opposition and its associated vehement protests is not supported by the affected local population living in the surrounding of a nuclear facility, but by more or less professional nuclear opponents.

The general public is often exposed to protest actions and to the arguments of nuclear opponents, without having any real information on the subject. Thus, a vital objective of the Asse Salt Mine, as well as of DWK and other companies and authorities dealing with nuclear power, is to provide the public with factual information.

Besides carrying out research and development programmes and tests of waste disposal, about 6000 visitors in 1977 were shown through the Asse Salt Mine and obtained firsthand information. DWK is not only represented at the Gorleben site to provide detailed factual information about the *Entsorgungszentrum* to all interested parties, particularly the local authorities, but it also tries to inform the general public with all the relevant facts on the broadest possible basis. (See also the recent IAEA booklet[17] on this subject.)

From 28 March until 3 April 1979 the 'Gorleben Hearings' took place in Hanover, during which international opponents of nuclear power, the so-called 'critics', and international nuclear scientists, the so-called 'countercritics'—both invited by the state government of Lower Saxony—discussed the German integrated concept of nuclear waste management. The Prime Minister of Lower Saxony, Herr Albrecht, attended the whole week of discussions.

On the basis of these discussions he delivered on 16 May 1979 a governmental declaration to his parliament. In this declaration he acknowledged that the disposal of radioactive waste in a suitable salt dome is combined with only a very small risk for future generations and in fact that there is no risk for the living and directly following generation.

However, concerning the other above-ground facilities at the *Entsorgungszentrum*, namely the spent fuel storage and reprocessing facility, he requested DWK to consider some technical alterations; these were accepted by DWK.

In spite of this, his conclusion was that the nuclear *Entsorgungszentrum*, although perfectly feasible from a technological and safety point of view, cannot at present be carried out for political reasons. Thus, he proposed the following actions:

◇ the immediate construction of safe long-term spent-fuel storage facilities with passive cooling systems;
◇ an increase of research and development activities for the safe disposal of radioactive waste;
◇ the exploration of the salt dome at Gorleben by deep drillings and, in the case of positive drilling results, by mining and underground workings. If drillings prove negative, then exploration of other sites suitable for a repository;
◇ that a decision about reprocessing should not be made until a settled concept for future energy policy existed.

DWK accepted the technical changes requested by Herr Albrecht and has already altered its concept in order to meet the requirements of reprocessing capacity of 1400 tU/year. On 28 September 1979 the Prime Ministers of the German Federal States met Chancellor Herr Schmidt and discussed the waste management issue. The results summarized below give the present state of the political discussions:

It was confirmed that reprocessing of spent fuel, recycling of nuclear fuel and disposal of nuclear wastes is perfectly feasible today from a safety and technological point of view and that it can guarantee the *Entsorgung* of nuclear power plants. Therefore, activities to bring about the integrated waste mangement concept are to continue unchanged, based on planning, research and development results already obtained.

In parallel to these, other waste management techniques, such as the direct disposal of spent fuel without reprocessing, will be investigated to decide by the middle of the 1980s if such techniques have substantial safety advantages. A commission of the Federal Government and the State Governments will be set up which, under the leadership of the Federal Government, will supervise these investigations.

The government of the Federal State of North Rhine-Westphalia offered to license a spent fuel storage facility under the condition that at the time of its commissioning a decision on the waste management technique had been made. The Federal State of Lower Saxony was also ready to take over a spent fuel storage facility and in addition its government declared its willingness to license a waste repository in the salt dome of Gorleben if the drilling and mining investigations confirmed its suitability. Investigations will therefore continue uninterrupted so that a decision on the suitability of the Gorleben salt dome can be made in the second half of the 1980s. Finally, it was decided that by AD 2000 at the very latest, aboveground facilities for the waste management technique be chosen and that the repository for the disposal of radioactive wastes shall then be ready for operation.

NOTES AND LITERATURE CITED

1. National Academy of Science–National Research Council, *The Disposal of Radioactive Waste on Land.* NAS–NRC Publication 519. Washington DC (1957).
2. J. H. Martini, Bericht zur Frage der Möglichkeiten der Endlagerung radioaktiver Abfälle im Untergrund der Bundesrepublik, Unpublished expert report, Bundesanstalt für Bodenforschung, Hannover (1963).
3. H. Venzlaff, Tieflagerung radioaktiver Abfälle aus geologischer Sicht. *Atomwirtsch. Atomtech.* (July–August 1978).
4. K. Kühn and J. Hamstra, Geologic isolation of radioactive wastes in the Federal Republic of Germany and the respective program of The Netherlands. In *Proceedings of International Symposium of USERDA on Management of Wastes from the LWR Fuel Cycle*, Denver, Colorado 11–16 July 1976).
5. E. Albrecht and F. Perzl, The research and development program on waste disposal in the Federal Republic of Germany. In *Proceedings of Symposium on Waste Management*, Tucson, Arizona (6–8 March 1978).
6. R. Proske, Zur Endlagerung radioaktiver Abfälle. *Chem. Ing.-Tech.* Nr. 1, 9–13 (1978).
7. J. Wolf, Endlagerung verbrauchter Brennelemente des AVR-Versuchskernkraftwerkes im Salzbergwerk Asse. Jül-1163 (1976).
8. P. Ploumen and G. Strickmann, Berechnung der zeitlichen und räumlichen Temperaturverteilung bei der säkularen Lagerung hochradioaktiver Abfälle in Salzstöcken. RWTH Aachen (February 1977).
9. Forschungs- und Entwicklungsarbeiten zur Endlagerung radioaktiver Abfälle. GSF-T70, KfK 2522. Annual Report 1976 of the Entwicklungsgemeinschaft Tieflagerung (EGT), München (August 1977).
10. R. L. Bradshaw and W. C. McClain (eds.), Project salt vault—A demonstration of disposal of high-activity solidified wastes in underground salt mines. ORNL-4555 (April 1971).
11. Gesellschaft für Strahlen- und Umweltforschung mbH, Bedingungen für die Lagerung von schwachaktiven Abfällen im Salzbergwerk Asse. München (December 1975).
12. Gesellschaft für Strahlen- und Umweltforschung mbH, Bedingungen für die Versuchseinlagerung von mittelradioaktiven Abfällen in die Kammer 8a der 511 m-Sohle des Salzbergwerkes Asse. München (December 1976).
13. The term *Entsorgungszentrum* cannot be translated. In the German language the word *Entsorgung* has been created recently and is now widely used as an antonym to *Versorgung*; this means supply or procurement. In the nuclear context *Entsorgung* is the integrated concept which includes handling and treatment of spent fuel elements and all other radioactive wastes from that moment onwards when they are taken out of the reactor, until their final disposal as wastes. Therefore the *Entsorgungszentrum* includes all facilities for the reprocessing of spent nuclear fuel elements as well as those used for the safe and final disposal of radioactive wastes.
14. C. Salander, Konzept der deutschen Electrizitätswirtschaft für die Entsorgung der Kernkraftwerke. *Kerntechnik* Nr. 5, 229–237 (1978).
15. Reaktorsicherheitskommission, Strahlenschutzkommission, Grundsätzliche sicherheitstechnische Realisierbarkeit des Entsorgungszentrum. Deutscher Bundestag-8, Wahlperiode, Drucksache 8/1281.
16. R. A. Kerr, Geological disposal of nuclear wastes: salt's lead is challenged. *Science* **204,** 603 (1979).
17. International Atomic Energy Agency, *Radioactive Wastes.* IAEA, Vienna (1978).

The complete manuscript was received 30 October 1979

Risks and Consequences in Energy Production

DR HERBERT INHABER
Oak Ridge National Laboratories, Oak Ridge, Tennessee, USA

The word *risk* is often used in association with energy production and in some cases it is equated with *consequences*. The distinction is important and has implications for the study of energy systems. Failure in a given energy system, such as nuclear power or hydroelectric dams, can give rise to human consequences, such as deaths, injuries or disease. The extent of these consequences varies from one system to another. Furthermore, the probability of such failures is a characteristic of the given system. For example, a nuclear reactor or hydroelectric dam may have a probability of one catastrophic failure in x years of operation, where x is a large number. The probability multiplied by the number of deaths, injuries and illnesses caused by the failures—the consequences—can be taken as a measure of the risk, or risk index.

Society tends to think of consequences of energy systems only in terms of failures. Catastrophes, such as possible accidents in nuclear reactors, dam failures or liquid natural gas explosions, inevitably attract more attention that the day-to-day accidents and illnesses required to fabricate, construct and operate these facilities. However, day-to-day deaths and losses of health in producing energy may far outweigh the consequences from the more spectacular catastrophic events.

The difference between risk and consequence in terms of energy systems is analogous to what takes place in a lottery. In that form of gambling, there are usually one, or more, large and pleasant consequences – the major prizes. The great number of tickets which do not win a prize tend to be ignored, even though the amount spent on them is always greater than the total of prize money.

Much the same can apply when we think of risk and consequences to human health in energy production. Failures of hydroelectric dams, mishaps at nuclear reactors, explosions at natural gas terminals – these tend to draw most public attention. However, it can be shown that these large-scale events produce a relatively small fraction of the total risk or consequences of energy systems. The day-to-day effects of mining the raw materials, processing them, building the energy installations and operating them, tend to be greater than the more spectacular events of which we hear. In terms of the lottery mentioned above, it is as if the prizes had the same values as one or two tickets, instead of many more.

RISK AND CONSEQUENCE

We must begin by defining risk and consequence, since the two are confused in the literature. Consequences to health can be defined as the number of deaths, injuries or diseases attributable to a given phase of energy production. These consequences have a finite probability of occurring, where the probability ranges between 0 and 1. The consequences in deaths, injuries or disease can be multiplied by the probability to yield the risk, sometimes known as the risk index. Since the probability is unitless, the units of risk are the same as that of consequences.

To take a hypothetical example, consider a hydroelectric dam. The consequences of its failure may be estimated to be y deaths. Dams of a particular type may fail, on average, once in x years, yielding a probability of $1/x$. The total annual risk is then the product of the two values, or y/x deaths.

The above calculation is applicable to the more spectacular events associated with energy production which, as will be shown below, form a relatively small proportion of overall risk. However, the same reasoning can be applied to the day-to-day consequences as well. For example, producing a given number of tons of coal or uranium may produce z deaths, on average. In reality, the exact number of consequences will depend on such factors as the ore grade, the type of mine, and so on. A fixed value is assumed only for illustrative purposes. The probability of this number of deaths occurring may be taken as close to 1, providing working conditions are not changed. The expected number of deaths is again equal to the consequences times the probability, or z.

SET OF BASIC RULES

The field of risk assessment has grown rapidly in the last decade. By now, there are risk studies on many of the conventional energy systems, defined as those

CCC–0308–0188/80/0005–0304$04.00/1

which presently produce large quantities of energy.[2,3] A start has recently been made on some non-conventional sources, such as solar space heating, windmills and methanol.[1] While no risk study can yet be deemed to be perfectly definitive, those dealing with conventional systems are generally thought to be more accurate than those on non-conventional systems, primarily because less is known about the latter group.

While papers on the subject have grown in quantity, their conclusions have sometimes differed. To some extent this is because of varying ground rules. While no set of rules can be applicable to all risk studies, the following were used in a recent work:[1]

- Present-day energy systems were evaluated, not those of the future. This stipulation is important in making fair comparisons. For example, fossil-fueled systems of the future may have better pollution controls than at present, thus reducing the risk. It would be invalid to compare the risk of these hypothetical future systems to that of present systems. As systems change in the future, risk evaluations should be revised.
- For many energy systems, there are factors relating to risk which at present cannot be quantified. For example, in terms of nuclear power there are questions of proliferation and sabotage; for some fossil fuels, there is the question of increasing concentrations of carbon dioxide in the atmosphere and its effect on climate. Because these questions cannot yet be quantified does not, of course, mean they should be ignored. Rather, they are so important that they deserve papers or even books of their own. If and when these long-term or risk-related questions can be quantified, they should be included in risk studies.
- The actual risk incurred in producing a unit of energy should be shown, not the incremental risk. By *incremental risk* we mean the extra risk that would be incurred by workers in producing energy above and beyond what they would incur in non-energy-related activities. For example, coal miners have a relatively high risk per man-year of labor. If they had been office workers this risk would have been lower. The difference between the two is the incremental risk.

  There are at least two reasons why this system was not adopted in my previous paper.[1] First, there is no way of telling what occupations coal miners, oil drillers, reactor operators and others associated with energy production would have chosen if they were not employed in their present jobs. Second, writers of major treatises on risk, such as Comar and Sagan[2] and Smith *et al.*[3] have generally adopted the approach of evaluating total risk, not incremental risk. However, it must be admitted that this requirement is only a philosophical assumption, and the calculations could be done either way. There are few, if any, studies on determining incremental risks.
- Public and occupational risk should be shown separately. These categories are sometimes called 'involuntary' and 'voluntary' risk, although it is unclear whether a coal or uranium miner 'volunteers' for risk in the ordinary sense. Public risk is sometimes catastrophic but sometimes not; members of the public can suffer accidents when coal is moved by railroad to a generating plant, for example.
- The risk attributable to obtaining the raw and finished materials for energy systems should be included in the total risk. That is, the risk in producing not only the coal, oil and uranium, but the steel, copper and aluminum for turbines, boilers and reactors should also be included. The argument can be made that the steel, copper and aluminum would be produced in any case, and so the risk attributable to them should not be counted. This again is a philosophical choice. Generally speaking, writers on the subject have included the risk attributable to fuel-gathering, so it seems logical to include risk attributable to non-fuels. This question can be important for energy systems, such as hydroelectricity, where there is no fuel in the ordinary sense.
- Environmental effects, as opposed to health effects, probably deserve separate study. Of course many environmental effects, such as air or water pollution, are related to health. However, combining the two topics in one study would likely lead to confusion among readers.

There are other rules which could be adopted in producing risk studies, and those mentioned above

DR HERBERT INHABER is a former scientific advisor with the Atomic Energy Control Board of Canada. Educated at McGill University and in America, he holds three degrees in physics and mathematics. Prior to joining the Control Board, he worked at the Canadian Federal Department of the Environment on measuring the state of environmental quality. Dr Inhaber has been a visiting lecturer in the Department of History of Science and Medicine at Yale University and an advisor to the Science Council of Canada. Author of about 50 scientific papers in mechanical engineering, physics, sociology of science and pollution, he has written two books: *Environmental Indices* and *Physics of the Environment*. Dr Inhaber started his comparison of the relative risk of energy systems in 1977 and has published the Canadian Control Board report and other papers on the subject.
Address: Health and Safety Research Division, Oak Ridge National Laboratories, PO Box X, Oak Ridge, Tennessee 37830, USA.

are not comprehensive. They do give some idea of the choices which have to be made before calculations begin. With these assumptions, the risk analysis can proceed. As mentioned above, the relative risk of catastrophic accidents is small compared to day-to-day risk. This contention can be explored in depth.

A recent excellent article by Reissland and Harries[15] evaluated some of these questions in detail. Readers are referred to it for a useful discussion, primarily in terms of nuclear power.

CATASTROPHIC ACCIDENTS

Some energy systems can have catastrophic accidents affecting the public. For example, hydroelectricity and nuclear power have had the consequences of public accidents studied in detail. On the other hand, some energy systems seem to have little or no chance of catastrophic accidents. Examples of this type might be solar space heating, windmills and solar photovoltaic systems. While these technologies are not risk-free due to the relatively large amounts of materials and construction labor they require, it may be assumed that they have negligible risk in terms of public catastrophic accidents.

Of the eleven systems studied in a recent paper,[1] only three – hydroelectricity, natural gas and nuclear power – were assumed to have potential large accidents affecting the public.

HYDROELECTRICITY

Let us now consider in detail the risk of hydroelectricity. The total risk may be divided into a number of parts: material acquisition, construction, operation and maintenance, transportation and dam failures. One way of showing the sources of risk for any energy system is indicated in Fig. 1. Dam failures may be included in public health risk. For some energy systems, such as coal and nuclear power, there may be a risk from waste disposition. It is zero for hydroelectricity.

It should be noted that, as in all energy systems, different models will produce different results. For example, all hydroelectric dams differ in design and construction and only a comprehensive treatment could evaluate all the variables. What follows is only applicable to some dams in Canada.

The materials required for an average 1000 MW power is shown in Table 1. From these values one can estimate the total occupational risk per unit energy produced if other data are known. The data required are the load factor, or ratio of average energy output to the maximum output of a dam, and the average lifetime of dams. This lifetime is generally taken in the literature as the economic as opposed to the physical lifetime. The load factor chosen is 0.60; the lifetime is about 50 years.

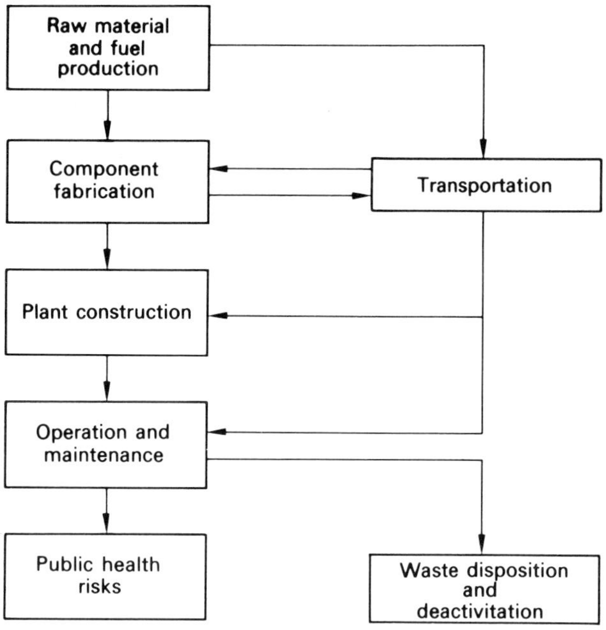

Figure 1. Sources of risk in energy production. The relative importance of the components depends on the energy system. For example, there will be no fuel requirements for systems like solar or wind. Public health risk from coal will be primarily due to air pollution, and that from nuclear power from the possibility of reactor accidents. Transportation plays a part in a number of components.

From the materials requirements, one can go on to calculate the occupational risk per unit energy (one megawatt-year) of this system; only the methodology and not the data is discussed here. For each ton of steel, we can assume that a given number of tons of iron ore and coal are required. (In practice, other materials will be needed as well, but this is a simplified model.) One can use industrial accident rates to determine that, on average, x deaths and y accidents occur for each ton of iron ore and coal produced. Of course, these rates will vary from mine to mine and country to country, but again only averages are being considered here. The same considerations apply to the rock, earth and concrete mentioned in Table 1.

A schematic diagram of how the calculations are performed for materials like concrete, steel and fabricated parts is shown in Fig. 3. The risk incidence per unit weight is multiplied by the unit energy output of the particular material. These intermediate values are then summed to yield the total number of occupational deaths, accidents and disease.

Table 1. Material required for 1000 MW hydroelectric system in 10^5 Metric Tons[4 and a]

Reinforcing steel	Structural steel	Rock	Earth	Concrete
0.14	0.068	117	44	27.5

[a] Steel for turbines and generators is apparently not included in these estimates.

Figure 2. Hydroelectric Dam. The largest accidents affecting the public have occurred in hydroelectric dams, yet in spite of this their total risk is relatively small. Here a typical dam, the Stewartville Generating Station on the Madawaska River in Ontario, Canada, is shown. It has a total capacity of 153 000 kW and was completed in 1969. The dam is 455 m long and 72 m high. (Courtesy Ontario Hydro.)

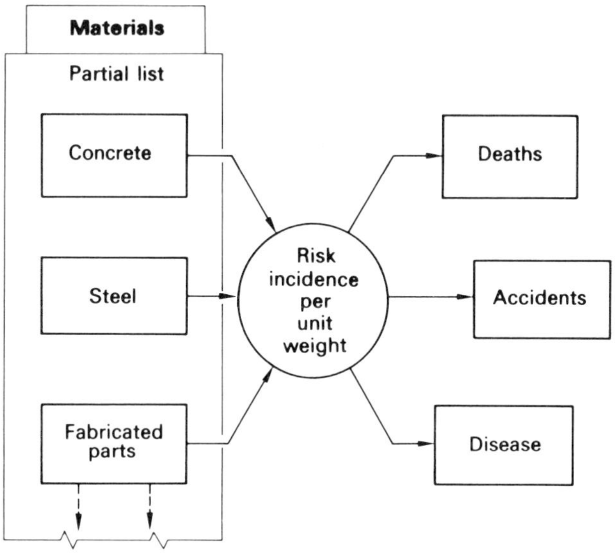

Figure 3. Risk from material acquisition. Each of the raw materials which go into an energy system has an associated calculable risk. The risk depends on the accident, illness and death rate per unit weight produced in the appropriate industry. The dashes indicate that other materials are used.

Morison[4] has estimated the number of construction fatalities in Ontario per unit power capacity. Other authors[5] have estimated the average number of accidents and deaths produced in operation and maintenance of hydro dams.

The final item bearing on occupational risk is that of transportation. The materials for all energy systems must be moved to intermediate and final locations. For some systems like coal, with large amounts to be moved, the transportation risk can be large. For hydroelectricity, most of the materials, as shown in Table 1, are earth and rock. It can safely be assumed that this type of material will be acquired near the site of the dam. As a result, the major transportation risk will be that of the steel and other materials required to produce it. It can be shown that the transportation risk for the relatively small amount of steel will be slight compared with other sources of occupational risk.

Results are shown in Table 2. There are around 20–30 man-days lost per unit energy for hydroelectricity production. However, these numbers will mean very little unless they are compared with something else. As mentioned above, it is of interest to determine the risk from dam failures.

Table 2. Hydroelectricity occupational risk per Megawatt-year net Electrical Output[5 and a]

Accidental and disease	Material acquisition and construction	Operation and maintenance	Total
Deaths[b]	$8.9–20 \times 10^{-4}$	6.3×10^{-4}	$15–26 \times 10^{-4}$
Injury	0.34	0.013	0.35
Man-days lost[b]	15.9–22.5	4.8	21–27

[a] Deaths and disabilities from disease are small.
[b] Deaths counted as 6000 man-days lost (About 30 years of working at about 200 days per year with a 5-day working week and 4–6 weeks holidays).

Hydroelectric Dam Failures

Although there is much information available on dams throughout the world, there appears to be no master list of deaths and injuries associated with accidents related to hydroelectric sites. The events listed in Table 3 form only a partial list of possible incidents which have occurred.

The next question to be answered is: what is the risk of hydroelectricity dam failures? There are a number of ways in which this question can be answered. One can say that there is one chance in x per year, where x is a large number, of dam failure. However, if one is to compare catastrophic risk to routine or occupational risk, the units must be the same. From an extension of Table 1, one can estimate the total number of people who have been killed due to hydroelectric dam accidents. An analogous calculation can be made for accidents. These two quantities can then be combined into total man-days lost.

The denominator of the fraction 'man-days lost per unit energy' is still lacking. One approach to supplying it is to estimate the total amount of hydroelectricity produced in history. This would be analogous to the historical or actuarial data displayed in Table 3. Using the values of Darmstadter,[6] one can estimate that the total amount of hydroelectricity produced up to 1978 was about 3.7 million megawatt-years.

Results for Hydroelectricity

The relative sizes of catastrophic and non-catastrophic risk can now be compared. As noted above, non-catastrophic (or occupational) risk for this energy source is about 20–27 man-days per unit energy. The catastrophic risk is about 7–10 man-days lost per unit energy. While not negligible, it constitutes only about 26% of the total risk from this energy source.

Table 3. Partial List of Major Disasters Associated with Hydroelectricity Generation[5]

Place	Date	Number dead
Gleno, Italy	1923	600
St Francis, USA	1926	426–450
Kiev, USSR	1961	145
Vajont, Italy	1963	2600–3000
Koyna, India	1967	180

NUCLEAR POWER

The risk of nuclear power has probably been studied more than any other type. Results will depend on the type of reactor, the degree of reprocessing, assumptions about the frequency and severity of catastrophic accidents, and so on. A recent study[7] evaluated the total risk of the light-water type of reactor, used widely in the USA. While there exist other types of reactor, such as the Canadian CANDU, their risk has not been studied as extensively as that of light-water reactors.

NUCLEAR OCCUPATIONAL RISK

Estimates for nuclear occupational risk seem to vary widely. Some, such as that of the US Council on Environmental Quality,[8] are comparatively low; others are higher. In Table 4, a range of values are shown. In general, the lower values are the lower part of the range indicated by Comar and Sagan,[2] and the higher value is either their maximum value or that of Smith *et al.*[3] Both of these are in turn extensive data compilations.

Results are shown in Table 4. As in the case of hydroelectricity, each death is counted as 6000 man-days lost. Most of the risk apparently comes from gathering and handling of fuel, namely mining, refining and enrichment. While the object of this paper is not to compare the risk between systems, it is, however, of interest to note that nuclear power seems to have a lower occupational risk than hydroelectricity. Table 4 indicates a factor of three between the lower and upper end of the range in terms of total

Table 4. Nuclear Occupational Risk per Megawatt-year[a]

Accidental and disease	Gathering and handling fuel	Transportation	Operation, maintenance and waste management	Total
Deaths	$1.0–12 \times 10^{-4}$	$2.7–16 \times 10^{-6}$	1.2×10^{-4}	$4–15 \times 10^{-4}$[b]
Injury and disability	$3.4–16 \times 10^{-3}$	$6.0–21 \times 10^{-5}$	2×10^{-3}	$1.6–2.9 \times 10^{-2}$[b]
Man-days lost	0.8–7.8	$1.9–11 \times 10^{-2}$	0.8	3.1–10[b]

[a] From Refs 2 and 3. Light-water reactors only.
[b] Includes risk for acquisition of non-nuclear material and construction of reactors, turbines and so forth.

man-days lost. If one takes the average as a likely value, there are about six occupational man-days lost per unit energy.

NUCLEAR PUBLIC RISK

There has been considerable controversy about appropriate values to use for nuclear public risk, primarily because of the relative lack of experience with this energy form. This paper mentions the important points briefly. The Rasmussen report,[9] a theoretical study of public risk, has often served as the starting point for discussion of this subject. In an extensive review article, Smith *et al.*[3] used values substantially lower than that of Rasmussen, while another long review by Comar and Sagan[2] had values comparable to those of Rasmussen. In order to avoid possible pro-nuclear bias, the highest values in these two review articles were chosen in my previous work:[1] deaths equal $3–23 \times 10^{-5}$, disabilities equal $7–160 \times 10^{-7}$, and man-days lost are 0.2–1.4, all per unit energy. This group of values is used here.

Some estimates for nuclear public risk of light-water reactors have been higher than that of Rasmussen. Comparable studies for other type of reactors have apparently not been made. For example, one study, commissioned by the Ford Foundation,[10] noted that under the most pessimistic assumptions the values of public risk could be substantially higher than that of Rasmussen. However, this is an example of a 'worst case', and not necessarily representative of the likely value of risk. An analogous situation might be considering the risk of Italian dams in 1963. As noted in Table 3, it was then the place and year of what apparently was the largest accident associated with hydroelectricity. Considering only this 'worst case' would not give an adequate picture of the overall public risk of hydroelectricity. This question is discussed more fully at the end of this paper.

The values for public risk noted above form a relatively small fraction of occupational risk. They are 8–15% of the deaths, 0.004–0.09% of the accidents and disabilities, and 6–14% of the man-days lost. It could then be concluded, if the estimates for public risk are reasonably valid, that catastrophic risk of nuclear power seems to be smaller than non-catastrophic risk.

Another approach to studying the validity of public risk estimates for nuclear power is to put actual data on accidents on a historical or actuarial basis, as was done in the case of hydroelectricity. The accident at

Figure 4. Nuclear Power Station. A general ground level view of the well-known nuclear power station of Three Mile Island in Pennsylvania where on 28 March 1979 the most serious accident to date at any nuclear power station occurred. However, this article suggests that accidents affecting the public will form only a small part of the total nuclear risk; at the TMI accident, radiation exposure above normal was not great. (Courtesy Metropolitan Edison.)

Three Mile Island in early 1979 was probably the most serious event to date affecting public health in the history of civilian nuclear reactors. Apparently about 3500–4000 man-rems of radiation dose was incurred by the public.[13] This value is subject to change as more information is made public, and the following calculations should be regarded only as preliminary. Using a recent report of the US Committee on the Biological Effects of Ionizing Radiation, this exposure corresponds to 0.22–1.4 excess fatal cancer cases.

The total amount of energy produced to the end of 1978 in reactors in the Western world was about 170 000 megawatt-years, according to *Nucleonics Week* [29 April 1979]. If one divides the estimated number of man-days lost due to the Three Mile Island accident by the total amount of nuclear energy produced, one obtains $0.8–5.1 \times 10^{-2}$ man-days lost per megawatt-year. These values are about 4% of the corresponding theoretical quantities mentioned above. In turn, these 'experimental' values of public risk are about 0.25–0.5% of the occupational risk.

These calculations do not prove what will happen in the future. However, since the public risk of hydroelectricity was measured on an historical basis, it is of interest to do the same for nuclear power.

NATURAL GAS OCCUPATIONAL RISK

Occupational risk for natural gas seems to be fairly small on the basis of published studies. Results are shown in Table 5. The number of man-days lost per unit energy is 1.9–6.5; this value is lower than either nuclear power or hydroelectricity. The reasons for this result are complicated, but are probably due to the fact that natural gas needs little or no refining of fuel, in contrast to nuclear power, and seems to require fewer materials and less construction labor than hydroelectricity. Note that these values apply only to natural gas, not to liquefied or synthetic gas, which probably have different values both for occupational and catastrophic risk.

NATURAL GAS PUBLIC RISK

Public risk from natural gas tends to be of a less potentially catastrophic nature than from hydroelectricity or nuclear power. That is, the number of people affected tends to be smaller. The 1970–1975 period saw 13 deaths and 69 injuries to the public due to gas pipeline accidents in the USA.[11] This is for transmission and gathering only, not distribution into homes. The distribution of energy is also excluded from the risk analysis of hydroelectricity and nuclear power. If all the natural gas used in the USA had been transformed into electricity, it can be calculated that the number of deaths, injuries and man-days lost per unit energy would be 9×10^{-6}, 4.7×10^{-5} and 0.06, respectively, assuming 50 days lost per injury. The number of man-days lost due to catastrophic risk is then 1–3% of the total occupational risk. If distribution of natural gas had been included, the fraction would have been 10–27%.

Table 5. Natural Gas Occupational Risk per Megawatt-year[14]

Accidental and disease	Gathering and handling fuels	Transportation	Electricity production	Total
Deaths	$3.9–31 \times 10^{-5}$	$2.9–3.4 \times 10^{-5}$	$1.4–5.3 \times 10^{-5}$	$1.6–4.8 \times 10^{-4a}$
Injury and disability	0.004–0.31	0.0017–0.0019	$0.9–2.1 \times 10^{-3}$	0.009–0.36[a]
Man-days lost	0.6–5.0	0.34–0.39	0.17–0.52	1.9–6.5[a]

[a] Includes risk for acquisition of non-fuel material and construction of gas-fired plant and so forth.

These values are historical. No prediction of future risk can be made using them. As stated above, only natural gas, and not liquefied gas, is considered here. Concern has been expressed that the latter form of energy could have large public risk. However, there have been few studies on the subject. An incident related to liquefied gas resulted in the death of a few dozen people in Cleveland, Ohio during the 1940s, but there has been comparatively little analysis of its historical record beyond this. Presumably the public risk of distribution would be comparable to natural gas.

WORST CASES AND AVERAGES

In considering public risk due to catastrophes, it should be noted that worst cases can be hypothesized, in contrast to representative or average values. The example of Italian hydroelectric dams in 1963 was given above.

The Rasmussen report[9] and a predecessor hypothesized large nuclear accidents, which in their estimation would occur extremely infrequently. Using these large hypothetical accidents as a substitute for the representative or historical values used in this paper would be invalid. Morris[12] notes that a comparison between an upper limit, or worst case, for one technology and a representative value for another system would not be a fair one.

CONCLUSIONS

The proportion of catastrophic to non-catastrophic risk for the three energy systems considered appears to be low. Results are shown in Fig. 5. The hatched section of each bar is the proportion of catastrophic risk. Both types of risk are measured in man-days

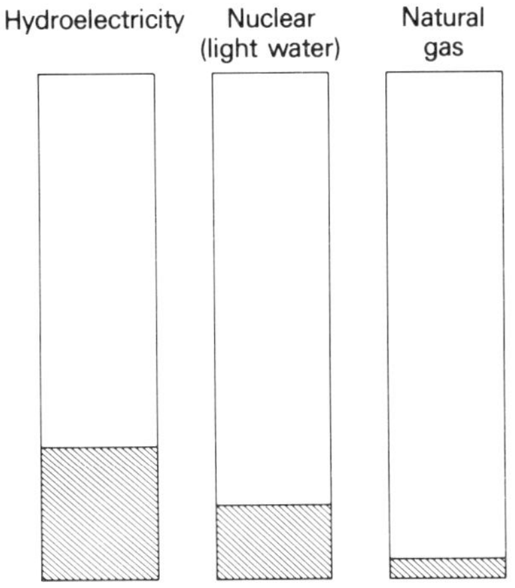

Figure 5. Catastrophic and non-catastrophic risk for three energy systems. The clear portion of each bar represents the proportion of non-catastrophic or occupational risk; the hatched part represents the catastrophic proportion. Each energy system has a range of values for the two aspects of risk. Only the maxima of the catastrophic and non-catastrophic risk were considered for this graph. Fairly similar results would be produced if the minima were considered. Note that the absolute values of risk for the three systems are different. Only normalized values are shown here. Risk for nuclear power deals only with light-water reactors.

lost. As noted in the caption the data refers to the maxima of the bands of uncertainty for each energy system.

The other eight systems, coal, oil, solar thermal electric, solar photovoltaic, solar space heating, methanol, ocean thermal and windpower, whose risk was considered in a recent study[1] appear to have little or no risk of public catastrophe. In some cases, their overall risk can be substantial, but this is primarily due to occupational risk in gathering materials, construction, air pollution from burning fossil fuels, and so forth.

It can then be concluded that for most major conventional or non-conventional energy systems, the proportion of catastrophic to non-catastrophic risk is small or negligible. This, of course, is directly contrary to the impression one would get from reading some newspapers. How what appears to be fact and what is reported can be reconciled is not clear at this point.

The study of risk and consequences is almost by definition an interdisciplinary effort. One needs some knowledge of air pollution, occupational statistics, the engineering of energy systems, history of these systems, radiology and other fields. Combining pieces of information from many sources naturally leads to some uncertainties in these conclusions.

It is likely that the relative risk of energy will become a more important field in the future. While the risk of some non-conventional systems has been studied to a limited extent, that of systems like geothermal, fusion and solar power satellites still await detailed analysis. Study of relative risk will give us some idea of what energy systems are doing *to* us, as well as *for* us.

LITERATURE CITED

1. H. Inhaber, *Risk of Energy Production* Report AECB 1119, Atomic Energy Control Board, Ottawa, Canada (1978). This work contains many references and readers are referred to this document.
2. C. L. Comar and L. A. Sagan, Health effects of energy production and conversion. *Ann. Rev. Energy* **1**, 581–599 (1976). This and Ref. 3 are studies of conventional systems.
3. K. R. Smith, J. Weyant and J. P. Holdren, *Evaluation of Conventional Power Plants*, Report ERG 75–5. Energy and Resources Program, University of California, Berkeley, California (July 1975).
4. W. G. Morison, Director, Design and Development, Ontario Hydro, Toronto, personal communication (December 1977).
5. Ref. 1, Appendix K.
6. J. Darmstadter, P. D. Teitelbaum and J. G. Polach, *Energy in the World Economy*, Johns Hopkins Press, Baltimore (1971).
7. Ref. 1, Appendix D. This was based in large part on Refs 2 and 3.
8. *Energy and the Environment*, Council on Environmental Quality, Washington, DC (1973).
9. *Reactor Safety Study: An Assessment of Accident Risks in US Commercial Nuclear Power Plants*, Report WASH-1400, Nuclear Regulatory Commission, Washington, DC (1975).
10. Ford Foundation, *Nuclear Power Issues and Choices*, Report of the Nuclear Energy Policy Study Group, Ballinger Publishers, Cambridge, Massachusetts (1977).
11. *Accidents and Unscheduled Events Associated with Non-Nuclear Energy Resources and Technology*, Report EPA-600/7-77-016, p. 141. Environmental Protection Agency, Washington, DC (February 1977).
12. S. C. Morris, Comparative Effects of Coal and Nuclear Fuel on Mortality, presented at 137th Annual Meeting, American Statistical Association, Chicago (August 1977).
13. R. Gotchy, US Nuclear Regulatory Commission, private communication (May 1979).
14. Ref. 1, Appendix C.
15. J. Reissland and V. Harries, A scale for measuring risks. *New Sci.* 809–811 (13 September 1979).

The manuscript was received 20 January 1980

Infection
From Worms to Nucleotides

N. W. PIRIE, FRS
Formerly Head of Biochemistry Department, Rothamsted Experimental Station, Harpenden, Hertfordshire, UK

Grant-giving bodies often use the word *timely*, sometimes coupled with the word *promise*, to describe the principles used in allocating money. Research cannot be undertaken on certain themes until technical restraints have been removed, namely until it becomes possible to attain certain levels of cold, heat, magnification or pressure. Even a superficial survey of some aspects of the history of biology suggests that, apart from such physical constraints, *timeliness* is less a statement about a research project, than about the vision and receptiveness of the grant-giving body to which the project is submitted. A symposium with the title 'Of timeliness and promise – sterile and fruitful approaches in the development of biochemistry' (December 1978), gave an opportunity for assembling a few examples, in the general domain of infection, of misjudged timeliness. This is a slightly expanded version of the paper given at the symposium.

'I would use (these weapons) and avenge Patroclus, were it not that flies might settle on his wounds during my absence, breed maggots, and cause decay.' That is Robert Graves' translation of a conversation at the beginning of Book 19 of the *Iliad*. Thetis, who had given her son, Achilles, the weapons, then reassures him that she will embalm the corpse with 'red nectar and ambrosia' so that he need not sit by it to keep off flies, but can take part in the fighting outside Troy. Homer, or the syndicate that goes by that name, made a perfectly clear statement that has remained in the *Iliad* for nearly 3000 years because it accords with common experience in farm, kitchen and tomb.

We may wonder therefore what was new in the experiments published by Redi in 1688.[1] He found that uncovered pieces of meat, from a score of different animal species, produced maggots of the same type, but he found none on covered meat. He claimed no originality; butchers and hunters usually covered meat at that time to keep off flies. In spite of the near-universality of classical education, the point had escaped the attention of scientists and philosophers. They concocted the idea of spontaneous generation. Having bedevilled biology for a millennium, it still caused trouble for two centuries after Redi.

Many points meriting discussion arise from that story. Why have maggots, or worms to use a general term common even after Linnaeus, played such an important role in discussions about the causation of disease, the preservation of food, and the origins of life? What is the nature of discovery? Were Homer and Redi discoverers, because they traced a rational connection between protection from flies and the elimination of maggots, whereas butchers and hunters may merely have been following a ritual or superstition? People who are not experimentally minded 'know' many things – only a fraction of them true. Separating fact from fiction is as truly discovery as noticing something that no one had suspected before.

As Josh Billings remarked 'It ain't what a man dont know that makes a fool of him, it's what he do know that ain't so.' Finally, why do perfectly definite observations, and not just suggestions and interpretations that are later found to have merit, so often lie neglected for decades or centuries? The many examples that crop up in the history of our knowledge of various aspects of even a single subject, such as infection, make it legitimate to wonder whether the concept of 'timeliness and promise', enshrined in many statements from Research Councils, is not bogus. Within broad limits, any project which is within the compass of the physical equipment of the time, is timely if there is an enthusiast eager to pursue it.

WORMS

Until about the beginning of the 19th century, movement was the usual criterion separating living from non-living structures. This attitude is shown by the word *viviparous*, the phrase 'the quick and the dead' and Coleridge's comment that scientists 'could not hear the life of metals asserted with a more contemptuous surprise than they themselves incur from the vulgar, when they speak of Life in mould or mucor'.[2] Worms are indubitably alive because they wriggle; and they are obviously connected with mortality.

In the confused history of the development of ideas about infection – the colonization of an organism by something that multiplies within it – worms have an important place. Fracastor[3] seems to have been the

CCC–0308–0188/80/0005–0312$04.00/1

first to emphasize the significance of multiplication as a means of distinguishing infection from poisoning. Aulus Cornelius Celsus, Rome's greatest medical writer (about 50 AD) may have had a glimmer of this idea; when writing of rabies he used the word *virus*, but for snake poison he used *venenum*. Fracastor introduced the word *fomes* for the material that he suggested was the transmitter of disease from person to person; he recognized that each infection had a specific agent and that this multiplied in the host, but he did not explicitly say that the agent was alive. That generalization had to wait for the 17th century. Then, magnifying lenses, and the urge to use them, became timely.

Reliance on unaided vision is now amazing to us, when so many even before old age makes it hard to read small print, have a lens among our pocket paraphernalia and use it when anything interesting is noticed. Shakespeare probably did not use a lens. One version of his comment on the scabies mite runs:

> . . . a little worme,
> Pickt from the lasie finger of a maide,
> (Romeo and Juliet, Act I, sc. iv).

But Redi used one to study mite anatomy in detail. Extensive use was advocated by Glanville,[4] who wrote in his *Plus ultra* (1668) 'There is an inexhaustible variety of Treasure which Providence hath lodged in Things, that to the World's end will afford fresh Discoveries' and went on to mention the '. . . little Threds and Springs . . .' in an organism specifically.

EARLY MICROSCOPISTS

Athanasius Kircher[5] was already an enthusiastic microscopist. He saw worms everywhere and was largely responsible for what Singer called the 'vermicular obsession' of the 17th century. Some went so far as to attribute all disease, indeed even 'natural' death in man and animals, to worms. Bearing in mind how much of early medicine was developed in warm countries where maggots develop quickly, this was not wholly unreasonable. Nevertheless, Avicenna thought that worms could be beneficial – an interesting forerunner of the occasional use during the present century of fly maggots for removing dead tissue from a wound.

Whether or not worms were responsible for disease, Redi had shown that maggots were not spontaneously generated; others doubted the spontaneous generation of worms elsewhere. Thus Boyle, in *The sceptical chymist* (1661), makes Carneades say '. . . common water . . . will putrifie and stink, and then perhaps too produce moss and little worms, or other insects, according to the nature of the seeds that were lurking in it'. Nevertheless, the idea remained untimely for a further 200 years. Because of a rich accretion of philosophical, political and theological associations, it was discussed with a level of acrimony unusual in scientific debate. Tapeworms dominated an entertaining phase of the wrangle. Ecclesiastical dogma seems to have been clear. No new species were created after the Sixth Day; the spontaneous generation of anything as elaborate as a tapeworm was therefore contrary to Holy Writ; Adam and Eve were perfect before the Fall and therefore could not have had worms or even a tendency to develop them for later transmission. The problem seemed insoluble until the essential intermediate hosts of these parasites were discovered in the 1840s. Interest in the place of worms as the cause or consequence of disease spread into literary circles; Addison, Defoe and Swift comment on worms seriously or humorously.[6]

At about the time when belief in worms and insects generally as causes of disease was beginning to wane, the idea started that they could be vectors. Malaria was associated with insects by Lancisi,[7] and diseases that were probably bartonellosis and leishmaniasis were associated with sand flies by Bueno.[8] These associations were more specific than the vague condemnation of insects by such Roman writers as Varro and Columella – the latter listed biting insects, along with visitors, among the hazards of living beside the main road! These 18th century correlations were about as good as those that started Reed and Ross on the control of yellow fever and malaria. The long interval is interesting. Perhaps it was needed for insects to recover timeliness after the demise of the 'vermicular obsession', perhaps delay was caused by the intellectual difficulty of separating the roles of vector and infective agent.

GERMS

As improvements in microscopy enabled more of the internal structure of the supposedly simple organisms

N. W. PIRIE, FRS, worked in Sir F. G. Hopkins' laboratory in Cambridge, UK, for 11 years on the isolation, metabolism and behaviour of glutathione, methionine and other S compounds; on the antigens of *Brucella*; and on the isolation of plant viruses. He moved to Rothamsted in 1940 primarily to facilitate virus research, but the move enabled work on the extraction of edible protein from leaves to be conducted on a larger scale than was possible in Cambridge. From 1947 to 1973 he was head of the Biochemistry Department at Rothamsted. On retirement, work on plant viruses ended. Work on leaf protein continues – especially on the design of simple and robust equipment that could be used for making it by relatively unskilled people.
Address: 42 Leyton Road, Harpenden, Hertfordshire, UK.

to be seen, it became increasingly improbable that such elaborate structures could simply assemble themselves. By the beginning of the 18th century, most biologists had come to agree with Homer and Redi. But fermentation remained a problem. Leeuwenhoek asserted that the 'beasties' he studied arose from other organisms only; he saw yeast cells, but did not assign them a role in fermentation. Boerhaave[9] 'discovered' that fermentation started more quickly if material from an old brew was added to the wort – brewers had, of course, known this for millennia. Yeast cells were not admitted to the category of living organisms until Cagniard-Latour[10] described their reproduction by budding. Chemists, like Liebig and Wöhler, rejected the idea with contumely and ridicule, but it led to the well-known series of experiments, notably by Pasteur, on the precautions needed to keep fermentable fluids from fermenting. The fact that air carried the particles needed for the initiation of fermentations and putrefactions, started Pasteur on the search for identifiable agents, rather than vague concepts such as contagion and miasma, as causes of disease. In this he was following Bassi who, 30 years earlier,[11] had transmitted *muscardine* to silk worms with a needle, and elaborated from his experiments a reasonably coherent germ theory of disease.

To a large extent many of us still depend on the sewage systems and clean water supplies installed in the middle of the 19th century. Tentative moves toward slum clearance were also made then. Those responsible for these improvements, Simon, Snow,[12] Nightingale, Virchow and others attached little importance to micro-organisms though they were aware of their existence. There is a tendency now to attach too much importance to the purely microbiological consequences of the reforms. The reformers had a more Hippocratic approach and were concerned with the general phenomena of dirt, air, food and the standard of living. As Feuerbach – usually remembered only because Karl Marx attacked him – put it 'Do you want to improve people? Then, instead of preaching against sin, give them better food'.

As so often in the history of science, trouble was caused by false antitheses and the difficulty of distinguishing a necessary, from a necessary and sufficient, condition. Instead of seeing that personal constitution, the nature of the environment and the presence of an infective agent all played a part in the final appearance of an infectious disease, many people argued and still argue, for one factor to the exclusion of the others. As recently as 1847, Semmelweiss was ostracized for telling doctors to wash more often, and washing was counted as one more bit of heterodoxy in Paracelsus' (1493–1541) turbulent career; it has now become timely.

When Lister and Tyndall[13] published articles in the *British Medical Journal* on the suppuration of wounds and the role of airborne germs in human disease, they felt it wise to point out that there was no conflict between their preoccupation with the immediate environment of the patient, and the sanitary reformer's interest in the environment as a whole. They stressed, as Fracastor had done, the importance of multiplication. Lister tried to avoid conflict by saying that it did not really matter whether the 'septic particles' were organisms. However, he thought they were (see Figs. 1 and 2). In the course of a long account of operations done aseptically, he said 'That they are self-propagating, like living beings, and their energy is extinguished by precisely the same agencies as extinguish vitality . . . is certain, and is of the utmost practical importance'.

Tyndall was more doctrinaire and wrote 'Now it is in the highest degree important to know whether the parasites in question are spontaneously developed, or are wafted from without to those afflicted with the disease. The means of prevention, if not of cure, would be widely different in (the) two cases'. The medical profession remained, for a time, unimpressed. At an International Sanitary Conference in 1885, Koch's recent discovery of the cholera vibrio was not mentioned. However, in spite of the scepticism of Pettenkofer and Metchnikoff, who ostentatiously drank cultures supposedly containing the vibrio, the infectivity of cholera was accepted in 1892.

There are few better examples of compartmentalized thinking than the almost complete absence from all this discussion of any mention of what was going on in the kitchen. Sausages were preserved with spices less exotic than Thetis' 'red nectar and ambrosia', but sausages needed only temporary preservation! The use of thin pastry casings is more relevant to the general theme of infection. Every cook knew that an unopened pie or Cornish pasty could be kept for longer than an open stew. Cooks did not, obviously, realize that this was because the dry casing excluded micro-organisms from the moist contents; it is odd that scientists did not think about the matter.

In 1732, C. Carter published 'The Compleat City and Country Cook' (see Fig. 3). This memorably alliterative work describes in some detail what is in essence the modern method of bottling fruit. An anonymous book of receipts had described the bottling of gooseberries in 1680. As a result, 'Gooseberry bottles' were made until late in the 18th century in Europe when restrictions on certain raw materials made it impossible to produce in Britain glass of suitable quality. At the end of the 18th century the experiments of Spallanzani provided an explanation for the efficacy of these techniques. In 1808, Saddington received a five guinea prize from the Society of Arts for exhibiting twelve fruits preserved for two to three years in wide-mouthed wine bottles. In the light of all that, it is hard to see what was so novel about Appert's book[14], which is usually cited as having started the food preservation industry. It is even harder to see why the demise of spontaneous generation did not become timely until after the later work of Schwann, Pasteur and Tyndall. It can be

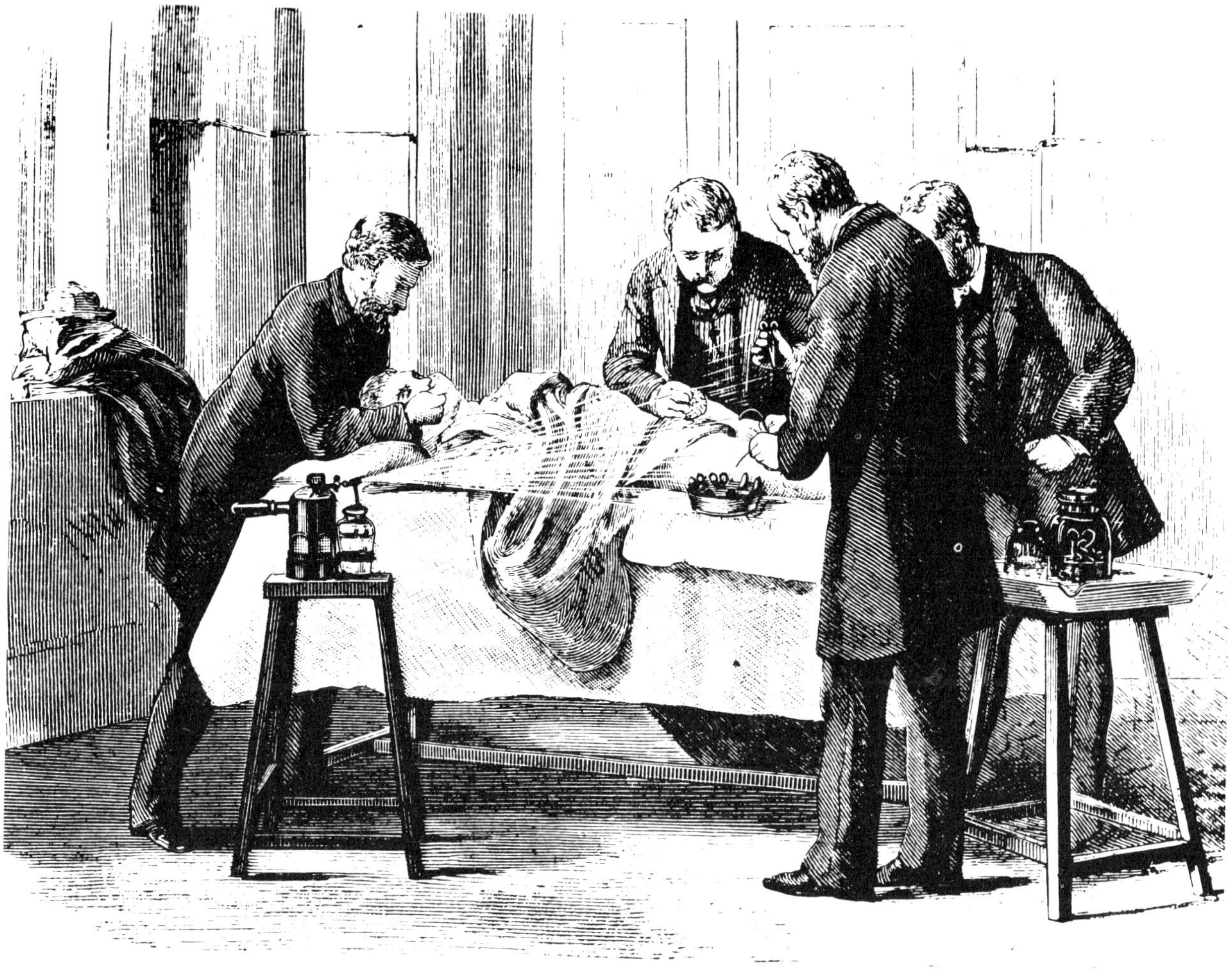

Figure 1. The Lister carbolic acid spray in use. From W. Watson Cheyne, *Antiseptic Surgery*, London (1882). Courtesy of the Wellcome Trustees.

argued that Pasteur and the others did little more than teach scientists what housewives already knew.

VIRUSES

Pasteur, like his contemporaries, used the word virus loosely; when in 1889 his microscope revealed nothing in a fluid transmitting rabies, he postulated an organism similar to those he was familiar with – but smaller. Filters that were supposed to retain bacteria were already used at that time so Loeffler and Frosch,[15] who were trying to make a vaccine to protect cattle against foot and mouth disease, were surprised by the infectivity of a filtrate. By simple arithmetic they demonstrated that they were dealing with an agent that multiplied, and not with a toxin. In the same year, Beijerinck came to the same conclusion about tobacco mosaic virus. Although he was a lifelong friend of the physical chemist van't Hoff, he confused matters by calling the agent 'fluid' and thus set pathologists off on a metaphysical wild goose chase about 'soluble substances' and 'particles' from which some have not yet returned. By accident these agents, and others found soon afterwards, were called ultramicroscopic or filterable viruses rather than bacteria: the designation, without the prefix, has stuck.

Although these viruses multiplied in susceptible hosts, they did not multiply in cell-free culture media. Beijerinck stressed this correlation, as did Landsteiner[16] who dismissed, on the grounds that the medium contained nucleated erythrocytes, the supposed cell-free cultivation of fowl plague virus. The generalization that viruses, because of their absolute dependence on a host, cannot usefully be classed as organisms, did not become timely for a further 40 years. The implications of the generalization have not yet been fully appreciated: viruses are still sometimes introduced into discussions on the origins of life. The dependence of viruses on a host was an untimely idea; equally untimely observations were being made at that time. Copeman and Centanni cultivated viruses in embryonated eggs, and Landsteiner agglutinated erythrocytes with fowl plague, a form of influenza, virus. Both techniques were commercialized a generation later (see Fig. 4).

Bacteria, like plants and animals, vary in composition according to their age and state of nutrition. It is reasonable to expect that the scope for variability will

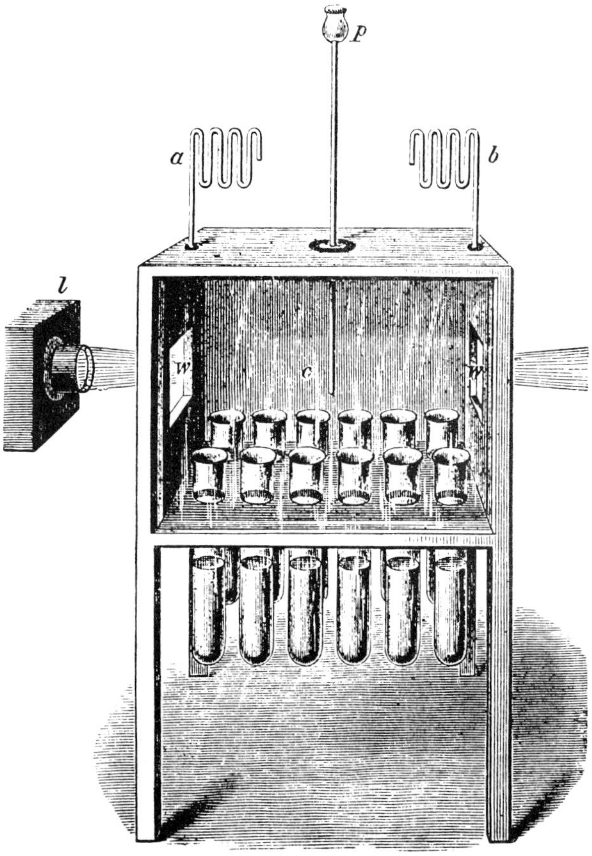

Figure 2. Tyndall's box in which he studied the relationship between contamination of the bacterial culture medium in the glass tubes and the presences of particles in the air. The box contained a glass front window (C) as well as two smaller windows (W) through which a strong light could be projected thus making the particles visible. (l) is the lens mount, (a) and (b) glass tubes to prevent entrance into the box of unwanted particles and (p) funnel to fill the tubes. Reproduced from *Essays on the Floating Matter of Air* London (1883) courtesy of Ann Ronan Picture Library; this illustration was first published by Tyndall in *Philos. Trans. R. Soc. London* **27**, 166 Part 1 (1876).

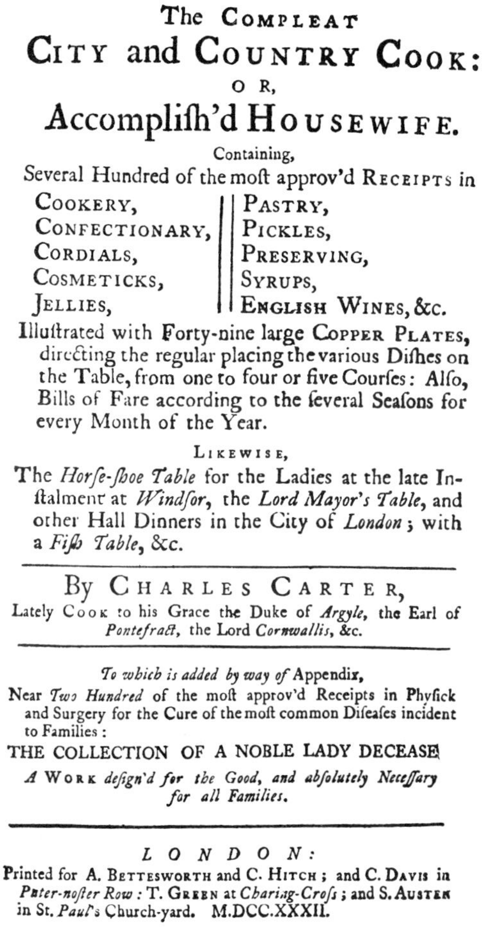

The COMPLEAT
CITY and COUNTRY COOK:
OR,
Accompliſh'd HOUSEWIFE.
Containing,
Several Hundred of the moſt approv'd RECEIPTS in

COOKERY,	PASTRY,
CONFECTIONARY,	PICKLES,
CORDIALS,	PRESERVING,
COSMETICKS,	SYRUPS,
JELLIES,	ENGLISH WINES, &c.

Illuſtrated with Forty-nine large COPPER PLATES, directing the regular placing the various Diſhes on the Table, from one to four or five Courſes: Alſo, Bills of Fare according to the ſeveral Seaſons for every Month of the Year.

LIKEWISE,

The *Horſe-ſhoe Table* for the Ladies at the late Inſtalment at *Windſor*, the *Lord Mayor's Table*, and other Hall Dinners in the City of *London*; with a *Fiſh Table*, &c.

By CHARLES CARTER,
Lately COOK to his Grace the Duke of *Argyle*, the Earl of *Pontefract*, the Lord *Cornwallis*, &c.

To which is added by way of Appendix,
Near *Two Hundred* of the moſt approv'd Receipts in Phyſick and Surgery for the Cure of the moſt common Diſeaſes incident to Families:
THE COLLECTION OF A NOBLE LADY DECEASED.
A WORK *deſign'd for the Good, and abſolutely Neceſſary for all Families.*

LONDON:
Printed for A. BETTESWORTH and C. HITCH; and C. DAVIS in *Pater-noſter Row*: T. GREEN at *Charing-Croſs*; and S. AUSTEN in St. *Paul's* Church-yard. M.DCC.XXXII.

Figure 3. Title page of Carter's famous book *The Compleat City and Country Cook*, London (1732). Courtesy of The British Library.

be smaller, the smaller the infective agent. When it is an obligate parasite, in a sense a metabolic product of the host, it is still more likely to have constant composition. These expectations, coupled with the tendency of viruses to be more resistant to chemical insult than bacteria, made it reasonable to try to purify viruses in the chemical sense. Fowl plague virus was called a globulin by Mrowka,[17] and fowl pox virus a nucleoprotein by Sanfelice.[18] The evidence that Schlesinger[19] published for the nucleoprotein nature of a bacteriophage was much more convincing and prompted Bawden and Pirie[20] to make the cautious comment on Potato virus X '. . . protein is an essential part of virus 'X', but there is no evidence that other equally important substances may not also be present'.

Tobacco mosaic virus was called a globulin by Stanley;[21] we found[22] that it was a nucleoprotein and went on to show that a dozen other plant viruses, or virus strains, were also nucleoproteins. The presence of nucleic acid was first disputed; it was then neglected because of the widespread illusion that nucleic acids were tetranucleotides and so too small to carry much specificity. Our suggestion[23] that the nucleic acid was just as likely as the protein to be important, was interpreted as our nucleic acid obsession in spite of the evidence, marshalled by Avery and his colleagues[24] that the pneumococcal transforming factor was nucleic acid. Nucleic acids began to be considered timely when Hershey and Chase[25] found that more nucleic acid than protein went into a bacterium when it was infected by a bacteriophage. The job was completed when Fraenkel-Conrat and Schramm infected plants with the nucleic acid separated from tobacco mosaic virus. The inoculum had to be stronger than usual, but this relegated protein in viruses to a protective rather than a primary function and gave nucleic acids their present dominant role in biochemistry.

SOME REASONS FOR DELAY

Throughout this account of the growth of knowledge about processes of infection, I have commented on

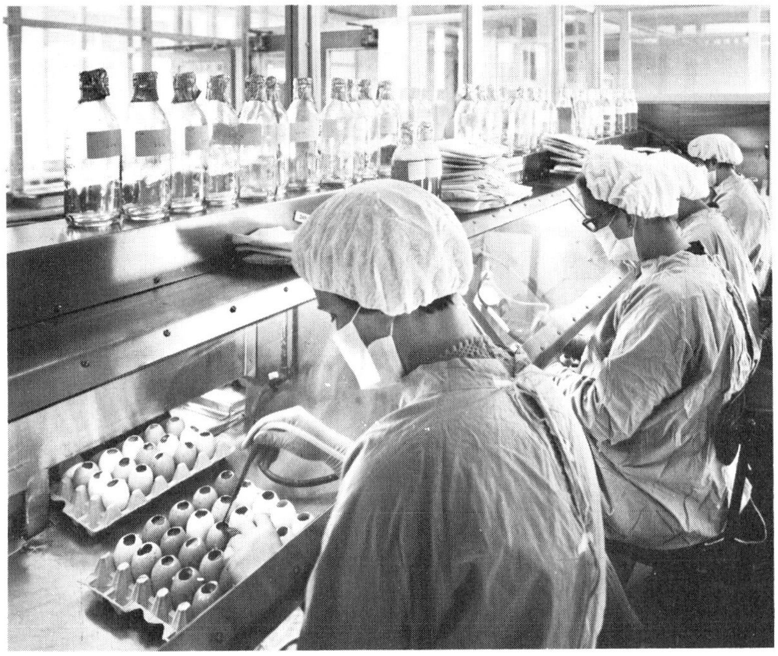

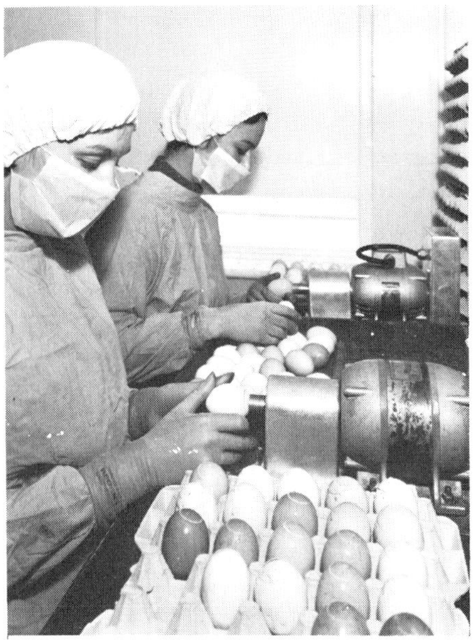

Figure 4. The cultivation of viruses in embryonated eggs. The eggs are inoculated with 0.2 ml of diluted virus suspension into the allantoic cavity, an automatic syringe being used (Left). After 2–3 days incubation, the eggs are transferred to a cold room to chill the embryos. After a suitable period of time, the eggs are swabbed with a meth/iodine solution and a circular cut is made in the shell over the air sac with a trephaning machine (Right). Courtesy of Duncan Flockhart and Company, London E2.

the fact that several lines of research would have been timely decades or centuries before they were pursued. This review may therefore end with some comments on the general topic of delay. Acceptance of ideas and observations may, understandably, be delayed until developments in related sciences have supplied a congenial intellectual environment. To a casual observer it seems that Mendel's observations on inheritance had to await the rise of biometry, Waterston's approach to the gas laws had to await unequivocal evidence for atomicity, and Wegener's suggestions about geodynamics had to await better bathymetric and magnetic surveys. The list could easily be extended.

Popular histories of science contain so many stories of vigorous opposition to ideas which later gained general acceptance, that it is tempting to think that sound novel ideas are sure to be opposed. The thought is sometimes inverted: opposition or neglect has been regarded as evidence for the soundness of an idea. But most novel ideas are wrong. The situation is different with an observation. It may be misinterpreted, but it cannot be wholly wrong. Thus, to take an example connected with the theme of infection, in some communities a priest blesses a new building in which cheese will be made. Part of the ritual involves rubbing it with old cheese. Presumably the ritual works: which is the effective part of it is a matter of opinion and for later experiment. When a technique is observed to work, it is hard to understand why the attempt to find out the reason is so often delayed.

Louis Racine (*fils*) remarked '. . . in order better to humiliate those who cultivate the sciences, God has permitted the finest discoveries to be made by chance, and by those least apt to make them. The sea compass was not invented by a mariner, nor the microscope by a physicist, nor printing by a writer, nor gunpowder by a soldier'.[26] We can add that Priestley was a non-conformist minister, Boussingault a mining engineer, Pasteur a crystallographer, and the first observations on current electricity were made by biologists. This sort of thing annoys specialists. They think, naturally, that advances in their subject should be made by them. The argument goes back a long way in both time and epistemology. Posidonius, who was a philosopher, said that innovations were made by philosophers: Seneca (90th Epistle) said they were made by workmen. In *The fable of the Bees*, Bernard de Mandeville discussed the issue and agreed with Seneca, so did Adam Smith in *The Wealth of Nations*.

When outsiders make suggestions or observations, opposition or neglect are understandable. Disregard for observations made by people with orthodox qualifications, and on subjects later given great importance, is harder to understand. The 'Non-conservation of Parity' is an excellent example with biological implications. The Nobel Prize for physics was awarded to Lee and Yang within a year of their observation of chirality in some atomic processes. Research on the subject was then undertaken with such enthusiasm, or mania, that another physicist had to withdraw five papers because his dream world had intruded on his experiments! Franklin,[27] in a thoughtful discussion of the whole matter, points out that similar observations had been reported, and neglected, 29 years earlier.

Fortuitously, newspaper accounts of Lee and Yang's observation coincided with a New York Academy of Sciences symposium on 'Modern ideas on spontaneous generation' held at the end of 1956. Pasteur regarded the preferential use of only one of the isomers of a chiral molecule as a peculiarly biological phenomenon. During the 1930s it became clear that, although individual members of a group of molecules such as the amino acids rotated the plane of polarization of light in either direction, all members of the group had the same configuration in 3-dimensional space. The origin and significance of chiral preference was therefore regularly discussed in papers on biopoesis. The possibility that it was the consequence of some fundamental atomic asymmetry was therefore an important subject of conversation of those attending the symposium. The sudden excitement about parity in physics laboratories can hardly be explained by this biological interest because that did not emerge in print until 1960.[28] All that Franklin feels able to conclude is that '... not only the physical and logical content of an experiment determines whether it is crucial or not.'

From one point of view, the present-day influence of think-tanks and the increasing control of research by civil servants could be beneficial. It is often easier to get a non-specialist than a specialist to understand the essence of a new idea, or the significance of an old observation – fewer layers of habit and conventional wisdom have to be peeled away. On the other hand, the non-specialist usually submits the matter to a specialist committee and, unless very obstinate, accepts condemnation by the committee. The basic fallacy underlying the British customer/contractor principle of financing research is the assumption that the customers know enough about what it might be possible for them to get, to know what to ask for. That is the sort of knowledge that people who actually handle material and equipment are most likely to have. The point was well put by Mees, who directed research for Eastman Kodak.

'The best person to decide what research shall be done is the man who is doing the research. The next best is the head of a department. After that you leave the field of best persons and meet increasingly worse groups. The first of these is the research director, who is probably wrong more than half the time. Then comes a committee which is wrong most of the time. Finally there is a committee of company vice-presidents, which is wrong all the time.'[29]

That may be an extreme point of view. Nevertheless, it recognizes the fact that developments are most likely to come from people with practical experience, especially if that experience extends to what is happening in factory, farm and kitchen.

NOTES AND LITERATURE CITED

1. F. Redi, *Esperienze intorno alla generazione degl'insetti* (1668).
2. S. T. Coleridge, *Hints towards the formation of a more comprehensive theory of life*, ed. by S. B. Watson (1848).
3. H. Fracastor, *De sympathia et antipathia rerum, liber unus. De contagione et contagiosis morbis et curatione* (1546).
4. J. Glanville, *Plus ultra: or the advancement of knowledge since the days of Aristotle. In an account of some of the most remarkable late improvements of practical, useful learning: to encourage philosophical endeavours. Occasioned by a conference with one of the notional way* (1668).
5. A. Kircher, *Scrutinium physico-medicum contagionsae etc.* (1658).
6. For example: D. Defoe in *A journal of the Plague Year* (1722), J. Addison in various articles in *The Tatler* and *The Spectator* about 1710, and J. Swift, in *Remarks upon a book, intitled, the Rights of the Christian Church asserted, against the Romish, and all other Priests, who claim an Independent Power over it* (1708). Swift wrote of a man who believed that '... all we eat and drink, all the whole consistence of human bodies, and those of every other animal, the very air we breathe; in short, all Nature throughout was nothing but worms: and by that system, he solved all difficulties, and from thence all cases in philosophy'.
7. G. M. Lancisi, *De noxiis paludum effluviis eorumque remediis* (1718).
8. A. Herrer and H. A. Christensen, Implication of *Phlebotomus* sand flies as vectors of Bartonellosis and Leishmaniasis as early as 1764. *Science* **190**, 154 (1975).
9. H. Boerhaave, *Elementa chemiae* (1732).
10. C. Cagniard-Latour, Mémoire sur la fermentation vineuse. *C. R. Acad. Sci.* **4**, 905 (1837).
11. A Bassi, *Del mal del segno calcinaccio o moscardino, malattia che affligge i bachi da seta e sul modo di liberarne le bigattaje anche le piu infestate* (1835).
12. Dr John Snow is best known for the part he played in stopping the cholera outbreak in Broad Street, London, in 1854. A pub on the corner of Broadwick and Lexington Streets in London's West end was renamed 'The John Snow' to celebrate the centenary of that outbreak. It is clear from the historical material displayed in that pub that the local residents disliked the squalor and stench intensely. For example, they wrote to the editor of *The Times* 'Sur, – May we beg and beseach your proteckshion and power, We are Sur, as it may be, living in a Wilderness, so far as the rest of London knows anything of us, or as the rich and great people care about. We live in muck and filth. We aint got no Priviy, no dust bins, no drains no water-splies and no drain or suer in the hole place ...'.
13. J. Lister, Address in surgery. *Brit. Med. J.* 227 (26/8/1871); J. Tyndall, Dust and disease. *Brit. Med. J.* 661 (24/6/1871).
14. N. F. C. Appert, *L'art de conserver pendant plusieurs anneés toutes les substances animales et végétales* (1810).
15. F. Loeffler and P. Frosch, Berichte der Kommission zur Erforschung der Maul–und Klauenseuche bei dem Institut für Infektionskrankheiten in Berlin. *Zentralbl. Bakteriol. Parasitenkd.* **23**, 371 (1898).
16. K. Landsteiner and M. Berliner, Ueber die Kultivierung des Virus der Hühnerpest. *Zentralbl. Bakteriol. Parasitenkd.* **67**, 165 (1912).
17. Mrowka, Das Virus der Hühnerpest ein Globulin. *Zentralbl. Bakteriol. Parasitenkd. Abt. 1 Orig.* **67**, 249 (1912).
18. F. Sanfelice, Untersuchungen über das Epithelioma contagiosum der Tauben. *Z. Hyg. Infektionskr.* **76**, 258 (1914).

19. M. Schlesinger, The Feulgen reaction of the bacteriophage substance. *Nature* (*London*) **138**, 508 (1936) and earlier papers.
20. F. C. Bawden and N. W. Pirie, Experiments on the chemical behaviour of potato virus 'X'. *Br. J. Exp. Pathol.* **17**, 64 (1936).
21. W. M. Stanley, Chemical studies on the virus of tobacco mosaic. VI. The isolation from diseased Turkish tobacco plants of a crystalline protein possessing the properties of tobacco mosaic virus, *Phytopathology* **26**, 305 (1936).
22. F. C. Bawden, N. W. Pirie, J. D. Bernal and I. Fankuchen, Liquid crystalline substances from virus-infected plants, *Nature* (*London*) **138**, 1051 (1936).
23. F. C. Bawden and N. W. Pirie, Virus multiplication considered as a form of protein synthesis in *The Nature of virus multiplication* ed. by P. Fildes and W. E. van Heyningen. Cambridge University Press, Cambridge (1953).
24. O. T. Avery, C. M. McLeod and M. McCarty, Studies on the chemical nature of the substance inducing transformation of pneumococcal types. Induction of transformation by a desoxyribonucleic acid fraction isolated from pneumococcus type III. *J. Exp. Med.* **79**, 137 (1944).
25. A. D. Hershey and M. Chase, Independent functions of viral protein and nucleic acid in growth of bacteriophage. *J. Gen. Physiol.* **36**, 39 (1952).
26. Quoted from *Nat. Resour.* **10**(3), 2 (1974).
27. A. Franklin, The discovery and non-discovery of parity non-conservation. *Stud. Hist. Philos. Sci.* **10**, 201 (1979).
28. J. B. S. Haldane, Pasteur and cosmic asymmetry. *Nature* (London) **187**, 879 (1960).
29. Quoted from *Biograph. Mem. Fellows R. Soc.* **7**, 182 (1961).

The manuscript was received 6 June 1980

Fumigation of the special artist of the *Illustrated London News* when covering the Hamburg cholera epidemic in 1892. From *Illustrated London News*, 17 September 1892. (Courtesy Ann Ronan Picture Library.)

The Sahelian Problem

DR GOBIND T. NANKANI

Development Economics Department, World Bank, Washington, DC, USA

The economic dimensions of the Sahelian Problem, and in particular the case for induced migration and population redistribution, are presented. A list of priorities in the development of the region follows but these are subject to social and institutional constraints.

The Sahelian problem[1] is defined here as a problem of low, and highly variable, agricultural production levels. Beginning with an empirically defined problem one is compelled to confront head-on the enormous complexity of the social world.[2] The study then becomes field-determined, rather than discipline-determined, and therefore an interdisciplinary approach is warranted.

In analysing the Sahelian problem, one can distinguish four major aspects: climatological – is the long-term climate changing; technological – can the ecosystem in these countries support[3] their populations now and in the future; economic – in what directions should efforts be made to utilize the regional resources economically to sustain the basic needs of its population; and institutional – what constraints and opportunities arise from the particular configuration of institutions among the peoples of the region?[4]

We can accept the conclusions of earlier studies[5] that the climate of the region is not changing fundamentally and that technologically it is possible, given sufficient investment, for the region to support its projected population on a sustained basis. That the socio-cultural and institutional limits to otherwise suitable policies are very real is also amply indicated from other studies.[6] Here the entire economic dimension and one specific institutional trait, that of migration, are examined. The need cannot be overemphasized for a more complete study of the institutional limits to the economic policies that are tentatively suggested here.

SAHELIAN ECOLOGY

Sahelian countries are the seven West African nations having land in the Sahel ecological zone: Senegal, Mauritania, Mali, Niger, Upper Volta, Chad and the Gambia. [See Map] The Sahel is one of the four agro-climatic zones found in the Sahelian countries, the others being the sub-desert and desert zones, the Sudan zone and the Guinea zone. The basic distinction is between the sub-desert and Sahel zones on the one hand, and the Sudan and Guinea zones on the other. The former are primarily characterized by a livestock economy carried out by nomadic and semi-nomadic herdsmen, while the latter are inhabited mainly by sedentary farmers.

The salient characteristics of the physical environment in the seven countries are that, due to the climatic, vegetational and soil quality characteristics, the environment becomes increasingly harsh and less productive from south to north; that the ecological balance is fragile; and that, due to the high variation in annual rainfall levels, the environment is highly variable from year to year. In conjunction with the demographic characteristics of the region,[7] these aspects of the physical environment define the Sahelian problem as being one of low and highly variable agricultural production levels. Thus, the most immediate objectives for these countries must be to raise their level of agricultural production, and to offset its variability.

Studies abound[5] of the Sahelian problem and strategies for its resolution. They adopt some common elements in their choice of objectives and of policy variables; much emphasis is given to food self-sufficiency as a major objective, and the major policy variable noted is investment.

However, such a conceptualization of the problem is questionable on at least two counts. Not only has the definition of food self-sufficiency remained ambiguous, but very little attention has been given to an assessment of the trade-off, if any, between food self-sufficiency through domestic production and through trade: the region's comparative advantage is implictly assumed to reside in food production.[8] Furthermore, resource development is enthroned as the *primum mobile* and little attention is given to the possible role of induced migration, the potential for which is enhanced by the traditional mobility of the rural population in the Sahelian countries.[10]

CCC–0308–0188/80/0005–0320$02.00/1

POTENTIAL FOR RESOURCE DEVELOPMENT

The potential, centred primarily on an agriculture and livestock economy, can be briefly summarized. The evidence suggests that there is a substantial potential for livestock development in the Sudano-Guinean livestock area; but a reduction of livestock numbers in the Sahelian livestock area is necessary if further desertification[10] is to be avoided; additionally, a substantial increase in production in dry-land farming is possible both through increases in acreage and the use of more intensive techniques.

Although the potential for irrigated agriculture is significant, its high cost implies that it must, for the present, play a secondary role in agriculture until the yields associated with expanded dry-land production rise to similar levels. Furthermore, settlement possibilities in disease eradicated areas, despite current optimism, seem to be difficult to implement. Finally, a careful study is needed of the economics of the various alternative mechanisms for offsetting the effects of variable rainfall. In the light of these findings, the study of induced migration as a complement to resource development, and the determination of priorities in resource development activities become especially important.

THE CASE FOR INDUCED MIGRATION

The suggestion that induced migration be considered in addition to financial investment in the choice of policy variables cannot only rest on the observation that migration has a long tradition in the area. In particular, estimates of 1970 per capita income in sedentary agriculture by country-zone cell may be compared with the technological resource potential per capita in these cells. The comparison is made by examining the degree of divergence in each country-zone cell, between present resource-utilization, as reflected in income per capita, and potential resource utilization per capita. Such divergence was found to be greatest for Senegal and Chad, lowest for Mali, Mauritania and the Gambia, and moderate for Niger and Upper Volta.

This suggests that accordingly in sedentary agricultural activity, the degree of population redistribution, through migration and/or differential growth,[11] called for between country-zone cells is greatest for Senegal and Chad, lowest for Mali, Mauritania and the Gambia, and moderate for Niger and Upper Volta.

This latter analysis, although crude,[12] reflects an attempt to make migration and population measures endogenous to the planning exercise, and clearly suggests the desirability of further research in this direction.

DEVELOPMENT PRIORITIES

On the basis of an analysis of the land and human resources of the region, as shown by the FAO and AID[5] studies of the Sahelian problem, and from the evidence of the present income levels in the region, some tentative implications for project selection and their priorities may be inferred. The tentative nature of these suggestions must be emphasized, given the paucity of data on the qualitative characteristics of the region's land and human resources. Given these qualifications, the order of priority in the selection and execution of projects, as determined by the logic of maximizing the productivity of investment funds over time, of laying the foundation for longer-term projects, and of meeting the most important basic needs first, would tentatively be as follows:

◇ Reorganization of the livestock economy in the Sahelian livestock area, with the encouragement of marginal migration out of the area.
◇ Extensive dry land agriculture and related rural development projects.
◇ Surveys and studies of river basins, water resources and disease-eradicated areas.
◇ Intensive dry land agriculture, small-scale irrigated agriculture (under partial control) and related rural development projects.
◇ Livestock development projects in the Sahelo-Sudanian Livestock area.
◇ Large-scale irrigation projects.
◇ Agro-industrial projects.

CONCLUDING NOTE

The above observations on the selection and sequencing of projects in the Sahel region suggest the broad directions in which, on grounds of efficiency and equity, future efforts ought to be channelled. However, two points must be made. First, given the high order of magnitude of the costs of the suggested

DR GOBIND T. NANKANI is an economist with the Development Economics Department at the Headquarters of the World Bank. He was a teaching fellow at Harvard University and before then an economist with Data Resources Inc. of Lexington, Massachusetts. He received his BSc from the University of Ghana in 1970 and his PhD in Economics from Harvard University in 1976.
Address: The World Bank, 1818 H Street, Washington, DC 20433, USA.

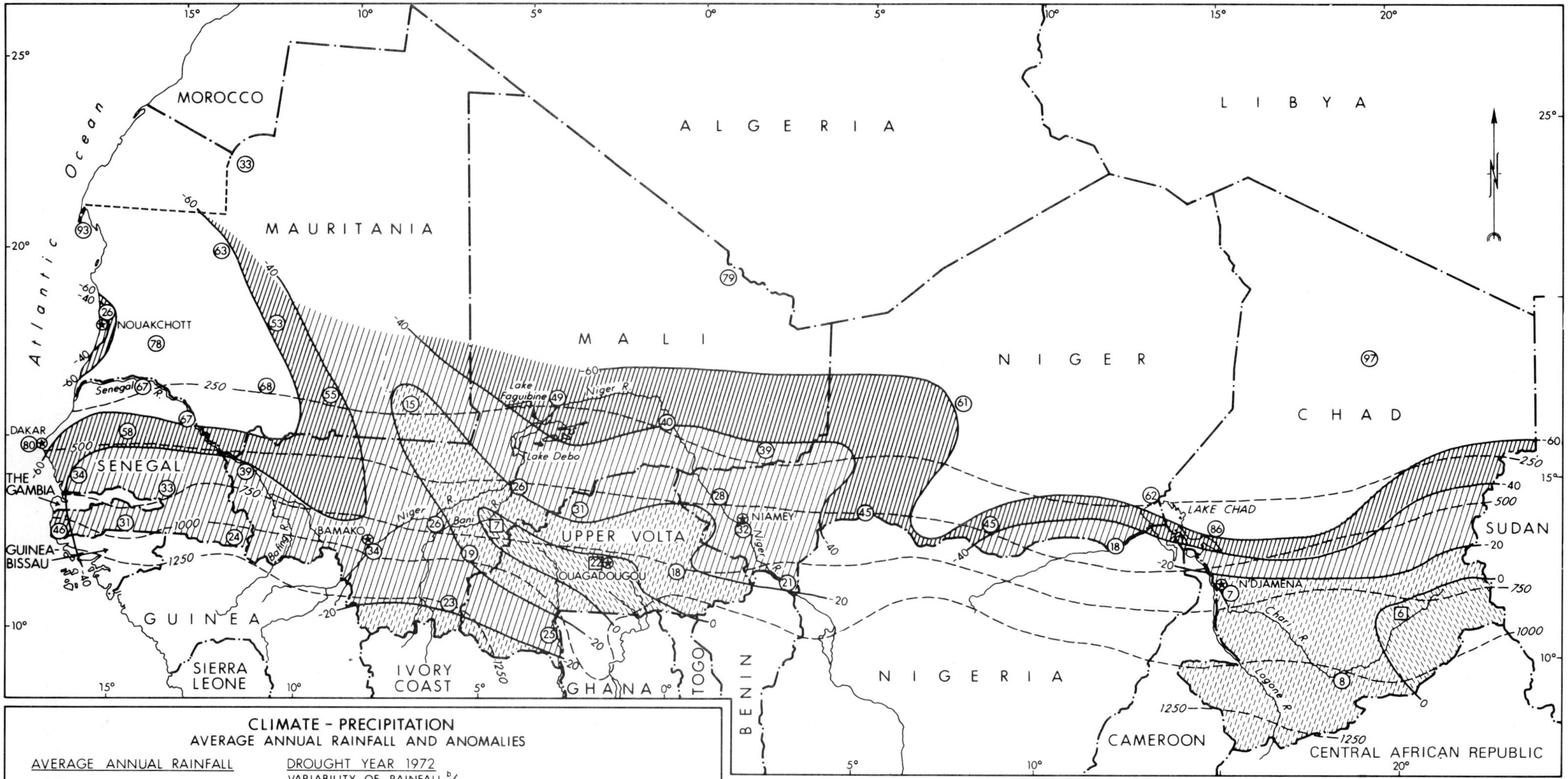

CLIMATE - PRECIPITATION
AVERAGE ANNUAL RAINFALL AND ANOMALIES

AVERAGE ANNUAL RAINFALL

— — — Isohyets, 250 mm interval

NORMAL YEAR

VARIABILITY OF RAINFALL [a]

Area of lowest normal variability from the average rainfall (less than 20%)

DROUGHT YEAR 1972

VARIABILITY OF RAINFALL [b]

(26) Deficit percent from average annual rainfall as recorded by climate station

[7] Excess percent from average annual rainfall as recorded by climate station

Area of greatest deficit but with low population density

-60%
-40%
-20%
0%
+25%

Percent of 1972 deficit or excess from the norm

SOURCES

[a] World Meteorological Organization, 'Surface Distribution of Coefficient of Variability' (Fig. 15), Technical Note No. 86, An Agroclimatology Survey of a Semiarid Area in Africa South of the Sahara, (WMO No. 210 TP 110) Geneva (1967).

[b] The percent values from the normal precipitation, from Yveline Poncet's Map 3, 'Provisional Assessment', Center of Development, Organization for Economic Cooperation (May 1973).

WEST AFRICA - SAHEL REGION

⊛ CAPITALS

RIVERS

INTERNATIONAL BOUNDARIES

INDICATES A PORTION OF THE TERRITORY OF THE FORMER SPANISH SAHARA (WESTERN SAHARA)

0 200 400 600 800 KILOMETRES

0 100 200 300 400 500 MILES

development programme for the region,[13] it is clearly worthwhile to pursue a study of the Sahelian problem under a broadened agenda of available range of choices. This must include particularly the degree of food self-sufficiency and the possible role of induced migration.[14]

Second, it must be emphasized that all the above suggestions are subject to various socio-cultural and institutional constraints which have yet to be studied carefully. As Bauer and Walters state: '(the) repercussions of changes in variables on factors ignored in conventional analysis or treated parametrically raise issues of much intellectual interest and practical significance . . . (the) interactions may differ widely in different contexts but are nevertheless susceptible to observation, analysis and prediction.'[15]

NOTES AND LITERATURE CITED

1. The present note is a highly condensed version of a much larger unpublished report. I am much indebted to Ralph Hofmeister and Shahid Yusuf for their comments. The views expressed here are solely those of the author and do not necessarily reflect those of the World Bank.
2. As Hayek notes, 'the social sciences, like much of biology but unlike most fields of the physical sciences, have to deal with structures of essential complexity, i.e. with structures whose characteristic properties can be exhibited only by models made up of relatively large numbers of variables.' My agreement with Hayek on this point does not, of course, constitute endorsement of any of his other views. See F. A. von Hayek, The pretence of knowledge. *Swed. J. Econ.* 176–185 (June 1975).
3. Defined, say, in terms of a 'core of basic needs'.
4. The region is inhabited by ethnic groups that differ markedly in their socio-cultural traits: using a relatively broad classification one may still distinguish 22 ethnic groups each of which comprises at least 5% of the population of at least one country in the region. The livestock-rearing peoples, comprising 25% of the region's population, belong to six of these ethnic groups, while most of the rest are sedentary farmers.
5. These studies are: FAO, *A Perspective Study on Agricultural Development in the Sahelian Countries, 1975–1990,* Vols. I–III. FAO, Rome (1976); *AID, Proposal for a long-term Comprehensive Development Program for the Sahel,* Vols. I–II. AID, Washington, DC (1976); M.I.T. Center for Policy Alternatives, *A Framework for Evaluating Long-Term Strategies for the Development of the Sahel–Sudan Region,* Vols. I–II M.I.T. Center for Policy Alternatives, Cambridge, Massachusetts (1974).
6. For example, see A. C. Picardi and W. W. Seifert, A tragedy of the commons *Technol. Rev.* 42–51 (May 1976); R. Baker, Innovation technology transfer and nomadic pastoral societies, in *The Politics of Natural Disaster – Case of the Sahel Drought,* ed. by M. Glantz, pp. 176–185 Praeger, New York, (1976); J. Adams, The economic development of African pastoral societies, *Kyklos* 852–865 (May 1975); R. M. A. van Zwanenberg and A. King, *An Economic History of Kenya and Uganda 1800–1970,* pp. 70–109 Humanities Press, Atlantic Highlands, New Jersey (1975); L. Taylor, The misconstrued crisis: Lester Brown and world food *World Dev.* 827–837 (November 1975).
7. The Sahelian countries have a total population of around 25 million, of which approximately 6 million are nomadic and transhuman herdsmen, and most of the rest are sedentary farmers. The rate of population growth is expected to increase over the next 25 years from about 1.5% to 2.5%, resulting in an increase from 25 to about 42 million by the year 2000. Relative to present trends in food production, these figures portend an increasing food deficit. See J. C. Caldwell, *The Sahelian Drought and its Demographic Implications,* Overseas Liaison Committee Paper No. 8, Washington, DC (December 1975).
8. The objective of food self-sufficiency can probably be explained on political and sociological grounds. However, to the extent that they are susceptible to trade-offs, the economic costs involved ought to be considered. The political implications of international trading patterns are excellently analysed in: A. O. Hirschman, *National Power and the Structure of Foreign Trade,* University of California Press, Berkeley (1945); the sociological reasons underlying the desire for food self-sufficiency at the micro-level are well described in: J. C. Scott, *The Moral Economy of the Peasant,* pp. 1–90 Yale University Press, New Haven (1976).
9. A. L. Mabogunje, *Regional Mobility and Resource Development in West Africa,* McGill-Queens University Press, Montreal (1972) and references cited therein.
10. J. Schechter, Desertification processes and the search for solutions. *Interdiscip. Sci. Rev.* **2**, 36 (1977).
11. Population control measures, it is believed, are unlikely to have significant effects before the year 2000; see J. C. Caldwell, Ref. 7, pp. 15–17. Hence the greater emphasis here on migration.
12. It is crude because of data limitations: it does not take account of quality differences in the land resources within the various country-zone cells.
13 That the implied investment costs are high is suggested by the projected costs of the FAO and AID programs: in 1975 prices US $3.7 thousand million or $292 per capita and US $15 thousand million or $600 per capita, respectively.
14. Induced migration is here emphasized because it has become generally recognized that guided settlements have fared better than planned and directed ones, based on the criteria of employment and economic viability. See R. Chambers, *Settlement Schemes in Tropical Africa,* pp. 1–15 Praeger, New York (1969).
15. P. T. Bauer and A. A. Walters, The state of economics, *J. Law Econ.* 1–23 (April 1975).

The manuscript was received 25 May 1978

Mushrooms
Their Consumption, Production and Culture Development

ASSISTANT PROFESSOR DANIEL J. ROYSE and
PROFESSOR LEE C. SCHISLER
Department of Plant Pathology, The Pennsylvania State University, University Park, Pennsylvania, USA

In the last 20 years, mushroom consumption and production have increased at a faster rate than almost any other agricultural food product. The consumption and production of Agaricus (*A. bisporus* and *A. bitorquis*) and the Shiitake (*Lentinus edodes*) mushrooms account for 74% and 14%, respectively, of the total world production and demand for cultivated mushrooms. Since the origin of Agaricus cultivation in France, much progress has been made and the bulk of the scientific and engineering progress has been achieved since the beginning of this century. Mushroom science, a composite science of many disciplines including plant pathology, entomology, horticulture, food science, agricultural engineering, agricultural economics and rural sociology, has contributed a great deal to the worldwide growth of the industry. Future improvements in mushroom culture will come not only from a more thorough understanding of single facets within the overall process, but from a more complete understanding of the interrelationships of these facets with each other.

The mushroom industry has grown at a rapid rate since the 1950s. Mushroom consumption and production have increased not only as the population increased but also on a *per capita* basis, especially in the marketing of fresh mushrooms. These gains have come at a relatively rapid rate due to promotional efforts and increase in the 'private expenditures' sector of the Gross National Product (GNP).

The farm value of the mushroom crop in the USA, for example, reached $360 million in 1979 according to Porter.[1] Ten years ago the comparable figure was only $68 million. In 1969 the index of producer prices was about half the 1979 figure. The value of the US crop increased fivefold while most standards for measuring economic change merely doubled. Production at 205 million kg increased 2.5 times in the same period. These figures give a relative picture of how great has been the demand for mushrooms in the past 10 years. It is a performance seldom duplicated in agriculture today.

CONSUMPTION

Consumption of cultivated mushrooms has increased steadily during the last 30 years. Agaricus (*A.*

Figure 1. The common mushroom, *Agaricus bisporus*, accounts for over 70% of all edible mushrooms produced and consumed. The United States, France, and the Federal Republic of Germany together are responsible for over 50% of the total world consumption.

Figure 2. The Shiitake mushroom, *Lentinus edodes*, is commonly produced on logs. Japan is the primary producer and consumer of this species. The Shiitake accounts for approximately 14% of the total world production of edible mushrooms.

CCC–0308–0188/80/0005–0324$04.50/1

bisporus and *A. bitorquis*) (Fig. 1) consumption accounts for nearly 75% of the total world demand. The Shiitake (*Lentinus edodes*) (Fig. 2) mushroom accounts for an additional 14% of total world consumption. These two mushrooms together make up the bulk of world consumption, accounting for nearly 90% of the world total (Fig. 3).

During the last 14 years, consumption of Agaricus has increased considerably in several countries (Table 1). *Per capita* rate of increase per year varies from a low of 24 g in New Zealand to a high of 134 g per person per year in West Germany. The 16 countries listed in Table 1 consume over 80% of the total world demand for both processed and fresh Agaricus.

The USA, West Germany and France together are responsible for over 50% of the total world consumption (Table 2). West Germany's consumption has increased at a rate of nearly twice that of other major countries. The reasons for this phenomenon are not clear although it is thought to be

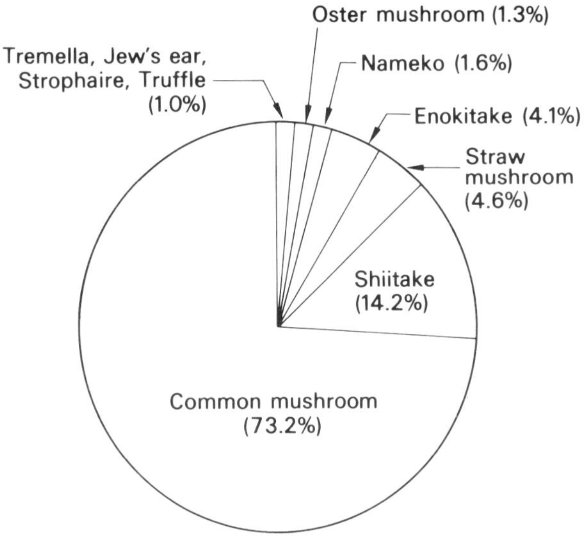

Figure 3. World production of edible mushrooms according to common name.

Table 1. Total Consumption (Fresh Equivalent) of Agaricus in grams *per capita*[a] and Rate of Increase in Consumption per person per year

Country	1960	1965	1970	1972	1974	1980[b]	Rate of increase per year
West Germany	150	530	1130	1700	2020	2550	134
Belgium	440	730	880	930	1560	2000	64
Denmark	390	780	960	1070	1350	1800	62
France	700	850	1020	1420	1440	2100	60
UK	300	470	740	880	980	1380	50
Italy	40	140	370	590	700	1050	48
Netherlands	110	500	560	860	950	1350	57
Sweden	220	410	550	840	1000	1460	53
Switzerland	330	460	920	1140	1340	2250	74
Spain	20	80	120	180	400	1180	30
Austria	140	360	440	590	760	—[c]	40
Ireland	100	130	360	460	620	—[c]	36
Australia	140	200	400	480	660	1000	36
New Zealand	90	200	340	350	440	—[c]	24
Canada	250	570	750	1290	1450	2000	84
USA	290	330	510	680	730	1200[d]	34

[a] Ref. 2.
[b] Estimate.
[c] Data not available.
[d] Actual consumption in 1979.

related to the 'private expenditure' sector of the Gross National Product.[2] According to Engel,[3] the demand for deluxe food products follows a sigmoid curve when consumption is plotted against private expenditure (Fig. 4).

Delcaire[2] found that by actually plotting the *per capita* 'private expenditure' against the total consumption of Agaricus, a curve could be drawn. He concluded that: (i) until \$1000 the *per capita* consumption is always lower than 200 g per year; (ii) from \$1000–2500 consumption varies between 200 and 600 g per year; and (iii) over \$2500 consumption varies from 600–2000 g per year. Apparently this finding does not apply to Japan, as this country mainly consumes the Shiitake mushroom.

DR DANIEL J. ROYSE is Assistant Professor of Plant Pathology at the Pennsylvania State University. Born in Illinois in 1950, he studied zoology and plant pathology at universities in his home state and after teaching there, he accepted his present position in 1978. He has lectured and published widely in his field and has received grants, awards and honors from a number of American institutions.

DR LEE C. SCHISLER is Professor of Plant Pathology at the Pennsylvania State University and has been Director of its Mushroom Research Center since 1968. He has worked on many mycological research projects and lectured about them in Canada, the U.K. and Australia as well as holding many honors of American mushroom institutions. His principal areas of research now include mushroom composts, diseases of mushrooms and mechanized systems of mushroom production; he has taught and published extensively in these fields.

Address: Mushroom Research, Department of Plant Pathology, College of Agriculture, The Pennsylvania State University, 211 Buckhout Laboratory, University Park, Pennsylvania 16802, USA.

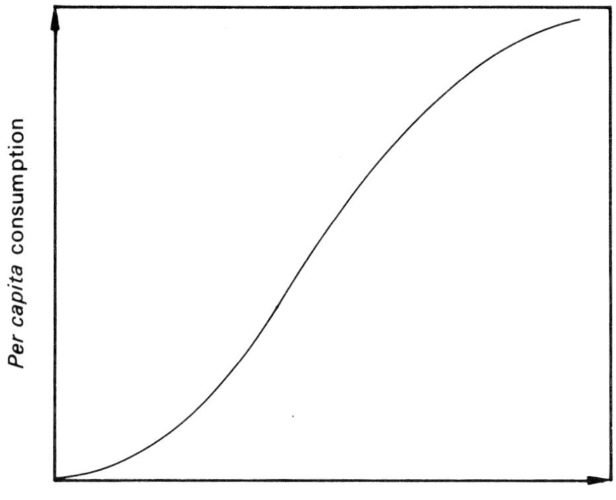

Figure 4. *Per capita* consumption of cultivated mushrooms vs *per capita* private expenditures portion of the Gross National Product.

NUTRITIVE VALUE

A limited number of studies of the mushroom's nutritional value have been conducted; the majority have resulted in general statements that compare mushroom protein with meat and other vegetables.[4] Food analysts have tended to dismiss the mushroom as unworthy of serious consideration as a dietary factor in nutrition.[5] In fact, most textbooks and tables of nutritive values class mushrooms with spices and condiments used primarily for flavoring purposes. We contend this is unfortunate, and that the increasing availability and use of mushrooms justifies a closer look at their potential place in the average diet.

In the absence of animal feeding studies, the most reliable method of estimating the nutritive value of mushrooms appears to be a measure of the content of essential amino acids,[4] and several methods can be used for determining the nutritional value of foods based on their content of essential amino acids.[6,7] Crisan and Sands proposed the use of a Nutritional Index (NI) calculated as:

$$\text{Nutritional Index} = \frac{(\text{Essential Amino Acid Index} \times \text{percentage protein})}{100}$$

where the Essential Amino Acid Index is calculated as a ratio of the essential amino acids contained in a particular food relative to the essential amino acid pattern based on known adult human dietary requirements.[6,7]

The Nutritional Indices (NI) for several foods in comparison with two species of mushrooms are presented in Table 3. *Agaricus bisporus* ranks above all vegetables except soybeans and spinach. Even the lesser nutritive Shiitake mushroom ranks above corn, potatoes, turnips, tomatoes and carrots in its NI score. *Agaricus bisporus* can therefore be considered to provide nutritive value comparable to several high protein foods. Between 70 and 90% of the protein present can be easily digested from mushrooms.

Table 2. Percentage of Total World Consumption of Agaricus for the 16 Richest Countries[a]

Country	Percentage of total consumption
USA	23.4
West Germany	18.9
France	11.3
UK	8.2
Italy	5.8
Canada	4.9
Belgium	2.3
Spain	2.1
Netherlands	1.9
Ireland	1.3
Sweden	1.3
Switzerland	1.3
Denmark	1.0
Austria	0.8
Australia	0.3
New Zealand	0.2
Total other countries	15.0

[a] Ref. 2.

PRODUCTION

Fresh and Processed Mushrooms

Fresh mushrooms, Agaricus, account for a larger percentage of the total world production than do processed mushrooms (Fig. 5). Europe, which produces 31% of the world total, has a larger portion of production sold as fresh product than as canned product. The Far East, however, which produces

Table 3. Comparison of Nutritive Value of Two Species of Mushrooms with Various Foods[a]

Nutritional indices	
59	Chicken
43	Beef
35	Pork
31	Soybeans
26	Spinach
25	Milk
22	Mushrooms (*Agaricus bisporus*)
21	Kidney beans
20	Peanuts
17	Cabbage
14	Cucumbers
13	Mushrooms (*Lentinus edodes*)
11	Corn
10	Turnips
9	Potatoes
8	Tomatoes
6	Carrots

[a] Ranking based on calculated values against the Food and Agriculture Organization Reference Protein Pattern.[4]

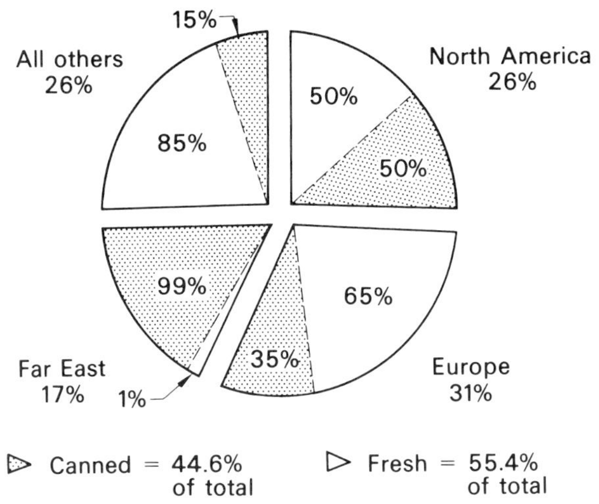

Figure 5. World production of fresh and canned Agaricus (*A. bisporus* and *A. bitorquis*).

approximately 17% of the world total, cans 99% of all mushrooms grown. These mushrooms are exported primarily from China and Korea with a large percentage going to the USA.

As a result of mushroom imports, the US mushroom industry is steadily changing from processing mushrooms to producing mushrooms for the fresh vegetable market. In 1966–1967 for example, fresh market sales accounted for only 25% of total sales in the USA. In 1978–1979 50% of the total mushroom sales were for the fresh market. This trend will undoubtedly continue for the next few years.

History of Mushroom Production

The cultivation of mushrooms (*Agaricus bisporus*) originated in the Paris region in France. Melon growers in this region discovered how mushrooms could be grown and started cultivating them in 1650.[8]

They observed that mushrooms were produced on spent compost used for producing melons. Hot beds used for melon cultivation contained manure which provided the heat necessary for forcing melons. These melon producers found that more mushrooms grew if the manure was sprayed with rinse water used to clean mature mushrooms. They also observed that only mushrooms coming from the melon beds could be grown in the garden on a pile of donkey or mule manure watered with mushroom rinse water. 'The mushrooms from the meadow cannot be made to grow in the garden,' according to Bonnefons.[8] It was then established that there was a distinct difference between mushrooms growing wild in meadows and cultivated mushrooms.

Nearly 60 years later Tournefort, a French botanist, described the first method of culture of mushrooms. He concluded that mushrooms could be shown as 'originating from the horse' and believed that mushroom spores were present in their natural state in horse manure. His cultivation method consisted of placing portions of horse manure covered with fluffy mold into a bed of horse manure and covering it with a layer of soil. In most cases mushrooms eventually appeared and were harvested.

In the 1780s Chambry, a French gardener, observed that *A. bisporus* could grow without light, and that very favorable conditions for growing mushrooms prevailed in subterranean tunnels and caves. As a result of this discovery successful culture was undertaken inside the numerous caves that were excavated for building stones and for gypsum. The caves presented, from the climatic point of view, several advantages over the previous growing conditions in open air. Factors such as temperature and relative humidity were much more constant in caves compared with above ground conditions.

From France, mushroom culture spread to other parts of the world. By 1825, the first mushroom crops were being produced in caves in Holland. In 1865, mushroom culture entered the USA via England and the first mushrooms were grown on a small scale; by 1870, the mushroom industry had begun to develop.

Development of the Mushroom Industry

In the early 1900s mushroom growing started to receive attention from horticultural specialists and mycologists.[9] Progress from the primitive culture to the more highly technical controlled methods of today has increased most rapidly since World War II, and especially since 1960 (Fig. 6). Laboratories for research on mushroom growing were established by Government experimental stations in the USA as well as in many countries of Europe, and in Japan, Taiwan and Southern Korea. Research is also conducted in facilities funded and operated by the larger growers and several spawn companies.

While the French continue to grow in underground quarries, most growers in other countries, with a few notable exceptions, have built above-ground houses.

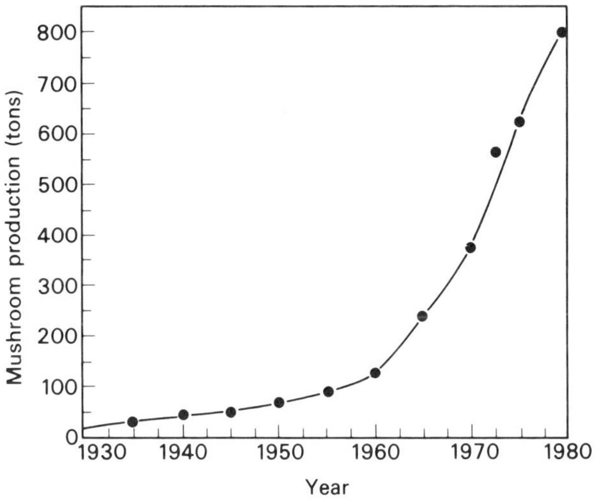

Figure 6. World production of Agaricus (fresh equivalent) since 1935.

Figure 7. A fully mature, open mushroom showing the gills on the underside of the cap.

These mushroom houses have evolved from simple uninsulated buildings heated with dry heat and ventilated with convection currents, to insulated vapor-proofed buildings with air conditioning equipment designed to provide controlled temperature, relative humidity and air exchange. Mushroom growing has become nearly independent of outside climatic conditions, with modern structures and equipment providing conditions most favorable to the mushrooms during the various phases of the growing cycle.

SPAWN

To understand spawn, spawning and spawn growth, the mushroom itself must be understood. Mushrooms are fruits of the fungus plant and consist of two main parts, the cap and the stem. As the mushroom matures the cap opens and the gills are exposed (Fig. 7). Mushroom spores are produced in the gills. Spores are microscopic spheres roughly comparable with the seeds of higher plants (Fig. 8). These spores are produced in large numbers in the gills. A 7.5 cm mushroom produces as many as 40 million spores an hour. Since spores germinate and grow into a mycelium rather unpredictably, they are not used to 'seed' mushroom compost. Spores will germinate and grow into a thread-like mycelium, which a laboratory producing spawn uses. Mycelium from the spores is placed onto steam-sterilized grain and, in time, the mycelium completely grows through the grain. This mixture of grain fully grown through by the mycelium is spawn, and spawn is used to 'seed' mushroom compost.

Spawn was first obtained by digging up wild mycelium from meadows where wild agarics grew and horses were active. The early methods of finding, digging up and perpetuating the wild mycelium were uncertain and unreliable. The first manufactured spawn combined a mixture of horse and cow manure pressed into bricks, dried and inoculated with mycelium from similar bricks, the original source being the wild mycelium (Fig. 9). The bricks were not sterilized, however, and could harbor pathogens and weed molds.

In 1902, Ferguson[10] described the process of spore germination in *A. bisporus* and Duggar[11] reported culturing mushroom pileus tissue on sterile media. Even so, it was 1915 before pure culture spawn was produced on sterilized horse manure compost in bottles. The development of horse manure spawn, however, did not eliminate the problem of obtaining productive, reliable spawn, free from pests. In 1930, The Pennsylvania State College (University) employed Dr James W. Sinden to work on mushroom production problems. During one of his first series of experiments Dr Sinden 'was struck with the variation in yields and feared that this might be a result of the spawn planted'.[12] He realized that in order to carry on further experimental work on mushroom growing, it was essential to eliminate any variation from this cause. Therefore, Dr Sinden sought a medium on which the mycelium would grow more vigorously and

Figure 8. Scanning electron micrograph showing mushroom spores (magnified 5000 times) borne on basidia. The basidia line the gills of the mushroom.

Figure 9. Mushroom spawn of *Agaricus bisporus* used to seed growing beds and trays: (a) first manufactured brick spawn used prior to 1930; (b) grain spawn developed in 1930 and used almost exclusively today.

that would provide a more uniform product. One of the first of these used was grain, specifically wheat, which was placed in flasks with a small amount of water and heat sterilized. On introduction of the mycelium he found that 'it grew very vigorously and in a manner entirely different than anything I had previously seen.' Thus, grain spawn was born and is used almost exclusively today.

During the same time period, Sinden began a culture collection of *A. bisporus* at Penn State now known as The Pennsylvania State University Mushroom Culture Collection. Most of the strains deposited in the collection are selections from an on-going testing program using mushroom tissue, single spore or multispore germinations.[13] Some are vegetative transfers of cultures obtained from elsewhere. From this culture bank 500–600 transfers a year are sent to domestic and overseas spawn laboratories in response to specific requests. It is estimated that over half of the commercial mushrooms produced in the US and Canada and a substantial quantity in a dozen other countries originate directly or indirectly from spawn multiplied from these stock cultures.

Also, 425 isolates of 203 different species of wild fleshy fungi are maintained by periodic transfer or in liquid nitrogen storage. Tissue isolations of mushrooms collected in British Columbia and Pennsylvania form the basis of this culture collection. In some instances, vegetative cultures are backed up by spore prints. Major collection emphasis is on members of Agaricaceae, Boletaceae and Morchella. Cultures are available for research on vegetative growth and fruiting initiation with an aim toward potential commercial cultivation.

COMPOST AND COMPOST SUPPLEMENTS

All commercial mushroom growing has depended on a composted material. Although the time and nature of the composting process has been modified, no one has been successful in eliminating it from the preparation of a suitable medium for the mushroom bed. In compost suitable for mushroom growing, other fungi with few exceptions either do not grow at all or only slowly in comparison to the mushroom. The objectives of composting are therefore to create a compost which has sufficient nutrients in it to grow a good crop of mushrooms and at the same time provide little or no nutrition for other fungi and competitor organisms.

The nutrients for mushroom growth are provided in composts which are traditionally prepared from horse manure and wheat straw. However, compost may have as its base pig manure, hay and straw mixtures, or hay and rough ground corn cobs, or any combination of these. The finished compost ready to plant should have a nitrogen (N) content of 2–2.5%, and to reach such a level, N-rich supplements must be added. Inorganic N-supplements can be added, but only to provide part of the necessary amount. Chicken manure is a common source and other organic sources include brewers' grain, malt sprouts and oilseed meal. Potassium may be deficient in hay–straw combinations and can be added as KCl or K_2SO_4. The only other essential addition is gypsum which is added at 3–5% of the dry weight; this prevents greasiness and improves the compost's physical characteristics.

Composting is a procedure involving two interdependent processes: phase I (Fig. 10) where the raw ingredients begin to break down, and phase II, the cookout as some growers call it, or pasteurization process.

Phase I occurs outside, or in a covered area. It can also occur in a drum or composting device but commercial applications of such have been limited. This phase is characterized by building the raw ingredients into piles. These piles are then periodically turned, watered and formed. This mixing and watering tends to equalize the environment so as to promote a uniform breakdown of the ingredients.

Compost ingredients harbor a great number of microorganisms. With the addition of water to the dry ingredients and the building of a pile, these microbes grow and reproduce and their growth requires: (a) suitable temperature; (b) adequate moisture; (c) sufficient oxygen; and (d) available food, which is supplied by the bulk compost ingredients and the compost supplements. The addition of nutrients at this time is directed therefore toward the feeding of microbial populations in the compost rather than direct nutrient addition for the mushroom. Microbial growth causes the temperature of the pile to increase and much of the nitrogen present is ammonified. After the temperature within the pile reaches 68 °C carmelization takes place, thus eliminating water from carbohydrates and thereby concentrating this

Figure 10. Composting machine used at The Pennsylvania State University's Mushroom Test Demonstration Facility. The machine is used to turn, water and supplement raw materials in the initial stage (phase I) of the compost process.

valuable food for the mushroom. This process is responsible for the darkening in color of the compost. The microbes use water, readily available carbon, nitrogen and oxygen and as long as enough of each of these is present, the pile will continue to generate heat.

The first phase is complete as soon as the raw ingredients have become pliable, are capable of holding water, the odor of ammonia is sharp and the dark brown color indicates carmelization has occurred.

Phase II of composting is carried out inside an insulated building where the conditions in the compost and in the ambient atmosphere can be controlled. Phase II has two main purposes: the final conditioning of compost so that it becomes mushroom specific through the elimination of ammonia and readily available carbohydrates, and pasteurization. Through the proper manipulation of temperature and ventilation these two primary objectives can be realized.

Once the building is filled, the temperature of the compost increases through microbial growth. The heating of the compost causes cool air surrounding it to be drawn towards the heat and this action assures adequate air movement for microbial growth. The increasing temperature of the compost causes pasteurization which requires an air and compost temperature of 60 °C for a minimum of two hours. Once the compost temperature has increased to this, either steam is introduced into the room to raise the air temperature to the same point or the room is closed tightly to allow the compost to heat the air.

If the compost and all other exposed surfaces are not subjected to this temperature, pasteurization will not be completed, leaving harmful organisms to reproduce and grow and subsequently these can cause difficulty in the mushroom crop. Once pasteurization is accomplished, cool air is introduced into the room to: (a) assure adequate oxygen; (b) attempt to hold compost below 60 °C; (c) help dissipate ammonia. Conversion of ammonia to microbial protein is best accomplished by thermophilic organisms, such as Actinomyces (*Thermomonspora*, *Streptomyces* and *Thermoactinomyces* spp.), fungi (*Humicola*, *Torula*, *Chaetomium*, and *Mucor*), thermotolerant mesophilic fungi – *Aspergillus fumigatus*. This process incorporates the ammonia into microbial cells and ultimately makes it available to the mushroom via enzymatic activity. These microbes grow best at temperatures between 46–60 °C and the longer the compost is kept within this range, and all factors are available in sufficient quantities, the faster the ammonia will be converted. Air is introduced into the house or room at a rate sufficient to cool the compost to this range. Once the compost has cooled to below 60 °C, the amount of ventilation must be regulated so as to keep the rate of cooling no greater than 2 °C per 24 hours. This will assure at least 96 h in the desired temperature range of 46–60 °C. The thermophiles should be able to convert the ammonia in 72 h in this temperature range. Once the smell of ammonia is gone, the compost temperature can be dropped to 24 °C for spawning.

DELAYED RELEASE NUTRIENT SUPPLEMENTS

In the early 1960s yield increases were observed when compost was supplemented with noncomposted protein- and/or lipid-rich materials at spawning, at casing and later. Schisler and Sinden[14] obtained a 10% crop increase when small amounts of protein supplements were added to the compost at spawning. Excessive heating during the spawn growth period and strong stimulation of competitor molds present in the compost drastically limited the rate of supplementation and corresponding benefit. Overheating during the spawn growth period may cause mycelial injury and render it nonproductive. To circumvent this competition, supplementation was tried at casing.[15]

Results showed that much greater amounts of nutrients could be added and yield increases were almost proportional to the amount added. Although yield increases as high as 100% have been realized, certain potential problems and limitations exist for supplementation at casing. Weed molds, nematodes, mites and disease-causing organisms must not be present in the compost when supplementing at casing. They would be dispersed throughout the compost when it is fragmented prior to supplementation and could multiply very rapidly before the mushroom mycelium recovered its growth. Nematode infestations are especially to be avoided.

The supplemented compost has a tendency to produce heat in amounts far above that of the undisturbed compost. The increased thermogenesis is probably a result of the increased metabolic rate of the mushroom mycelium in the presence of the supplement. Unless the air around the compost can absorb the heat as rapidly as it is produced the temperature can rapidly rise high enough to injure or kill the mushroom mycelium. Maximum thermogenesis occurs usually about the third or fourth day after supplementation and casing. Most commercial mushroom growing operations are not equipped with air-moving facilities sufficient to cope with removal of the heat developed.

Carroll and Schisler[16] reasoned that an ideal supplement for use at spawning would be unavailable to the compost microflora and the mushroom for about 18 days and then would gradually become available to the mushroom mycelium. The competition by foreign molds should be minimized with supplement that is unavailable until the compost is colonized by the mushroom mycelium. Gradual availability would minimize excessive heat production by the mycelium allowing higher quantities of supplementation and extend the period of yield enhancement past the first few flushes. It was these objectives that led Carroll and Schisler[16] to develop delayed release nutrient supplementation. The disadvantages associated with the supplementation of noncomposted nutrients to mushroom compost at spawning were largely overcome by encapsulating microdroplets of vegetable oil within a protein coat that was denatured with formaldehyde. Increases in mushroom yield of 60% were obtained. Delayed nutrient release was indicated by prolonged stimulation of yields beyond the first few flushes. Additional research on compost supplementation and mushroom nutritional requirements will undoubtedly lead to increased yields in the future.

FUTURE OUTLOOK

Consumption and Production

In many countries the *per capita* consumption of mushrooms is increasing steadily. Where mushrooms are considered as part of the national diet, for example Canada and some European countries, consumption exceeds 2 kg per person per annum. In the UK it is estimated that the figure is about 1.5 kg and in the USA about 1 kg per person per annum.

In 1973, the National Economic Development Organization on UK farming and the Common Market, published a report which indicated that labor is the major single item of cost in the production of mushrooms in the UK.[17] Labor accounts for 40% of the total costs, half being for harvesting and packing. Compost preparation and materials are estimated to cost 20–25% of the total expenditures. These figures suggest that the industry will need continued evolution of culture systems to stay competitive with other vegetables. Improvements in composting and harvesting will be necessary to contribute not only to standardization and control but also to the economics of culture.

Role of a Computer

The main use of a computer in the future of mushroom production will undoubtedly involve process control. There are a number of sensors used in mushroom production that can feed their signals into a computer. These sensors include thermocouples, hygrometers, pressure gauges, position sensors, gas sensors, flow meters and so forth. The values assembled by the translation of sensor signals can be processed according to programs written by the operator or by 'canned' programs that may become available through a 'master library'. From a computer a number of control signals may be generated for other equipment controlling motors, valves, lights, dampers and others.

The strength of a computer is its capability to handle large amounts of data or to process the data at great speed. Due to the energy intensive nature of mushroom production the use of the computer to control various production processes may well determine the economic status of a mushroom grower and the mushroom industry or both in the future.

The use of computers in the mushroom industry will definitely increase in the years to come. A number of medium-sized growers will use the computer to assist them in better financial and statistical work to help stay on top of the economic situation. By carefully processing the large amount of continuously updated figures to supply statistical data, information on important parameters in management and finance will give the grower early notice on developments in his business which require his attention.

Mushroom science, a composite science of many different disciplines has contributed a great deal to the worldwide growth of the industry. One example of this is the Mushroom Test Demonstration Facility (MTDF) of The Pennsylvania State University constructed in 1969. The MTDF, which constitutes a systems approach to mushroom growing, was an interdisciplinary research project involving four departments in the College of Agriculture: Agricultural Economics and Rural Sociology, Plant Pathology, Entomology, and Agricultural Engineering.[18] The MTDF has a flow type design which allows the introduction of the composting material at one end and removal of the harvested mushrooms at the other end. Automatic controls for the entire facility are located in panels in a central control room.

Environmental control, mechanization and automation provide a means of improving efficiency of mushroom production and reducing the labor requirement. Some of the unique features of the MTDF include: pressurized zones or areas that provide for crop protection, specially designed cooling sources and insulating capacities to permit supplementation at casing, refined program control for phase II composting, and controlled ventilation in production rooms based on monitored gas analysis. The first completely automatic system of phase II composting developed by research at the MTDF was widely adopted by the industry. The first automatic system for carbon dioxide control of ventilation also developed at the MTDF is rapidly gaining acceptance by the commercial industry. This system not only optimizes yield but since only essential fresh air is introduced and conditioned, the energy conservation effected is substantial.

Future improvements in mushroom culture will come not only from a more thorough understanding of single facets within the overall process, but from a more complete understanding of the interrelationships of these facets with each other.

Acknowledgements

Contribution No. 1166, Department of Plant Pathology, The Pennsylvania Agricultural Experiment Station. Authorized for publication 2 April 1980 as journal series paper 5953.

LITERATURE CITED

1. C. W. Porter, Economic developments in the US mushroom industry. Some problems some prospects. *Mushroom News* **27** (11), 4–7 (1979).
2. J. R. Delcaire, Economics of cultivated mushrooms, in *The Biology and Cultivation of Edible Mushrooms*, ed. by S. T. Chang and W. R. Hayes, pp. 727–793. Academic Press, New York (1978).
3. J. Engel, Food and Agriculture Organization, Publication on Methods of Forecast for Agriculture, p. XL. FAO, UN, Rome, Italy (1974).
4. E. V. Crisan and A. Sands, Nutritional value, in *The Biology and Cultivation of Edible Mushrooms*, ed. by S. T. Chang and W. A. Hayes, pp. 137–168. Academic Press, New York (1978).
5. K. Morgareidge, The nutritional value of mushrooms. Presentation before the Third Annual Mushroom Industry Short Course, Pennsylvania State University, (1958). Reprint 1972.
6. Food and Agriculture Organization, *Amino-acid Content of Foods and Biological Data on Proteins*, Nutrition Study No. 24. Food Policy and Food Science Service, Nutrition Division, FAO UN, Rome, Italy (1970).
7. Food and Agriculture Organization, *Energy and Protein Requirements*, Report of a Joint FAO/WHO *ad hoc* Expert Committee. Food Nutrition Meeting Report Series No. 52. FAO UN, Roma, Italy (1973).
8. P. J. C. Vedder, *Modern Mushroom Growing*, p. 420. Hauser Champignonkulturen AG, Gossau-Zürich (1978).
9. J. W. Sinden, Ecological control of pathogens and weed molds in mushroom culture, *Ann. Rev. Phytopathol.* **9**, 411–432 (1971).
10. M. C. Ferguson, A preliminary study of the spores of *Agaricus campestris* and other basidiomycetous fungi. *US Dep. Agric. Bur. Plant Ind. Bull.* **16**, 11–43 (1902).
11. B. M. Duggar, The principles of mushroom growing and mushroom spawn making, *US Bull. Dept. Agric. Bur. Plant. Ind.* **85**, 1–60 (1905).
12. R. Snetsinger, Mushrooms and Penn State, past, present, future. *Pa. Agric. Exp. Stn. Bull.* **767**, 34 (1970).
13. M. H. Jodon and D. J. Royse, Care and handling of cultures of the cultivated mushroom. *Pa. Agric. Exp. Stn. Bull.* **258**, 4 (1979).
14. L. C. Schisler and J. W. Sinden, Nutrient supplementation of mushroom compost at spawning. *Mushroom Sci.* **5**, 150–164 (1962).
15. J. W. Sinden and L. C. Schisler, Nutrient supplementation of mushroom compost at casing. *Mushroom Sci.* **5**, 267–280 (1962).
16. A. D. Carroll and L. C. Schisler, Delayed release nutrient supplement for mushroom culture. *Appl. Environ. Microbiol.* **31**, 499–503 (1976).
17. W. A. Hayes and N. G. Nair, The Cultivation of *Agaricus bisporus* and other edible mushrooms, in *The Filamentous Fungi*, Vol. 1, *Industrial Mycology*, ed. by J. E. Smith and D. R. Berry, p. 340. (1975).
18. M. E. Schroeder, W. L. Barr, L. C. Schisler and R. Snetsinger, A systems approach to mushroom growing. A preview of the mushroom test demonstration facility. *Pa. Agric. Exp. Stn. Prog. Rep.* **307**, 4 (1970).

The manuscript was received 15 April 1980

Mycological Medal
The scientific study of fungi is less than 300 years old, but the manifestations of these organisms have been known for centuries. Fungi play a crucial role in the slow but constant changes taking place in our environment because of their ubiquity and their astonishingly large numbers.
Here the French sculptor, A. Bloc, has beautifully summarized this idea, 'For the Best and for the Worst' in a 90 mm diameter copper medal, struck by the Paris Mint.

Book Reviews

PROFESSOR FRANZ SONDHEIMER FRS
Reviews
The Politics of Contraception

Carl Djerassi: The Politics of Contraception. W. W. Norton & Company, New York, 1979. pp. 274, $10.95, £6.50. ISBN 0 393 01264 6

This interesting book covers a far wider range than appears from the title. Undoubtedly, *The Politics of Contraception* has been chosen in order to distinguish it from so many other existing books on contraception which deal with the more mundane aspects of the subject. Today it is the politics, not the techniques, of contraception which are the most important aspect of the subject.

The book is a mine of information on contraception, and presents many thought-provoking ideas on the subject. It should be read by all those interested in this important field, including the intelligent non-scientist who is concerned with the 'population explosion', one of the most challenging problems facing mankind today. Although a large number of scientific facts are given, the material is presented deliberately in such a way that it can easily be understood by the layman.

Professor Djerassi is uniquely qualified to write this book, since, as he points out, he is one of the few individuals who has been engaged both in the basic research leading to the development of the oral contraceptive at the academic level, as well as an industrialist who has found world-wide practical application for his discoveries. At the same time, and this is perhaps the most fascinating aspect of the book, he has become deeply involved in the politics of contraception. This has brought him into close contact with politicians who have had to make decisions of tremendous importance, often with little or contradictory information. The ignorance and prejudice of some of these politicians, for which ample evidence is given, is quite frightening.

Although Professor Djerassi gives a thorough survey of all the different types of contraceptives available, and points out that abortion is the most widely practised form of birth control used in the world today, he pays most attention to the development and use of the oral contraceptive. This is the 'pill', an expression first coined for this particular form of medication by Aldous Huxley in *Brave New World Revisited* in 1958; i.e. two years before the pill was introduced into medicine! In a period of less than 20 years, the pill has become the most significant contraceptive hardware. Between 50 and 80 million women are taking the pill today, and cost, about 15 US cents per woman per month, is no longer a major factor.

No modern medicine has been studied as extensively as the steroid oral contraceptive. This is certainly justified, since the continuous use of a potent drug by a healthy person for a prolonged period of time poses fundamental questions. Unfortunately, the climate of over-regulation in Western Society has become such that much of the progress in the development of new contraceptive agents made in the 20 year period starting in the early 1950s, has almost dried up. The effect of the regulations in force in the US today is such that the production of a female 'once-a-month' pill, which would have many advantages over the presently available ones, would take $10\frac{1}{2}$–$17\frac{1}{2}$ years to develop, and cost US$ 15–46 million. The disincentives to the pharmaceutical industry to develop a male pill are of the same orders of magnitude. In fact, mainly because of the use of the female beagle as an experimental animal, which is highly prone to develop cancerous growths through administration of a variety of chemicals, it is very unlikely that the presently used oral contraceptives would have been approved had the present climate of regulations been operative 20 years ago.

In the period 1965–1969, just five US pharmaceutical companies spent $68 million dollars on contraceptive research, and several leading companies have now decided to discontinue research in this area. There is no doubt that toxicology and testing requirements will become even more stringent and time-consuming, and costs will continue to escalate. The problems associated with developing new contraceptive agents will continue to increase.

It is here and in related areas of contraceptive research and usage that the politics of contraception becomes of such importance. A suggestion made is that there must be a collaborative effort between the pharmaceutical industry and government, i.e. the tax-payer, to develop new contraceptive agents, since each organization is no longer in a position to undertake this task individually. Another suggestion is the formation of a new US government agency, whose main aim is to stimulate and facilitate new contraceptive research, and which will act as a buffer between the pharmaceutical industry and the Food and Drug Administration, which has less to lose by saying 'no' than 'yes'. Suggestions of great sociological significance are concerned with 'persuading' populations to use oral contraceptives, such as placing these into the drinking water supply, or into rice, in developing nations. Although many objections can and will be made to these suggestions, they are certainly thought-provoking.

There are many other interesting subjects discussed, such as the irresponsibility of the press, the prospects for the development of new female and male contraceptive agents – very small in the near future – the manufacture and use of oral contraceptives in China and the early history of research on the 'pill' in Mexico.

In summary, this is a fascinating book. My only criticisms, and these are quite minor ones, concern the poor quality of the printing, and the irrelevance of many of the illustrations. These show that Professor Djerassi possesses an excellent collection of pre-Columbian statues which, however, usually bear little relation to the text.

Professor Carl Djerassi is a Member of the Editorial Board of ISR

Professor F. Sondheimer, FRS, is a Royal Society Research Professor at University College, London. In the 1950s he was responsible for some important discoveries which led to the development of the contraceptive pill. More recently, his syntheses of annulenes have brought him distinction and world-wide recognition.

PROFESSOR ROMILA THAPAR

Reviews

Magic, Reason and Experience

G. E. R. Lloyd: Magic, Reason and Experience, Cambridge University Press, Cambridge, 1979. pp. xii+335. Hard cover, £20.00, ISBN 0 521 22373 3. Paperback, £7.95, ISBN 0 521 29641 2

Magic, Reason and Experience is at one level the study of the origins of Greek scientific ideas and at another level an attempt at investigating pre-modern science. It begins therefore with a discussion of some recent work on the modes of thought of what are often regarded as 'primitive' or early societies. This provides some of the methodology in tracing the changes in Greek scientific ideas from the traditional and non-rational to a perceptible increase in rational analyses, although not to the exclusion of the other, since magic and the irrational remain a part of Greek thought.

One of the areas in which this change can be more easily perceived is in the evolution of Greek medical knowledge. Epilepsy, for example, was traditionally regarded as 'the sacred disease'. Gradually by demonstrating the fraudulent nature of this concept its medical basis came to be recognized. Advances in knowledge drew primarily on the dialectical method of reasoning and, wherever possible, on the use of demonstration. Lloyd argues that even the practice of Rhetoric was not merely an oratorical device since it involved the formal analysis of arguments and did appeal, implicitly if not explicitly, to both reason and logic. He examines further the difference between those areas of knowledge in which observation was the substance of information, such as Greek ideas on geography, and those where observation was aimed towards extending knowledge thereby resulting in research, as in medicine and astronomy. Astronomy was particularly important, since, apart from its seemingly catering to a popular demand for astrology, it incorporated a continuing tradition of knowledge from the Babylonians and the Egyptians; with advances in astronomical instruments during the fifth and fourth centuries B.C. it linked observation with theory. The latter assisted also in a noticeable development of mathematics.

Lloyd cautions against too great an emphasis on scientific ideas *per se* in Greece at this time and points to the continuity of popular and traditional beliefs as part of the corpus of knowledge. In assessing the reasons for the emergence of Greek scientific ideas he suggests that in addition to the existence of literacy, of Greek familiarity with other societies and belief systems indicating a catholicity of interests and also of the existence of an economic surplus, the contemporary political developments also played a significant part. Not only the variety of political forms but the fact that they were frequently submitted to radical examination provided an atmosphere conducive to the required innovation of thought, debate and scrutiny, so essential to a scientific process of thinking.

Magic, Reason and Experience, in analysing the origins of Greek scientific thought, will be useful to those interested in such origins in other societies as well. It provides a welcome addition to the literature seeking to understand the roots of rationality in what are often dismissed as 'primitive' or irrational traditions.

Romila Thapar is Professor of Ancient Indian History at the Jawaharlal Nehru University in New Delhi, a Senior Fellow of Lady Margaret Hall, Oxford and a Distinguished Service Professor of Cornell University, Ithaca, New York.

PROFESSOR MELVIN KRANZBERG

Reviews

A History of Technology

Trevor I. Williams (Editor): Volume VI, The 20th Century, *c* 1900 to *c* 1950, Part I, pp. 690; Vol. VII, The 20th Century, *c* 1900 to *c* 1950, Part II, pp. 691–1482. Oxford University Press, 1979. £30 and £32 respectively. ISBN 019 8581 513 and 556 respectively.

These two tomes extend the original five volume Oxford *History of Technology* (published 1954–1958), edited by Charles Singer, E. J. Holmyard, A. R. Hall and Trevor Williams, from 1900 until approximately 1950. The first five volumes heralded the beginning of the scholarly study of the hitherto neglected field of the history of technology. Focusing on 'how things are done and made,' they presented straightforward accounts of technical developments and they still remain a basic reference work. Despite their pioneering nature, the original volumes were criticized on several counts. For the most part, they treated technology as a *Ding an sich*, developing without regard to socioeconomic or human factors. For another, by stopping at 1900, they omitted the great technical developments which helped shape the modern world. A third major criticism was that the volumes were largely British-centered, ignoring the great technological contributions of other countries and civilizations.

These final two volumes possess all the virtues of the preceding five and, alas, they also possess many of the same shortcomings. True, in his preface Dr Williams claims that more emphasis will be placed on economic, social and political factors because they 'cannot be ignored' in 20th century technological development. They should not have been ignored in the previous volumes either, but be that as it may, the attempt to incorporate them in 20th century history is so clumsily done that the reader will emerge without any clear insights regarding the interplay among non-technical factors and technological developments.

Volume VI begins promisingly enough, with seven chapters dealing with socioeconomic-political matters. But T. K. Derry's first chapter, a short review of world history from 1900 to 1950, is so sketchy as to be meaningless, especially since there is little interweaving of political with technical developments. Then follow chapters on 'The Sources of Innovation' (David Sawers), 'The Economics of Technological Development' (F. R. Bradbury), 'Management' (Glenn Porter), 'Trade Unions' (Harold Pollins), 'The Role of Government' (Alexander King) and 'Education in Industrialized Societies' (David Layton). Sawers, Bradbury, Porter and King provide excellent essays, but the chapters in the rest of the book focus on technical developments in specific fields and do little to relate those developments to these introductory chapters. The result is that, as in the previous volumes, the technical developments are treated in a vacuum, even though some of the technical chapters are themselves prefaced by brief economic summaries. Yet the volumes as a whole fail to contribute a great deal to understanding the industrial and managerial infrastructure which played such a great role in 20th century technical developments.

Of course, all such generalizations regarding the 58 chapters in these massive two volumes are subject to exceptions. For example, Roland J. Mainstone, writing on 'Building and Architecture,' shows very well the interplay between design considerations and construction methods; similarly, Sir Colin Buchanan shows the interaction between non-technical and strictly technical factors in his discussion of planning. Frank Greenaway's chapter on 'Instruments' is one of the few to bring out the important interplay between science and industrial development during the 20th century. But these are exceptions. For the most part, the authors of the technical chapters ignore Williams's dictum that economic, political and social forces 'cannot be ignored.'

The argument made in the original set of five volumes for stopping at 1900, namely, the difficulty in sorting out the historically significant from the trivial when treating recent developments, is also utilized to justify stopping these two volumes at 1950, so as to have the 'benefit of the lapse of a quarter of a century . . . in which to make some sort of assessment.' But surely the task of the historian is to sift out what is significant from what is trivial, today and in the past. And one does not have to be much of an historian to recognize, for example, that the recent development of minicomputers is of obvious technological and societal significance.

Ah, one may say, minicomputers developed after 1950; but the mid-century limit is not rigorously observed throughout. As Dr Williams observes, the time period 'has been treated fairly elastically,' so the chapters on space flight and atomic power perforce go beyond the 1950 limit. Other chapters, however, stop abruptly at 1950, thereby ignoring major developments in the third quarter of the century. Thus, M. L. Pearl, writing on 'Iron and Steel,' states that 'the most important development of the whole half century [oxygen method of steelmaking] lies outside our period,' as does the continuous casting of steel, another major technological development. Similarly, discussing the world transport market, K. W. Guilliam, ends his story with the Boeing 707, so jumbo jets and the supersonic transport do not appear. P. S. A. Berridge's discussion of road building omits any mention of the United States Interstate Highway System, the largest roadbuilding and public works project in the history of mankind, presumably because it began after 1950. Robert S. Woodbury's chapter on 'Machine Tools' also observes the 1950 limit, so there is virtually no discussion of computer controlled machine tools leading to robotics and cybernation. And Helmut Gernsheim's otherwise excellent chapter on 'Photography' mentions but does not deal with Polaroid 'instant' photography, though certainly one could have discerned its importance in photographic history in a chapter written in the 1970s.

Of course, even when a chapter carries its story into the 1970s, the author can guess wrong about historical trends. Thus T. A. Margerison, writing on computers, states that 'it now appears that the very rapid technical development of computers is stabilizing.' Insamuch as there has been a 16-fold increase in chip memory capacity in just the past three years, so that a \$20 hand computer today has more computing power than did the world's largest computer a quarter century ago, it is obvious that the 'very rapid technical development' has not yet stabilized. To Margerison's credit, however, it should be noted that he carries the story of the computer back to the 17th century, because the subject was not treated in the earlier volumes.

Criticism that the earlier volumes centered overmuch on Britain was vitiated by the fact that British leadership in the Industrial Revolution necessitated concentration on British developments. In the 20th century, however, when first Germany and then the USA came to dominate the industrial scene, it is hard to justify the British emphasis. Thus, in his chapter on

'Trade Unions,' Harold Pollins focuses on England and deals mainly with the response of trade unions to technological change; there is no mention of Karl Marx, an omission perhaps appropriate in dealing with British trade unionism, but certainly not so in dealing with trade unions elsewhere, especially on the Continent.

Yet, except when history is distorted thereby, I find nothing wrong with the British emphasis in these volumes. In equivalent general histories of technology published in France, the USA, and the Soviet Union, there is particular emphasis on developments within those nations, and in the Russian history virtually everything is home grown! If the British do not tell of their own role in their history of technology, who will do it for them?

Of course, not all the chapters are parochial. Frank Greenaway, writing on 'The Chemical Industry,' correctly shows the international character of developments, as does Anthony Michaelis in his neat (14 pages) summary of 'Space Technology,' a chapter that is noteworthy in (rightly) stressing the managerial factor in the American space program.

The fact is that the chapters are very uneven, as is perhaps inevitable in a multi-authored collection of this nature. Although most chapters are mere recitals of names and dates, a dry listing of technical developments in chronological order, others are outstanding. Thus Lynwood Bryant provides us with 'internal,' or 'hardware,' history of technology at its best by treating the development of the internal combustion engine as a series of problems to be solved and showing how they were solved.

As might be expected in a comprehensive, multi-author work of this nature, there are some startling oversights and omissions. Henry Ford's moving assembly-line, the major development in mass production during the 20th century, receives only two paragraphs. Some of the great giants of modern architecture – Le Corbusier, Saarinen, and Frank Lloyd Wright – do not appear in the Index. Elton Mayo's famous Hawthorne experiments, which led to the development of industrial sociology and the 'human relations' movement in the factory, are well described in Porter's fine chapter on 'Management,' but find no place in the chapters dealing with the actual production mechanism. And Norbert Weiner, whose books *Cybernetics* and *The Human Use of Human Beings* first concentrated attention on the social impact of computing devices and control systems, is strangely missing – perhaps indicating again the failure to tie technological developments to the social context throughout the work. On the other hand, there are some additional technologies not treated in the earlier volumes, such as Audrey B. Davis's chapter on medical technology and Margerison's on the computer.

At the conclusion of these two volumes, the editor evaluates technology's contributions to modern civilization in a chapter entitled 'Technology and the Quality of Life.' Williams makes a good case for technology in providing food, clothing, shelter – the necessities of life. He is fair enough to mention some shortcomings of the modern urban environment; he also takes a few swipes at the environmentalist movement – 'the pendulum has swung too far' – and he points out that the environment and energy are 'embarrassing legacies' to the future. But outside of several paragraphs denying that technology is to be blamed for the decline of religion, Williams has little to say about the 'quality of life' when measured in other than material terms. What about technology's contributions to human dignity, freedom, individualism, social justice? What has technology done for the arts, education, the humanities in general? Williams does not bother to ask such questions, let alone answer them.

These last two volumes of *A History of Technology* do not go much beyond the narrow concept of technological development which limited their predecessors a quarter century ago. But in the meantime, the history of technology has grown – and along with it the sociology and philosophy of technology – and the interactions between technology and the socio-cultural matrix have become major concerns of historians and other scholars. Because these final two volumes provide little in the way of synthetical treatment or historical interpretation, they do not truly advance the history of technology, although these massive tomes will find use as valuable encyclopedic references.

The first five volumes of *A History of Technology* started the scholarly study of the history of technology. Ever since then historians of technology have been engaged in a joyous race to uncover new facts, new relationships, new interpretations. But Dr Williams and his cohorts have remained at the starting gate.

Professor Melvin Kranzberg is the Callaway Professor of the History of Technology at the Georgia Institute of Technology, Atlanta, Georgia, USA. He is the author of many books in his field and enjoys a world reputation.

PROFESSOR ANDRE GUILCHER

Reviews

Coastal Research in the North Sea

Deutsche Forschungsgemeinschaft: Sandbewegung im Küstenraum: Rückschau, Ergebnisse und Ausblick. Harald Boldt Verlag, Boppard, 1979. 416 pp.

This book is a collection of papers on the subject of coastal sand and sandy mud flats, with special consideration of the East and North Frisian coast and the Dithmarschen in the German Federal Republic. It is the result of researches carried out under the supervision of the German Centre for Scientific Research, and it gives excellent examples of what can be done by interdisciplinary teams in the field of earth sciences.

Ulrich points out on page 348, in his study of ripple marks as indicators of sand movement in tidal channels, that such researches and measurements demanded the co-operation of seven working groups based in Kiel, Tönning, Helgoland and Tondern (the latter in Denmark), and dealing with sedimentology, seismology, various aspects of marine hydrology and marine biology. As Führböter says in his introductory remaks (pages 12–13), a full understanding of the present situations on sand flats requires the consideration of many different time scales, from geological periods of $\pm 10^4$ years duration, to secular periods and tidal periods of half-day duration; for this purpose, various techniques and teams working in association are necessary. A good example of the relationship between residual currents and sediment transport, between the Weser estuary and Amrum Island, North Friesland (Göhren), appears on pages 104 and 108. Theoretical studies must always be checked with field observations, since differences may be large between the former and the latter; see Oebius and Fenner, page 271.

I should like to add that all papers are extremely accurate and all observations most precise. The western German coast is very short indeed, but the German scientists have always been attracted by the sea; they have studied their shores, foreshores and the Wadden Sea flats as carefully as their Dutch neighbours have done.

Professor André Guilcher – a Member of our Editorial Board – is Professor of Geography at the Université de Bretagne, Occidentale.

PROFESSOR L. L. CAVALLI-SFORZA

Reviews

Human Genetics

Human Genetics: Possibilities and Realities, Ciba Foundation Symposium No. 66. (New Series); in honour of Sir Gordon Wolstenholme. Excerpta Medica, Amsterdam, Oxford, New York, 1979. $51.25. ISBN 90 219 4072 8.

For at least six participants in the Symposium on 'Human Genetics: Possibilities and Realities' that took place in London at the Ciba Foundation in June 1978, the meeting had a very special flavour. Nineteen years earlier the same people had taken part in a very similar Ciba Symposium which had been convened in the most pleasant and unlikely location of a Naples *fin de siècle* hotel. The earlier Symposium had collected a group of scientists who were mostly engaged in human biochemical genetics, an almost new, but already very active and exciting field at the time. Revisited almost twenty years later, human genetics again seems to be on the verge of a new take-off for a period of even more sustained growth.

Speakers on DNA included : P. M. B. Walker on repetitive DNA, the only true selfish gene according to Bodmer; C. Weissmann on methods of generating point mutations at predetermined sites; J. B. Gordon on the demonstration of action of specific cloned genes by injection into oocytes; P. E. Polani on the four classical DNA repair syndromes; and D. J. Weatherall on human hemoglobin genetics, where the analysis at the aminoacid sequence has now been coupled very effectively with studies at the DNA and RNA level. This last paper was placed, quite appropriately, in a somewhat intermediate position between those on DNA and those on proteins.

Henry Harris followed on multilocus polymorphic enzymes, then W. F. Bodmer on HLA and clusters of presumably duplicated genes in the HLA region, a theme to which he returned in the summary he gave of the Symposium. P. J. Lachmann spoke on the genetics of complement proteins; and C. Milstein on cell surface antigens and the possibilities of analysis generated by the powerful new technique of monoclonal antibodies.

The third group of lectures ranged from cancer biology papers by M. Siniscalco, Henry Harris and G. Klein, to a review of genetics of learning disabilities by B. Childs, and to the present and future importance of genetics in health care by C. Scriver.

Discussion at the meeting has been printed verbatim, and much of it is of special interest considering how many well known scientists attended the meeting without contributing papers. Syd Brenner was a spirited and multifaceted chairman.

The Symposium was dedicated to Sir Gordon Wolstenholme, retiring as director of the Ciba Foundation, under whose enlightened direction the Ciba Foundation has reached its remarkable scientific usefulness, quality and prestige. All those who have been guests of the Foundation, or attended one of its many Symposia, are well aware of how uniquely attractive and efficient that environment has always been. Here I would also like to propose special thanks to Sir Gordon for promoting certain scientific areas in which I have a particular interest. For example, under his aegis the Foundation held a Symposium on Medical Biology and the origin of the Etruscans. This may sound far-fetched to many, but I do not believe so. I hope that my introduction to the Symposium on areas of overlap between human genetics and archaeology and culture, in general, may help to prove the validity of this viewpoint.

Professor L. L. Cavalli-Sforza – a member of our Editorial Board – is Professor of Genetics at the Medical School, Stanford University, California. He is most eminent in the field of human population genetics and cultural evolution. See *Interdiscip. Sci. Rev.* 3, 99 (1978).

DR FRANCES MICHAELIS
Reviews
Pond Littoral Ecosystems

Dagmar Dykyjová and Jan Květ (Editors): Pond Littoral Ecosystems. (Ecological Studies; Volume 28). Springer-Verlag, Berlin, Heidelberg, New York, 1978. xiv+464 pp., 183 Figs. DM 98.00, $49.60. ISBN 3-540-08569-6

This book presents the Czechoslovakian contribution to the wetland studies of the International Biological Programme (IBP). It is concerned with fishponds managed primarily for the production of carp (*Cyprinus carpio* L.) but also for the harvesting of duck- and goose-meat and the common reed (*Phragmites communis* Trin.). Two areas of Czechoslovakia were studied: Třeboň and the well known region of Lednice, a Project Aqua site under the IBP. The work on the ecology of the 'littoral' or shoreline was carried out from 1965–1974 by a team of 27 scientists.

The 'Introduction to the Ecology of Fishpond Littorals' provides a fascinating background to the art of fish-rearing in Czechoslovakia since about 1450 and would be useful reading for the graduate ecologist. Section 7 on Fishpond Management and Section 8 on Conservation of Wetlands in Czechoslovakia are both readable and generally applicable. The rest of the book (Sections 1–6) is for the specialist, as it ranges in detail from pollen analyses of peatbog profiles through spatial variation of dissolved organic phosphate to the mean annual biomass of spiders. The section on Primary Production in the Littoral Plant Community is by far the largest and contains many methods which would be useful in other locations. One shortcoming of the book is that there is no concluding chapter. This is one of the few large-scale limnological studies ever undertaken and is worthy of an attempt at a summary.

The general presentation is good with an attractive cover, both subject and taxonomic indices and a long reference list at the end of each Section. The book is unified in spite of the many contributors. According to the Editors, the volume attempts to give 'condensed quantitative data' for 'synthesis and ecosystem modeling'. This is exactly what it does, with careful maps, tables and species lists. Some of the graphs contain too much detail to be clear. A striking feature is the high standard of monochrome photography, making it attractive to the non-expert. Hemispherical photography is used to estimate radiation fluxes.

The language leaves much to be desired. Many of the technical words appear to be used incorrectly e.g. 'inventarization' (p. 11) for 'inventory' and 'deponia' (p. 410) presumably meaning 'detritus' or 'debris'. Some of the word groups are verbose, e.g. 'long-persisting perennial macrophyte communities' (p. 23), or difficult to interpret, e.g. 'stenoecious cenoses of ephemers' (p. 26) and the sentence 'A specimen vegetation relevé documents the association' (p. 34). This simply highlights the need for rigorous editing by a native English speaker, well versed in 'Basic English'.

Unfortunately the book will be limited to people familiar with the complex technical vocabulary of the English language, which will preclude much of the developing world. This is a small criticism of a fine volume that attempts so much in the way of synthesis and largely succeeds.

Dr Frances B. Michaelis works at the Fisheries Laboratory, Rotorua, New Zealand and currently holds a Research Fellowship with the New Zealand National Parks Authority. She has a wide range of freshwater experience, ranging from Alpine springs to estuaries.

DR ULI WEIDMANN
Reviews
Vergleichende Verhaltensforschung: Grundlagen der Ethologie

Konrad Lorenz: Vergleichende Verhaltensforschung: Grundlagen der Ethologie, Springer-Verlag, Wien–New York, 1978. pp. xiv+308. $19.50, DM 39.00.

In the early 1950s, I was privileged to work among a group of students under Professor Lorenz; at intervals he would treat us, while watching ducks and geese, to an *ad hoc* lecture on key issues in ethology (as the 'new' comparative study of animal behaviour is called). These invariably exciting and stimulating lectures were based on chapters of a manuscript for a book he hoped to publish. This is the book!

It hardly needs to be stated that a work which took so long to mature is not simply yet another text on ethology. Indeed, in many ways it is quite unrepresentative of modern developments and current concerns: most citations of literature refer to works published 10 or 20 years ago and many of the currently topical research lines in animal behaviour are not mentioned.

What the book *does* bring are the thoughts of one of the 'fathers' of ethology on its philosophical and methodological foundations. In addition it gives a succinct and masterly review especially of the early history of this science and how its major concepts were developed.

As Section One explains, modern ethology started with the discovery, by Whitman and Heinroth, that many behaviour patterns of related species reflect systematic (evolutionary) relatedness just as do their morphological features. Hence, it would appear that such behaviour patterns, no less than morphological structures, develop under the influence of 'genetic blueprints'; a conclusion which can often be corroborated by experiments in which a given behaviour pattern develops even in an animal deprived of many of the normal environmental influences. Section Two covers motivational aspects of these patterns: releasing mechanisms, instinct hierarchies, and so forth, as well as the types of spatial orientation behaviour. A final section deals with adaptive modifications in behaviour, learning phenomena and curiosity.

Amongst brilliant insights there is much that is speculative, controversial and in need of experimental confirmation. While some views expressed may be currently 'unfashionable', their presentation in one book (hopefully soon available in English) is likely to stimulate renewed interest and research.

Young ethologists, psychologists, philosophers of science and, indeed, anyone who may have heard of Lorenz mainly as a proponent of rigid views on aggression and instinct is here given a chance to follow the development of his ideas and to appreciate why he was deemed worthy of a Nobel prize.

Dr Uli Weidmann is Senior Lecturer in Psychology at Leicester University, UK.

PROFESSOR JOHN J. McDERMOTT

Reviews

Genesis and Development of a Scientific Fact

Ludwig Fleck: Genesis and Development of a Scientific Fact, The University of Chicago Press, Chicago; 1979. $22.75. ISBN 0 226 25324 4

This book is a translation from German, of a classic work in the philosophy and methodology of science, which was published originally in 1935. Although comparatively unsung for many years, due partially to the onset of World War II, Fleck's work has received increasing attention of late. Thomas Kuhn, in his Foreword, views the book as a forerunner of his own influential book, *Structure of Scientific Revolutions.*

The fundamental theme pursued by Fleck, in Kuhn's judgment, is 'the personal tentative and incoherent character of journal science together with the essential and creative act of the individuals who add order and authority by selective systematization within a vademecum.' To that end, he tracks the genesis and the empirical reliability of the concept of syphilis and also of the Wassermann reaction. In effect, Fleck casts the tradition of scientific inquiry into a sociological context, thereby raising important questions about data, verification and the alleged unity of the history of science.

Considerable work has been done in this area since that of Fleck but it is salutary to have an English edition of this important forerunner to contemporary positions in the nature and activity of science.

Professor John J. McDermott is Head of the Department of Philosophy, Texas A+M University, College Station, Texas, USA.

DR NAYLOR FIRTH

Reviews

Environmental Geochemistry and Health

Environmental Geochemistry and Health: Proceedings of a Royal Society Discussion 15–16 March 1978.
The Royal Society, London, 1980. viii+216 pp. £21.50 UK, £22.55 overseas. ISBN 0 85403 114 6

During the 20th century two significant developments have contributed to the emergence of environmental geochemistry as a major scientific discipline. The first has been the control and, in many cases, the virtual elimination of communicable diseases as major causes of morbidity and mortality. Since we have all to die, this has resulted in the promotion of diseases of the heart and cancer and other malignant neoplasms to the top of the league. The suspicion that many of these could be caused by imbalances in trace element concentrations had to await the second development before detailed studies were possible. As Lenihan said at the first 'Trace Substances in Environmental Health' conference in Missouri in 1967, 'Unless we can detect an element and measure its distribution in the tissues of the body, it is unlikely that we shall ever be able to say anything meaningful about its biological purpose'. The ability to analyse more and more for less and less has been acquired at an alarming, if expensive, rate over the last decade so that conference proceedings such as those contained in this book can say something 'meaningful' about the contribution of a number of elements to diseases.

The volume contains 17 papers by authors from Australia, the UK, Switzerland, the USA and the USSR. Six papers deal specifically with trace elements, three with the techniques of geochemical surveying and four with the relationship between geochemistry and other environmental parameters. Separate chapters are devoted to the relationship between geochemistry and cardiovascular disease as well as between lead and human health. The remaining two chapters deal with epidemiological surveys and with a review of environmental geochemical research in the USA.

The study of the relationships between environment and disease involves workers from many different disciplines, all of whom use their own particular scientific jargon. There is thus a danger that communications between them can be difficult. However, nearly all the authors in this publication succeed in putting their views in language that should be understood by anyone with an interest in the subject; however, there are one or two lapses.

The volume has not been edited and consequently there is a variation in the degree of detail encountered in the individual chapters. This ranges from attention to the minutiae of sampling, preparation and analysis in the Plant and Moore chapter on 'Regional Geochemical Mapping in Britain' to a very broad brush approach by Wixson in his chapter on 'Geochemistry and Pollution'.

The chapters cover an exceptionally wide range of science but several common themes are worth highlighting. The analysis of a substrate for a particular element is of limited use unless the chemical species of the element is also identified: the effects of barium meals containing pure barium sulphate and barium sulphate contaminated with barium nitrate will be very different, even though the total barium content of each is the same. In spite of the advances made in analytical techniques, the determination of speciation is still difficult and it is in this field that a large effort is still needed.

Mills, in his chapter on 'Trace Elements in Animals', questions the selection of tissues such as blood and liver for analysis, and suggests that for the sake of convenience of accessibility we may be omitting to analyse those tissues most susceptible to, and indicative of, the effects of trace element imbalance. There is also particular urgency to study the health of those few remaining communities with simple staple diets originating from indigenous crops and waters, features which are no longer characteristic of the 'developed' world.

Selenium emerges as an element with many disguises, on some occasions protective – as in its antagonism to mercury, lead, silver and cadmium toxicity – and on others toxic – as in the selenium plant accumulators. Part of the disguise is the ability of the element to exist as different chemical species and the importance of differentiating between selenate and selenite anions in seleniferous soils and plants is emphasized by Peterson in his paper on 'Geochemistry and Ecology'.

The links between the hard facts of mortality and morbidity, and the multitude of environmental variables that could be contributors, are rarely well defined. In the UK, as Thornton and Webb point out in their excellent chapter on 'Geochemistry and Health in the UK', actual causal relations only exist between iodine and goitre as well as fluorine and dental caries, with an empirical relation between cardiovascular disease and water hardness. Even in the many cases where good correlations exist between an element and a disease, Underwood raises the question 'When does a correlation or association become a causation?' and therein lies the fascination and challenge of environmental geochemistry.

The standard of production of this volume is excellent, with eight colour plates, photographs and line diagrams. I counted only three minor typographic errors. However, the lack of an index will be a disadvantage for many readers. The papers cover such a wide range of topics, elements, substrates and geographic areas that it seems a pity not to have devoted a little more time to the compilation of an index for such a useful survey.

The study of environmental geochemistry and health – or should it be disease? – demands an interdisciplinary approach and regular meetings, such as this Royal Society discussion and the annual University of Missouri Trace Substances Conferences. They do much to ensure that contact is maintained between the diverse disciplines. This excellent publication should ensure that many workers as yet unassociated with the problems will realize that they too have a contribution to make.

Living systems have presumably evolved on our planet because of, and not in spite of, the environment that it provided, so that the knowledge of anyone studying animal, vegetable or mineral systems will have a contribution to make to our understanding of health and disease. *Environmental Geochemistry and Health* can be recommended for the initiated and the uninitiated alike.

Dr Naylor Firth is Scientific Adviser to the Welsh Office in Cardiff; he has worked extensively in environmental geochemistry.

SUBJECT INDEX